New Frontiers in Extracellular Vesicles

New Frontiers in Extracellular Vesicles

Editor: Erina Jimenez

FOSTER ACADEMICS

www.fosteracademics.com

www.fosteracademics.com

FOSTER ACADEMICS

Cataloging-in-Publication Data

New frontiers in extracellular vesicles / edited by Erina Jimenez.
p. cm.
Includes bibliographical references and index.
ISBN 978-1-64646-562-0
1. Extracellular space. 2. Coated vesicles. 3. Cell organelles.
4. Cell membranes. 5. Cytology. I. Jimenez, Erina.
QH603.E93 E98 2023
571.65--dc23

Foster Academics,
118-35 Queens Blvd., Suite 400,
Forest Hills, NY 11375, USA

ISBN 978-1-64646-562-0 (Hardback)

Contents

Preface

Extracellular vesicles (EVs) are lipid bilayer-delimited particles which are released naturally from nearly all types of cells but cannot replicate like cells. There are three types of EVs that are categorized on the basis of their size and synthesis route. These include apoptotic bodies, exosomes, and microvesicles. They transport a cargo of lipids, proteins, metabolites, nucleic acids and even parent cell organelles. The communication mechanism among cells relies heavily on EVs. They are also capable of carrying disease biomarkers, including CD24, glypican-1, CD63, and colon cancer-associated transcript 2 (CCAT2). Current research around EVs focuses on their role in various diseases and conditions such as cancer, neurodegeneration, metabolic disorders, and intercellular communication. Owing to their cargo delivery potential, EVs may be used as drug delivery platforms for cancer therapy. Furthermore, engineered EVs may find application in cancer combination therapy. This book explores the new frontiers in the study of extracellular vesicles. It is appropriate for students seeking detailed information on extracellular vesicles, as well as for experts.

After months of intensive research and writing, this book is the end result of all who devoted their time and efforts in the initiation and progress of this book. It will surely be a source of reference in enhancing the required knowledge of the new developments in the area. During the course of developing this book, certain measures such as accuracy, authenticity and research focused analytical studies were given preference in order to produce a comprehensive book in the area of study.

This book would not have been possible without the efforts of the authors and the publisher. I extend my sincere thanks to them. Secondly, I express my gratitude to my family and well-wishers. And most importantly, I thank my students for constantly expressing their willingness and curiosity in enhancing their knowledge in the field, which encourages me to take up further research projects for the advancement of the area.

Editor

1

Salivary Extracellular Vesicle-Associated exRNA as Cancer Biomarker

Giulia Chiabotto [1], Chiara Gai [1], Maria Chiara Deregibus [2] and Giovanni Camussi [1,*]

[1] Department of Medical Sciences, University of Torino, Torino 10126, Italy
[2] 2i3T Business Incubator and Technology Transfer, University of Torino, Torino 10126, Italy
* Correspondence: giovanni.camussi@unito.it

Abstract: Extracellular vesicles (EVs) secreted in biological fluids contain several transcripts of the cell of origin, which may modify the functions and phenotype of proximal and distant cells. Cancer-derived EVs may promote a favorable microenvironment for cancer growth and invasion by acting on stroma and endothelial cells and may favor metastasis formation. The transcripts contained in cancer EVs may be exploited as biomarkers. Protein and extracellular RNA (exRNA) profiling in patient bio-fluids, such as blood and urine, was performed to identify molecular features with potential diagnostic and prognostic values. EVs are concentrated in saliva, and salivary EVs are particularly enriched in exRNAs. Several studies were focused on salivary EVs for the detection of biomarkers either of non-oral or oral cancers. The present paper provides an overview of the available studies on the diagnostic potential of exRNA profiling in salivary EVs.

Keywords: miRNA; non coding RNA; exosomes; microvesicles; cancer; saliva

---

1. Introduction

The aim of liquid biopsy is to identify biomarkers with diagnostic, predictive and prognostic values in bio-fluids, to avoid more invasive approaches. Researchers focused on different types of biomarkers, including proteins, circulating DNA fragments and cells, and extracellular RNAs (exRNAs). ExRNAs are more sensitive and specific biomarkers than proteins and better reflect the cell dynamic than DNA does [1]. However, several limitations in the use of exRNA as biomarkers still remain, related to their heterogeneity, the incomplete definition of their multiple targets and functions, and their stability in different biological fluids [2].

Nowadays, the recently developed techniques of sequencing allow for an accurate evaluation of RNA expression, which reflects cellular genetic and functional states. Different types of RNA biomarkers have been considered in cancer. Differential mRNA expression profiles may reflect the positive and negative regulation of tumor-associated genes in several cancers and may provide suitable biomarkers for monitoring the clinical outcome of patients [3–5]. Non-coding RNAs, such as microRNAs (miRNAs), piwi-interacting RNA (piRNA), small nucleolar RNA (snoRNA), circular RNA (circRNA) and long non-coding RNAs (lncRNAs), have also been investigated as potential biomarkers in cancer [1]. Moreover, the detection of chimeric RNAs may allow for the identification of chromosomal aberrations [6,7]. The stability of different exRNAs in the biological fluids depends on protection from exonucleases, provided either by RNA binding proteins, such as those of the Argonaute family, and high- and low-density lipoproteins, and by encapsulation in membrane vesicles [8–10].

Membrane vesicles released by cells in the extracellular space have recently emerged as a good evolutionarily preserved mechanism of inter-cellular communication. The vesicles are able to share genetic information among cells by delivering proteins, bio-active lipids and nucleic acids protected from degrading enzymes [11–13]. These vesicles, termed extracellular vesicles (EVs), are abundant in all biological fluids and can be exploited for searching biomarkers since they retain the molecular signature of the cell of origin.

One challenge of liquid biopsy is the choice of the bio-fluid that better reflects the occurrence of cancer. Most studies have focused on blood, but several other bio-fluids are now gaining attention, including saliva. Saliva is enriched in EVs and may represent a bio-fluid suitable for searching for markers of oral and systemic diseases.

2. EVs as Carriers of exRNA

EVs are a heterogeneous population, which includes membrane vesicles of different sizes and biogenesis. The three main categories of EVs include exosomes, ectosomes, and apoptotic bodies [14–17]. Exosomes are nano-sized vesicles (35–100 nm), which originate from the multivesicular bodies and are secreted by a process of exocytosis. This process requires the inward budding of multivesicular bodies-membrane, followed by fusion with plasma membrane and release in the extracellular space. The endosomal sorting complex required for transport (ESCRT) machinery, and several components of the Ras GTPases (RAB) family [18,19] and of the tetraspanin family [18], are involved in such processes. Vesicles generated by the budding of surface plasma membrane with the inclusion of cytoplasmic constituents have been termed microvesicles. This term is misleading as these vesicles include a large population of vesicles within the nano-range (60–250 nm), such as those released from healthy cells. It has been therefore suggested that one should name these vesicles ectosomes or shedding vesicles [14]. Shedding vesicles also include larger vesicles that may reach 1000 nm, and some of them may derive from cells in a pre-apoptotic phase. Microvesicle formation is related to the modification of plasma membrane curvature due to changes in lipid and protein interactions involving the arrestin domain-containing protein-1 (ARRDC1) and the late endosomal protein tumor susceptibility gene 101 (*TSG101*). The cytoskeleton rearrangements controlled by the signaling cascade of Ras-related GTPase ADP-ribosylation factor 6 (ARF6) promote vesiculation and release [20]. The apoptotic bodies released by cells undergoing programmed death are vesicles with a diameter of 1000–5000 nm and may contain nuclear fragments and intact chromosomes [21].

Most of the studies on the use of EVs as potential biomarkers have been performed on exosomes and microvesicles, as both types of vesicles may encapsulate fragments of genomic and mitochondrial DNA origin [22–25], and different classes of RNA, such as mRNA, miRNA, lncRNA, mitochondrial RNA, transfer RNA, and ribosomal RNA [26–29]. Furthermore, nano-sized vesicles may be released by the same cell by exocytosis or by surface membrane budding, and it may be difficult to discriminate vesicles discharged by non-apoptotic cells on the basis of mechanisms of origin. In fact, some molecules constituent of the endosomal sorting complex required for transport (ESCRT) and some ancillary proteins such as *TSG101*, Alix and Vacuolar protein sorting-associated protein 4 (VPS4) implicated in the formation of exosomes, are also reported in the literature to also be shared by shedding vesicles [30]. The discharge of exosomes may involve some constituents of the RAB family of GTPase proteins implicated in the MVBs/plasma membrane interaction [31,32]. Furthermore, the biogenesis of shedding vesicles may depend on a reorganization of the proteins of the cytoskeleton myosin and actin under the control of the ARF6 signaling [30]. Some tetraspanins and some ESCRT proteins are often reported as common exosome and shedding vesicle markers and cannot represent a peculiarity principle [33]. However, CD9, CD63 and CD81 tetraspanins are reported to be enhanced in exosomes [18], while annexin A1 is considered a marker for microvesicles [34]. Due to the heterogeneity of EVs produced by different cell types and present in the biological fluids, the protocols used for EV purification assume a critical relevance. For this reason, the public available databases [35–37] take into account the procedures used for the purification of EVs when describing the lipid, protein and nucleic acid composition. Of interest, the comparative lipidomic, proteomic and genomic analyses between the cells of origin and their released EVs highlight the presence of qualitative and quantitative differences, in both basal and stimulated condition. These data suggest that the EV cargo is actively modulated [38–40]. The EV mediated transfer of their cargo into recipient cells can induce epigenetic and functional changes into the recipient cells [41]. Some studies indicate that genetic materials encapsulated in EV include mitochondrial [24,42] and genomic [22] DNAs. Other studies performed on exosomal sub-fractions of EVs suggest that the DNA release is not related to the small vesicle release but to the autophagy- and multivesicular-endosome-dependent mechanism [34]. Several RNA species were found to be associated with EVs. EVs contain intact mRNA that can be translated into proteins in the recipient cells [27,28], but also many fragments of 200 nucleotides [43] that may have a biological role as scavengers and/or values as biomarkers. The exRNA enriched in EVs include miRNAs, ribosomal RNAs, tRNA fragments piRNA, snoRNA, Y-RNA, circRNA and lncRNAs [44–48].

Little is known about the process of nucleic acid compartimentalization into EVs [39,49–52]. Some proteins involved in EV biogenesis are potential candidates for RNA encapsulation in EVs. For instance, it has been shown that in EVs purified by differential ultracentrifugation from liver stem cells, Alix coprecipitate with Argonate 2 (Ago2) protein and miRNAs. The significant reduction of EV-associated miRNAs in Alix knock-down cells suggests that Alix can have a role as a component of ESCRT in the export of the Ago2-miRNA complex [53]. Through a high-resolution density gradient fractionation coupled with an immunoaffinity capture of exosomes, Argonaute proteins were detected in the non-vesicular compartment [34], which may contain components of the multivesicular body membranes. In breast cancer-derived EVs, miRNAs associated with Ago2 were shown to induce an alteration in the transcriptome of the recipient cells [54]. By regulating the Ago2 secretion [55], GTPase KRas (KRAS) has been involved in the miRNA compartmentalization into EVs released by colorectal cancer cells [56]. Moreover, miRNA packing into EVs depends on the interaction with the heterogeneous nuclear ribonucleoprotein A2B1 (hnRNPA2B1) [57] and with the RNA-binding protein Y-box protein I (YBX1) [58].

3. EVs in Cancer Biology

EVs released by cancer cells may act both locally, contributing to create a favorable microenvironment for cancer growth, and at distance, promoting the metastatic niche formation. Several studies have shown that cancer EVs contribute to the induction of chemo-resistance [59–63], to the remodeling of extracellular matrix [64,65], to tumor vascularization [66] and to epithelial-mesenchymal transition with a consequent enhanced migration/invasion and metastasis formation [67–69]. EVs also participate as active players in the bi-directional crosstalk between cancer cells and cells present in the microenvironment, such as fibroblasts [70,71], which may secrete EVs conferring chemo-resistance [72–74] and invasiveness to cancer cells [70,71].

Several mechanisms of action involving the EV-mediated transfer of proteins and exRNA have been described and exploited as diagnostic markers. In particular, the miRNA-mediated effects have been extensively studied. Several miRNAs present in EVs released from breast cancer (miR-100, miR-222, miR30a and miR-17), lung cancer (miR-100-5p), and ovarian cancer (miR-21) or released from stromal cells (miR-21 and miR-146a) were shown to confer chemo-resistance [60–62,73,74] (Table 1). Cancer EVs may contribute to new blood vessel formation by transferring to recipient fibroblasts and endothelial cells pro-angiogenic miRNAs such as miR-155, miR-210 and miR-494, which are under the regulation of the hypoxia-inducible factor (HIF) 1α [75–79] (Table 2). Moreover, several studies on EVs released by cancer cells indicate that they promote the development of a pre-metastatic niche by transferring either proteins or oncogenic miRNAs [80–83]. For instance, some miRNAs (miR-125b, miR-130b and miR-155) present in prostate cancer and released by EVs have been shown to confer a protumorigenic phenotype to adipose-derived mesenchymal stem cells [84].

Table 1. The role of EVs in chemo-resistance and immune-modulation. Several proteins and exRNAs have been described to be involved in tumor chemo-resistance and immune-modulation.

Biological Effect	Mechanism of Action	Cell Source	Target	References
Resistance to chemotherapy	Transfer of MDR-1/P-gp	Docetaxel-resistant prostate cancer	Docetaxel-sensitive prostate cancer	[59]
	Transfer of miR-100, miR-222, miR-30a and miR-17	Adriamycin and docetaxel-resistant breast cancer	Adriamycin and docetaxel-sensitive breast cancer	[60]
	Transfer of miR-21	Platinum-resistant ovarian cancer	Platinum-sensitive ovarian cancer	[61]
	Transfer of miR-100-5p, miR-21 and miR-133b	Cisplatin-resistant lung cancer	Cisplatin-sensitive lung cancer	[62,63]
	Transfer of miR-21, which downregulates APAF1	Stroma	Ovarian cancer	[74]
	Transfer of miR-146a with Snail mRNA	Cancer-Associated Fibroblasts	Pancreatic cancer	[73]
	Activation of the antiviral/ NOTCH3 signaling pathway	Stroma	Breast cancer	[72]
Tumor immune-escape	Release of pro-inflammatory cytokines by macrophages, possibly mediated by miR-21 and miR-29a	Breast and lung cancer, melanoma	Tumor cells, fibroblasts, endothelial cells, and immune cells	[85–87]
	Inhibition of dendritic cell maturation and functions, by delivering specific miRNAs (e.g., miR-203, miR-212-3p)	Renal carcinoma, pancreatic cancer, melanoma	Dendritic and T cells	[87–89]
	MDSCs activation, which leads to TGF-β-mediated suppression of T cell activity	Melanoma and colorectal carcinoma	$CD14^+$ monocytes	[90,91]
	Suppression of the T-cell activity mediated by PDL-1, TGF-β, Fas ligand and TRAIL	Melanoma, colorectal, gastric and prostate cancer, head and neck squamous cell carcinoma	$CD8^+$T cells	[92–96]
	Inhibition of NK cell cytotoxic activity, possibly mediated by MIC A ligand of NKG2D receptor	Mammary carcinoma, melanoma, cervical, head and neck, liver cancer	NK cells	[97–99]
Enhancement of immune response	Activation of a tumor antigen-specific immune response in humans	Melanoma and non-small cell lung cancer patients-derived dendritic cells	systemic administration	[100,101]

EVs: extracellular vesicles, MDR: multidrug resistance protein, APAF1: apoptotic protease-activating factor 1, MDSCs: myeloid-derived suppressor cells, TGF-β: trasforming growth factor-β, PDL-1: programmed death-ligand 1, TRAIL: tumor necrosis factor-related apoptosis-inducing ligand, NK: natural killer, MIC: MHC class I–related chain, NKG2D: NKG2-D type II integral membrane protein.

Table 2. Role of EVs as biomarkers of tumor progression. Cancer-derived EV content has been proposed as a tumor biomarker and has been related to several processes involved in tumor aggressiveness.

Biological Effect	Mechanism of Action	Cell Source	Target	References
Tumor biomarkers	Transfer of miR-21, miR-141, miR-200a, miR-200b, miR-200c, miR-203, miR-205 and miR-214	Ovarian cancer	Serum	[102]
	Transfer of miR-17-3p, miR-21, miR-29a, miR-106a, miR-146 miR-155, miR-191, miR-192, miR-203, miR-205, miR-210, miR-212 and miR-214	Lung cancer	Serum	[85,103]
	Transfer of miR-18a, miR-221 and miR-224	Hepatocellular carcinoma	Serum	[104]
Pro-angiogenic effect	Transfer of proangiogenic miRNAs, mostly regulated by HIF-1α (miR-155-5p, miR-210 and miR-494)	Melanoma, hepatocellular, lung and renal adenocarcinoma	CAFs and endothelial cells	[75,77–79,81]
Decrease cell-to-cell adhesion	Reduction of E-cadherin, let-7i and β-catenin expression, and increase of Snail1-2, Twist1-2, Sip1, vimentin, ZEB2 and N-cadherin expression, activation of MAPK pathway	Breast and bladder cancer, melanoma	Mammary and urothelial cells epithelial cells, primary melanocytes	[67–69]
Increase in cell migration/invasion	Lipids and proteins (e.g., CD81)-dependent stimulation of the cancer cell motility via Wnt signaling	Cancer Associated Fibroblasts	Melanoma, breast and prostate cancer	[70,71]
Development of premetastatic niche	Delivery of TYRP2, VLA4, HSP70, an HSP90 isoform and the MET oncoprotein	Melanoma	Bone marrow progenitor cell	[83]
	Exosomal expression of tumor-specific integrin patterns	Osteosarcoma, rhabdomyosarcoma, Wilms tumor, skin and uveal melanoma, breast, colorectal, pancreatic and gastric cancer	Brain, lung and liver epithelium	[82]
	Delivery of MIF	Pancreatic ductal adenocarcinoma	Kupffer cell	[80]
	Delivery of specific oncogenic miRNAs, e.g., miR-125b, miR-130b and miR-155, which induce a neoplastic reprogramming of recipient cells	Prostate, renal cancer	Adipose-derived stem cells, lung epithelium	[81,84]

HIF-1a: hypoxia inducible factor 1α, HSP90: heat shock protein 90, MET: hepatocyte growth factor receptor., TYRP2: tyrosinase-related protein-2, VLA4: very late antigen 4, HSP70: heat shock protein 70, MIF: macrophage migration inhibitory factor.

A contribution in favoring the tumor immune-escape of EVs released by cancer cells has been also suggested [105,106] (Table 1). The mechanisms involved the activation of tumor-associated macrophages [85–87], suppressor myeloid cells [90,91] and the inhibition of NK cell activity [97–99]. By expressing PDL1 [92,93], the transforming growth factor (TGF) beta [94], the tumor necrosis factor–related apoptosis-inducing ligand (TRAIL) and the Fas ligand [95,96], cancer EVs exhibit an immunosuppressive activity on T cells. Moreover, cancer EVs inhibit the maturation of dendritic cells through a mechanism involving the expression of HLA-G [88] and specific miRNAs, such as miR-203 and miR-212-3p [89]. Despite tumor EVs have been mainly implicated in the tumor immune escape, they can also be exploited to cross-present tumor antigens to the antigen presenting cells eliciting an antigen-specific cytotoxic lymphocyte anti-tumoral response [100,101]. However, clinical trials based on this assumption have provided conflicting results [107–109].

At present, most of the studies looking for exRNAs as cancer biomarkers have been performed on whole blood, urine and cerebrospinal fluids. Recently, several studies have explored the detection of exRNAs associated with the EVs. Despite the fact that the quantitative and stoichiometric analyses revealed that many miRNAs are present in less than one single copy per single exosome [110], several studies indicate a potential utility as cancer biomarkers [102,103]. For instance, in hepatocellular carcinoma, the expression by serum EVs of miR-18a, miR-221and miR-224 has been suggested as potential diagnostic biomarkers [104]. Fabbri et al. demonstrated that EV-associated miR-21 and miR-29a bound to a Toll-like receptor family favoring an inflammatory pro-metastatic response in lung [85]. On the other hand, the expression of miR-21 in serum EVs in patients with breast cancer correlates with a favorable outcome [111]. By comparing the miRNA signature of ovarian cancer EVs with that of EVs from normal subjects, Taylor and colleagues suggested a potential utility to screening asymptomatic patients [102]. A significant similarity of EV-associated miRNAs was observed with tumor-derived miRNAs in lung adenocarcinoma [103]. Moreover, the miRNA patterns of patients were clearly distinct from those of normal controls, suggesting that circulating EV-associated miRNAs might be useful as a non-invasive screening test [103].

4. Salivary EVs as Biomarkers

EVs are particularly enriched in saliva, which in respect to blood does not undergo coagulation. This is an important issue because many studies have been performed on serum. Coagulation induces a consistent release of EVs from platelets, thus modifying the composition of circulating EVs [112]. Salivary EVs should derive in part from salivary glands and in part from circulation: indeed, about a 30% similarity of salivary and plasma proteome has been described by a few studies [113–115]. In particular, using liquid chromatography and mass spectrometry, 19,474 unique peptides have been isolated from whole saliva in a multicenter study [113]. Protein annotation was assessed by matching the identified peptides with a recently published dataset of the human plasma proteome [116], and 1939 different proteins were identified as commonly expressed in blood and saliva. However, a puzzling aspect is the expression of neuronal markers in salivary EVs with significant changes in the miRNA pattern and in the proteomic profile after a head concussion [117] and in neurological diseases [118,119]. Moreover, the EV composition may be affected by the presence in saliva of viruses, including the human papillomavirus (HPV) [120–124] and the neurotropic human herpesviruses (e.g., HHV-6), which are detectable in the saliva of infected subjects [125].

A critical aspect in the use of salivary EVs as biomarkers is the purification technique that is used (Table 3). In fact, results may vary depending on the purified subpopulations and the presence of contaminants, such as bacterial flora. Therefore, accurate mouth washing, careful standardization on saliva collection and sample filtration are recommended to abate the bacterial load.

Table 3. Biomarkers detected in salivary EVs. Salivary EVs can be purified using different EV isolation techniques and can be exploited as biomarkers because they contain disease-related proteins and exRNA.

Disease	Isolation Method	EV Biomarkers	Type of Biomarker	References
Brain injury and neurological disorders	Differential ultracentrifugation	CDC2, CSNK1A1, and CTSD	mRNA	[117]
	XYCQ EV Enrichment KIT	α-synuclein	protein	[119]
Oral squamous cell carcinoma	Differential ultracentrifugation	CD63	protein	[126,127]
	Differential ultracentrifugation	PPIA	protein	[128]
	Charge-based precipitation	miR-412-3p, miR-512-3p, miR-27a-3p, miR-494-3p, miR-302b-3p, miR-517b-3p	miRNA	[129]
Lung cancer	Affinity chromatography column combined with filter system (ACCF)	Annexin A1, A2, A3, A5, A6, A11; NPRL2; CEACAM1; MUC1; PROM1; HIST1H4A; TNFAIP3	protein	[130]
	Affinity chromatography column combined with filter system (ACCF)	BPIFA1, CRNN, MUC5B, IQGAP	protein	[131]
Head and neck carcinoma	Differential ultracentrifugation	miR-486-5p, miR-486-3p, miR-10b-5p, miR-122	miRNA	[132]
Pancreatic cancer	Total Exosome Isolation Reagent (Invitrogen)	miR-1246, miR-4644	miRNA	[133]
	Differential ultracentrifugation	Apbb1ip, Aspn, BCO31781, Daf2, Foxp1, Gng2, Incenp	mRNA	[134]

CDC2: Cyclin-dependent kinase A-1, CSNK1A1: Casein Kinase 1 Alpha 1, CTSD: Cathepsin D, PPIA: Peptidyl-prolyl cis-trans isomerase A, NPRL2: GATOR complex protein NPRL2, CEACAM1: Carcinoembryonic antigen-related cell adhesion molecule 1, MUC1: Mucin 1, PROM1: Prominin 1, HIST1H4A: Histone H4, TNFAIP3: Tumor necrosis factor alpha-induced protein 3, BPIFA1: BPI fold-containing family A member 1, CRNN: Cornulin, MUC5B: Mucin 5b, IQGAP: Ras GTPase-activating-like protein IQGAP1.

Differential ultracentrifugation or density gradient ultracentrifugation are considered the gold standard for the purification of EV subpopulations. These techniques have been further implemented with the combined use of the immune-affinity capture of exosomes [34]. To improve the separation of vesicles from non-vesicular components a floating technique has been proposed, based on gradient fractioning centrifugation, with samples applied to the bottom of tubes [135]. However, the standardization of these techniques may be difficult, as the results are influenced not only by the centrifugal radius of the rotor and g force type, but also by the viscosity of the starting solution. In addition, due to mechanical damage, membrane debris are generated, as seen by electron microscopy. Moreover, the difficult detection of proteins and RNAs has been described [136–139]. To avoid shear stress due to ultracentrifugation, size exclusion chromatography has been employed with the aim to separate small vesicles from protein contaminants [140–142]. Immuno-affinity purification allows for the recovery of sub fractions of EVs based on the expression of surface markers [139,143–145], and several kits are commercially available. Microfiltration has also been used with membranes with appropriate pore sizes to remove cell debris and apoptotic bodies [143]. However, this technique is limited by EV adhesion to membranes and pore clogging. In addition, to isolate small biological samples, all these techniques may have a low efficient recovery of EVs. Another approach for isolating EVs from biological liquids is based on polymeric precipitation [146–151]. This approach allows for a rapid precipitation of EVs, but it is limited by the co-precipitation of proteins of a non-vesicular origin such as lipoproteins [136,152,153]. Recently, a new technique based on electric field-induced release and measurement has been successfully applied to liquid biopsy in saliva [154]. Using this technique, the mutation of epidermal growth factor receptor (EGFR) in patients with lung cancers was detected and matched with biopsy genotyping [154,155]. Moreover, the electric field-induced release has been combined with the magnetic beads immune-capturing of exosomes [156,157], resulting in a highly sensitive and specific method of exRNA extraction and analysis. Compared to polymeric precipitation and differential centrifugation, this approach is less time consuming, requires smaller sample volumes and does not involve sample lysis that may reduce exRNA yield. However, for each EV extraction, the capture probe that is attached to the magnetic beads allows for the isolation of only those EVs containing the exosome-specific surface marker used for capturing EVs [156]. In fact, EVs are a heterogeneous population of vesicles, and individual EV analyses show that not all EVs co-express the same tetraspanin. Therefore, this technique may not include the whole pattern of EV-associated exRNA.

By quantitative nano-structural and single molecule force spectroscopy, Sharma et al. [126] performed a bio-molecular analysis of exosomes present in the saliva from patients with oral cancer. They demonstrated that exosomes were augmented in number and size, displayed a dissimilar morphology and showed an increased expression of CD63. Similarly, Zlotogorski-Hurvitz et al. [127] described a bigger salivary exosome concentration and size in patients with oral cancers in comparison with healthy subjects, a higher expression of CD63 and a decreased expression of CD9 and CD81. Few other studies performed a proteomic analysis of salivary exosomes in search of potential biomarkers of oral [128] and lung carcinomas [130]. A higher expression of the CD63 molecule was observed in EVs from the saliva of patients with oral cancers in respect to normal subjects [126]. Sun et al. performed a comparative proteomic analysis of salivary EVs in normal subjects and lung cancer patients [131]. In this study, several proteins were found to be dysregulated, and four of them were present in both salivary microvesicles and exosomes, suggesting their potential use for the detection of lung cancer.

It has been reported that in saliva, the bulk of miRNAs is packaged in exosomes [13]. In fact, miRNAs are easily detectable in EVs present in saliva [158,159]. Several studies focused on the possibility of exRNA isolation from saliva and oral samples [160–162] and in particular on salivary EV associated miRNAs in patients with oral cancer [160–164].

Langevin et al. performed a comprehensive miRNA sequence analysis of EVs derived from the saliva of patients with head and neck carcinomas and identified a distinct pattern of secretion and, in particular, miRNAs secreted only by cancer cells [132]. Some miRNAs, such as miR-486-5p, miR-486-3p and miR-10b-5p, were specifically overexpressed in the EVs of a subset of head and neck carcinomas. Machida and colleagues showed that miR-1246 and miR-4644 present in salivary EVs are potential biomarkers of cancers of the pancreato-biliary tract [133]. Taken together, these analyses may provide the bases for the development of new tumor biomarkers (Table 3).

A transcriptomic signature specific for pancreatic [165] and ovarian cancers [166] and proteomic signature modifications in lung cancer [167] have been described in whole saliva. Zhang et al. [165] demonstrated that the combination of KRAS, metyl CpG binding domain protein 3 like 2 (MBD3L2), acrosomal vesicle 1 (ACRV1), and dolichyl-phosphate mannosyltransferase subunit 1 (DPM1) mRNAs in saliva may differentiate patients with pancreatic carcinomas from patients with chronic pancreatitis and healthy subjects with a high sensitivity and specificity. Moreover, a transcriptomic analysis of salivary EVs by next generation sequencing showed the presence of many coding and non-coding RNAs, such as mRNAs for several proteins, miRNAs, snoRNAs, piRNAs, and lncRNAs [48,168]. Palanisamy et al. [169] found, in exosomes isolated from saliva, 509 mRNA transcripts, which once incorporated in keratinocytes were able to modify the protein expression in these cells. Moreover, exosomes from adenocarcinoma of the pancreatic ducts were able to modify the biology of exosomes derived from the salivary gland and induce changes in the salivary biomarker profiles [134]. Similarly, they showed an interaction between exosomes derived from the human metastatic mammary gland epithelial adenocarcinoma cell line MDA-MB-231 cells and exosomes derived from the human submandibular gland (HSG) cells. This interaction induced an activation of the HSG cell transcriptional machinery with an increase of total cellular RNA and transcriptomic and proteomic changes [170]. Salivary EVs derived from patients with pancreatic carcinoma were shown to inhibit NK cell activation, thus favoring tumor immune escape [171].

We analyzed the zeta potential of salivary EVs and, based on their negative charge, we developed a charge-based precipitation protocol. This technique allows for the efficient recovery of exRNA from a salivary EV population with a very homogeneous size and shape [159] (Figure 1). In a recent work [129], we used the charge-based precipitation method to isolate EVs from the saliva of patients with oral squamous cell carcinoma (OSCC) to investigate the presence of exRNAs suitable as biomarkers. Our aim was to assess whether this quick, simple and efficient technique could be useful for detecting exRNA in the salivary EVs of patients with OSCC. The diagnosis of OSCC is based on oral examination and histological analysis. However, the identification of salivary biomarkers may have potential prognostic and therapeutic values. To exclude misleading results due to a different exposition to risk factors, at the time of patients' recruitment, subjects included in our study were checked for their habits regarding smoking and alcohol consumption. In fact, smoke and alcohol consumption has been described as potentially affecting the composition of EV-associated exRNA [172,173]. Therefore, patients and controls were matched to obtain a similar distribution of risk factors among the two groups to reduce this bias. Moreover, 5 out 21 patients were positive to the Human Papilloma Virus (HPV). To avoid the detection of exRNA of viral origin, we screened EVs for the presence of about 800 miRNAs with human-specific primers. Although HPV infection may alter the EV release and cargo, we did not observe any significant change in the size and concentration of EVs from HPV-positive patients compared to negative patients. A differential expression of the EV miRNA signature in OSCC cells infected or not by HPV has been previously shown [124]. In this study, the authors observed that HPV infected cells released EVs enriched with 14 miRNAs, whereas non-infected cells overexpressed 19 miRNAs. The cohorts of patients we studied were too small to draw any conclusions. However, we did not observe the differential expression of miRNAs, which has been previously described for EVs released by in vitro infected OSCC cells.

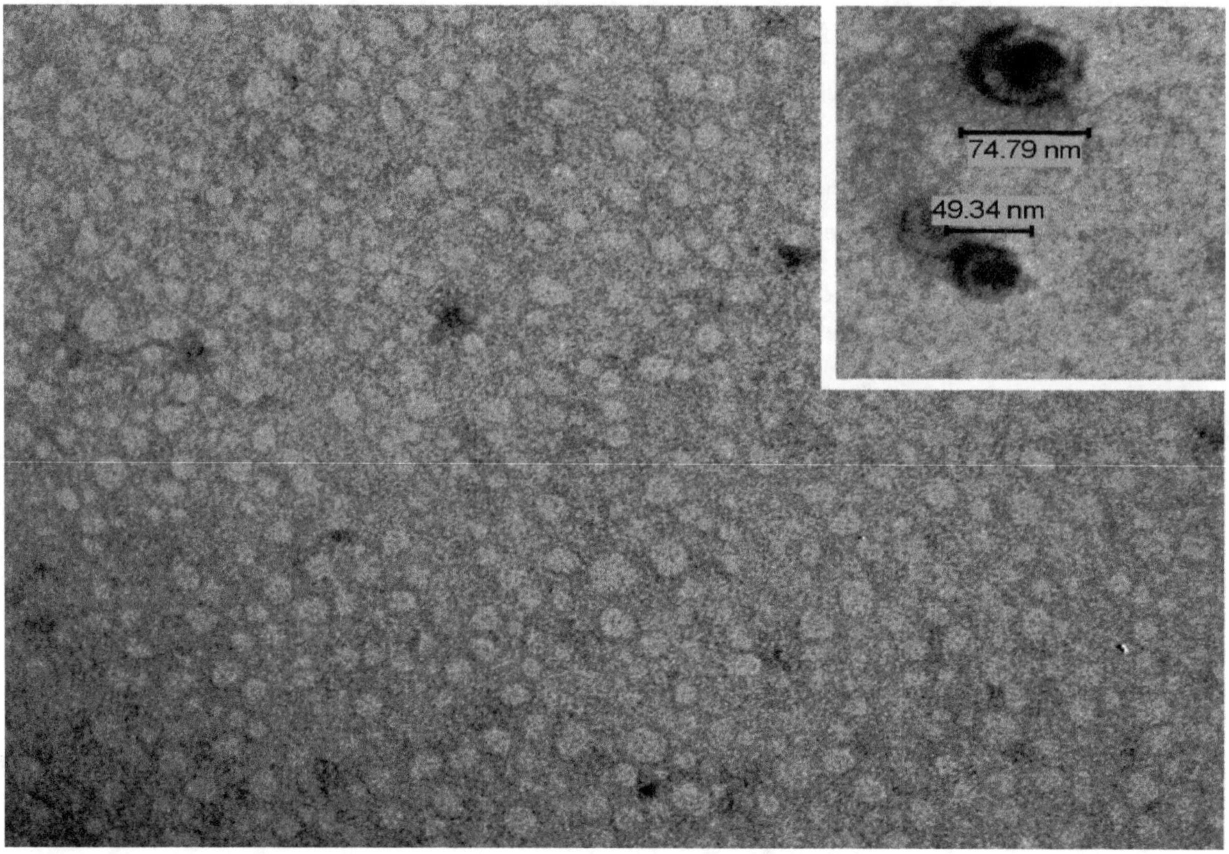

Figure 1. Salivary EVs characterization. A representative transmission electron microscopy image of EVs isolated by a charge-based precipitation method, showing a carpet of vesicles in the nano-range. In the inset, the bars indicate the size of the extracellular vesicles (EVs). The preparation was stained with NanoVan (JEOL Jem-1010 electron microscope, original magnification ×75,000; inset ×150,000).

By comparing the miRNA expression of cancer patients and matched controls, we observed an up-regulation of miR-412-3p, miR-512-3p, miR-27a-3p and miR-494-3p in patients with oral squamous cell carcinoma. MiR-512-3p and miR-412-3p were also potentially sensitive and specific biomarkers, as indicated by the high AUC values (0.847 and 0.871 respectively, with p values < 0.02) and a maximum Youden's Index. Interestingly, we also observed an exclusive expression of miR-302b-3p and miR-517b-3p in cancer EVs. Moreover, we performed a bio-informatic analysis to better understand whether the tumor-enriched miRNAs could be functionally related to the tumor. We observed that eight tumor-related pathways were potentially targeted by these miRNAs. In particular, miR-512-3p and miR-27a-3p may target 7 and 20 genes, respectively, of the ErbB signaling pathway, which is known to promote cell proliferation and survival in cancer [174] and is activated in oral carcinomas [175–177]. MiR-512-3p, miR-27a-3p, and miR-302b-3p could potentially target proteoglycan genes and CD44 involved in c-Fos-mediated cell invasion and migration [178], ERK1/2 phosphorylation [179] and the phenotype of oral cancer stem cells [180]. Moreover, miR-512-3p, miR-412-3p, miR-27a-3p, and miR-302b-3p reduced the expression of TGFβR2, frequently reduced in cancer and stroma cells in patients with oral squamous carcinomas [181]. Increased levels of the oncogenic miR-27a-3p has also been detected in EVs obtained from the plasma of OSCC patients [182]. In this study, a comparable miRNA signature was observed between plasma EVs and EVs released by OSCC cells in vitro.

Recent studies have shown that EVs also contain lncRNAs [183]. The expression of lncRNAs has not been investigated in salivary EVs. However, salivary lncRNAs may represent a potential marker for OSSC [184]. In fact, a subset of lncRNAs was correlated with high metastatic OSCC. In particular, the lncRNA HOTAIR was found to be highly expressed in the saliva of patients with lymph node metastasis. Therefore, besides miRNAs, the search for lncRNAs in salivary EVs could be a valuable diagnostic and prognostic tool for OSCC.

5. Conclusions

Taken together, these studies suggest that EVs derived from cancer cells may modulate the function and may induce epigenetic changes in neighboring or distant cells. These biological effects are related to the delivery of transcripts that are specific of the originator cells. Several studies have shown a prominent role of exRNAs associated with vesicles. Since EVs may retain the molecular signature of the cell of origin, it has been suggested that they are a potential diagnostic exploitation. The salivary EV composition may reflect the presence of local or systemic diseases and has been investigated as a potential biomarker for both oral and non-oral cancers. Changes in the molecular composition of the EVs of non-oral cancers may either depend on their derivation from blood (since salivary glands are vascularized) or be the consequence of phenotypic changes occurring in gland cells (as the results of the stimulation by circulating cancer EVs). However, so far, available studies are relatively few and include a low number of patients. Further studies are necessary to optimize the protocol of EV isolation from saliva in order to obtain reproducible results. Moreover, the use of the EV content as a biomarker should take into account that this may be influenced by a number of cancer-associated risk factors, such as viral infections, smoking, alcohol abuse, as well as a number of non-cancer-associated factors related to concomitant diseases. However, these limitations in the use of EVs as biomarkers are not restricted to saliva, but may influence EVs derived from any biological fluid. Since saliva is an easily obtainable non-invasive bio-fluid particularly enriched in EVs, it may represent a new approach for cancer biomarker discovery. However, to define whether salivary EVs have a real clinical diagnostic and prognostic potential would require comparative studies between EVs derived from tumor cells, blood and saliva, which are not at present available.

Author Contributions: All authors equally contributed to the conceptualization of the article. The research of the pertinent literature was performed by M.C.D., G.C.; writing—original draft preparation, G.C.; review and editing, C.G., M.C.D., and G.C.

References

1. Xi, X.; Li, T.; Huang, Y.; Sun, J.; Zhu, Y.; Yang, Y.; Lu, Z.J. RNA Biomarkers: Frontier of Precision Medicine for Cancer. *NonCoding RNA* **2017**, *3*, 9. [CrossRef] [PubMed]
2. Lan, H.; Lu, H.; Wang, X.; Jin, H. MicroRNAs as Potential Biomarkers in Cancer: Opportunities and Challenges. *Biomed. Res. Int.* **2015**, *2015*, 125094. [CrossRef] [PubMed]
3. Martinez-Ledesma, E.; Verhaak, R.G.W.; Treviño, V. Identification of a Multi-Cancer Gene Expression Biomarker for Cancer Clinical Outcomes Using a Network-Based Algorithm. *Sci. Rep.* **2015**, *5*, 11966. [CrossRef] [PubMed]
4. Cancer Genome Atlas Network. Comprehensive Molecular Portraits of Human Breast Tumours. *Nature* **2012**, *490*, 61–70. [CrossRef] [PubMed]
5. Cuzick, J.; Swanson, G.P.; Fisher, G.; Brothman, A.R.; Berney, D.M.; Reid, J.E.; Mesher, D.; Speights, V.O.; Stankiewicz, E.; Foster, C.S.; et al. Prognostic Value of an RNA Expression Signature Derived from Cell Cycle Proliferation Genes in Patients with Prostate Cancer: A Retrospective Study. *Lancet Oncol.* **2011**, *12*, 245–255. [CrossRef]
6. Asmann, Y.W.; Necela, B.M.; Kalari, K.R.; Hossain, A.; Baker, T.R.; Carr, J.M.; Davis, C.; Getz, J.E.; Hostetter, G.; Li, X.; et al. Detection of Redundant Fusion Transcripts as Biomarkers or Disease-Specific Therapeutic Targets in Breast Cancer. *Cancer Res.* **2012**, *72*, 1921–1928. [CrossRef] [PubMed]
7. Attard, G.; Clark, J.; Ambroisine, L.; Fisher, G.; Kovacs, G.; Flohr, P.; Berney, D.; Foster, C.S.; Fletcher, A.; Gerald, W.L.; et al. Duplication of the Fusion of TMPRSS2 to ERG Sequences Identifies Fatal Human Prostate Cancer. *Oncogene* **2008**, *27*, 253–263. [CrossRef] [PubMed]

8. Arroyo, J.D.; Chevillet, J.R.; Kroh, E.M.; Ruf, I.K.; Pritchard, C.C.; Gibson, D.F.; Mitchell, P.S.; Bennett, C.F.; Pogosova-Agadjanyan, E.L.; Stirewalt, D.L.; et al. Argonaute2 Complexes Carry a Population of Circulating MicroRNAs Independent of Vesicles in Human Plasma. *Proc. Natl. Acad. Sci. USA* **2011**, *108*, 5003–5008. [CrossRef]
9. Vickers, K.C.; Palmisano, B.T.; Shoucri, B.M.; Shamburek, R.D.; Remaley, A.T. MicroRNAs Are Transported in Plasma and Delivered to Recipient Cells by High-Density Lipoproteins. *Nat. Cell Biol.* **2011**, *13*, 423–433. [CrossRef]
10. Guallar, D.; Wang, J. RNA-Binding Proteins in Pluripotency, Differentiation, and Reprogramming. *Front. Biol. (Beijing)* **2014**, *9*, 389–409. [CrossRef]
11. Quesenberry, P.J.; Aliotta, J.M. Cellular Phenotype Switching and Microvesicles. *Adv. Drug Deliv. Rev.* **2010**, *62*, 1141–1148. [CrossRef] [PubMed]
12. Chen, X.; Liang, H.; Zhang, J.; Zen, K.; Zhang, C.-Y. Secreted MicroRNAs: A New Form of Intercellular Communication. *Trends Cell Biol.* **2012**, *22*, 125–132. [CrossRef] [PubMed]
13. Gallo, A.; Tandon, M.; Alevizos, I.; Illei, G.G. The Majority of MicroRNAs Detectable in Serum and Saliva Is Concentrated in Exosomes. *PLoS ONE* **2012**, *7*, e30679. [CrossRef] [PubMed]
14. Cocucci, E.; Meldolesi, J. Ectosomes and Exosomes: Shedding the Confusion between Extracellular Vesicles. *Trends Cell Biol.* **2015**, *25*, 364–372. [CrossRef] [PubMed]
15. Bobrie, A.; Colombo, M.; Krumeich, S.; Raposo, G.; Théry, C. Diverse Subpopulations of Vesicles Secreted by Different Intracellular Mechanisms Are Present in Exosome Preparations Obtained by Differential Ultracentrifugation. *J. Extracell. Vesicles* **2012**, *1*, 18397. [CrossRef] [PubMed]
16. Ratajczak, M.Z.; Ratajczak, J. Extracellular Microvesicles as Game Changers in Better Understanding the Complexity of Cellular Interactions-From Bench to Clinical Applications. *Am. J. Med. Sci.* **2017**, *354*, 449–452. [CrossRef]
17. Ratajczak, J.; Wysoczynski, M.; Hayek, F.; Janowska-Wieczorek, A.; Ratajczak, M.Z. Membrane-Derived Microvesicles: Important and Underappreciated Mediators of Cell-to-Cell Communication. *Leukemia* **2006**, *20*, 1487–1495. [CrossRef]
18. Kowal, J.; Tkach, M.; Théry, C. Biogenesis and Secretion of Exosomes. *Curr. Opin. Cell Biol.* **2014**, *29*, 116–125. [CrossRef]
19. Hyenne, V.; Apaydin, A.; Rodriguez, D.; Spiegelhalter, C.; Hoff-Yoessle, S.; Diem, M.; Tak, S.; Lefebvre, O.; Schwab, Y.; Goetz, J.G.; et al. RAL-1 Controls Multivesicular Body Biogenesis and Exosome Secretion. *J. Cell Biol.* **2015**, *211*, 27–37. [CrossRef]
20. Tricarico, C.; Clancy, J.; D'Souza-Schorey, C. Biology and Biogenesis of Shed Microvesicles. *Small Gtpases* **2017**, *8*, 220–232. [CrossRef]
21. Hristov, M.; Erl, W.; Linder, S.; Weber, P.C. Apoptotic Bodies from Endothelial Cells Enhance the Number and Initiate the Differentiation of Human Endothelial Progenitor Cells in Vitro. *Blood* **2004**, *104*, 2761–2766. [CrossRef] [PubMed]
22. Balaj, L.; Lessard, R.; Dai, L.; Cho, Y.-J.; Pomeroy, S.L.; Breakefield, X.O.; Skog, J. Tumour Microvesicles Contain Retrotransposon Elements and Amplified Oncogene Sequences. *Nat. Commun.* **2011**, *2*, 180. [CrossRef] [PubMed]
23. Lázaro-Ibáñez, E.; Sanz-Garcia, A.; Visakorpi, T.; Escobedo-Lucea, C.; Siljander, P.; Ayuso-Sacido, A.; Yliperttula, M. Different GDNA Content in the Subpopulations of Prostate Cancer Extracellular Vesicles: Apoptotic Bodies, Microvesicles, and Exosomes. *Prostate* **2014**, *74*, 1379–1390. [CrossRef] [PubMed]
24. Guescini, M.; Genedani, S.; Stocchi, V.; Agnati, L.F. Astrocytes and Glioblastoma Cells Release Exosomes Carrying MtDNA. *J. Neural Transm. (Vienna)* **2010**, *117*, 1–4. [CrossRef] [PubMed]
25. Sansone, P.; Savini, C.; Kurelac, I.; Chang, Q.; Amato, L.B.; Strillacci, A.; Stepanova, A.; Iommarini, L.; Mastroleo, C.; Daly, L.; et al. Packaging and Transfer of Mitochondrial DNA via Exosomes Regulate Escape from Dormancy in Hormonal Therapy-Resistant Breast Cancer. *Proc. Natl. Acad. Sci. USA* **2017**, *114*, E9066–E9075. [CrossRef] [PubMed]
26. Ratajczak, J.; Miekus, K.; Kucia, M.; Zhang, J.; Reca, R.; Dvorak, P.; Ratajczak, M.Z. Embryonic Stem Cell-Derived Microvesicles Reprogram Hematopoietic Progenitors: Evidence for Horizontal Transfer of MRNA and Protein Delivery. *Leukemia* **2006**, *20*, 847–856. [CrossRef] [PubMed]

27. Deregibus, M.C.; Cantaluppi, V.; Calogero, R.; Lo Iacono, M.; Tetta, C.; Biancone, L.; Bruno, S.; Bussolati, B.; Camussi, G. Endothelial Progenitor Cell Derived Microvesicles Activate an Angiogenic Program in Endothelial Cells by a Horizontal Transfer of MRNA. *Blood* **2007**, *110*, 2440–2448. [CrossRef] [PubMed]
28. Valadi, H.; Ekström, K.; Bossios, A.; Sjöstrand, M.; Lee, J.J.; Lötvall, J.O. Exosome-Mediated Transfer of MRNAs and MicroRNAs Is a Novel Mechanism of Genetic Exchange between Cells. *Nat. Cell Biol.* **2007**, *9*, 654–659. [CrossRef] [PubMed]
29. Fatima, F.; Nawaz, M. Vesiculated Long Non-Coding RNAs: Offshore Packages Deciphering Trans-Regulation between Cells, Cancer Progression and Resistance to Therapies. *Noncoding RNA* **2017**, *3*, 10. [CrossRef] [PubMed]
30. van Niel, G.; D'Angelo, G.; Raposo, G. Shedding Light on the Cell Biology of Extracellular Vesicles. *Nat. Rev. Mol. Cell Biol.* **2018**, *19*, 213–228. [CrossRef] [PubMed]
31. Baietti, M.F.; Zhang, Z.; Mortier, E.; Melchior, A.; Degeest, G.; Geeraerts, A.; Ivarsson, Y.; Depoortere, F.; Coomans, C.; Vermeiren, E.; et al. Syndecan-Syntenin-ALIX Regulates the Biogenesis of Exosomes. *Nat. Cell Biol.* **2012**, *14*, 677–685. [CrossRef] [PubMed]
32. D'Souza-Schorey, C.; Chavrier, P. ARF Proteins: Roles in Membrane Traffic and Beyond. *Nat. Rev. Mol. Cell Biol.* **2006**, *7*, 347–358. [CrossRef] [PubMed]
33. Nabhan, J.F.; Hu, R.; Oh, R.S.; Cohen, S.N.; Lu, Q. Formation and Release of Arrestin Domain-Containing Protein 1-Mediated Microvesicles (ARMMs) at Plasma Membrane by Recruitment of *TSG101* Protein. *Proc. Natl. Acad. Sci. USA* **2012**, *109*, 4146–4151. [CrossRef]
34. Jeppesen, D.K.; Fenix, A.M.; Franklin, J.L.; Higginbotham, J.N.; Zhang, Q.; Zimmerman, L.J.; Liebler, D.C.; Ping, J.; Liu, Q.; Evans, R.; et al. Reassessment of Exosome Composition. *Cell* **2019**, *177*, 428–445. [CrossRef] [PubMed]
35. Kalra, H.; Simpson, R.J.; Ji, H.; Aikawa, E.; Altevogt, P.; Askenase, P.; Bond, V.C.; Borràs, F.E.; Breakefield, X.; Budnik, V.; et al. Vesiclepedia: A Compendium for Extracellular Vesicles with Continuous Community Annotation. *PLoS Biol.* **2012**, *10*, e1001450. [CrossRef] [PubMed]
36. Kim, D.-K.; Kang, B.; Kim, O.Y.; Choi, D.-S.; Lee, J.; Kim, S.R.; Go, G.; Yoon, Y.J.; Kim, J.H.; Jang, S.C.; et al. EVpedia: An Integrated Database of High-Throughput Data for Systemic Analyses of Extracellular Vesicles. *J. Extracell. Vesicles* **2013**, *2*, 20384. [CrossRef] [PubMed]
37. Mathivanan, S.; Simpson, R.J. ExoCarta: A Compendium of Exosomal Proteins and RNA. *Proteomics* **2009**, *9*, 4997–5000. [CrossRef] [PubMed]
38. Li, Y.; Zheng, Q.; Bao, C.; Li, S.; Guo, W.; Zhao, J.; Chen, D.; Gu, J.; He, X.; Huang, S. Circular RNA Is Enriched and Stable in Exosomes: A Promising Biomarker for Cancer Diagnosis. *Cell Res.* **2015**, *25*, 981–984. [CrossRef]
39. Li, C.C.Y.; Eaton, S.A.; Young, P.E.; Lee, M.; Shuttleworth, R.; Humphreys, D.T.; Grau, G.E.; Combes, V.; Bebawy, M.; Gong, J.; et al. Glioma Microvesicles Carry Selectively Packaged Coding and Non-Coding RNAs Which Alter Gene Expression in Recipient Cells. *RNA Biol.* **2013**, *10*, 1333–1344. [CrossRef]
40. Villarroya-Beltri, C.; Baixauli, F.; Gutiérrez-Vázquez, C.; Sánchez-Madrid, F.; Mittelbrunn, M. Sorting It out: Regulation of Exosome Loading. *Semin. Cancer Biol.* **2014**, *28*, 3–13. [CrossRef]
41. Quesenberry, P.J.; Aliotta, J.; Deregibus, M.C.; Camussi, G. Role of Extracellular RNA-Carrying Vesicles in Cell Differentiation and Reprogramming. *Stem Cell Res.* **2015**, *6*, 153. [CrossRef] [PubMed]
42. Guescini, M.; Guidolin, D.; Vallorani, L.; Casadei, L.; Gioacchini, A.M.; Tibollo, P.; Battistelli, M.; Falcieri, E.; Battistin, L.; Agnati, L.F.; et al. C2C12 Myoblasts Release Micro-Vesicles Containing MtDNA and Proteins Involved in Signal Transduction. *Exp. Cell Res.* **2010**, *316*, 1977–1984. [CrossRef] [PubMed]
43. Batagov, A.O.; Kurochkin, I.V. Exosomes Secreted by Human Cells Transport Largely MRNA Fragments That Are Enriched in the 3′-Untranslated Regions. *Biol. Direct* **2013**, *8*, 12. [CrossRef] [PubMed]
44. Cheng, L.; Sun, X.; Scicluna, B.J.; Coleman, B.M.; Hill, A.F. Characterization and Deep Sequencing Analysis of Exosomal and Non-Exosomal MiRNA in Human Urine. *Kidney Int.* **2014**, *86*, 433–444. [CrossRef] [PubMed]
45. Crescitelli, R.; Lässer, C.; Szabó, T.G.; Kittel, A.; Eldh, M.; Dianzani, I.; Buzás, E.I.; Lötvall, J. Distinct RNA Profiles in Subpopulations of Extracellular Vesicles: Apoptotic Bodies, Microvesicles and Exosomes. *J. Extracell. Vesicles* **2013**, *2*, 20677. [CrossRef] [PubMed]
46. Hill, A.F.; Pegtel, D.M.; Lambertz, U.; Leonardi, T.; O'Driscoll, L.; Pluchino, S.; Ter-Ovanesyan, D.; Nolte-'t Hoen, E.N.M. ISEV Position Paper: Extracellular Vesicle RNA Analysis and Bioinformatics. *J. Extracell. Vesicles* **2013**, *2*, 22859. [CrossRef] [PubMed]

47. Huang, X.; Yuan, T.; Tschannen, M.; Sun, Z.; Jacob, H.; Du, M.; Liang, M.; Dittmar, R.L.; Liu, Y.; Liang, M.; et al. Characterization of Human Plasma-Derived Exosomal RNAs by Deep Sequencing. *BMC Genom.* **2013**, *14*, 319. [CrossRef] [PubMed]
48. Ogawa, Y.; Taketomi, Y.; Murakami, M.; Tsujimoto, M.; Yanoshita, R. Small RNA Transcriptomes of Two Types of Exosomes in Human Whole Saliva Determined by next Generation Sequencing. *Biol. Pharm. Bull.* **2013**, *36*, 66–75. [CrossRef]
49. Abels, E.R.; Breakefield, X.O. Introduction to Extracellular Vesicles: Biogenesis, RNA Cargo Selection, Content, Release, and Uptake. *Cell. Mol. Neurobiol.* **2016**, *36*, 301–312. [CrossRef]
50. Zhang, Y.; Liu, D.; Chen, X.; Li, J.; Li, L.; Bian, Z.; Sun, F.; Lu, J.; Yin, Y.; Cai, X.; et al. Secreted Monocytic MiR-150 Enhances Targeted Endothelial Cell Migration. *Mol. Cell* **2010**, *39*, 133–144. [CrossRef]
51. Collino, F.; Deregibus, M.C.; Bruno, S.; Sterpone, L.; Aghemo, G.; Viltono, L.; Tetta, C.; Camussi, G. Microvesicles Derived from Adult Human Bone Marrow and Tissue Specific Mesenchymal Stem Cells Shuttle Selected Pattern of MiRNAs. *PLoS ONE* **2010**, *5*, e11803. [CrossRef] [PubMed]
52. Goldie, B.J.; Dun, M.D.; Lin, M.; Smith, N.D.; Verrills, N.M.; Dayas, C.V.; Cairns, M.J. Activity-Associated MiRNA Are Packaged in Map1b-Enriched Exosomes Released from Depolarized Neurons. *Nucleic Acids Res.* **2014**, *42*, 9195–9208. [CrossRef] [PubMed]
53. Iavello, A.; Frech, V.S.L.; Gai, C.; Deregibus, M.C.; Quesenberry, P.J.; Camussi, G. Role of Alix in MiRNA Packaging during Extracellular Vesicle Biogenesis. *Int. J. Mol. Med.* **2016**, *37*, 958–966. [CrossRef] [PubMed]
54. Melo, S.A.; Sugimoto, H.; O'Connell, J.T.; Kato, N.; Villanueva, A.; Vidal, A.; Qiu, L.; Vitkin, E.; Perelman, L.T.; Melo, C.A.; et al. Cancer Exosomes Perform Cell-Independent MicroRNA Biogenesis and Promote Tumorigenesis. *Cancer Cell* **2014**, *26*, 707–721. [CrossRef] [PubMed]
55. McKenzie, A.J.; Hoshino, D.; Hong, N.H.; Cha, D.J.; Franklin, J.L.; Coffey, R.J.; Patton, J.G.; Weaver, A.M. KRAS-MEK Signaling Controls Ago2 Sorting into Exosomes. *Cell Rep.* **2016**, *15*, 978–987. [CrossRef] [PubMed]
56. Cha, D.J.; Franklin, J.L.; Dou, Y.; Liu, Q.; Higginbotham, J.N.; Demory Beckler, M.; Weaver, A.M.; Vickers, K.; Prasad, N.; Levy, S.; et al. KRAS-Dependent Sorting of MiRNA to Exosomes. *eLife* **2015**, *4*, e07197. [CrossRef] [PubMed]
57. Villarroya-Beltri, C.; Gutiérrez-Vázquez, C.; Sánchez-Cabo, F.; Pérez-Hernández, D.; Vázquez, J.; Martin-Cofreces, N.; Martinez-Herrera, D.J.; Pascual-Montano, A.; Mittelbrunn, M.; Sánchez-Madrid, F. Sumoylated HnRNPA2B1 Controls the Sorting of MiRNAs into Exosomes through Binding to Specific Motifs. *Nat. Commun.* **2013**, *4*, 2980. [CrossRef]
58. Shurtleff, M.J.; Temoche-Diaz, M.M.; Karfilis, K.V.; Ri, S.; Schekman, R. Y-Box Protein 1 Is Required to Sort MicroRNAs into Exosomes in Cells and in a Cell-Free Reaction. *ELife* **2016**, *5*, e19276. [CrossRef]
59. Corcoran, C.; Rani, S.; O'Brien, K.; O'Neill, A.; Prencipe, M.; Sheikh, R.; Webb, G.; McDermott, R.; Watson, W.; Crown, J.; et al. Docetaxel-Resistance in Prostate Cancer: Evaluating Associated Phenotypic Changes and Potential for Resistance Transfer via Exosomes. *PLoS ONE* **2012**, *7*, e50999. [CrossRef]
60. Chen, W.; Liu, X.; Lv, M.; Chen, L.; Zhao, J.; Zhong, S.; Ji, M.; Hu, Q.; Luo, Z.; Wu, J.; et al. Exosomes from Drug-Resistant Breast Cancer Cells Transmit Chemoresistance by a Horizontal Transfer of MicroRNAs. *PLoS ONE* **2014**, *9*, e95240. [CrossRef]
61. Crow, J.; Atay, S.; Banskota, S.; Artale, B.; Schmitt, S.; Godwin, A.K. Exosomes as Mediators of Platinum Resistance in Ovarian Cancer. *Oncotarget* **2017**, *8*, 11917–11936. [CrossRef] [PubMed]
62. Qin, X.; Yu, S.; Zhou, L.; Shi, M.; Hu, Y.; Xu, X.; Shen, B.; Liu, S.; Yan, D.; Feng, J. Cisplatin-Resistant Lung Cancer Cell-Derived Exosomes Increase Cisplatin Resistance of Recipient Cells in Exosomal MiR-100-5p-Dependent Manner. *Int. J. Nanomed.* **2017**, *12*, 3721–3733. [CrossRef] [PubMed]
63. Xiao, X.; Yu, S.; Li, S.; Wu, J.; Ma, R.; Cao, H.; Zhu, Y.; Feng, J. Exosomes: Decreased Sensitivity of Lung Cancer A549 Cells to Cisplatin. *PLoS ONE* **2014**, *9*, e89534. [CrossRef] [PubMed]
64. Nawaz, M.; Shah, N.; Zanetti, B.R.; Maugeri, M.; Silvestre, R.N.; Fatima, F.; Neder, L.; Valadi, H. Extracellular Vesicles and Matrix Remodeling Enzymes: The Emerging Roles in Extracellular Matrix Remodeling, Progression of Diseases and Tissue Repair. *Cells* **2018**, *7*, 167. [CrossRef] [PubMed]
65. Sung, B.H.; Ketova, T.; Hoshino, D.; Zijlstra, A.; Weaver, A.M. Directional Cell Movement through Tissues Is Controlled by Exosome Secretion. *Nat. Commun.* **2015**, *6*, 7164. [CrossRef] [PubMed]
66. Song, W.; Yan, D.; Wei, T.; Liu, Q.; Zhou, X.; Liu, J. Tumor-Derived Extracellular Vesicles in Angiogenesis. *Biomed. Pharm.* **2018**, *102*, 1203–1208. [CrossRef] [PubMed]

67. Galindo-Hernandez, O.; Serna-Marquez, N.; Castillo-Sanchez, R.; Salazar, E.P. Extracellular Vesicles from MDA-MB-231 Breast Cancer Cells Stimulated with Linoleic Acid Promote an EMT-like Process in MCF10A Cells. *Prostaglandins Leukot. Essent. Fat. Acids* **2014**, *91*, 299–310. [CrossRef] [PubMed]
68. Franzen, C.A.; Blackwell, R.H.; Todorovic, V.; Greco, K.A.; Foreman, K.E.; Flanigan, R.C.; Kuo, P.C.; Gupta, G.N. Urothelial Cells Undergo Epithelial-to-Mesenchymal Transition after Exposure to Muscle Invasive Bladder Cancer Exosomes. *Oncogenesis* **2015**, *4*, e163. [CrossRef] [PubMed]
69. Xiao, D.; Barry, S.; Kmetz, D.; Egger, M.; Pan, J.; Rai, S.N.; Qu, J.; McMasters, K.M.; Hao, H. Melanoma Cell-Derived Exosomes Promote Epithelial-Mesenchymal Transition in Primary Melanocytes through Paracrine/Autocrine Signaling in the Tumor Microenvironment. *Cancer Lett.* **2016**, *376*, 318–327. [CrossRef]
70. Santi, A.; Caselli, A.; Ranaldi, F.; Paoli, P.; Mugnaioni, C.; Michelucci, E.; Cirri, P. Cancer Associated Fibroblasts Transfer Lipids and Proteins to Cancer Cells through Cargo Vesicles Supporting Tumor Growth. *Biochim. Biophys. Acta* **2015**, *1853*, 3211–3223. [CrossRef]
71. Luga, V.; Zhang, L.; Viloria-Petit, A.M.; Ogunjimi, A.A.; Inanlou, M.R.; Chiu, E.; Buchanan, M.; Hosein, A.N.; Basik, M.; Wrana, J.L. Exosomes Mediate Stromal Mobilization of Autocrine Wnt-PCP Signaling in Breast Cancer Cell Migration. *Cell* **2012**, *151*, 1542–1556. [CrossRef] [PubMed]
72. Boelens, M.C.; Wu, T.J.; Nabet, B.Y.; Xu, B.; Qiu, Y.; Yoon, T.; Azzam, D.J.; Twyman-Saint Victor, C.; Wiemann, B.Z.; Ishwaran, H.; et al. Exosome Transfer from Stromal to Breast Cancer Cells Regulates Therapy Resistance Pathways. *Cell* **2014**, *159*, 499–513. [CrossRef] [PubMed]
73. Richards, K.E.; Zeleniak, A.E.; Fishel, M.L.; Wu, J.; Littlepage, L.E.; Hill, R. Cancer-Associated Fibroblast Exosomes Regulate Survival and Proliferation of Pancreatic Cancer Cells. *Oncogene* **2017**, *36*, 1770–1778. [CrossRef] [PubMed]
74. Au Yeung, C.L.; Co, N.-N.; Tsuruga, T.; Yeung, T.-L.; Kwan, S.-Y.; Leung, C.S.; Li, Y.; Lu, E.S.; Kwan, K.; Wong, K.-K.; et al. Exosomal Transfer of Stroma-Derived MiR21 Confers Paclitaxel Resistance in Ovarian Cancer Cells through Targeting APAF1. *Nat. Commun.* **2016**, *7*, 11150. [CrossRef] [PubMed]
75. Cui, H.; Seubert, B.; Stahl, E.; Dietz, H.; Reuning, U.; Moreno-Leon, L.; Ilie, M.; Hofman, P.; Nagase, H.; Mari, B.; et al. Tissue Inhibitor of Metalloproteinases-1 Induces a pro-Tumourigenic Increase of MiR-210 in Lung Adenocarcinoma Cells and Their Exosomes. *Oncogene* **2015**, *34*, 3640–3650. [CrossRef] [PubMed]
76. Dang, K.; Myers, K.A. The Role of Hypoxia-Induced MiR-210 in Cancer Progression. *Int. J. Mol. Sci.* **2015**, *16*, 6353–6372. [CrossRef] [PubMed]
77. Mao, G.; Liu, Y.; Fang, X.; Liu, Y.; Fang, L.; Lin, L.; Liu, X.; Wang, N. Tumor-Derived MicroRNA-494 Promotes Angiogenesis in Non-Small Cell Lung Cancer. *Angiogenesis* **2015**, *18*, 373–382. [CrossRef]
78. Matsuura, Y.; Wada, H.; Eguchi, H.; Gotoh, K.; Kobayashi, S.; Kinoshita, M.; Kubo, M.; Hayashi, K.; Iwagami, Y.; Yamada, D.; et al. Exosomal MiR-155 Derived from Hepatocellular Carcinoma Cells Under Hypoxia Promotes Angiogenesis in Endothelial Cells. *Dig. Dis. Sci.* **2019**, *64*, 792–802. [CrossRef] [PubMed]
79. Zhou, X.; Yan, T.; Huang, C.; Xu, Z.; Wang, L.; Jiang, E.; Wang, H.; Chen, Y.; Liu, K.; Shao, Z.; et al. Melanoma Cell-Secreted Exosomal MiR-155-5p Induce Proangiogenic Switch of Cancer-Associated Fibroblasts via SOCS1/JAK2/STAT3 Signaling Pathway. *J. Exp. Clin. Cancer Res.* **2018**, *37*, 242. [CrossRef]
80. Costa-Silva, B.; Aiello, N.M.; Ocean, A.J.; Singh, S.; Zhang, H.; Thakur, B.K.; Becker, A.; Hoshino, A.; Mark, M.T.; Molina, H.; et al. Pancreatic Cancer Exosomes Initiate Pre-Metastatic Niche Formation in the Liver. *Nat. Cell Biol.* **2015**, *17*, 816–826. [CrossRef]
81. Grange, C.; Tapparo, M.; Collino, F.; Vitillo, L.; Damasco, C.; Deregibus, M.C.; Tetta, C.; Bussolati, B.; Camussi, G. Microvesicles Released from Human Renal Cancer Stem Cells Stimulate Angiogenesis and Formation of Lung Premetastatic Niche. *Cancer Res.* **2011**, *71*, 5346–5356. [CrossRef] [PubMed]
82. Hoshino, A.; Costa-Silva, B.; Shen, T.-L.; Rodrigues, G.; Hashimoto, A.; Tesic Mark, M.; Molina, H.; Kohsaka, S.; Di Giannatale, A.; Ceder, S.; et al. Tumour Exosome Integrins Determine Organotropic Metastasis. *Nature* **2015**, *527*, 329–335. [CrossRef] [PubMed]
83. Peinado, H.; Alečković, M.; Lavotshkin, S.; Matei, I.; Costa-Silva, B.; Moreno-Bueno, G.; Hergueta-Redondo, M.; Williams, C.; García-Santos, G.; Ghajar, C.; et al. Melanoma Exosomes Educate Bone Marrow Progenitor Cells toward a Pro-Metastatic Phenotype through MET. *Nat. Med.* **2012**, *18*, 883–891. [CrossRef] [PubMed]
84. Abd Elmageed, Z.Y.; Yang, Y.; Thomas, R.; Ranjan, M.; Mondal, D.; Moroz, K.; Fang, Z.; Rezk, B.M.; Moparty, K.; Sikka, S.C.; et al. Neoplastic Reprogramming of Patient-Derived Adipose Stem Cells by Prostate Cancer Cell-Associated Exosomes. *Stem Cells* **2014**, *32*, 983–997. [CrossRef] [PubMed]

85. Fabbri, M.; Paone, A.; Calore, F.; Galli, R.; Gaudio, E.; Santhanam, R.; Lovat, F.; Fadda, P.; Mao, C.; Nuovo, G.J.; et al. MicroRNAs Bind to Toll-like Receptors to Induce Prometastatic Inflammatory Response. *Proc. Natl. Acad. Sci. USA* **2012**, *109*, E2110–E2116. [CrossRef] [PubMed]
86. Chow, A.; Zhou, W.; Liu, L.; Fong, M.Y.; Champer, J.; Van Haute, D.; Chin, A.R.; Ren, X.; Gugiu, B.G.; Meng, Z.; et al. Macrophage Immunomodulation by Breast Cancer-Derived Exosomes Requires Toll-like Receptor 2-Mediated Activation of NF-KB. *Sci. Rep.* **2014**, *4*, 5750. [CrossRef] [PubMed]
87. Marton, A.; Vizler, C.; Kusz, E.; Temesfoi, V.; Szathmary, Z.; Nagy, K.; Szegletes, Z.; Varo, G.; Siklos, L.; Katona, R.L.; et al. Melanoma Cell-Derived Exosomes Alter Macrophage and Dendritic Cell Functions in Vitro. *Immunol. Lett.* **2012**, *148*, 34–38. [CrossRef]
88. Grange, C.; Tapparo, M.; Tritta, S.; Deregibus, M.C.; Battaglia, A.; Gontero, P.; Frea, B.; Camussi, G. Role of HLA-G and Extracellular Vesicles in Renal Cancer Stem Cell-Induced Inhibition of Dendritic Cell Differentiation. *BMC Cancer* **2015**, *15*, 1009. [CrossRef]
89. 101 Ding, G.; Zhou, L.; Qian, Y.; Fu, M.; Chen, J.; Chen, J.; Xiang, J.; Wu, Z.; Jiang, G.; Cao, L. Pancreatic Cancer-Derived Exosomes Transfer MiRNAs to Dendritic Cells and Inhibit RFXAP Expression via MiR-212-3p. *Oncotarget* **2015**, *6*, 29877–29888. [CrossRef]
90. Chalmin, F.; Ladoire, S.; Mignot, G.; Vincent, J.; Bruchard, M.; Remy-Martin, J.-P.; Boireau, W.; Rouleau, A.; Simon, B.; Lanneau, D.; et al. Membrane-Associated Hsp72 from Tumor-Derived Exosomes Mediates STAT3-Dependent Immunosuppressive Function of Mouse and Human Myeloid-Derived Suppressor Cells. *J. Clin. Investig.* **2010**, *120*, 457–471. [CrossRef]
91. Valenti, R.; Huber, V.; Filipazzi, P.; Pilla, L.; Sovena, G.; Villa, A.; Corbelli, A.; Fais, S.; Parmiani, G.; Rivoltini, L. Human Tumor-Released Microvesicles Promote the Differentiation of Myeloid Cells with Transforming Growth Factor-Beta-Mediated Suppressive Activity on T Lymphocytes. *Cancer Res.* **2006**, *66*, 9290–9298. [CrossRef] [PubMed]
92. Chen, G.; Huang, A.C.; Zhang, W.; Zhang, G.; Wu, M.; Xu, W.; Yu, Z.; Yang, J.; Wang, B.; Sun, H.; et al. Exosomal PD-L1 Contributes to Immunosuppression and Is Associated with Anti-PD-1 Response. *Nature* **2018**, *560*, 382–386. [CrossRef] [PubMed]
93. Theodoraki, M.-N.; Yerneni, S.S.; Hoffmann, T.K.; Gooding, W.E.; Whiteside, T.L. Clinical Significance of PD-L1+ Exosomes in Plasma of Head and Neck Cancer Patients. *Clin. Cancer Res.* **2018**, *24*, 896–905. [CrossRef] [PubMed]
94. Yen, E.-Y.; Miaw, S.-C.; Yu, J.-S.; Lai, I.-R. Exosomal TGF-B1 Is Correlated with Lymphatic Metastasis of Gastric Cancers. *Am. J. Cancer Res.* **2017**, *7*, 2199–2208. [PubMed]
95. Abusamra, A.J.; Zhong, Z.; Zheng, X.; Li, M.; Ichim, T.E.; Chin, J.L.; Min, W.-P. Tumor Exosomes Expressing Fas Ligand Mediate CD8+ T-Cell Apoptosis. *Blood Cells Mol. Dis.* **2005**, *35*, 169–173. [CrossRef] [PubMed]
96. Huber, V.; Fais, S.; Iero, M.; Lugini, L.; Canese, P.; Squarcina, P.; Zaccheddu, A.; Colone, M.; Arancia, G.; Gentile, M.; et al. Human Colorectal Cancer Cells Induce T-Cell Death through Release of Proapoptotic Microvesicles: Role in Immune Escape. *Gastroenterology* **2005**, *128*, 1796–1804. [CrossRef] [PubMed]
97. Ashiru, O.; Boutet, P.; Fernández-Messina, L.; Agüera-González, S.; Skepper, J.N.; Valés-Gómez, M.; Reyburn, H.T. Natural Killer Cell Cytotoxicity Is Suppressed by Exposure to the Human NKG2D Ligand MICA*008 That Is Shed by Tumor Cells in Exosomes. *Cancer Res.* **2010**, *70*, 481–489. [CrossRef] [PubMed]
98. Liu, C.; Yu, S.; Zinn, K.; Wang, J.; Zhang, L.; Jia, Y.; Kappes, J.C.; Barnes, S.; Kimberly, R.P.; Grizzle, W.E.; et al. Murine Mammary Carcinoma Exosomes Promote Tumor Growth by Suppression of NK Cell Function. *J. Immunol.* **2006**, *176*, 1375–1385. [CrossRef]
99. Ludwig, S.; Floros, T.; Theodoraki, M.-N.; Hong, C.-S.; Jackson, E.K.; Lang, S.; Whiteside, T.L. Suppression of Lymphocyte Functions by Plasma Exosomes Correlates with Disease Activity in Patients with Head and Neck Cancer. *Clin. Cancer Res.* **2017**, *23*, 4843–4854. [CrossRef]
100. Escudier, B.; Dorval, T.; Chaput, N.; André, F.; Caby, M.-P.; Novault, S.; Flament, C.; Leboulaire, C.; Borg, C.; Amigorena, S.; et al. Vaccination of Metastatic Melanoma Patients with Autologous Dendritic Cell (DC) Derived-Exosomes: Results of Thefirst Phase I Clinical Trial. *J. Transl. Med.* **2005**, *3*, 10. [CrossRef]
101. Morse, M.A.; Garst, J.; Osada, T.; Khan, S.; Hobeika, A.; Clay, T.M.; Valente, N.; Shreeniwas, R.; Sutton, M.A.; Delcayre, A.; et al. A Phase I Study of Dexosome Immunotherapy in Patients with Advanced Non-Small Cell Lung Cancer. *J. Transl. Med.* **2005**, *3*, 9. [CrossRef] [PubMed]
102. Taylor, D.D.; Gercel-Taylor, C. MicroRNA Signatures of Tumor-Derived Exosomes as Diagnostic Biomarkers of Ovarian Cancer. *Gynecol. Oncol.* **2008**, *110*, 13–21. [CrossRef] [PubMed]

103. Rabinowits, G.; Gerçel-Taylor, C.; Day, J.M.; Taylor, D.D.; Kloecker, G.H. Exosomal MicroRNA: A Diagnostic Marker for Lung Cancer. *Clin. Lung Cancer* **2009**, *10*, 42–46. [CrossRef] [PubMed]
104. Sohn, W.; Kim, J.; Kang, S.H.; Yang, S.R.; Cho, J.-Y.; Cho, H.C.; Shim, S.G.; Paik, Y.-H. Serum Exosomal MicroRNAs as Novel Biomarkers for Hepatocellular Carcinoma. *Exp. Mol. Med.* **2015**, *47*, e184. [CrossRef] [PubMed]
105. Taylor, D.D.; Gercel-Taylor, C. Exosomes/Microvesicles: Mediators of Cancer-Associated Immunosuppressive Microenvironments. *Semin. Immunopathol.* **2011**, *33*, 441–454. [CrossRef]
106. Jabalee, J.; Towle, R.; Garnis, C. The Role of Extracellular Vesicles in Cancer: Cargo, Function, and Therapeutic Implications. *Cells* **2018**, *7*, 93. [CrossRef] [PubMed]
107. Bu, N.; Wu, H.; Sun, B.; Zhang, G.; Zhan, S.; Zhang, R.; Zhou, L. Exosome-Loaded Dendritic Cells Elicit Tumor-Specific CD8+ Cytotoxic T Cells in Patients with Glioma. *J. Neurooncol.* **2011**, *104*, 659–667. [CrossRef]
108. Graner, M.W.; Alzate, O.; Dechkovskaia, A.M.; Keene, J.D.; Sampson, J.H.; Mitchell, D.A.; Bigner, D.D. Proteomic and Immunologic Analyses of Brain Tumor Exosomes. *FASEB J.* **2009**, *23*, 1541–1557. [CrossRef]
109. Kunigelis, K.E.; Graner, M.W. The Dichotomy of Tumor Exosomes (TEX) in Cancer Immunity: Is It All in the ConTEXt? *Vaccines (Basel)* **2015**, *3*, 1019–1051. [CrossRef]
110. Chevillet, J.R.; Kang, Q.; Ruf, I.K.; Briggs, H.A.; Vojtech, L.N.; Hughes, S.M.; Cheng, H.H.; Arroyo, J.D.; Meredith, E.K.; Gallichotte, E.N.; et al. Quantitative and Stoichiometric Analysis of the MicroRNA Content of Exosomes. *Proc. Natl. Acad. Sci. USA* **2014**, *111*, 14888–14893. [CrossRef]
111. Anfossi, S.; Giordano, A.; Gao, H.; Cohen, E.N.; Tin, S.; Wu, Q.; Garza, R.J.; Debeb, B.G.; Alvarez, R.H.; Valero, V.; et al. High Serum MiR-19a Levels Are Associated with Inflammatory Breast Cancer and Are Predictive of Favorable Clinical Outcome in Patients with Metastatic HER2+ Inflammatory Breast Cancer. *PLoS ONE* **2014**, *9*, e83113. [CrossRef] [PubMed]
112. Gemmell, C.H.; Sefton, M.V.; Yeo, E.L. Platelet-Derived Microparticle Formation Involves Glycoprotein IIb-IIIa. Inhibition by RGDS and a Glanzmann's Thrombasthenia Defect. *J. Biol. Chem.* **1993**, *268*, 14586–14589. [PubMed]
113. Yan, W.; Apweiler, R.; Balgley, B.M.; Boontheung, P.; Bundy, J.L.; Cargile, B.J.; Cole, S.; Fang, X.; Gonzalez-Begne, M.; Griffin, T.J.; et al. Systematic Comparison of the Human Saliva and Plasma Proteomes. *Proteom. Clin. Appl.* **2009**, *3*, 116–134. [CrossRef] [PubMed]
114. Bandhakavi, S.; Stone, M.D.; Onsongo, G.; Van Riper, S.K.; Griffin, T.J. A Dynamic Range Compression and Three-Dimensional Peptide Fractionation Analysis Platform Expands Proteome Coverage and the Diagnostic Potential of Whole Saliva. *J. Proteome Res.* **2009**, *8*, 5590–5600. [CrossRef]
115. Zhao, M.; Yang, Y.; Guo, Z.; Shao, C.; Sun, H.; Zhang, Y.; Sun, Y.; Liu, Y.; Song, Y.; Zhang, L.; et al. A Comparative Proteomics Analysis of Five Body Fluids: Plasma, Urine, Cerebrospinal Fluid, Amniotic Fluid, and Saliva. *Proteom. Clin. Appl.* **2018**, *12*, e1800008. [CrossRef]
116. Omenn, G.S.; States, D.J.; Adamski, M.; Blackwell, T.W.; Menon, R.; Hermjakob, H.; Apweiler, R.; Haab, B.B.; Simpson, R.J.; Eddes, J.S.; et al. Overview of the HUPO Plasma Proteome Project: Results from the Pilot Phase with 35 Collaborating Laboratories and Multiple Analytical Groups, Generating a Core Dataset of 3020 Proteins and a Publicly-Available Database. *Proteomics* **2005**, *5*, 3226–3245. [CrossRef]
117. Cheng, Y.; Pereira, M.; Raukar, N.; Reagan, J.L.; Queseneberry, M.; Goldberg, L.; Borgovan, T.; LaFrance, W.C.; Dooner, M.; Deregibus, M.; et al. Potential Biomarkers to Detect Traumatic Brain Injury by the Profiling of Salivary Extracellular Vesicles. *J. Cell. Physiol.* **2019**, *234*, 14377–14388. [CrossRef]
118. Saeedi, S.; Israel, S.; Nagy, C.; Turecki, G. The Emerging Role of Exosomes in Mental Disorders. *Transl. Psychiatry* **2019**, *9*, 122. [CrossRef]
119. Cao, Z.; Wu, Y.; Liu, G.; Jiang, Y.; Wang, X.; Wang, Z.; Feng, T. α-Synuclein in Salivary Extracellular Vesicles as a Potential Biomarker of Parkinson's Disease. *Neurosci. Lett.* **2019**, *696*, 114–120. [CrossRef]
120. Jin, Y.; Guan, Z.; Wang, X.; Wang, Z.; Zeng, R.; Xu, L.; Cao, P. ALA-PDT Promotes HPV-Positive Cervical Cancer Cells Apoptosis and DCs Maturation via MiR-34a Regulated HMGB1 Exosomes Secretion. *Photodiagn. Photodyn.* **2018**, *24*, 27–35. [CrossRef]
121. Harden, M.E.; Munger, K. Human Papillomavirus 16 E6 and E7 Oncoprotein Expression Alters MicroRNA Expression in Extracellular Vesicles. *Virology* **2017**, *508*, 63–69. [CrossRef] [PubMed]
122. Honegger, A.; Leitz, J.; Bulkescher, J.; Hoppe-Seyler, K.; Hoppe-Seyler, F. Silencing of Human Papillomavirus (HPV) E6/E7 Oncogene Expression Affects Both the Contents and the Amounts of Extracellular Microvesicles Released from HPV-Positive Cancer Cells. *Int. J. Cancer* **2013**, *133*, 1631–1642. [CrossRef] [PubMed]

123. Chiantore, M.V.; Mangino, G.; Iuliano, M.; Zangrillo, M.S.; De Lillis, I.; Vaccari, G.; Accardi, R.; Tommasino, M.; Columba Cabezas, S.; Federico, M.; et al. Human Papillomavirus E6 and E7 Oncoproteins Affect the Expression of Cancer-Related MicroRNAs: Additional Evidence in HPV-Induced Tumorigenesis. *J. Cancer Res. Clin. Oncol.* **2016**, *142*, 1751–1763. [CrossRef] [PubMed]
124. Peacock, B.; Rigby, A.; Bradford, J.; Pink, R.; Hunter, K.; Lambert, D.; Hunt, S. Extracellular Vesicle MicroRNA Cargo Is Correlated with HPV Status in Oropharyngeal Carcinoma. *J. Oral Pathol. Med.* **2018**, *47*, 954–963. [CrossRef] [PubMed]
125. Hukin, J.; Farrell, K.; MacWilliam, L.M.; Colbourne, M.; Waida, E.; Tan, R.; Mroz, L.; Thomas, E. Case-Control Study of Primary Human Herpesvirus 6 Infection in Children with Febrile Seizures. *Pediatrics* **1998**, *101*, E3. [CrossRef] [PubMed]
126. Sharma, S.; Gillespie, B.M.; Palanisamy, V.; Gimzewski, J.K. Quantitative Nanostructural and Single-Molecule Force Spectroscopy Biomolecular Analysis of Human-Saliva-Derived Exosomes. *Langmuir* **2011**, *27*, 14394–14400. [CrossRef] [PubMed]
127. Zlotogorski-Hurvitz, A.; Dayan, D.; Chaushu, G.; Salo, T.; Vered, M. Morphological and Molecular Features of Oral Fluid-Derived Exosomes: Oral Cancer Patients versus Healthy Individuals. *J. Cancer Res. Clin. Oncol.* **2016**, *142*, 101–110. [CrossRef] [PubMed]
128. Winck, F.V.; Prado Ribeiro, A.C.; Ramos Domingues, R.; Ling, L.Y.; Riaño-Pachón, D.M.; Rivera, C.; Brandão, T.B.; Gouvea, A.F.; Santos-Silva, A.R.; Coletta, R.D.; et al. Insights into Immune Responses in Oral Cancer through Proteomic Analysis of Saliva and Salivary Extracellular Vesicles. *Sci. Rep.* **2015**, *5*, 16305. [CrossRef] [PubMed]
129. Gai, C.; Camussi, F.; Broccoletti, R.; Gambino, A.; Cabras, M.; Molinaro, L.; Carossa, S.; Camussi, G.; Arduino, P.G. Salivary Extracellular Vesicle-Associated MiRNAs as Potential Biomarkers in Oral Squamous Cell Carcinoma. *BMC Cancer* **2018**, *18*, 439. [CrossRef] [PubMed]
130. Sun, Y.; Xia, Z.; Shang, Z.; Sun, K.; Niu, X.; Qian, L.; Fan, L.-Y.; Cao, C.-X.; Xiao, H. Facile Preparation of Salivary Extracellular Vesicles for Cancer *Proteom. Sci. Rep.* **2016**, *6*, 24669. [CrossRef] [PubMed]
131. Sun, Y.; Huo, C.; Qiao, Z.; Shang, Z.; Uzzaman, A.; Liu, S.; Jiang, X.; Fan, L.-Y.; Ji, L.; Guan, X.; et al. Comparative Proteomic Analysis of Exosomes and Microvesicles in Human Saliva for Lung Cancer. *J. Proteome Res.* **2018**, *17*, 1101–1107. [CrossRef] [PubMed]
132. Langevin, S.; Kuhnell, D.; Parry, T.; Biesiada, J.; Huang, S.; Wise-Draper, T.; Casper, K.; Zhang, X.; Medvedovic, M.; Kasper, S. Comprehensive MicroRNA-Sequencing of Exosomes Derived from Head and Neck Carcinoma Cells in Vitro Reveals Common Secretion Profiles and Potential Utility as Salivary Biomarkers. *Oncotarget* **2017**, *8*, 82459–82474. [CrossRef] [PubMed]
133. Machida, T.; Tomofuji, T.; Maruyama, T.; Yoneda, T.; Ekuni, D.; Azuma, T.; Miyai, H.; Mizuno, H.; Kato, H.; Tsutsumi, K.; et al. MiR-1246 and MiR-4644 in Salivary Exosome as Potential Biomarkers for Pancreatobiliary Tract Cancer. *Oncol. Rep.* **2016**, *36*, 2375–2381. [CrossRef] [PubMed]
134. Lau, C.; Kim, Y.; Chia, D.; Spielmann, N.; Eibl, G.; Elashoff, D.; Wei, F.; Lin, Y.-L.; Moro, A.; Grogan, T.; et al. Role of Pancreatic Cancer-Derived Exosomes in Salivary Biomarker Development. *J. Biol. Chem.* **2013**, *288*, 26888–26897. [CrossRef] [PubMed]
135. Kowal, J.; Arras, G.; Colombo, M.; Jouve, M.; Morath, J.P.; Primdal-Bengtson, B.; Dingli, F.; Loew, D.; Tkach, M.; Théry, C. Proteomic Comparison Defines Novel Markers to Characterize Heterogeneous Populations of Extracellular Vesicle Subtypes. *Proc. Natl. Acad. Sci. USA* **2016**, *113*, E968–E977. [CrossRef] [PubMed]
136. Momen-Heravi, F.; Balaj, L.; Alian, S.; Trachtenberg, A.J.; Hochberg, F.H.; Skog, J.; Kuo, W.P. Impact of Biofluid Viscosity on Size and Sedimentation Efficiency of the Isolated Microvesicles. *Front. Physiol.* **2012**, *3*, 162. [CrossRef] [PubMed]
137. Jeppesen, D.K.; Hvam, M.L.; Primdahl-Bengtson, B.; Boysen, A.T.; Whitehead, B.; Dyrskjøt, L.; Orntoft, T.F.; Howard, K.A.; Ostenfeld, M.S. Comparative Analysis of Discrete Exosome Fractions Obtained by Differential Centrifugation. *J. Extracell. Vesicles* **2014**, *3*, 25011. [CrossRef] [PubMed]
138. Cvjetkovic, A.; Lötvall, J.; Lässer, C. The Influence of Rotor Type and Centrifugation Time on the Yield and Purity of Extracellular Vesicles. *J. Extracell. Vesicles* **2014**, *3*, 23111. [CrossRef] [PubMed]
139. Tauro, B.J.; Greening, D.W.; Mathias, R.A.; Ji, H.; Mathivanan, S.; Scott, A.M.; Simpson, R.J. Comparison of Ultracentrifugation, Density Gradient Separation, and Immunoaffinity Capture Methods for Isolating Human Colon Cancer Cell Line LIM1863-Derived Exosomes. *Methods* **2012**, *56*, 293–304. [CrossRef]

140. Müller, G. Novel Tools for the Study of Cell Type-Specific Exosomes and Microvesicles. *J. Bioanal. Biomed.* **2012**, *4*, 46–60. [CrossRef]
141. Taylor, D.D.; Lyons, K.S.; Gerçel-Taylor, C. Shed Membrane Fragment-Associated Markers for Endometrial and Ovarian Cancers. *Gynecol. Oncol.* **2002**, *84*, 443–448. [CrossRef] [PubMed]
142. Böing, A.N.; van der Pol, E.; Grootemaat, A.E.; Coumans, F.A.W.; Sturk, A.; Nieuwland, R. Single-Step Isolation of Extracellular Vesicles by Size-Exclusion Chromatography. *J. Extracell. Vesicles* **2014**, *3*, 23430. [CrossRef] [PubMed]
143. Taylor, D.D.; Shah, S. Methods of Isolating Extracellular Vesicles Impact Down-Stream Analyses of Their Cargoes. *Methods* **2015**, *87*, 3–10. [CrossRef] [PubMed]
144. Caby, M.-P.; Lankar, D.; Vincendeau-Scherrer, C.; Raposo, G.; Bonnerot, C. Exosomal-like Vesicles Are Present in Human Blood Plasma. *Int. Immunol.* **2005**, *17*, 879–887. [CrossRef] [PubMed]
145. Zarovni, N.; Corrado, A.; Guazzi, P.; Zocco, D.; Lari, E.; Radano, G.; Muhhina, J.; Fondelli, C.; Gavrilova, J.; Chiesi, A. Integrated Isolation and Quantitative Analysis of Exosome Shuttled Proteins and Nucleic Acids Using Immunocapture Approaches. *Methods* **2015**, *87*, 46–58. [CrossRef] [PubMed]
146. Szatanek, R.; Baran, J.; Siedlar, M.; Baj-Krzyworzeka, M. Isolation of Extracellular Vesicles: Determining the Correct Approach (Review). *Int. J. Mol. Med.* **2015**, *36*, 11–17. [CrossRef] [PubMed]
147. Alvarez, M.L.; Khosroheidari, M.; Kanchi Ravi, R.; DiStefano, J.K. Comparison of Protein, MicroRNA, and MRNA Yields Using Different Methods of Urinary Exosome Isolation for the Discovery of Kidney Disease Biomarkers. *Kidney Int.* **2012**, *82*, 1024–1032. [CrossRef]
148. Alvarez, M.L. Isolation of Urinary Exosomes for RNA Biomarker Discovery Using a Simple, Fast, and Highly Scalable Method. *Methods Mol. Biol.* **2014**, *1182*, 145–170. [CrossRef]
149. Rekker, K.; Saare, M.; Roost, A.M.; Kubo, A.-L.; Zarovni, N.; Chiesi, A.; Salumets, A.; Peters, M. Comparison of Serum Exosome Isolation Methods for MicroRNA Profiling. *Clin. Biochem.* **2014**, *47*, 135–138. [CrossRef]
150. Zlotogorski-Hurvitz, A.; Dayan, D.; Chaushu, G.; Korvala, J.; Salo, T.; Sormunen, R.; Vered, M. Human Saliva-Derived Exosomes: Comparing Methods of Isolation. *J. Histochem. Cytochem.* **2015**, *63*, 181–189. [CrossRef]
151. Kanchi Ravi, R.; Khosroheidari, M.; DiStefano, J.K. A Modified Precipitation Method to Isolate Urinary Exosomes. *J. Vis. Exp.* **2015**, *95*, e51158. [CrossRef] [PubMed]
152. Witwer, K.W.; Buzás, E.I.; Bemis, L.T.; Bora, A.; Lässer, C.; Lötvall, J.; Nolte-'t Hoen, E.N.; Piper, M.G.; Sivaraman, S.; Skog, J.; et al. Standardization of Sample Collection, Isolation and Analysis Methods in Extracellular Vesicle Research. *J. Extracell. Vesicles* **2013**, *2*, 20360. [CrossRef]
153. Yuana, Y.; Levels, J.; Grootemaat, A.; Sturk, A.; Nieuwland, R. Co-Isolation of Extracellular Vesicles and High-Density Lipoproteins Using Density Gradient Ultracentrifugation. *J. Extracell. Vesicles* **2014**, *3*, 23262. [CrossRef] [PubMed]
154. Wei, F.; Lin, C.-C.; Joon, A.; Feng, Z.; Troche, G.; Lira, M.E.; Chia, D.; Mao, M.; Ho, C.-L.; Su, W.-C.; et al. Noninvasive Saliva-Based EGFR Gene Mutation Detection in Patients with Lung Cancer. *Am. J. Respir. Crit. Care Med.* **2014**, *190*, 1117–1126. [CrossRef] [PubMed]
155. Pu, D.; Liang, H.; Wei, F.; Akin, D.; Feng, Z.; Yan, Q.; Li, Y.; Zhen, Y.; Xu, L.; Dong, G.; et al. Evaluation of a Novel Saliva-Based Epidermal Growth Factor Receptor Mutation Detection for Lung Cancer: A Pilot Study. *Thorac. Cancer* **2016**, *7*, 428–436. [CrossRef]
156. Wei, F.; Yang, J.; Wong, D.T.W. Detection of Exosomal Biomarker by Electric Field-Induced Release and Measurement (EFIRM). *Biosens. Bioelectron.* **2013**, *44*, 115–121. [CrossRef] [PubMed]
157. Cheng, J.; Nonaka, T.; Wong, D.T.W. Salivary Exosomes as Nanocarriers for Cancer Biomarker Delivery. *Materials (Basel)* **2019**, *12*, 654. [CrossRef]
158. Michael, A.; Bajracharya, S.D.; Yuen, P.S.T.; Zhou, H.; Star, R.A.; Illei, G.G.; Alevizos, I. Exosomes from Human Saliva as a Source of MicroRNA Biomarkers. *Oral Dis.* **2010**, *16*, 34–38. [CrossRef]
159. Deregibus, M.C.; Figliolini, F.; D'Antico, S.; Manzini, P.M.; Pasquino, C.; De Lena, M.; Tetta, C.; Brizzi, M.F.; Camussi, G. Charge-Based Precipitation of Extracellular Vesicles. *Int. J. Mol. Med.* **2016**, *38*, 1359–1366. [CrossRef]
160. Li, Y.; St John, M.A.R.; Zhou, X.; Kim, Y.; Sinha, U.; Jordan, R.C.K.; Eisele, D.; Abemayor, E.; Elashoff, D.; Park, N.-H.; et al. Salivary Transcriptome Diagnostics for Oral Cancer Detection. *Clin. Cancer Res.* **2004**, *10*, 8442–8450. [CrossRef]

161. Zhou, Y.; Kolokythas, A.; Schwartz, J.L.; Epstein, J.B.; Adami, G.R. MicroRNA from Brush Biopsy to Characterize Oral Squamous Cell Carcinoma Epithelium. *Cancer Med.* **2017**, *6*, 67–78. [CrossRef]
162. Yap, T.; Vella, L.J.; Seers, C.; Nastri, A.; Reynolds, E.; Cirillo, N.; McCullough, M. Oral Swirl Samples—A Robust Source of MicroRNA Protected by Extracellular Vesicles. *Oral Dis.* **2017**, *23*, 312–317. [CrossRef]
163. Mestdagh, P.; Van Vlierberghe, P.; De Weer, A.; Muth, D.; Westermann, F.; Speleman, F.; Vandesompele, J. A Novel and Universal Method for MicroRNA RT-QPCR Data Normalization. *Genome Biol.* **2009**, *10*, R64. [CrossRef]
164. Kaczor-Urbanowicz, K.E.; Martin Carreras-Presas, C.; Aro, K.; Tu, M.; Garcia-Godoy, F.; Wong, D.T. Saliva Diagnostics-Current Views and Directions. *Exp. Biol. Med. (Maywood)* **2017**, *242*, 459–472. [CrossRef]
165. Zhang, L.; Farrell, J.J.; Zhou, H.; Elashoff, D.; Akin, D.; Park, N.-H.; Chia, D.; Wong, D.T. Salivary Transcriptomic Biomarkers for Detection of Resectable Pancreatic Cancer. *Gastroenterology* **2010**, *138*, e1–e7. [CrossRef]
166. Lee, Y.-H.; Kim, J.H.; Zhou, H.; Kim, B.W.; Wong, D.T. Salivary Transcriptomic Biomarkers for Detection of Ovarian Cancer: For Serous Papillary Adenocarcinoma. *J. Mol. Med.* **2012**, *90*, 427–434. [CrossRef]
167. Xiao, H.; Zhang, L.; Zhou, H.; Lee, J.M.; Garon, E.B.; Wong, D.T.W. Proteomic Analysis of Human Saliva from Lung Cancer Patients Using Two-Dimensional Difference Gel Electrophoresis and Mass Spectrometry. *Mol. Cell Proteom.* **2012**, *11*, M111.012112. [CrossRef]
168. Ogawa, Y.; Tsujimoto, M.; Yanoshita, R. Next-Generation Sequencing of Protein-Coding and Long Non-Protein-Coding RNAs in Two Types of Exosomes Derived from Human Whole Saliva. *Biol. Pharm. Bull.* **2016**, *39*, 1496–1507. [CrossRef]
169. Palanisamy, V.; Sharma, S.; Deshpande, A.; Zhou, H.; Gimzewski, J.; Wong, D.T. Nanostructural and Transcriptomic Analyses of Human Saliva Derived Exosomes. *PLoS ONE* **2010**, *5*, e8577. [CrossRef]
170. Lau, C.S.; Wong, D.T.W. Breast Cancer Exosome-like Microvesicles and Salivary Gland Cells Interplay Alters Salivary Gland Cell-Derived Exosome-like Microvesicles in Vitro. *PLoS ONE* **2012**, *7*, e33037. [CrossRef]
171. Katsiougiannis, S.; Chia, D.; Kim, Y.; Singh, R.P.; Wong, D.T.W. Saliva Exosomes from Pancreatic Tumor-Bearing Mice Modulate NK Cell Phenotype and Antitumor Cytotoxicity. *FASEB J.* **2017**, *31*, 998–1010. [CrossRef]
172. Serban, K.A.; Rezania, S.; Petrusca, D.N.; Poirier, C.; Cao, D.; Justice, M.J.; Patel, M.; Tsvetkova, I.; Kamocki, K.; Mikosz, A.; et al. Structural and Functional Characterization of Endothelial Microparticles Released by Cigarette Smoke. *Sci. Rep.* **2016**, *6*, 31596. [CrossRef]
173. Kodidela, S.; Wang, Y.; Patters, B.J.; Gong, Y.; Sinha, N.; Ranjit, S.; Gerth, K.; Haque, S.; Cory, T.; McArthur, C.; et al. Proteomic Profiling of Exosomes Derived from Plasma of HIV-Infected Alcohol Drinkers and Cigarette Smokers. *J. Neuroimmune Pharmacol.* **2019**, 1–19. [CrossRef]
174. Appert-Collin, A.; Hubert, P.; Crémel, G.; Bennasroune, A. Role of ErbB Receptors in Cancer Cell Migration and Invasion. *Front. Pharm.* **2015**, *6*, 283. [CrossRef]
175. Li, X.; Sun, R.; Geng, X.; Wang, S.; Zen, D.; Pei, J.; Yang, J.; Fan, Y.; Jiang, H.; Yang, P.; et al. A Comprehensive Analysis of Candidate Gene Signatures in Oral Squamous Cell Carcinoma. *Neoplasma* **2017**, *64*, 167–174. [CrossRef]
176. Ohnishi, Y.; Yasui, H.; Kakudo, K.; Nozaki, M. Lapatinib-Resistant Cancer Cells Possessing Epithelial Cancer Stem Cell Properties Develop Sensitivity during Sphere Formation by Activation of the ErbB/AKT/Cyclin D2 Pathway. *Oncol. Rep.* **2016**, *36*, 3058–3064. [CrossRef]
177. Li, S.-X.; Yang, Y.-Q.; Jin, L.-J.; Cai, Z.-G.; Sun, Z. Detection of Survivin, Carcinoembryonic Antigen and ErbB2 Level in Oral Squamous Cell Carcinoma Patients. *Cancer Biomark.* **2016**, *17*, 377–382. [CrossRef]
178. Dong, C.; Ye, D.-X.; Zhang, W.-B.; Pan, H.-Y.; Zhang, Z.-Y.; Zhang, L. Overexpression of C-Fos Promotes Cell Invasion and Migration via CD44 Pathway in Oral Squamous Cell Carcinoma. *J. Oral Pathol. Med.* **2015**, *44*, 353–360. [CrossRef]
179. Judd, N.P.; Winkler, A.E.; Murillo-Sauca, O.; Brotman, J.J.; Law, J.H.; Lewis, J.S.; Dunn, G.P.; Bui, J.D.; Sunwoo, J.B.; Uppaluri, R. ERK1/2 Regulation of CD44 Modulates Oral Cancer Aggressiveness. *Cancer Res.* **2012**, *72*, 365–374. [CrossRef]
180. Ghuwalewala, S.; Ghatak, D.; Das, P.; Dey, S.; Sarkar, S.; Alam, N.; Panda, C.K.; Roychoudhury, S. CD44(High)CD24(Low) Molecular Signature Determines the Cancer Stem Cell and EMT Phenotype in Oral Squamous Cell Carcinoma. *Stem Cell Res.* **2016**, *16*, 405–417. [CrossRef]

181. Meng, W.; Xia, Q.; Wu, L.; Chen, S.; He, X.; Zhang, L.; Gao, Q.; Zhou, H. Downregulation of TGF-Beta Receptor Types II and III in Oral Squamous Cell Carcinoma and Oral Carcinoma-Associated Fibroblasts. *BMC Cancer* **2011**, *11*, 88. [CrossRef]
182. Momen-Heravi, F.; Bala, S. Extracellular vesicles in oral squamous carcinoma carry oncogenic miRNA profile and reprogram monocytes via NF-κB pathway. *Oncotarget* **2018**, *9*, 34838–34854. [CrossRef]
183. Arantes, L.M.R.B.; De Carvalho, A.C.; Melendez, M.E.; Lopes Carvalho, A. Serum, Plasma and Saliva Biomarkers for Head and Neck Cancer. *Expert Rev. Mol. Diagn.* **2018**, *18*, 85–112. [CrossRef]
184. Tang, H.; Wu, Z.; Zhang, J.; Su, B. Salivary LncRNA as a Potential Marker for Oral Squamous Cell Carcinoma Diagnosis. *Mol. Med. Rep.* **2013**, *7*, 761–766. [CrossRef]

2

Tumor-Derived Extracellular Vesicles Inhibit Natural Killer Cell Function in Pancreatic Cancer

Jiangang Zhao [1,2], Hans A. Schlößer [1], Zhefang Wang [1], Jie Qin [1], Jiahui Li [1], Felix Popp [1], Marie Christine Popp [1], Hakan Alakus [1], Seung-Hun Chon [1], Hinrich P. Hansen [3], Wolfram F. Neiss [4], Karl-Walter Jauch [2], Christiane J. Bruns [1] and Yue Zhao [1,5,*]

1 Department of General, Visceral und Tumor Surgery, University Hospital Cologne, Kerpener Straße 62, 50937 Cologne, Germany; jiangang.zhao@uk-koeln.de (J.Z.); hans.schloesser@uk-koeln.de (H.A.S.); zhefang.wang@uk-koeln.de (Z.W.); jie.qin@uk-koeln.de (J.Q.); jiahui.li@uk-koeln.de (J.L.); felix.popp@uk-koeln.de (F.P.); marie.popp@uk-koeln.de (M.C.P.); hakan.alakus@uk-koeln.de (H.A.); seung-hun.chon@uk-koeln.de (S.-H.C.); christiane.bruns@uk-koeln.de (C.J.B.)

2 Department of General, Visceral und Vascular Surgery, Ludwig-Maximilian-University (LMU), 81377 Munich, Germany; Karl-Walter.Jauch@med.uni-muenchen.de

3 Department I of Internal Medicine, University Hospital of Cologne, Center for Integrated Oncology Cologne-Bonn, CECAD Center of Excellence on "Cellular Stress Responses in Aging-Associated Diseases", Center for Molecular Medicine Cologne, University of Cologne, 50931 Cologne, Germany; hinrich-peter.hansen@uk-koeln.de

4 Department of Anatomy I, Medical Faculty, University of Cologne, 50937 Cologne, Germany; wolfram.neiss@uk-koeln.de

5 Department of General, Visceral und Vascular Surgery, Otto von Guericke University, 39120 Magdeburg, Germany

* Correspondence: yue.zhao@uk-koeln.de

Abstract: Pancreatic ductal adenocarcinoma (PDAC) is one of the most lethal malignancies. Tumor-derived extracellular vesicles (EVs) induce pre-metastatic niche formation to promote metastasis. We isolated EVs from a highly-metastatic pancreatic cancer cell line and patient-derived primary cancer cells by ultracentrifugation. The protein content of EVs was analyzed by mass spectrometry. The effects of PDAC-derived EVs on natural kill (NK) cells were investigated by flow cytometry. The serum EVs' TGF-β1 levels were quantified by ELISA. We found that integrins were enriched in PDAC-derived EVs. The expression of NKG2D, CD107a, TNF-α, and INF-γ in NK cells was significantly downregulated after co-culture with EVs. NK cells also exhibited decreased levels of CD71 and CD98, as well as impaired glucose uptake ability. In addition, NK cell cytotoxicity against pancreatic cancer stem cells was attenuated. Moreover, PDAC-derived EVs induced the phosphorylation of Smad2/3 in NK cells. Serum EVs' TGF-β1 was significantly increased in PDAC patients. Our findings emphasize the immunosuppressive role of PDAC-derived EVs and provide new insights into our understanding of NK cell dysfunction regarding pre-metastatic niche formation in PDAC.

Keywords: extracellular vesicles; NK cells; pancreatic cancer; pre-metastatic niche

1. Introduction

Pancreatic ductal adenocarcinoma (PDAC) is one of the most lethal malignancies worldwide [1]. The majority of patients with pancreatic cancer are diagnosed at an advanced stage without the opportunity for curative surgery [2]. Even after R0 resection followed by adjuvant chemotherapy and/or radiotherapy, most patients will eventually suffer from recurrence [3]. The Surveillance, Epidemiology and End Results (SEER) Program (https://seer.cancer.gov/statfacts/) revealed that the five-year overall survival for PDAC patients with localized disease is 34.3%. For those who present with distant metastases, this drops to merely 2.7%. These daunting statistics emphasize the importance of improving our understanding of the metastatic process to reduce the incidence of metastasis and develop effective therapeutic strategies for PDAC patients.

As an important hallmark of cancer, metastasis is a complex process that propagates tumor cells from the site of origin to distant tissues, also known as the metastatic cascade [4,5]. The term "pre-metastatic niche" describes the microenvironment in a secondary organ that has been affected by the primary tumor to support tumor growth in advance of tumor cell entry [6]. Many studies have identified the existence of pre-metastatic niches in different organs, such as liver and lung [7,8]. Liu Y. et al. summarized six characteristics for the pre-metastatic niche, including immunosuppression, inflammation, lymphangiogenesis, angiogenesis/vascular permeability, organotropism, and reprogramming [9]. The liver pre-metastatic niche has been partly analyzed in orthotopic PDAC mouse models [7]. However, the underlying mechanisms of liver pre-metastatic niche formation remain unclear in human PDAC. Extracellular vesicles (EVs), like exosomes and microvesicles, can mediate intercellular communication, and thus affect physiological and pathological conditions through the transfer of various cargo molecules, including proteins, nucleic acids, and lipids [10,11]. As lipid bilayer membrane vesicles, EVs are one ideal carrier for drug delivery in cancer treatment [12,13]. Kamerkar S et al. have successfully modified exosomes to deliver short interfering RNA specific to KRAS mutation, which suppressed tumor growth in PDAC-bearing mice and significantly increased their overall survival [14]. Jang SC et al. synthesized nanovesicles by the breakdown of monocytes or macrophages, which had similar characteristics as the exosomes. These nanovesicles could deliver chemotherapeutics to inhibit tumor growth [15]. Saari H and team have loaded Paclitaxel into cancer cell-derived EVs. Paclitaxel-loaded EVs could bring the drug into prostate cancer cells and increase its cytotoxicity [16]. In addition, a study led by Garofalo et al. reported that human lung cancer cell-derived extracellular vesicles (EVs) formulations could specifically target the neoplasia. Administration of EVs with oncolytic virus alone or in combination with chemotherapeutics may serve as a novel strategy to treat cancer [17].

In addition, tumor-derived EVs can be released into the circulation and transferred to distant sites. Recent research has shed light on the role of tumor-derived EVs in pre-metastatic niche formation. In murine models of lung cancer and melanoma, tumor-derived exosomes delivered signals to lung epithelial cells and activated Toll-like receptor 3, which elicited chemokine production and promoted neutrophil infiltration [8]. PDAC-derived exosomes recruited macrophages and neutrophils to the liver and stimulated hepatic stellate cells to synthesize fibronectin to promote liver metastasis [7]. In addition, exosomal integrins are thought to determine organotropic metastasis [18]. In conclusion, tumor-derived exosomes are involved in pre-metastatic niche formation, such as angiogenesis, immunosuppression, and organotropism.

Natural killer (NK) cells are a group of innate lymphocytes with inherent capacities to recognize and eliminate virus-infected cells and malignant cells. The role of NK cells in PDAC has received less attention, but is increasingly being recognized. Gürlevik E et al. reported that after primary tumor resection, gemcitabine treatment triggered NK cell cytotoxicity against pancreatic cancer cells and decreased the incidence of local recurrence in orthotopic PDAC mouse models [19]. Ames E et al. found that NK cells preferentially killed pancreatic cancer stem cells (CSCs) in vitro. Intratumoral injection of NK cells in human pancreatic cancer-bearing NSG mice reduced the percentage of pancreatic CSCs and the tumor burden [20]. However, the ability to escape from immune surveillance has been established as a hallmark of tumor cells [5]. Increasingly, studies have revealed NK cells' exhaustion in tumor. Exhausted NK cells produce decreased cytokines, downregulate activating receptors, as well as exhibit impaired cytolytic activity [21–23]. In addition, dysregulated cellular metabolism has been observed in dysfunctional NK cells. Cong J et al. found that in lung cancer mouse models, the expression of gluconeogenesis enzyme fructose bisphosphatase 1 (FBP1) was upregulated in tumor-infiltrating NK cells, which mediates the dysfunction of NK cells by impairing glycolysis [24]. The mechanisms for NK cell dysfunction include direct inhibition via cell–cell contact and indirect inhibition via the production of inhibitory factors such as transforming growth factor beta 1 (TGF-β1), interleukin 10 (IL-10), prostaglandin E2 (PGE2), and indoleamine 2,3-dioxygenase (IDO) [25,26]. However, to date, the dynamic status of NK cells in the liver pre-metastatic niche of PDAC remains unclear. Tumor-derived EVs are emerging as a regulator in NK cell modulation in pancreatic cancer.

In this study, we have investigated the effect of tumor-derived EVs on NK cells in pancreatic cancer. Here, we provide evidence that pancreatic cancer-derived EVs carry abundant immunosuppressive factors and inhibit NK cell function, which contributes to pre-metastatic niche formation.

2. Results

2.1. Characterization of Pancreatic Cancer-Derived EVs

EVs were isolated from cell culture supernatants of a highly-metastatic pancreatic cancer cell line L3.6pl and a PDAC patient-derived primary cancer cell line TBO368 by differential centrifugation and ultracentrifugation (Figure S1a). In order to examine the morphology and measure the size of pancreatic cancer-derived EVs, we used transmission electron microscopy (TEM) and nanoparticle analysis (NTA). The TEM image showed that pancreatic cancer-derived EVs displayed features of membrane vesicles (Figure 1a andFigure S4). The result of NTA demonstrated that most of the pancreatic cancer-derived EVs had a diameter of around 110 nm (Figure 1b). EVs were further characterized by their expression of CD9, CD63, CD81, ALIX, flotillin-1, TSG101, and Rab5, which are commonly-used markers for EVs (Figure 1c,d and Figure S3. Intriguingly, we also detected mutant KRAS (G12D) in L3.6pl-derived EVs, which was consistent with that in genomic DNA (Figure S1b).

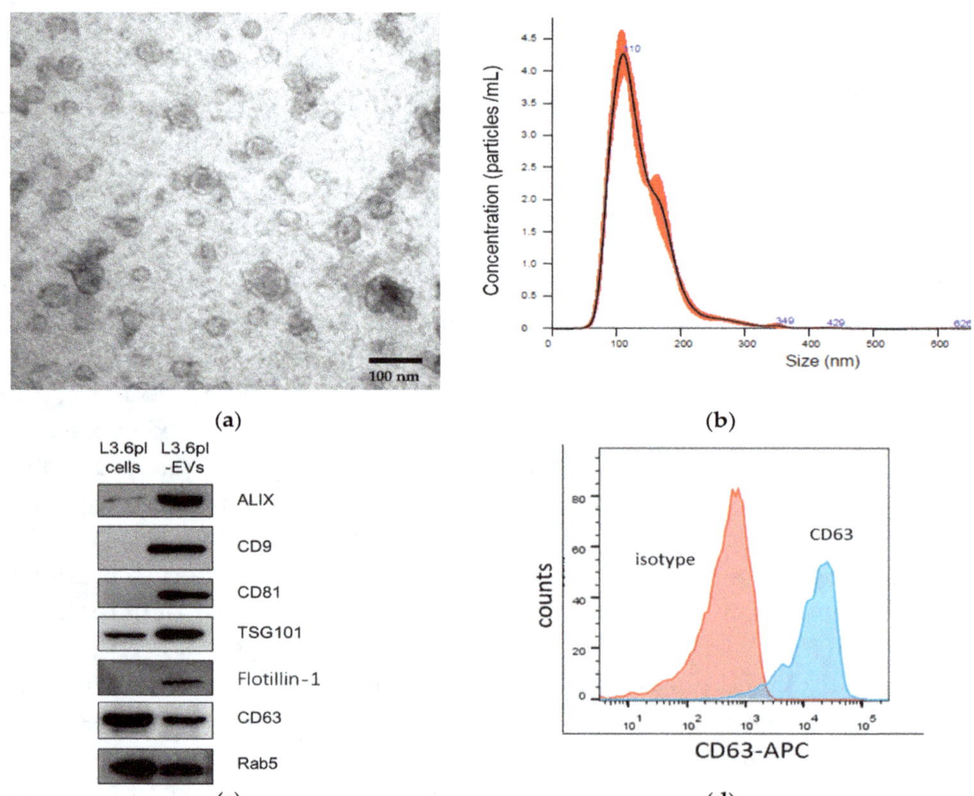

Figure 1. Characterization of pancreatic cancer cell-derived extracellular vesicles (EVs). (**a**) The representative image of pancreatic cancer-derived EVs by transmission electron microscopy (TEM). Scale bar, 100 nm. (**b**) The size of pancreatic cancer-derived EVs was determined by nanoparticle analysis (NTA). The size range was 136.1 ± 47.3 nm. (**c**) The expression of exosomal markers ALIX, flotillin-1, TSG101, CD9, CD63, CD81, and Rab5 for L3.6pl-derived EVs and parental cells was determined by Western blotting. (**d**) The expression of CD63 on L3.6pl-derived EVs coupled to carboxylate beads was analyzed by flow cytometry.

2.2. *Comprehensive Proteomic Analysis of Pancreatic Cancer-Derived EVs*

The proteomic profile of pancreatic cancer-derived EVs was analyzed by mass spectrometry. More than 2600 proteins were detected in both samples. A significant overlap was observed in L3.6pl-derived EVs and TBO368-derived EVs. (Figure S1c). Eighty-eight of the top 100 most frequently-identified proteins in EVs, according to the Exocarta database (www.exocarta.org), were identified in pancreatic cancer-derived EVs. The enrichment of exosomal markers verified the purity of EVs (Figure S1d).

To investigate the cellular component, molecular function, and biological process of proteins in pancreatic cancer-derived EVs, GO analysis was performed using the Gene Ontology Resource (http://geneontology.org/). Proteins were categorized according to their ontology as determined from their GO annotation terms. Based on the cellular component, around 40% of all the identified proteins were annotated to extracellular EVs (Figure 2a). The molecular function revealed the enrichment of proteins related to translation regulator activity (GO:0045182), transcription regulator activity (GO:0140110), molecular transducer activity (GO:0060089), binding (GO:0005488), structural molecule activity (GO:0005198), molecular function regulator (GO:0098772), catalytic activity (GO:0003824), and transporter activity (GO:0005215) (Figure 2b). The biologic process revealed that the proteins in pancreatic cancer-derived EVs were involved in cellular component organization or biogenesis (GO:0071840), cellular processes (GO:0009987), biological phases (GO:0044848), localization (GO:0051179), reproduction (GO:0000003), biological regulation (GO:0065007), response to stimulus (GO:0050896), developmental processes (GO:0032502), multicellular organismal processes (GO:0032501), biological adhesion (GO:0022610), metabolic processes (GO:0008152), cell proliferation (GO:0008283), and immune system processes (GO:0002376) (Figure 2c).

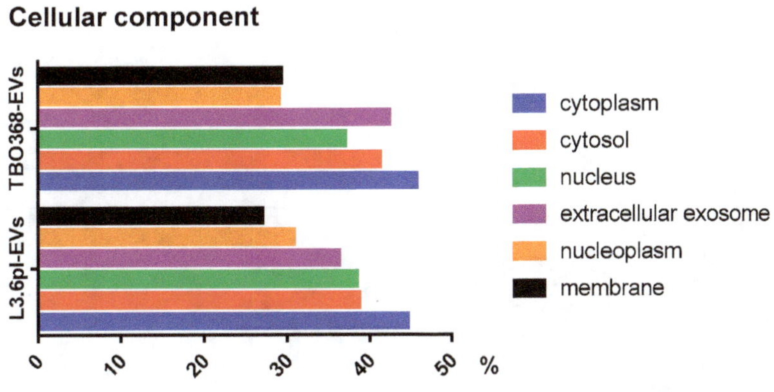

(a)

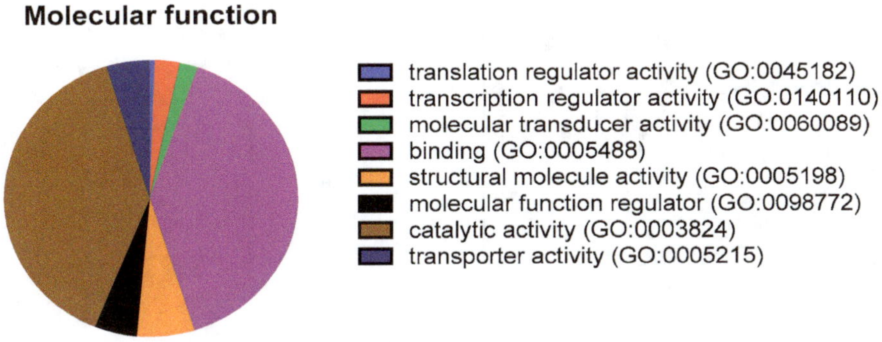

(b)

Figure 2. *Cont.*

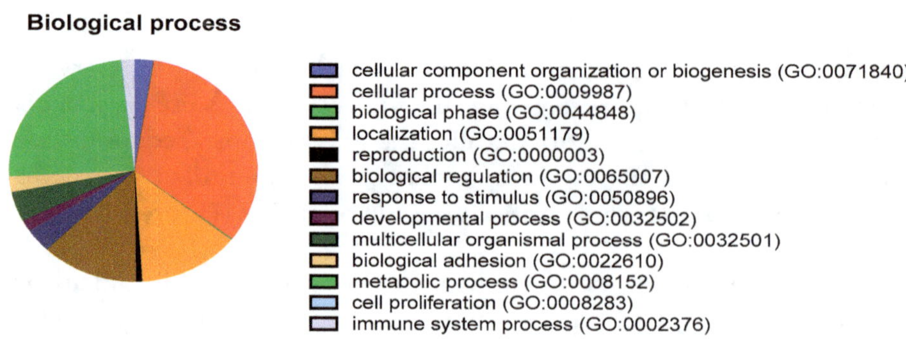

(c)

Figure 2. Proteomic analysis of pancreatic cancer-derived EVs. (**a**) The cellular component of proteins in TBO368-derived EVs and L3.6pl-derived EVs. (**b**) The molecular function of identified proteins in pancreatic cancer-derived EVs. (**c**) The biological process of identified proteins in pancreatic cancer-derived EVs.

2.3. *Pancreatic Cancer-Derived EVs Carry Adhesion Molecules*

To evaluate the role of pancreatic cancer-derived EVs in the pre-metastatic niche, GO analysis revealed that abundant cellular adhesion proteins exist in pancreatic cancer-derived EVs, particularly integrins, such as ITGA1, ITGA2, ITGA3, ITGA6, ITGAV, ITGB1, ITGB4, ITGB5, ITGB6, and ITGB8 (Figure 3a,b). We detected the expression of integrin alpha V (ITGAV) in L3.6pl-derived EVs by Western blotting (Figure 3c). To track the *in vivo* distribution of pancreatic cancer-derived EVs, we further injected PKH67-labelled L3.6pl-derived EVs intravenously into NOD-scid IL2rγ^{null} (NSG) mice. Twenty four hours after injection, PKH67-labelled EVs were detected by immunofluorescence in cryosections of mouse liver tissue, which indicated that pancreatic cancer-derived EVs reached the liver (Figure 3d and Figure S5).

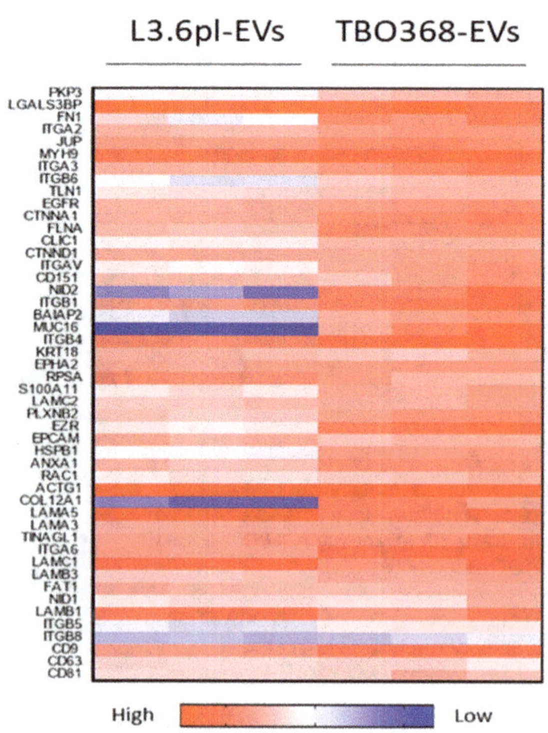

(a)

	L3.6pl-exo	TBO368-exo
ITGA1	+	+
ITGA2	+	+
ITGA3	+	+
ITGA4	-	+
ITGA5	-	+
ITGA6	+	+
ITGAV	+	+
ITGB1	+	+
ITGB4	+	+
ITGB5	+	+
ITGB6	+	+
ITGB8	+	+

(b)

Figure 3. *Cont.*

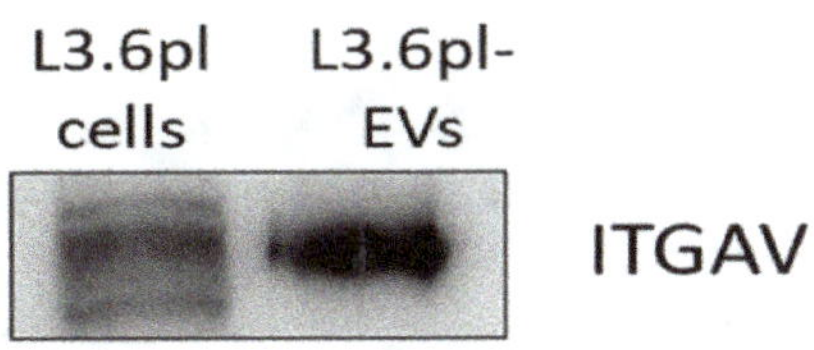

(c)

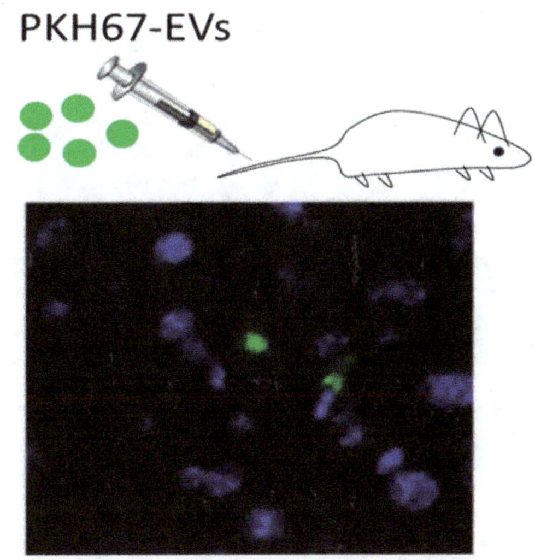

(d)

Figure 3. Pancreatic cancer-derived EVs carry adhesion molecules. (**a**) Heatmap of adhesion molecules in L3.6pl-derived EVs and TBO368-derived EVs, exosomal markers CD9, CD63, and CD81 as internal references. (**b**) Integrins in L3.6pl-derived EVs and TBO368-derived EVs. (**c**) Western blot analysis of ITGAV in L3.6pl-derived EVs. (**d**) Analysis of liver injected with PKH67-labeled L3.6pl-derived EVs (green) by confocal microscopy. Nuclei were stained with DAPI (blue).

2.4. Pancreatic Cancer-Derived EVs Carry Immune Regulatory Factors

To investigate the role of tumor-derived EVs in immune regulation, we first analyzed the expression pattern of immune regulatory factors in paired PDAC tumor tissues and adjacent non-tumor tissues based on the GSE28735 dataset ($n = 45$). Compared to non-tumor tissues (N), a variety of factors like TGF-β1, TGF-β2, HMGB1, PVR, nectin-2, galectin-9, PD-L1, PD-L2, and MICA/MICB were significantly higher in the tumor tissue (T) (Figure 4a). Interestingly, enrichment of some molecules, including TGF-β1, nectin-2, and PVR, was identified in pancreatic cancer-derived EVs by Western blotting (Figure 4b). TGFbRI and TGFbRII (TGF-β1 receptors), DNAM-1, TIGIT, and CD96 (nectin-2 and PVR receptors) are present on NK cells. These results support the hypothesis that pancreatic cancer-derived EVs potentially modulate NK cell function.

2.5. Pancreatic Cancer-Derived EVs Inhibit NK Cell Function

Subsequently, we determined whether NK cells could uptake pancreatic cancer-derived EVs. To address this issue, L3.6pl-derived EVs were stained with PKH67 (green). PKH67-labelled EVs were incubated with NK cells. After 24 h, we observed that PKH67-labelled EVs were present on the plasma membrane and in the cytoplasm of NK cells (Figure 4c). This result indicated that pancreatic cancer-derived EVs could be incorporated by NK cells, suggesting their potential role in the regulation of NK cell function. Thus, we examined the effects of pancreatic cancer-derived EVs on NK cells.

NKG2D is one of the most important activating receptors on NK cells, and the expression level of NKG2D correlates positively with their anti-tumor ability [27]. We co-cultured NK cells with L3.6pl-derived EVs or PBS for 24 h. After co-culture, the expression of NKG2D in NK cells was significantly downregulated (Figure 4d). CD107a is a functional marker for NK cells [28]. TNF-α and IFN-γ are two main cytokines produced by activated NK cells [29]. To measure the amount of CD107a, TNF-α, and IFN-γ, NK cells pre-treated with L3.6pl-derived EVs or PBS were co-cultured with L3.6pl cells at an effector:target cell ratio of 1:1 for 5 h. L3.6pl-derived EVs resulted in a significant decrease of CD107a, TNF-α, and IFN-γ in NK cells (Figure 4e). Nutrient uptake and glucose metabolism are essential for NK cell functionality [30]. CD71 (transferrin receptor), CD98 (large neutral amino acid transporter), and 2-NBDG incorporation ability are three commonly-used metabolic parameters in NK cells [31]. We found that L3.6pl-derived EVs significantly reduced the expression of CD71 and CD98 in NK cells. In addition, L3.6pl-derived EVs impaired the glucose uptake ability of NK cells (Figure 4f).

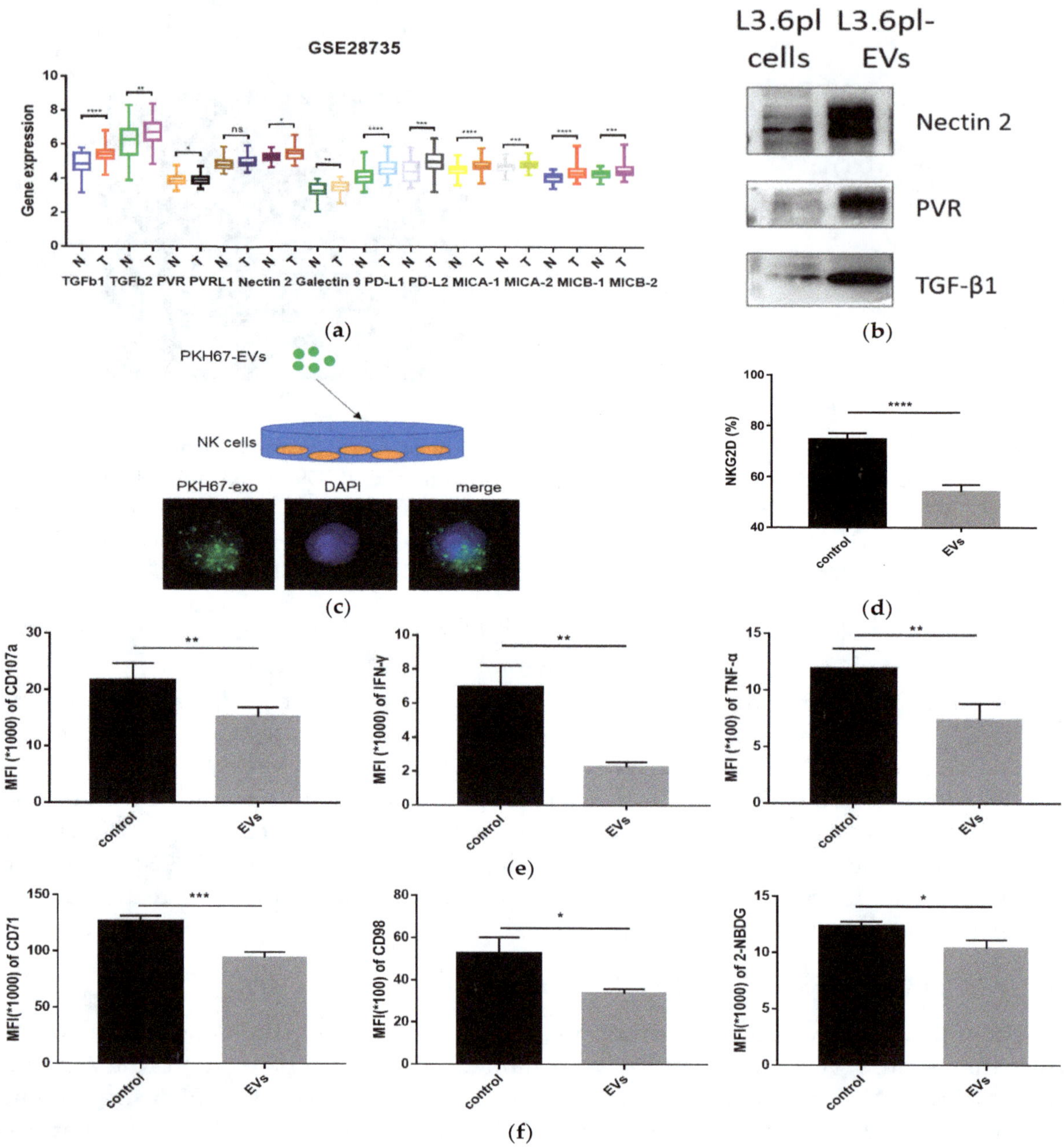

Figure 4. Pancreatic cancer-derived EVs impair natural kill (NK) cell function. (**a**) Relative mRNA expression of representative immune regulatory factors in tumor tissues (T) and non-tumor tissues (N) in pancreatic cancer from the GSE28735 dataset, $n = 45$. (**b**) The expression of nectin-2, PVR, and TGF-β1 was determined by Western blotting in L3.6pl-derived EVs and L3.6pl cells. (**c**) Analysis of pancreatic cancer-derived EVs' uptake by NK cells using confocal microscopy. L3.6pl-derived EVs were stained with PKH67 (green) and incubated with NK cells for 24 h. The nucleus was labeled with DAPI (blue). (**d**) NK cells were treated with PBS or L3.6pl-derived EVs for 24 h. The percentage of NKG2D-positive NK cells was analyzed by flow cytometry. (**e**) NK cells pre-treated with PBS or L3.6pl-derived EVs were co-cultured with L3.6pl cells at a 1:1 ratio for 5 h. The mean fluorescence intensity (MFI) of CD107a (left), IFN-γ (middle), and TNF-α (right) in NK cells was analyzed by flow cytometry. (**f**) NK cells were treated with PBS or L3.6pl-derived EVs for 24 h. NK cells were then analyzed by flow cytometry to determine the MFI of CD71 (left) and CD98 (middle) and 2-NBDG incorporation (right). Data are the means ± SD of four experiments. ns, no significant difference, * $p < 0.05$, ** $p < 0.01$, *** $p < 0.001$, **** $p < 0.0001$ by Student's t test.

2.6. *Pancreatic Cancer-Derived EVs Impair NK Cell Cytotoxicity*

It has been reported that NK cells have the capacity to kill CSCs [20]. We next examined whether pancreatic cancer-derived EVs impair NK cell cytotoxicity against pancreatic CSCs. After enrichment of CSCs using the sphere formation assay, we found higher mRNA expression levels of NKG2D ligands, MICB, and ULBP2 in the CSCs population (Figure 5a). Flow cytometric analysis confirmed a higher MICA/MICB expression in spheres than that in adherent cells (Figure 5b). This indicated that NK cells might prefer to recognize and eliminate pancreatic CSCs. NK cells were then pre-treated with L3.6pl-derived EVs in the presence of IL-2 (100 U/mL) for 24 h. Then, we co-cultured L3.6pl cells with untreated or L3.6pl-derived EV pre-treated NK cells. After 24-h killing, floating cells were washed away, and adherent cells were trypsinized for the sphere formation assay. We found that NK cells pre-treated with L3.6pl-derived EVs showed decreased cytotoxicity against pancreatic CSCs (Figure 5c,d).

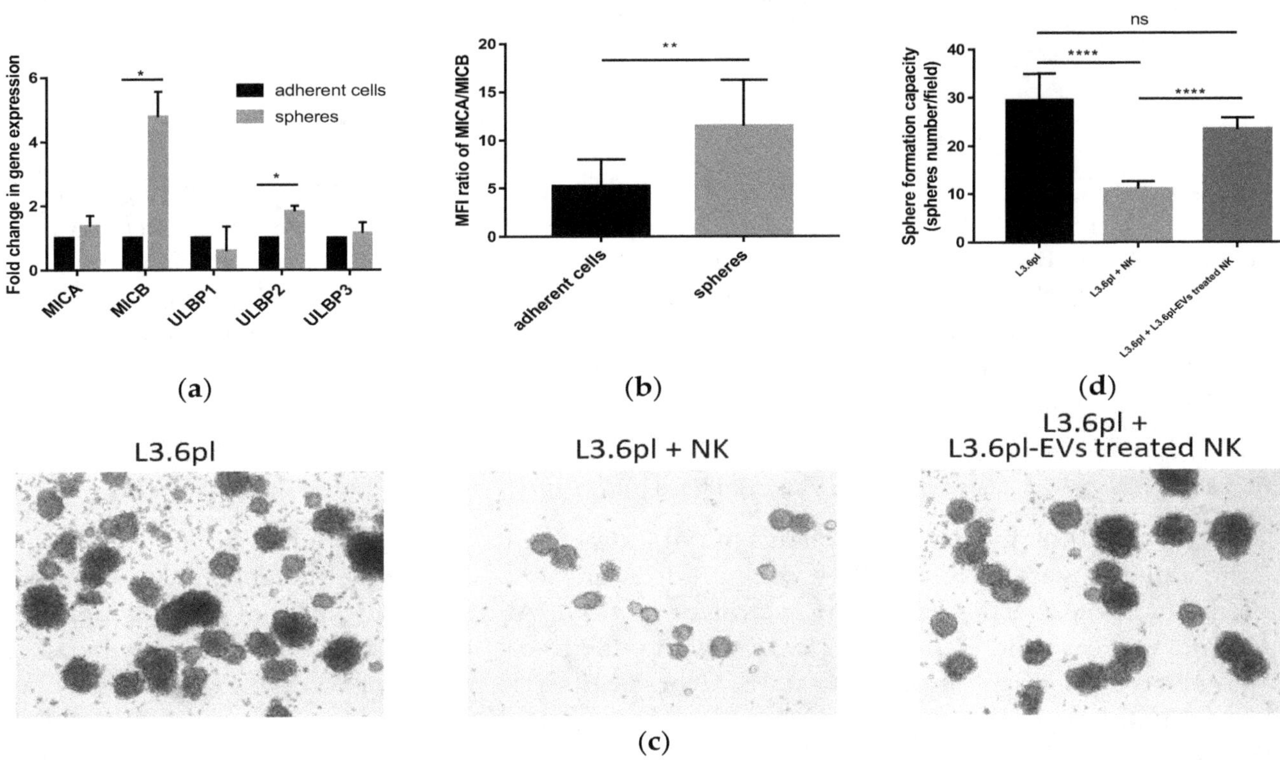

Figure 5. Pancreatic cancer-derived EVs suppress NK cell cytotoxicity against cancer stem cells (CSCs). (**a**) Gene expression of NKG2D ligands in adherent cells and spheres using qRT-PCR. Data were normalized to GAPDH and presented as fold change in comparison with genes in adherent cells. (**b**) The MFI of MICA/MICB in adherent cells and spheres was determined by flow cytometry. (**c**) Representative images of tumor spheres without NK cell killing (left), tumor spheres after untreated NK cell killing (middle), and tumor spheres after L3.6pl-derived EVs pre-treated NK cell killing (right). (**d**) The number of tumor spheres without NK cell killing, tumor spheres after untreated NK cell killing, and tumor spheres after L3.6pl-derived EV-pre-treated NK cell killing. Data are the means ± SD of four experiments. ns, no significant difference, * $p < 0.05$, ** $p < 0.01$, *** $p < 0.001$, **** $p < 0.0001$ by Student's t test.

2.7. *Pancreatic Cancer-Derived EVs Phosphorylate Smad2/3 in NK Cells*

As a major immunosuppressive cytokine, TGF-β1 inhibits the activation and function of NK cells through the TGFβ-Smad2/3 signaling pathway [32]. In our experiments, we observed that TGF-β1 attenuated the expression of the NKG2D, CD107a, IFN-γ, CD71, CD98, and 2-NBDG incorporation ability of NK cells (Figure S2a–c). As shown above, pancreatic cancer-derived EVs contained TGF-β1. Therefore, we investigated whether pancreatic cancer-derived EVs could activate the

TGFβ-Smad2/3 signaling pathway in NK cells. After incubation with TGF-β1 or L3.6pl-derived EVs, the phosphorylation level of Smad2/3 in NK cells was significantly elevated. However, in the presence of SB-431542 (an inhibitor of TGFβRI), the phosphorylation of Smad2/3 was reversed and returned to the original baseline level (Figure 6a). These findings suggest that pancreatic cancer-derived EVs deliver TGF-β1 to NK cells, induce Smad2/3 phosphorylation, and ultimately result in NK cell dysfunction.

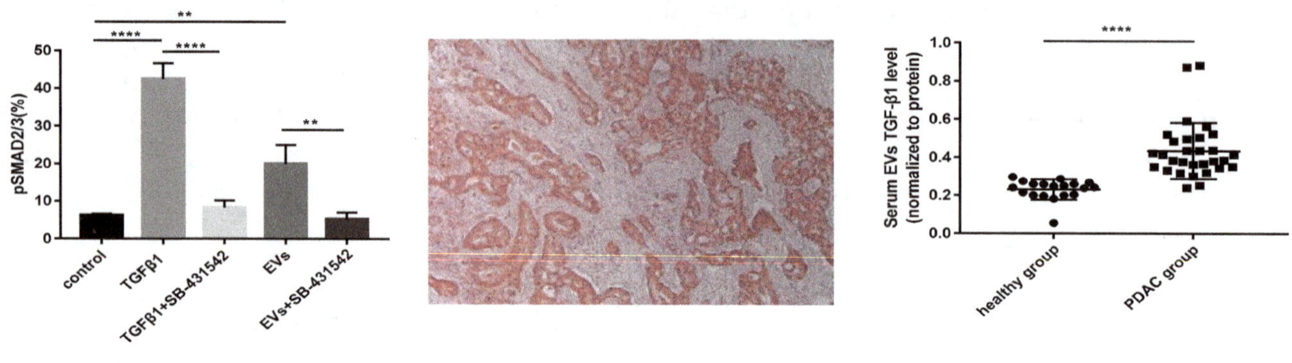

(a) (b) (c)

Figure 6. Pancreatic cancer-derived EVs inhibit NK cell function through the TGFβ-Smad2/3 signaling pathway, and serum EVs' TGF-β1 is increased in pancreatic cancer patients. (**a**) After co-culture with TGF-β1 or L3.6pl-derived EVs in the presence or absence of SB-431542, the phosphorylation level of SMAD2/3 in NK cells was measured by flow cytometry. Data are the means ± SD of four experiments. (**b**) The immunohistochemistry (IHC) result showed TGF-β1 overexpression in tumor tissue of PDAC. Magnification: 100×. (**c**) The amount of TGF-β1 per gram of serum EVs in the healthy control ($n = 19$) and PDAC group ($n = 30$). ns, no significant difference, * $p < 0.05$, ** $p < 0.01$, *** $p < 0.001$, **** $p < 0.0001$ by Student's t test.

2.8. Increased TGF-β1 Levels in Serum EVs of Patients with PDAC

The clinicopathological characteristics of patients with PDAC are listed in Table S1. The mean age of the patients was 66.1 y, and they were predominantly male. All the patients donated blood at the time point of diagnosis prior to any treatment ($n = 30$). At diagnosis, 53.3% of patients presented with a tumor stage T1 and T2, and 46.7% presented with T3 or T4. Seventy seven-point-seven percent of patients had a positive lymph node status. Three out of 30 patients had distant metastases (M1). Fifty six-point-seven percent of the patients were Union for International Cancer Control (UICC) I or II, and 43.3% were UICC III or IV. Nineteen healthy individuals were included in this study as controls ($n = 19$). TGF-β1 was overexpressed in PDAC (Figure 6b). The amount of TGF-β1 in serum EVs was determined by ELISA. The concentration of TGF-β1 per gram of EVs was calculated. The level of EVs' TGF-β1 in patients with pancreatic cancer ranged from 0.20–0.88 ng/g. In the healthy donors, it ranged from 0.05–0.30 ng/g. Compared to healthy donors, TGF-β1 in serum EVs was significantly elevated in patients with PDAC ($p < 0.0001$) (Figure 6c).

3. Discussion

Pancreatic ductal adenocarcinoma is one of the most lethal malignancies [1]. Metastasis accounts for a majority of cancer-related deaths in PDAC. Recently, the pre-metastatic niche has been proposed to elucidate the mechanisms of the organ-specific metastatic process in many cancer entities, such as melanoma, lung cancer, and pancreatic cancer [7,8,33]. Over the past few decades, EVs have attracted wide attention in early detection, diagnosis, and treatment of cancer. Moreover, as a mediator of intercellular communication, EVs released from tumor cells are found to interact with cells in distant organ sites and finally induce a pre-metastatic niche for future metastasis [34]. Despite tremendous advances, the underlying cellular and molecular events involved in pre-metastatic niche formation of PDAC have yet to be determined.

In the present study, tumor-derived EVs were isolated from two human pancreatic cancer cell lines, L3.6pl and TBO368, by differential centrifugation and ultracentrifugation to exclude dead cells, large debris, and microvesicles. Then, the morphology and size distribution of EVs were examined by TEM and NTA. Exosomal markers, including CD9, CD63, CD81, TSG101, Alix, glotillin-1, and Rab5, were identified by Western blotting. Comprehensive proteomic analysis is expected to elucidate the potential impact of tumor-derived EVs on pre-metastatic niche formation of PDAC. The proteomic profile of mouse PDAC cell line-derived EVs has been analyzed by a previous study. Yu Z et al. compared exosomes derived from Panc02 and Panc02-H7 cells using proteomic analyses. The differentially-expressed proteins in Panc02-H7-derived exosomes were thought to enhance tumor growth, invasion, and metastasis [35]. To the best of our knowledge, this is the first study to systematically analyze the protein content in human PDAC-derived exosomes. By mass spectrometry, more than 2600 proteins were detected in both samples. We found that about 90% of the protein identified in L3.6pl-derived EVs overlapped with those identified in TBO368-derived EVs. GO analysis of identified proteins was performed for cellular components, molecular functions, and biologic processes.

GO-based category clustering of the molecular functions of protein contents in pancreatic cancer-derived EVs suggested that there was a significant enrichment in localization and biological adhesion, which may facilitate the ability of EVs' to adhere to the surfaces of recipient cells, fuse with their membranes, and transfer exosomal components into the target cells to modulate their biological functions. The mechanisms of organ-specific homing and colonization of cancer cells are enormously complex. Cell adhesion to the extracellular matrix (ECM) determines the colonization of metastatic sites and facilitates the survival of circulating tumor cells in the new environment. Integrins can bind to fibronectin, vitronectin, laminin, and collagen in the ECM, thereby enhancing tumor cell motility and invasion ability [36]. Y Liu et al. reported that after either intravenous injection or intra-tumor injection, lung cancer-derived exosomes were detected in the lung and induced a lung pre-metastatic niche in mouse models [8]. GO analysis revealed that pancreatic cancer-derived exosomes exhibited abundant cellular adhesion molecules, especially integrins. Ayuko Hoshino et al. found that tumor-derived exosomal integrins determined organotropic metastasis. They reported that tumor-derived exosomes carrying integrins $\alpha6\beta4$ and $\alpha6\beta1$ were responsible for lung metastasis, while exosomes carrying integrin $\alpha v\beta5$ were associated with liver metastasis [18]. Our findings showed that integrin αv and integrin $\beta5$ were abundant in pancreatic cancer-derived EVs. After intravenous injection, PKH67-labeled pancreatic cancer-derived EVs reached the liver of the NSG mouse. Therefore, we proposed that tumor-derived EVs' tended to enter the liver, delivered cargos to the recipient cells, and induced a pre-metastatic niche for future metastasis in PDAC.

A key feature of the pre-metastatic niche is immunosuppression [9]. In our study, GO analysis revealed that the identified proteins in pancreatic cancer-derived EVs were involved in biological regulation and immune system processes. By Western blotting, we found that pancreatic cancer-derived EVs displayed a variety of immune regulatory molecules, such as TGF-$\beta1$, nectin-2, and PVR. Therefore, we speculated that pancreatic cancer-derived EVs might be involved in the modulation of immune cell functions. Next, we tried to investigate the effects of pancreatic cancer-derived EVs on NK cells. Firstly, we found that pancreatic cancer-derived EVs could be incorporated by NK cells. Nevertheless, it was still uncertain whether pancreatic cancer-derived EVs could mediate immune suppression upon co-incubation with NK cells.

The stress proteins MICA and MICB are widely expressed by cancer cells due to genomic damage [37]. NKG2D is a key activating receptor for NK cell cytotoxicity [27]. The binding of MICA/MICB to NKG2D receptors triggers NK cell-mediated cytotoxicity and enables them to eliminate cancer cells [38]. Our results indicated that the expression of NKG2D on NK cells was significantly downregulated by pancreatic cancer-derived EVs. After recognition and activation, NK cells synthesize and release effective cytokines into the tumor cells. For example, IFN-γ and TNF-α are two indispensable cytokines for NK cell cytotoxicity. Neutralization of IFN-γ and TNF-α significantly impaired NK cell activity [29]. We demonstrated that treatment with EVs led to less production of IFN-γ and TNF-α in NK cells. In addition, as a functional marker for NK cell activity, the expression of CD107a was significantly downregulated.

Recently, the importance of cellular metabolism of immune cells has gained increasing attention. Enough nutrients and energy are essential for NK effector functions [39]. Cong J et al. demonstrated that the tumor could reduce NK cell glycolytic capacity, which results in reduced cytotoxicity and NK cell dysfunction [24]. We found that NK cells exhibited less CD71 and CD98, as well as reduced glucose uptake ability after treatment with EVs. Dysregulated metabolism caused by pancreatic cancer-derived EVs affected multiple biological processes in NK cells such as interfered protein synthesis and impaired energy production. Our result was consistent with previous studies investigating the effects of tumor-derived EVs on immune cells [40–42]. Our findings suggest that pancreatic cancer-derived EVs induce a dysfunctional phenotype of NK cells, which ultimately contribute to an immunosuppressive microenvironment in the pre-metastatic niche.

Pancreatic cancer cells with high aldehyde dehydrogenase 1 (ALDH1) expression are considered as cancer stem cells (CSCs) [43]. Here, we found that after enrichment using the sphere formation assay, pancreatic CSCs exhibited high expression of ligands for NKG2D. However, pancreatic cancer-derived EVs impaired NK cell cytotoxicity against CSCs. CSCs are thought to be responsible for metastasis [44]. The inhibition of NK cell cytotoxicity allowed pancreatic CSCs to escape from NK cell immune surveillance and colonize in the target organ.

As a key inhibitory cytokine, TGF-β1 plays a dominant role in modulating NK cell function. For example, TGF-β1 attenuated NK cell responses by downregulating NKG2D expression in patients with advanced cancer [45]. Among various signals delivered by pancreatic cancer-derived EVs to NK cells, TGF-β1 was thought to be a candidate responsible for NK cell dysfunction. Our results showed that TGF-β1 impaired NK cell function, including downregulated expression of NKG2D, CD107a, CD71, and CD98, decreased production of cytokines, such as TNF-α and IFN-γ, as well as reduced glucose uptake ability. In general, activation of the TGF-β/Smad2/3 signaling pathway is implicated in NK cell dysfunction [32]. We found that either pancreatic cancer-derived EVs or TGF-β1 could induce the phosphorylation of Smad2/3 in NK cells. However, the phosphorylation level of Smad2/3 returned to baseline in the presence of SB-431542. Therefore, we proposed that pancreatic cancer-derived EVs inhibit NK cell function via the TGFβ1-Smad2/3 pathway. Pancreatic cancer-derived EVs delivered TGF-β1 to the surface of NK cells upon binding to the TGFβ receptors (TGFβRI/II). Activation of TGFβRI/II by TGF-β1 induced the phosphorylation of serine/threonine residues and triggered phosphorylation of Smad2/3. Phosphorylated-Smad2/3 then translocated to the nucleus and regulated gene transcription, thereby modulating NK cell function [46]. The TGFβRI inhibitor, SB-431542, exhibited cytotoxicity against pancreatic cancer cells in vitro [47]. Further investigations are needed to explore its anti-tumor effect in vivo, especially its influence on the phenotypic and functional diversity of NK cells.

Without specific symptoms, it is a major challenge to detect PDAC at early stages. EVs are promising to be developed as a liquid biopsy tool for early detection and diagnosis of PDAC [48]. Recently, it has been reported that the levels of exosomal PD-L1 in the plasma, rather than soluble PD-L1, were associated with disease progression in patients with head and neck squamous cell carcinomas (HNSCCs) [49]. In our study, we measured the levels of serum EVs' TGF-β1 in PDAC patients. Interestingly, compared to healthy donors, serum EVs' TGF-β1 was significantly elevated in the PDAC group. Therefore, serum EVs' TGF-β1 holds promise to be used as a diagnostic tool for detection of PDAC.

Apart from TGF-β1, pancreatic cancer-derived EVs also contained multiple other immune regulatory factors, such as PVR and nectin-2, which could be delivered as inhibitory signals to NK cells. Both PVR and nectin-2 can bind to inhibitory receptors on NK cell, including CD96, PVRIG, and TIGIT [50]. Therefore, in addition to TGF-β1, PVR and nectin-2 in pancreatic cancer-derived EVs could also impair NK cell function. It has been demonstrated that the PVR/nectin-2-TIGIT axis is involved in attenuated NK cell cytotoxicity [51]. As a checkpoint, receptor blockade of TIGIT prevented NK cell dysfunction and elicited NK cell anti-tumor responses in tumor-bearing mouse models [52]. Immune checkpoint inhibitors targeting CTLA-4, PD-1, and PD-L1 have shown clinical benefit for patients with non-small cell lung cancer (NSCLC), advanced melanoma and several other cancers [53–55]. However, these inhibitors are less effective in patients with PDAC [56]. Our results indicate that TIGIT offers a potential immunotherapeutic target for PDAC.

Nevertheless, a limitation of the present study is that in vivo effects of human pancreatic cancer-derived EVs on NK cells have not been investigated. To address this issue, humanized patient-derived xenograft mouse models, which can better recapitulate tumor heterogeneity and simulate the complexity of the immune system, serve as a better platform for further investigation [57].

In conclusion, we proposed a novel mechanism of immune escape in PDAC. Pancreatic cancer can establish a pre-metastatic niche in the liver via tumor-derived EVs. Pancreatic cancer-derived EVs carrying immunosuppressive cargos mediate NK cell dysfunction. Metastatic pancreatic cancer cells evade the immune surveillance of NK cells and ultimately generate metastases in the liver. Additionally, serum EVs' TGF-β1 may represent a promising non-invasive diagnostic tool for PDAC.

4. Materials and Methods

4.1. Cell Culture

The L3.6pl human pancreatic cancer cells were isolated and established after several cycles of *in vivo* selection in nude mice [58]. TBO368 was obtained with written consent from a pancreatic cancer patient and subsequently established as low-passage primary cell cultures. The study has been approved by the Ethics Committee of the University of Cologne (BIOMASOTA, (Biologische Material Sammlung zur Optimierung Therapeutischer Ansätze), ID: 13-091, approval in May 2016). L3.6pl was cultured in DMEM medium (Gibco, ThermoFisher Scientific, Waltham, MA, USA) supplemented with 10% FBS (*vol/vol*), 1% vitamin mixture, 1% sodium pyruvate, 1% nonessential amino acids, 1% L-glutamine and 1% penicillin/streptomycin (FBS from Capricorn Scientific, Ebsdorfergrund, Germany)(others from Invitrogen, ThermoFisher Scientific, Waltham, MA, USA). TBO368 was cultured in advanced DMEM medium (Gibco) supplemented with 10% FBS, 1% L-glutamine and 1% penicillin/streptomycin. Human NK cell line NK92 cells were obtained from the ATCC and cultured in MEM α, no nucleosides medium (Gibco) containing 12.5% FBS, 12.5% horse serum (Gibco), 100 IU/mL penicillin, 100 µg/mL streptomycin, 0.02 mM folic acid, 0.1 mM 2-mercaptoethanol, 0.2 mM Myo-inositol, 2 mM L-glutamine, and 100 U/mL recombinant human IL-2 (Peprotech, Hamburg, Germany).

4.2. EVs' Isolation

For EVs' preparation from cell culture supernatants, pancreatic cancer cells with a confluency of 70–80%, around 1.5×10^7 cells, were washed with Dulbecco's Phosphate-Buffered Saline (DPBS) 3 times and cultured in 25 mL of medium supplemented with 10% EVs-free FBS for an additional 24 h in a T-175 flask. EVs were isolated using a differential centrifugation and ultracentrifugation method. In brief, supernatants were centrifuged at 300× *g* for 10 min and 2000× *g* for 10 min at 4 °C to remove dead cells and cell debris. The supernatants were transferred into new tubes and centrifuged at 10,000× *g* for 30 min at 4 °C to remove large vesicles. One hundred and twenty milliliters of supernatant were then transferred into ultracentrifugation tubes and ultracentrifuged at 100,000× *g* for 70 min at 4 °C (Beckman Coulter, OptimaTM L-90K, Indianapolis, IN, USA). After the first round of ultracentrifugation, the supernatant was discarded. The pellet was resuspended in PBS and ultracentrifuged again at 100,000× *g* for 70 min at 4 °C. EVs were resuspended in 100 µL of PBS and stored in a −80 °C freezer for less than one year.

Serum EVs were isolated by a precipitation method using ExoQuick (System Biosciences, Palo Alto, CA, USA) according to the manufacturer's instructions. In brief, serum samples were centrifuged at 3000× *g* for 15 min at room temperature (RT) to remove remaining blood cells and cell debris.

After centrifugation, 250 µL of serum were put into a new tube and mixed with 63 µL of ExoQuick Exosome Precipitation Solution (System Biosciences, Palo Alto, CA, USA). The mixture was incubated at 4 °C for 30 min and then centrifuged at 1500× *g* at 4 °C for 30 min. The supernatant was aspirated. The tube was centrifuged at 1500× *g* for an additional 5 min to remove the residual ExoQuick solution. The pellet was resuspended completely in 100 µL of PBS. Serum EVs were stored in a −80 °C freezer for less than one year.

4.3. Nanoparticle Tracking Analysis

The size distribution of pancreatic cancer-derived EVs was examined by nanoparticle tracking analysis (NTA). Briefly, background measurements were performed with filtered PBS, which revealed the absence of any kinds of particles. EVs were diluted 1:1000 with PBS. After sample loading, five repeated measurements were recorded and then analyzed using a Nanosight NS300 with the NTA 3.0 software (Malvern Instruments, Malvern, UK).

4.4. Transmission Electron Microscopy

The morphology of pancreatic cancer-derived EVs was assessed by transmission electron microscopy (TEM). In brief, EVs were put onto formvar-carbon-coated electron microscopy grids for 10 min in a wet chamber. After brief blotting the grid edge with filter paper, the grids were placed on drops of 2% aqueous uranyl acetate for 1 min, removed, blotted again, and placed on H_2O drops for 1 min, removed, and blotted at the edge. After 24 h of air drying, the grids were inspected using a transmission electron microscope (Zeiss EM 912 Omega at 100 kV, Darmstadt, Germany).

4.5. Western Blot

Cells and EVs were lysed using complete lysis M buffer supplemented with phosphatase inhibitor cocktail (Sigma, St. Louis, MO, USA). Protein samples were centrifuged at 14,000× *g* for 15 min at 4 °C. After centrifugation, the supernatants were transferred into new tubes and stored in a −80 °C freezer. Protein concentrations were measured by the BCA Protein Assay (Thermo Scientific™, Waltham, MA, USA). Protein Samples were prepared in Pierce™ LDS Sample Buffer (Thermo Fisher Scientific), boiled for 10 min at 70 °C. Ten to fifteen micrograms of lysates were loaded and run in SDS polyacrylamide gels. Gels were then transferred onto PVDF membranes. The membranes were blocked in blocking buffer at RT for 1 h and incubated with specific primary antibodies at 4 °C overnight (Table S2). Blots were washed 3 × 5 min in PBST. Incubation with the appropriate horseradish peroxidase (HRP)-conjugated secondary antibodies was performed at RT for 1 h. Blots were again washed in TBST 3 × 5 min. Proteins were detected via chemiluminescence with SuperSignal™ West Pico PLUS Chemiluminescent Substrate (Thermo Fisher Scientific) using the Intas ChemoStar ECL Imager (Intas Science Imaging, Göttingen, Germany).

4.6. Flow Cytometry of EVs

Pancreatic cancer-derived EVs were incubated with polybead carboxylate microspheres (Polyscience, Niles, Illinois, USA) in PBS overnight at 4 °C. Samples were blocked with 2% BSA in PBS and centrifuged at 350× *g* for 2 min. The supernatants were discarded. The pellets were washed with PBS twice and resuspended in PBS. EVs were incubated with the anti-CD63 antibody at 4 °C for 20 min in the dark. The samples were analyzed by CytoFlex (Beckman Coulter, Indianapolis, IN, USA).

4.7. Flow Cytometry of Cells

For surface staining, cells were incubated with Human TruStain FcX™ (Biolegend, San Diego, CA, USA) on ice for 10 min. Then, conjugated fluorescent antibodies were added and incubated on ice for 20 min in the dark. After incubation, cells were washed in PBS twice and then analyzed by flow cytometry. For intracellular staining, surface antigen staining was performed as above.

Afterwards, cells were washed and fixed in fixation buffer (Biolegend) for 20 min at RT. Cells were washed and permeabilized with permeabilization wash buffer (Biolegend). Cells were incubated with appropriate conjugated fluorescent antibodies in the dark for 20 min at RT. After incubation, cells were washed with permeabilization wash buffer twice and analyzed by flow cytometry. For each sample, 30,000–50,000 events were acquired using CytoFlex (Beckman Coulter). Data were analyzed using FlowJo software (Treestar Inc., Ashland, USA).

4.8. *Mass Spectrometry of EVs*

Fifty micrograms of pancreatic cancer-derived EVs were used for mass spectrometry analyses. First, EVs were denatured using 8 mol/L urea with protease inhibitor cocktail. Samples were centrifuged for 15 min at 17,000× *g* to remove debris. The protein concentration was determined using the Direct Detect® Spectrometer (Darmstadt, Germany). Then, 50 μg per sample were transferred into a new 1.5-mL tube. Samples were reduced using 5 mM DTT at 25 °C for 1 h and incubated with 40 mM chloroacetamide in the dark for 30 min. This was followed by proteolytic digestion with lysyl endopeptidase (Lys-C) at an enzyme:substrate ratio of 1:75 at 4 °C for 4 h. The samples were diluted with 50 mM Triethylammoniumbicarbonate (TEAB) to achieve a final concentration of urea ≤2 M. A subsequent digestion with trypsin at an enzyme:substrate ratio of 1:75 and incubation at 25 °C overnight was performed. The digestion was terminated with formic acid at a final concentration of 1%. After digestion, the peptide mixtures were desalted using an in-house made StageTip per sample (containing 2 layers of SDB-RPS discs). StageTips were equilibrated as follows: 20 μL of methanol, centrifugation at 2600 rpm for 1 min; 20 μL of 0.1% formic acid in 80% acetonitrile, centrifugation at 2600 rpm for 1 min; 20 μL of 0.1% formic acid in water, centrifugation at 2600 rpm for 1.5 min; 20 μL of 0.1% formic acid in water, centrifugation at 2600 rpm for 2 min. The samples (acidified with formic acid) were centrifuged at full speed for 5 min and then loaded onto the equilibrated StageTips. After centrifugation at 2600 rpm for 5 min, the StageTips were washed according to the following protocol: 30 μL of 0.1% formic acid in water, centrifugation at 2600 rpm for 3 min; 30 μL of 0.1% formic acid in 80% acetonitrile, centrifugation at 2600 rpm for 3 min. The last wash step was performed twice. Finally, the StageTips were dried completely with a syringe and kept at 4 °C until LC-MS analysis. Nano LC-MS was performed using a gradient for 150 min and analyzed using the MAXQuant (Max Planck Institute of Biochemistry, Martinsried, Germany) and Perseus software (Max Planck Institute of Biochemistry, Martinsried, Germany).

4.9. *In Vitro EVs' Uptake Assay*

Pancreatic cancer cell-derived EVs were isolated as described above. The PKH67 Fluorescent Cell Linker kit (Sigma, St. Louis, MO, USA) was used to label EVs according to the manufacturer's instruction. In brief, after the first round of ultracentrifugation, the supernatant was discarded, and the pellet of EVs was resuspended in 750 μL of Diluent C. One microliter of PKH67 dye was dissolved in 250 μL of Diluent C. EVs and PKH67 dye were mixed gently and incubated at RT for 5 min. Nine milliliters of PBS with 1% BSA were added to bind excess PKH67 dye. EVs were ultracentrifuged at 100,000× *g* for 70 min at 4 °C and washed twice with PBS by ultracentrifugation. The PKH67-labeled EVs were then resuspended in PBS. NK cells were incubated with PKH67-labeled EVs for 24 h and put on polysine adhesion slides for 30 min at 37 °C. After fixation and permeabilization, NK cells were stained with DAPI. Uptake of PKH67-labeled EVs by NK cells was visualized by confocal microscopy.

4.10. *In Vivo Distribution of EVs*

To study the *in vivo* distribution of pancreatic cancer-derived EVs, EVs were fluorescently labeled as described above. Animal experiments were conducted according to protocols approved by the responsible national and local authority (81-02.04.2018.A139, LANUV NRW, approved on 20 September 2018). PKH67-labeled EVs were administered into the tail vein of two healthy 4–6-week-old NSG mice. One NSG mouse was injected with PBS as a negative control. Twenty four hours after injection, mice were euthanized. Organs were dissected and embedded in Tissue-Tek O.C.T.TM. Then samples were frozen and stored at −80 °C. For immunofluorescence, 10 μm of O.C.T.TM tissue cryosections were stained with DAPI. The distribution of PKH67-labeled EVs was analyzed by confocal microscopy.

4.11. *In Vitro NK Cell Cytotoxicity Assay*

L3.6pl cells (2×10^5) were plated with NK cells (effector:target = 5:1) in a 6-well plate. All wells contained 100 IU/mL rhIL-2 with 50% of NK cell medium and 50% of L3.6pl culture medium. After co-culture for 24 h, plates were washed with DPBS three times. Adherent cells were harvested and counted for subsequent experiments.

4.12. *Sphere Formation Assay*

L3.6pl cells were seeded as a single-cell suspension at a concentration of 2000 cells/well in 6-well ultra-low attachment plates (Corning Inc., Corning, NY, USA) in DMEM/F12 medium (Gibco) supplemented with 20 ng/mL EGF (Peprotech, Rocky Hill, NJ), 20 ng/mL bFGF (Peprotech, Rocky Hill, NJ), 5 μg/mL insulin (Sigma-Aldrich, MO, USA) and 1× B27 (Gibco). After 7 days, spheres were counted under microscopy. Spheres were collected for mRNA extraction and flow cytometric analysis.

4.13. *RNA Isolation, cDNA Synthesis, and qRT-PCR*

Total RNA was isolated from cultured cells according to the manufacturer's instructions using the RNeasy Mini Kit (Qiagen, Hilden, Germany). RNA was reverse transcribed according to the manufacturer's protocol using the High-Capacity cDNA Reverse Transcription Kit (Applied Biosystems, ThermoFisher Scientific, Waltham, MA, USA). The synthesized cDNA was then assessed for gene expression using the Fast SYBR green master mix (Invitrogen) with QuantStudio 7 flex (Applied Biosystems, Thermo Fisher Scientific, USA).

4.14. *Human Studies*

All human tissue and blood samples were collected ethically, and their research use was according to the terms of the informed consent (BIOMASOTA, ID: 13-091, approval in May 2016). Human peripheral blood samples were acquired from healthy subjects and PDAC patients at University Hospital of Cologne between October 2016 and December 2018. All patients with PDAC were pathologically confirmed. Blood was obtained and centrifuged at 2000 rpm for 10 min and 4000 rpm for 10 min at RT. Serum samples were aliquoted and stored in the −80 °C freezer. Serum EVs were isolated as described above in the "EVs' Isolation" section. The EVs' TGF-β1 levels in serum were determined using the TGF beta-1 Human/Mouse Uncoated ELISA Kit (Invitrogen, ThermoFisher Scientific, Waltham, MA, USA) and analyzed with an ELISA microplate reader at 450 nm.

4.15. *Statistical Analysis*

Differences between two groups were tested by the two-tailed Student's t-test. All statistical analyses were made using GraphPad Prism 8 (GraphPad Software, Inc., San Diego, USA). Data were considered statistically significant when the p-value was smaller than 0.05.

5. Conclusions

In this study, we showed that tumor derived-EVs are responsible for pre-metastatic niche formation in the liver for PDAC. The inhibitory effects of pancreatic cancer-derived EVs on NK cells represent a mechanism allowing metastatic tumor cells to escape from NK cell immune surveillance in the pre-metastatic niche. We also demonstrate that TGF-β1 in serum EVs was significantly increased in patients with PDAC. In conclusion, these findings emphasize the immunosuppressive role of pancreatic cancer-derived EVs and provide new insights into our understanding of NK cell dysfunction regarding pre-metastatic niche formation for PDAC.

Author Contributions: Conceptualization: J.Z., C.J.B., and Y.Z.; methodology: J.Z. and Z.W.; acquisition of data (provided animals, acquired and managed patients, provided facilities, etc.): F.P., M.C.P., H.A., S.-H.C., H.P.H., and W.F.N.; analysis and interpretation of data (e.g., statistical analysis, biostatistics, computational analysis): J.Z. and Y.Z.; writing, review, and/or revision of the manuscript: J.Z., Y.Z., and H.A.S.; administrative, technical, or material support (i.e., reporting or organizing data, constructing databases): Z.W., J.Q., and J.L.; Study supervision: K.-W.J, C.J.B., and Y.Z. All authors provided critical comments and approved the final version of the manuscript.

Acknowledgments: The authors thank to the technical assistance from Susanne Neiss, Michaela Heitmann, Anke Wienand-Dorweiler and Lisa Raatz. The authors thank Michael Korenkov and Lisa Raatz for the helpful suggestions and critical reading of the manuscript. We thank Tim van Beers for help with TEM imaging. We thank Sin Yuin Yeo for developing the EVs' isolation method. We thank Margarete Odenthal and Xiaojie Yu for useful discussions. Jiangang Zhao, Zhefang Wang, and Jiahui Li were financially supported by CSC scholarship (The China Scholarship Council).

References

1. Kamisawa, T.; Wood, L.D.; Itoi, T.; Takaori, K. Pancreatic cancer. *Lancet* **2016**, *388*, 73–85. [CrossRef]
2. Kommalapati, A.; Tella, S.H.; Goyal, G. Contemporary Management of Localized Resectable Pancreatic Cancer. *Cancers (Basel)* **2018**, *10*, 24. [CrossRef] [PubMed]
3. Suker, M.; Beumer, B.R.; Sadot, E.; Marthey, L.; Faris, J.E.; Mellon, E.A.; El-Rayes, B.F.; Wang-Gillam, A.; Lacy, J.; Hosein, P.J.; et al. FOLFIRINOX for locally advanced pancreatic cancer: A systematic review and patient-level meta-analysis. *Lancet Oncol.* **2016**, *17*, 801–810. [CrossRef]
4. Hanahan, D.; Weinberg, R.A. The hallmarks of cancer. *Cell* **2000**, *100*, 57–70. [CrossRef]
5. Hanahan, D.; Weinberg, R.A. Hallmarks of cancer: The next generation. *Cell* **2011**, *144*, 646–674. [CrossRef] [PubMed]
6. Peinado, H.; Zhang, H.; Matei, I.R.; Costa-Silva, B.; Hoshino, A.; Rodrigues, G.; Psaila, B.; Kaplan, R.N.; Bromberg, J.F.; Kang, Y.; et al. Pre-metastatic niches: Organ-specific homes for metastases. *Nat. Rev. Cancer* **2017**, *17*, 302–317. [CrossRef] [PubMed]
7. Costa-Silva, B.; Aiello, N.M.; Ocean, A.J.; Singh, S.; Zhang, H.; Thakur, B.K.; Becker, A.; Hoshino, A.; Mark, M.T.; Molina, H.; et al. Pancreatic cancer exosomes initiate pre-metastatic niche formation in the liver. *Nat. Cell Biol.* **2015**, *17*, 816–826. [CrossRef]
8. Liu, Y.; Gu, Y.; Han, Y.; Zhang, Q.; Jiang, Z.; Zhang, X.; Huang, B.; Xu, X.; Zheng, J.; Cao, X. Tumor Exosomal RNAs Promote Lung Pre-metastatic Niche Formation by Activating Alveolar Epithelial TLR3 to Recruit Neutrophils. *Cancer Cell* **2016**, *30*, 243–256. [CrossRef]
9. Liu, Y.; Cao, X. Characteristics and Significance of the Pre-metastatic Niche. *Cancer Cell* **2016**, *30*, 668–681. [CrossRef]
10. Thery, C.; Zitvogel, L.; Amigorena, S. Exosomes: Composition, biogenesis and function. *Nat. Rev. Immunol.* **2002**, *2*, 569–579. [CrossRef]
11. Yáñez-Mó, M.; Siljander, P.R.; Andreu, Z.; Zavec, A.B.; Borràs, F.E.; Buzas, E.I.; Buzas, K.; Casal, E.; Cappello, F.; Carvalho, J.; et al. Biological properties of extracellular vesicles and their physiological functions. *J. Extracell. Vesicles* **2015**, *4*, 27066. [CrossRef] [PubMed]
12. Luan, X.; Sansanaphongpricha, K.; Myers, I.; Chen, H.; Yuan, H.; Sun, D. Engineering exosomes as refined biological nanoplatforms for drug delivery. *Acta Pharmacol. Sin.* **2017**, *38*, 754. [CrossRef] [PubMed]
13. S, E.L.A.; Mager, I.; Breakefield, X.O.; Wood, M.J. Extracellular vesicles: Biology and emerging therapeutic opportunities. *Nat. Rev. Drug Discov.* **2013**, *12*, 347–357. [CrossRef]
14. Kamerkar, S.; LeBleu, V.S.; Sugimoto, H.; Yang, S.; Ruivo, C.F.; Melo, S.A.; Lee, J.J.; Kalluri, R. Exosomes facilitate therapeutic targeting of oncogenic KRAS in pancreatic cancer. *Nature* **2017**, *546*, 498–503. [CrossRef] [PubMed]
15. Jang, S.C.; Kim, O.Y.; Yoon, C.M.; Choi, D.S.; Roh, T.Y.; Park, J.; Nilsson, J.; Lotvall, J.; Kim, Y.K.; Gho, Y.S. Bioinspired exosome-mimetic nanovesicles for targeted delivery of chemotherapeutics to malignant tumors. *ACS Nano* **2013**, *7*, 7698–7710. [CrossRef]
16. Saari, H.; Lazaro-Ibanez, E.; Viitala, T.; Vuorimaa-Laukkanen, E.; Siljander, P.; Yliperttula, M. Microvesicle- and exosome-mediated drug delivery enhances the cytotoxicity of Paclitaxel in autologous prostate cancer cells. *J. Control. Release Off. J. Control. Release Soc.* **2015**, *220*, 727–737. [CrossRef] [PubMed]

17. Garofalo, M.; Villa, A.; Rizzi, N.; Kuryk, L.; Rinner, B.; Cerullo, V.; Yliperttula, M.; Mazzaferro, V.; Ciana, P. Extracellular vesicles enhance the targeted delivery of immunogenic oncolytic adenovirus and paclitaxel in immunocompetent mice. *J. Control. Release Off. J. Control. Release Soc.* **2019**, *294*, 165–175. [CrossRef]
18. Hoshino, A.; Costa-Silva, B.; Shen, T.L.; Rodrigues, G.; Hashimoto, A.; Tesic Mark, M.; Molina, H.; Kohsaka, S.; Di Giannatale, A.; Ceder, S.; et al. Tumour exosome integrins determine organotropic metastasis. *Nature* **2015**, *527*, 329–335. [CrossRef]
19. Gurlevik, E.; Fleischmann-Mundt, B.; Brooks, J.; Demir, I.E.; Steiger, K.; Ribback, S.; Yevsa, T.; Woller, N.; Kloos, A.; Ostroumov, D.; et al. Administration of Gemcitabine After Pancreatic Tumor Resection in Mice Induces an Antitumor Immune Response Mediated by Natural Killer Cells. *Gastroenterology* **2016**, *151*, 338–350. [CrossRef]
20. Ames, E.; Canter, R.J.; Grossenbacher, S.K.; Mac, S.; Chen, M.; Smith, R.C.; Hagino, T.; Perez-Cunningham, J.; Sckisel, G.D.; Urayama, S.; et al. NK Cells Preferentially Target Tumor Cells with a Cancer Stem Cell Phenotype. *J. Immunol.* **2015**. [CrossRef]
21. Mamessier, E.; Sylvain, A.; Thibult, M.L.; Houvenaeghel, G.; Jacquemier, J.; Castellano, R.; Goncalves, A.; Andre, P.; Romagne, F.; Thibault, G.; et al. Human breast cancer cells enhance self tolerance by promoting evasion from NK cell antitumor immunity. *J. Clin. Investig.* **2011**, *121*, 3609–3622. [CrossRef] [PubMed]
22. Platonova, S.; Cherfils-Vicini, J.; Damotte, D.; Crozet, L.; Vieillard, V.; Validire, P.; Andre, P.; Dieu-Nosjean, M.C.; Alifano, M.; Regnard, J.F.; et al. Profound coordinated alterations of intratumoral NK cell phenotype and function in lung carcinoma. *Cancer Res.* **2011**, *71*, 5412–5422. [CrossRef] [PubMed]
23. Sun, C.; Xu, J.; Huang, Q.; Huang, M.; Wen, H.; Zhang, C.; Wang, J.; Song, J.; Zheng, M.; Sun, H.; et al. High NKG2A expression contributes to NK cell exhaustion and predicts a poor prognosis of patients with liver cancer. *Oncoimmunology* **2017**, *6*, e1264562. [CrossRef] [PubMed]
24. Cong, J.; Wang, X.; Zheng, X.; Wang, D.; Fu, B.; Sun, R.; Tian, Z.; Wei, H. Dysfunction of Natural Killer Cells by FBP1-Induced Inhibition of Glycolysis during Lung Cancer Progression. *Cell Metab.* **2018**, *28*, 243–255. [CrossRef] [PubMed]
25. Jewett, A.; Tseng, H.C. Tumor induced inactivation of natural killer cell cytotoxic function; implication in growth, expansion and differentiation of cancer stem cells. *J. Cancer* **2011**, *2*, 443–457. [CrossRef] [PubMed]
26. Li, T.; Yang, Y.; Hua, X.; Wang, G.; Liu, W.; Jia, C.; Tai, Y.; Zhang, Q.; Chen, G. Hepatocellular carcinoma-associated fibroblasts trigger NK cell dysfunction via PGE2 and IDO. *Cancer Lett.* **2012**, *318*, 154–161. [CrossRef]
27. Sheppard, S.; Ferry, A.; Guedes, J.; Guerra, N. The Paradoxical Role of NKG2D in Cancer Immunity. *Front. Immunol.* **2018**, *9*, 1808. [CrossRef]
28. Alter, G.; Malenfant, J.M.; Altfeld, M. CD107a as a functional marker for the identification of natural killer cell activity. *J. Immunol. Methods* **2004**, *294*, 15–22. [CrossRef]
29. Wang, R.; Jaw, J.J.; Stutzman, N.C.; Zou, Z.; Sun, P.D. Natural killer cell-produced IFN-gamma and TNF-alpha induce target cell cytolysis through up-regulation of ICAM-1. *J. Leukoc. Biol.* **2012**, *91*, 299–309. [CrossRef]
30. O'Brien, K.L.; Finlay, D.K. Immunometabolism and natural killer cell responses. *Nat. Rev. Immunol.* **2019**. [CrossRef]
31. Viel, S.; Marcais, A.; Guimaraes, F.S.; Loftus, R.; Rabilloud, J.; Grau, M.; Degouve, S.; Djebali, S.; Sanlaville, A.; Charrier, E.; et al. TGF-beta inhibits the activation and functions of NK cells by repressing the mTOR pathway. *Sci. Signal.* **2016**, *9*, ra19. [CrossRef] [PubMed]
32. Rouce, R.H.; Shaim, H.; Sekine, T.; Weber, G.; Ballard, B.; Ku, S.; Barese, C.; Murali, V.; Wu, M.F.; Liu, H.; et al. The TGF-β/SMAD pathway is an important mechanism for NK cell immune evasion in childhood B-acute lymphoblastic leukemia. *Leukemia* **2016**, *30*, 800–811. [CrossRef] [PubMed]
33. Peinado, H.; Aleckovic, M.; Lavotshkin, S.; Matei, I.; Costa-Silva, B.; Moreno-Bueno, G.; Hergueta-Redondo, M.; Williams, C.; Garcia-Santos, G.; Ghajar, C.; et al. Melanoma exosomes educate bone marrow progenitor cells toward a pro-metastatic phenotype through MET. *Nat. Med.* **2012**, *18*, 883–891. [CrossRef] [PubMed]
34. Tkach, M.; Thery, C. Communication by Extracellular Vesicles: Where We Are and Where We Need to Go. *Cell* **2016**, *164*, 1226–1232. [CrossRef] [PubMed]
35. Yu, Z.; Zhao, S.; Ren, L.; Wang, L.; Chen, Z.; Hoffman, R.M.; Zhou, J. Pancreatic cancer-derived exosomes promote tumor metastasis and liver pre-metastatic niche formation. *Oncotarget* **2017**, *8*, 63461–63483. [CrossRef] [PubMed]
36. Hamidi, H.; Ivaska, J. Every step of the way: Integrins in cancer progression and metastasis. *Nat. Rev. Cancer* **2018**, *18*, 533–548. [CrossRef] [PubMed]

37. Gasser, S.; Orsulic, S.; Brown, E.J.; Raulet, D.H. The DNA damage pathway regulates innate immune system ligands of the NKG2D receptor. *Nature* **2005**, *436*, 1186–1190. [CrossRef]
38. Bauer, S.; Groh, V.; Wu, J.; Steinle, A.; Phillips, J.H.; Lanier, L.L.; Spies, T. Activation of NK cells and T cells by NKG2D, a receptor for stress-inducible MICA. *Science* **1999**, *285*, 727–729. [CrossRef]
39. Gardiner, C.M.; Finlay, D.K. What Fuels Natural Killers? Metabolism and NK Cell Responses. *Front. Immunol.* **2017**, *8*, 367. [CrossRef]
40. Berchem, G.; Noman, M.Z.; Bosseler, M.; Paggetti, J.; Baconnais, S.; Le Cam, E.; Nanbakhsh, A.; Moussay, E.; Mami-Chouaib, F.; Janji, B.; et al. Hypoxic tumor-derived microvesicles negatively regulate NK cell function by a mechanism involving TGF-beta and miR23a transfer. *Oncoimmunology* **2016**, *5*, e1062968. [CrossRef]
41. Chen, G.; Huang, A.C.; Zhang, W.; Zhang, G.; Wu, M.; Xu, W.; Yu, Z.; Yang, J.; Wang, B.; Sun, H.; et al. Exosomal PD-L1 contributes to immunosuppression and is associated with anti-PD-1 response. *Nature* **2018**, *560*, 382–386. [CrossRef] [PubMed]
42. Xia, Y.; Zhang, Q.; Zhen, Q.; Zhao, Y.; Liu, N.; Li, T.; Hao, Y.; Zhang, Y.; Luo, C.; Wu, X. Negative regulation of tumor-infiltrating NK cell in clear cell renal cell carcinoma patients through the exosomal pathway. *Oncotarget* **2017**, *8*, 37783–37795. [CrossRef] [PubMed]
43. Kim, M.P.; Fleming, J.B.; Wang, H.; Abbruzzese, J.L.; Choi, W.; Kopetz, S.; McConkey, D.J.; Evans, D.B.; Gallick, G.E. ALDH activity selectively defines an enhanced tumor-initiating cell population relative to CD133 expression in human pancreatic adenocarcinoma. *PLoS ONE* **2011**, *6*, e20636. [CrossRef] [PubMed]
44. Shiozawa, Y.; Nie, B.; Pienta, K.J.; Morgan, T.M.; Taichman, R.S. Cancer stem cells and their role in metastasis. *Pharmacol. Ther.* **2013**, *138*, 285–293. [CrossRef] [PubMed]
45. Lee, J.-C.; Lee, K.-M.; Kim, D.-W.; Heo, D.S. Elevated TGF-β1 Secretion and Down-Modulation of NKG2D Underlies Impaired NK Cytotoxicity in Cancer Patients. *J. Immunol.* **2004**, *172*, 7335–7340. [CrossRef]
46. Derynck, R.; Zhang, Y.E. Smad-dependent and Smad-independent pathways in TGF-β family signalling. *Nature* **2003**, *425*, 577. [CrossRef] [PubMed]
47. Halder, S.K.; Beauchamp, R.D.; Datta, P.K. A specific inhibitor of TGF-beta receptor kinase, SB-431542, as a potent antitumor agent for human cancers. *Neoplasia* **2005**, *7*, 509–521. [CrossRef]
48. Yadav, D.K.; Bai, X.; Yadav, R.K.; Singh, A.; Li, G.; Ma, T.; Chen, W.; Liang, T. Liquid biopsy in pancreatic cancer: The beginning of a new era. *Oncotarget* **2018**, *9*, 26900–26933. [CrossRef]
49. Theodoraki, M.N.; Yerneni, S.S.; Hoffmann, T.K.; Gooding, W.E.; Whiteside, T.L. Clinical Significance of PD-L1(+) Exosomes in Plasma of Head and Neck Cancer Patients. *Clin. Cancer Res. Off. J. Am. Assoc. Cancer Res.* **2018**, *24*, 896–905. [CrossRef]
50. Lopez-Soto, A.; Gonzalez, S.; Smyth, M.J.; Galluzzi, L. Control of Metastasis by NK Cells. *Cancer Cell* **2017**, *32*, 135–154. [CrossRef]
51. Stanietsky, N.; Simic, H.; Arapovic, J.; Toporik, A.; Levy, O.; Novik, A.; Levine, Z.; Beiman, M.; Dassa, L.; Achdout, H.; et al. The interaction of TIGIT with PVR and PVRL2 inhibits human NK cell cytotoxicity. *Proc. Natl. Acad. Sci. USA* **2009**, *106*, 17858–17863. [CrossRef] [PubMed]
52. Zhang, Q.; Bi, J.; Zheng, X.; Chen, Y.; Wang, H.; Wu, W.; Wang, Z.; Wu, Q.; Peng, H.; Wei, H.; et al. Blockade of the checkpoint receptor TIGIT prevents NK cell exhaustion and elicits potent anti-tumor immunity. *Nat. Immunol.* **2018**, *19*, 723–732. [CrossRef] [PubMed]
53. Borghaei, H.; Paz-Ares, L.; Horn, L.; Spigel, D.R.; Steins, M.; Ready, N.E.; Chow, L.Q.; Vokes, E.E.; Felip, E.; Holgado, E.; et al. Nivolumab versus Docetaxel in Advanced Nonsquamous Non-Small-Cell Lung Cancer. *N. Engl. J. Med.* **2015**, *373*, 1627–1639. [CrossRef] [PubMed]
54. Brahmer, J.; Reckamp, K.L.; Baas, P.; Crino, L.; Eberhardt, W.E.; Poddubskaya, E.; Antonia, S.; Pluzanski, A.; Vokes, E.E.; Holgado, E.; et al. Nivolumab versus Docetaxel in Advanced Squamous-Cell Non-Small-Cell Lung Cancer. *N. Engl. J. Med.* **2015**, *373*, 123–135. [CrossRef] [PubMed]
55. Hamid, O.; Robert, C.; Daud, A.; Hodi, F.S.; Hwu, W.J.; Kefford, R.; Wolchok, J.D.; Hersey, P.; Joseph, R.W.; Weber, J.S.; et al. Safety and tumor responses with lambrolizumab (anti-PD-1) in melanoma. *N. Engl. J. Med.* **2013**, *369*, 134–144. [CrossRef] [PubMed]
56. Morrison, A.H.; Byrne, K.T.; Vonderheide, R.H. Immunotherapy and Prevention of Pancreatic Cancer. *Trends Cancer* **2018**, *4*, 418–428. [CrossRef] [PubMed]
57. Olson, B.; Li, Y.; Lin, Y.; Liu, E.T.; Patnaik, A. Mouse Models for Cancer Immunotherapy Research. *Cancer Discov.* **2018**, *8*, 1358–1365. [CrossRef]

3

Role of Extracellular Vesicles (EVs) in Cell Stress Response and Resistance to Cancer Therapy

Clodagh P. O'Neill, Katie E. Gilligan and Róisín M. Dwyer *

Discipline of Surgery, Lambe Institute for Translational Research, National University of Ireland Galway (NUIG), Galway H91 YR71, Ireland; c.oneill15@nuigalway.ie (C.P.O.); k.gilligan3@nuigalway.ie (K.E.G.)

* Correspondence: roisin.dwyer@nuigalway.ie

Abstract: Extracellular vesicles (EVs) are nanosized particles released by all cells that have been heralded as novel regulators of cell-to-cell communication. It is becoming increasingly clear that in response to a variety of stress conditions, cells employ EV-mediated intercellular communication to transmit a pro-survival message in the tumor microenvironment and beyond, supporting evasion of cell death and transmitting resistance to therapy. Understanding changes in EV cargo and secretion pattern during cell stress may uncover novel, targetable mechanisms underlying disease progression, metastasis and resistance to therapy. Further, the profile of EVs released into the circulation may provide a circulating biomarker predictive of response to therapy and indicative of microenvironmental conditions linked to disease progression, such as hypoxia. Continued progress in this exciting and rapidly expanding field of research will be dependent upon widespread adoption of transparent reporting standards and implementation of guidelines to establish a consensus on methods of EV isolation, characterisation and nomenclature employed.

Keywords: extracellular vesicles; exosomes; cell stress; cancer; drug resistance; hypoxia; heat stress; biomarker

---

1. Introduction

As highlighted throughout this special issue, cancer cells develop mechanisms of response to conditions of stress to evade death and support tumour progression. These mechanisms support escape from apoptosis, endoplasmic reticulum (ER) stress, autophagy, hypoxia, anti-cancer therapies, heat stress and chemical stress. Understanding the mechanisms of escape from cell stress presents novel targets for therapeutic intervention and may unveil circulating biomarkers reflective of cell stress conditions and capable of predicting response to therapy. In recent years evidence has accumulated highlighting a role for extracellular vesicles (EVs) as novel mediators of these stress pathways in cancer.

EVs are nanosized particles released by all cells and first described as far back as 1946 [1], although it is only relatively recently that their importance in normal physiology and disease has become apparent. EVs were originally considered vehicles for waste removal by cells, but are now known to contain a range of functional nucleic acids, lipids and proteins that play a key role in cell-cell communication [2]. "EV" is now the recommended universal term used for nanosized vesicles released by cells including exosomes, microvesicles (or ectosomes) and apoptotic bodies [3]. Exosomes, the smallest of the vesicles, range from 30–120 nm and are formed through the endosomal pathway and predominantly released from cells through fusion with the plasma membrane. Microvesicles/ectosomes are slightly larger, ranging from 50 nm–1 μm and are formed through direct budding from the plasma membrane. Finally, during apoptosis a large range of vesicles are released ranging from 50 nm–5 μm. No purification method specific for one EV subgroup exists, with all current approaches resulting in a heterogenous population of EVs due to overlap in size of the vesicles, along with the fact that there are no specific markers to identify or select the particular vesicle of interest [1,4]. Based on this, throughout this article vesicles reported within this broad size range in all studies will be referred to simply as EVs.

EVs contain a range of factors including proteins, lipids, RNA (mRNA, microRNA, long non-coding (lnc) RNA) and genomic DNA, and act as a reservoir protecting the material from degradation until transfer to recipient cells. EV content is thought to reflect that of the cell from which it has been released. In recent years, EVs have been implicated as a driving force in intercellular communication in the primary tumor microenvironment, mediating crosstalk that supports cancer development, adaptation to conditions of hypoxia and nutrient deprivation, escape from apoptosis and anti-cancer therapies, immune evasion and disease progression [1,5]. For example, O'Brien et al [6], demonstrated that EVs released by triple negative breast cancer (TNBC) cells were capable of increasing migration, proliferation and invasion capacity of 3 separate recipient breast cancer cell lines. EVs derived from the sera of patients with TNBC were also capable of increasing invasion of recipient cells when compared to those from healthy controls [6]. Another study highlighted the role of EVs in ovarian cancer metastasis. Highly metastatic ovarian cancer cells were found to release EVs that could induce metastatic behaviour in recipient cancer cells, and apoptosis in mesothelial cells [7]. Matrix Metalloproteinase-1 (MMP1) expression in ovarian cancer is linked with poor prognosis and was found to be packaged into EVs and transferred to the mesothelial cells [7]. Studies such as these support the hypothesis that changes in cell phenotype during conditions of stress and development of resistance to therapy may result in a change in EV cargo and pattern of release. Further, EV release into the circulation raises potential for use as circulating biomarkers reflecting changes in microenvironmental conditions and predicting response to therapy [5].

2. Response to Anti-Cancer Therapy and Transfer of Drug Resistance

A number of studies have demonstrated a change in the content of EVs secreted by tumour cells in response to chemotherapy and radiation-induced cell stress, resulting in transfer of a resistant phenotype and immune evasion, and raising potential for this altered profile as a biomarker of response or resistance to therapy (Figure 1). Potential novel therapeutic targets have also been highlighted through study of altered EV content in this setting.

Evidence of EV-mediated transfer of drug resistance in response to cell stress in vitro continues to accumulate in a range of cancers including pancreatic [8], prostate [9], breast [10–12], ovarian [13], osteosarcoma [14], non-small cell lung cancer [15] melanoma [16] and acute myeloid leukaemia [17].

Recently, BRAF inhibitor resistant melanoma cells were shown to activate a novel truncated form of anaplastic lymphoma kinase (ALK), which was secreted in EVs that were then capable of transferring drug resistance through activation of the mitogen-activated protein kinase (MAPK) signalling pathway in recipient cells [16]. The ALK protein cargo in EVs following irradiation was also implicated in driving tumour growth and compromising therapeutic efficacy of ALK inhibitors in non-small-cell lung carcinoma (NSCLC) [18]. EVs isolated from ALK-containing H3122 cells following irradiation contained much higher levels of phosphorylated ALK (p-ALK) than control EVs, and activated AKT, STAT3, and the ERK pathway in recipient cells [18].

Following irradiation, GBM cells were also shown to promote proliferation and therapy resistance of surviving tumour cells by secreting apoptotic extracellular vesicles (apoEVs) enriched with various components of spliceosomes [19]. ApoEVs altered RNA splicing in recipient cells, partly through transfer of splicing factor RBM11, resulting in more oncogenic isoforms of MDM4 and Cyclin D1 thereby promoting therapy resistance and an aggressive migratory phenotype [19]. The protein cargo of EVs released from head and neck squamous cell carcinoma in vitro was also found to reflect changes in cellular processes induced by radiation e.g., transient suppression of transcription/translation and stress-induced signalling [20]. Indeed, investigation of the p53-mediated response following lung cancer cell irradiation revealed the protein TSAP6 enhanced EV production in cells, highlighting a novel function of the p53 protein in regulation of EV secretion [21].

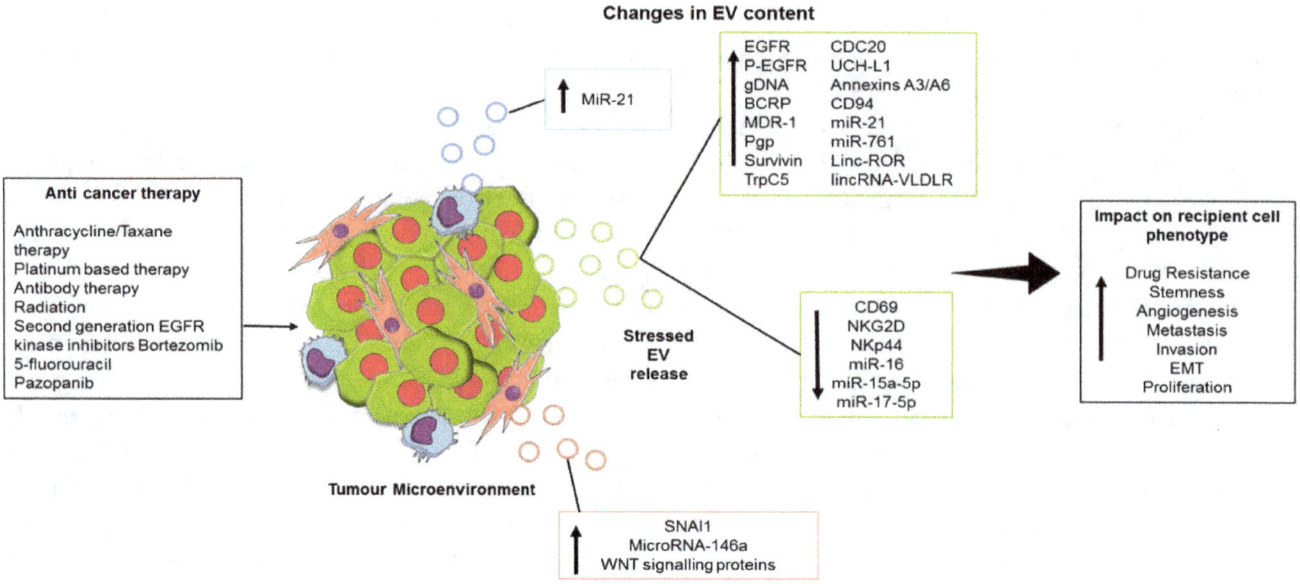

Figure 1. Impact of anticancer therapies on extracellular vesicle EV content and release from the tumour microenvironment. Anti-cancer therapy stress causes EV release from tumour cells and associated stromal cells including cancer-associated fibroblasts (CAFs) (pink) and tumour-associated macrophages (TAMs) (Blue). The resultant EVs have altered content including various bioactives that can have a phenotypic impact in recipient cells. This figure was created using MindtheGraph. Abbreviations: epidermal growth factor receptor (EGFR), phosphorylated-EGFR (P-EGFR), genomic DNA (gDNA), breast cancer resistance protein (BCRP), multi-drug resistance protein 1 (MDR-1), P-glycoprotein (Pgp), short transient receptor potential channel 5 (TrpC5), cell-division cycle protein 20 (CDC20), long intergenic non-protein coding RNA, regulator of reprogramming (Linc-ROR) LincRNA-very low density lipoprotein receptor (lincRNA-VLDLR), cluster of differentiation 69 (CD69), natural killer group 2 membrane D (NKG2D), activating NK receptor NKp44 (NKp44), snail family transcriptional repressor 1 (SNAI1).

Exposure to heat shock and chemotherapy drugs caused a significant increase in EV release by hepatocellular carcinoma (HCC) cells in vitro, with the secreted vesicles enriched with biologically active heat shock proteins (HSP60, HSP70, HSP90) and capable of stimulating an increase in cytotoxic activity of NK cells [22]. Also in HCC, changes in EV encapsulated long non-coding RNAs have been demonstrated in response to anticancer drugs, unveiling potential novel mechanisms involved in response to therapy and pathways towards acquired chemoresistance [23,24].

Breast cancer cells were shown to release EVs enriched with the cell survival protein Survivin following exposure to microtubule inhibiting drugs such as Paclitaxel, resulting in increased drug resistance and survival of recipient cells [25]. EVs from docetaxel-resistant prostate cancer cells were capable of transferring resistance to recipient cells, potentially due to EV Multidrug resistance-1 (MDR-1)/P-glycoprotein (P-gp) transfer preventing accumulation of drugs within cells through active drug efflux [9]. EVs from prostate cancer patients' sera were also demonstrated to induce increased cell proliferation and invasion, compared to EVs from age-matched controls [9].

Keklikoglou et al. [26] demonstrated an increase in the number of EVs released from cancer cells following exposure to two common breast cancer therapies, PTX and doxorubicin (DOX). Further, the EVs released from treated cells had an enhanced pro-metastatic capacity. Using immunocompetent animal models, the study demonstrated that injection of EVs derived from chemotherapy treated cells resulted in an increase in metastatic lung nodules. This pro-metastatic capacity was also confirmed in a zebrafish embryo model. Proteomic analysis revealed enrichment of the cell signalling protein annexin A6 (ANXA6) in the EVs, which was demonstrated to play a key role in the effects seen. ANXA6 was also detected in circulating EVs of breast cancer patients undergoing neoadjuvant chemotherapy, raising potential as a circulating biomarker of disease progression or response to therapy [26].

EVs released by cells within the tumour microenvironment in response to anti-cancer therapies have also been shown to mediate resistance. Following treatment with 5-fluorouracil, mesenchymal stem cell (MSC)-derived EVs had a protective effect on gastric cancer cells, reducing chemotherapy induced apoptosis through activation of the calcium/calmodulin-dependent protein kinases (CaM-Ks) and Raf/MEK/ERK kinase signalling cascade [27]. EVs derived from bone marrow mesenchymal stromal cells (BM-MSC) were also demonstrated to increase chronic lymphocytic leukemia B cell chemoresistance to several drugs, enhance migration and decrease apoptosis [28]. Gemcitabine treatment of carcinoma associated fibroblasts (CAFs) stimulated release of EVs capable of promoting proliferation and chemoresistance of pancreatic ductal adenocarcinoma cells [29]. The microRNA content of EVs released by gastric tumour-associated macrophages (TAMs) was also altered following cisplatin treatment [30], with increased EV-miR21 transferred to recipient gastric cancer cells, suppressing apoptosis and activating the PI3K/AKT signalling pathway through down regulation of PTEN [30]. These studies highlight the possibility that novel treatment strategies targeting EV-mediated intercellular communication prior to chemotherapy may be effective in reducing this pro-tumorigenic signalling.

In addition to mediating transfer of microRNA/mRNA/lncRNA and proteins to confer a drug resistant phenotype upon recipient cells, EVs have also been shown to act as drug decoys, thus shielding target tumour cells. B-cell lymphoma-released EVs were demonstrated to have elevated CD20 embedded in the membrane which acted as a decoy target, intercepting and binding to rituximab [31]. Transient receptor potential channel 5 (TrpC5) has also been shown to accumulate in EVs released by drug resistant breast cancer cells, resulting in sequestration of chemotherapeutic drugs [32]. Transfer of EV-TrpC5 to recipient cells was demonstrated to result in transfer of resistance. Further, the level of circulating EV-TrpC5 had a significant positive association with TRPC5 protein level in tissues of patients with advanced breast cancer prior to chemotherapy, suggesting a role for circulating EVs in evaluation of chemoresistance [33].

Further studies using patient samples have demonstrated that along with influencing a chemoresistant phenotype, EVs have potential as circulating biomarkers of resistance or response to chemotherapy, which may play an important role in the clinical decision making process. Breast cancer resistance protein (BRCP) was found to be upregulated at the mRNA and protein level in tumour derived circulating EVs from patients with a poor response to chemotherapy [34]. MUC1 was also co-expressed with BCRP, raising potential as a predictive biomarker of response to chemotherapy [34]. A strong negative correlation between the level of ubiquitin carboxyl terminal hydrolase-L1 (UCH-L1) in circulating EVs and the clinical outcome of chemotherapy has also been reported, highlighting potential for monitoring chemotherapeutic efficacy [10].

Yin et al. [35] reported increased annexin A3 in sera of platinum resistant ovarian cancer patients, raising potential as a biomarker of resistance to platinum-based therapies. In Vitro, ovarian cancer cells expressing higher levels of annexin A3 released increased numbers of EVs, with the protein detected in EVs from cisplatin-resistant cells [35]. In patients with multiple myeloma, downregulation of a panel of microRNAs (miR-16, miR-15a, miR20a and miR-17) was associated with bortezomib resistance [36]. Also focused on microRNA content, analysis of EVs released by pazopanib treated/untreated synovial sarcoma cells in vitro highlighted EV encapsulated miR-761 as a potential biomarker of pazopanib resistance [37]. Using cell lines from a range of cancer types, Montermini et al [38] demonstrated that treatment with second generation EGFR kinase inhibitors stimulated release of EVs containing EGFR, phosphorylated EGFR, and genomic DNA. The data presented suggests that the EV profile could be used as a surrogate indicator of drug-related stress response [38].

EVs have also been implicated in the mechanism of action of drugs in development. Shiau et al. [39], showed that the phytoagent deoxyelephantopin (DET) and its chemical derivative DETD-35 suppressed triple negative breast cancer (TNBC) in vitro and in vivo. DET and DETD-35 treatment induced mitochondrial dysfunction and downstream oxidative stress. Interestingly EVs released from the TNBC cells after DET/DETD-35 treatment had anti-tumour phenotypes i.e., reduced expression of proteins involved in cell migration, adhesion and angiogenesis. This study supports the suitability of DET/DETD-35 as a potential anti-cancer drug, to treat TNBC partly through emission of anti-cancer EVs [39].

Suberoylanilide hydroxamic acid (SAHA), is a histone deacetylase inhibitor that has anti-cancer drug potential. Campanella et al. [40], studied the mechanism of action of this chemical in lung carcinoma cells in vitro. Similar to DET/DETD-35 [39], treatment of cancer cells with SAHA, caused mitochondrial damage and oxidative stress. This chemical stimulated release of EVs containing high expression of a nitrated version of heat shock protein 60 (HSP60). It was postulated that EV-nitrated Hsp60 could interact with the immune system, but its anti-cancer properties are yet to be fully elucidated [40].

3. Hypoxia

Low oxygen tension or hypoxia is often present in tumour tissue, and this is associated with more aggressive cancers and poor patient prognosis. The tumour microenvironment is heterogeneous, with varying levels of hypoxic conditions throughout. Cancer cells have acquired many mechanisms to cope with this environmental stress, with the most well-known mediators being the hypoxia inducible factors (HIF) including HIFα and HIFβ. These proteins can activate downstream signalling pathways relating to tumour growth, metastasis and angiogenesis. The release of EVs in response to hypoxia has been reported in various cancers including breast, prostate and multiple myeloma [41–46]. EV release provides a novel hypoxia response mechanism facilitating transformation of the microenvironment into a cancer promoting utopia, with many studies showing EV involvement in transfer or regulation of HIF proteins (Table 1) [43–47].

Hypoxia resistant multiple myeloma (HR-MM) cells, generated through continuous exposure to hypoxic conditions in vitro, were shown to have a two fold increase in EV release compared to the parent MM cells [43]. Promotion of angiogenesis by EV-miRNA from HR-MM cells was also demonstrated, mediated through miR-135b suppression of the factor-inhibiting hypoxia-inducible factor 1 (FIH-1) [43]. EVs released from leukemia cells cultured in hypoxic conditions were also shown to stimulate a significant increase in human umbilical vein endothelial cells (HUVEC) tubule formation in vitro compared to EVs from control cells [46]. Kucharzewska et al. [48] showed that under hypoxia conditions GBM cell secreted EVs that induced growth factor and cytokine release by recipient endothelial cells, in turn activating PI3K/AKT signalling and migration of pericytes. A range of breast cancer cell lines exposed to varying conditions of hypoxia in vitro have also been shown to release increased numbers of EVs with increasing hypoxia [44,45].

EVs released from hypoxic oral squamous cell carcinoma (OSCC) cells initiated cell migration and metastasis in recipient cells through HIF-1α and HIF-2α expression [49]. miRNA analysis of the EVs highlighted miR-21 as being significantly elevated compared to EVs released in normoxic conditions. Functional analysis revealed that miR-21 promoted invasion and migration of recipient cells in a HIF-1α and HIF-2α dependent manner [49]. Further, EV release by breast cancer cells was reported to be HIF dependent and mediated by expression and co-localisation of shedding EVs with RAB22A, a small GTPase. Incubation of breast cancer cells with these EVs resulted in increased focal adhesion formation and ECM invasion in vitro and promoted extravasation in vivo. In hypoxic conditions prostate cancer cells were also shown to release EVs enriched with signalling molecules that were capable of promoting invasiveness of normal prostate cancer cells [50]. Long non-coding RNAs (lncRNAs) have in recent years been shown to play a role in regulation of gene expression. Long intergenic noncoding RNA, regulator of reprogramming (linc-ROR) was shown to be highly enriched in EV fractions from HCC cells after hypoxia exposure. Hypoxia-induced EVs promoted cell survival at least in part through Linc-ROR mediated regulation of the HIFα pathway in HCC cells [47].

Table 1. Impact of hypoxia on EV release and content. NTA: Nano Particle Tracking Analysis; TEM: Transmission Electron Microscopy; WB: Western Blot; HIF: Hypoxia Inducible Factor; TRPS: Tunable Resistive Pulse Sensing; TD-EV: Tumor Derived Extracellular Vesicles; TGF: Transforming Growth Factor.

Cancer Type	EV Isolation Technique	EV Characterization	EV Content Change under Stress	Study Outcome	Reference
Prostate	Ultracentrifugation Or commercially available kit	NTA	Prostate cancer (PCA) cells released EVs enriched in triglycerides due to activation of lipogenesis-related enzymes and signalling molecules	Significance of lipid accumulation in hypoxic PCA cells and EVs; therapeutic relevance in PCA.	[41]
Prostate	Ultracentrifugation Or commercially available kit	NTA TEM WB	Higher number of proteins in EV (Hypoxic) compared to EV (Normoxic), primarily associated with the remodelling of epithelial adherens junction pathway	EV (Hypoxic) are loaded with proteins that could enhance invasiveness, stemness and induce microenvironment changes, thereby promoting PCA aggressiveness	[50]
Multiple Myeloma (MM)	Commercially available kit	TEM NTA	miR-135b was significantly upregulated in EVs from Hypoxia resistant (HR) MM.	HR-MM cells may mimic in vivo bone marrow microenvironment. EV-miR-135b potential target for controlling angiogenesis	[43]
Leukaemia	Commercially available kit	TEM NTA WB	Range of miRNAs including miR210 significantly increased in EVs secreted from hypoxic K562 cell line	EV-miRNA derived from cancer cells under hypoxic conditions may influence angiogenic activity in endothelial cells	[46]
Breast	Differential centrifugation	NTA	Increased number of EV particles released in HIF-dependent manner. RAB22A protein co-localizes with budding EVs.	EVs can induce complex cytoskeletal alterations in a RAB22A-dependent manner promoting Breast cancer metastasis	[44]
Breast	Ultracentrifugation Or commercially available kit	TEM NTA WB	Increase in EV-miR-210 released from breast cancer cells	Hypoxia promoted the release of cancer EVs containing elevated miR-210	[45]
Hepatocellular cancer (HCC)	Ultracentrifugation	TEM NTA Protein yield	Long intergenic noncoding RNA (linc-RoR) was incorporated into EVs.	Linc-RoR enriched EVs released by tumour cells during hypoxia modulated recipient cell signaling and survival	[47]
Glioblastoma multiforme (GBM)	Ultracentrifugation	NTA WB	The proteome and mRNA profiles of EVs reflected the oxygenation status of GBM cells and patient tumors.	EV pathway constitutes a potentially targetable driver of hypoxia-dependent intercellular signaling. EVs signature holds biomarker potential.	[48]
Multiple Tumor Types	Ultracentrifugation	TRPS TEM	Hypoxic TD-EVs transferred TGF-β1. Increased EV-miR-210 and EV-miR-23a.	Novel mechanism of immune suppression mediated by hypoxic TD-EVs	[51]
Lung	Commercially available kit	WB	Increased levels of EV-miR-103a.	EV-miR-103a enhanced tumor progression and angiogenesis. Increased miR-103a levels seen in lung cancer patients.	[52]
Oral Squamous cell carcinoma (OSCC)	Commercially available kit	SEM WB	108 OSCC EV-miRNAs were differentially expressed, miR-21 significantly upregulated.	Hypoxic microenvironment may stimulate tumor cells to generate miR-21-rich EVs that are delivered to normoxic cells to promote prometastatic behaviors.	[49]

Given the mounting evidence for the role of EVs in cellular response to hypoxia, a number of groups have investigated the potential of circulating EVs as biomarkers of hypoxia. For example, miR-210 expression has been reported as upregulated in EVs released from multiple myeloma, leukaemia, and breast cancer cells in hypoxic conditions [43,45,46]. Li et al., [49] investigated the use of EV-miR-21 as a biomarker in OSCC. EVs were isolated from the plasma of OSCC patients (n = 108) and healthy controls (n = 108). Circulating EV-miR-21 was shown to be significantly higher in OSCC patients than healthy controls, and the levels were found to be associated with clinical characteristics such as tumor Stage and lymph node metastasis. Through HIF-1α and HIF-2α staining, EV-miR-21 was also linked to tumor hypoxia [49]. EVs released from glioblastoma (GBM) cells during hypoxia have been shown to be enriched in hypoxia regulated proteins and mRNA including caveolin 1 (CAV1), interleukin-8 (IL8), platelet-derived growth factor (PDGF) and MMPs [48]. The study raised the potential for these EV molecules as a signature of oxygenation status and aggressiveness of GBM tumours. Lipid accumulation in EVs derived from prostate cancer cells exposed to hypoxia has also been postulated to have biomarker potential to assess tumor oxygenation status and aggressiveness [41]. EVs from the hypoxic prostate cancer cells were found to have increased accumulation of triglycerides. After reoxygenation these lipids supported rapid prostate cell growth. Blockade of lipid formation by various drugs including the COX2 inhibitor celecoxib, reduced tumour growth and invasiveness after reoxygenation, suggesting a potential therapeutic target for prostate cancer treatment [41].

As previously highlighted, EVs play a key role in mediation of tumour-stroma interactions. In the hypoxic tumour environment, tumour-associated macrophages (TAMs) lose their anti-tumour phenotype and are linked with much poorer outcomes in cancer patients. Hsu et al. [52], showed a role for EVs in amplifying the macrophage oncogenic effects in lung cancer under hypoxic stress. Incubation of M2 macrophages with EVs released by hypoxic lung cancer cells reprogrammed the macrophages towards a pro-tumorigenic, immunosuppressive phenotype through EV-miR103a signalling [52]. Tumour-derived EVs have also been shown to interact with NK cells under hypoxic conditions. Through both in vitro and in vivo experiments EVs from hypoxic tumour cells were shown to impair NK cell cytotoxicity through the transfer of proteins including TGF-β1, and miRNA including miR-210 and miR-23a [51].

4. Nutrient Deprivation

Due to the rapid increase in cell growth in the tumour microenvironment, nutrient deprivation is a prevalent stress. MSCs are mass producers of EVs and have been shown to survive well under nutrient starvation stress [53,54]. Vallabhaneni et al., [54] investigated the cargo of EVs from serum-deprived MSCs (SD-MSCs) associated with the tumour environment. Proteomic, nucleic acid, and lipid analysis of the EV cargo was performed. EV lipid analysis confirmed the presence of bioactive lipids with pro-tumourgenic characteristics. MiRNA analysis identified miR21 and miR34a as key oncomiRs, with roles in tumour progression and proliferation confirmed in vitro and in vivo [54].

Following on from this study, in 2016 the same group studied the role of these nutrient deprived MSC-EVs in osteosarcoma (OC) [53]. OC cells incubated with EVs from SD-MSCS showed resistance to apoptosis and increased wound healing in vitro. The recipient cells were found to express miRNAs that could potentially target metabolism and metastasis associated genes. Alteration in expression of target genes including matrix metalloproteinase (MMP1) and focal adhesion kinase (PTK2) was validated by qPCR [53].

5. ER Stress and Apoptosis

The endoplasmic reticulum (ER) is essential in maintaining cell homeostasis, however under stressful conditions cells induce an unfolded protein response (UPR). ER Stress has been linked to multivesicular body (MVB) formation, and increased EV release. This increased EV release was only found in cells containing ER stress transducers inositol required enzyme 1 (IRE1) and PKR-like ER kinase (PERK) [55]. Interestingly ER stress has also been found to be induced by EVs. Tumour-derived

EVs containing miR-3091-3p internalised by hepatocytes suppressed autophagy-related protein 9b (Atg9b) expression. This led to ER stress-induced cell death by accumulation of ubiquitinated proteins [56]. Javeed et al., [57] found that pancreatic cancer shed adrenomedullin+/ CA1909+ EVs. The EVs then induced ER stress along with failure of the UPR causing paraneoplastic b-cell dysfunction resulting in inhibited insulin secretion.

If cell stress cannot be relieved by UPR and DNA damage response to maintain equilibrium, cell death is initiated [58]. Pavlyukov et al., [19] showed in GBM cells the paradoxical role of apoptotic cell derived EVs (ApoEVs) in promoting tumour growth and drug resistance in neighbouring cells, demonstrated in vitro and in vivo. Analysis of the ApoEVs uncovered that proteasomal and spliceosomal proteins were significantly enriched compared to control GBM EVs. These spliceosome related proteins were taken up by target cells and had an effect on mRNA splicing. Through this process a more aggressive and therapy-resistant phenotype was promoted in surviving cells [19].

6. Autophagy

Autophagy allows the breakdown of unnecessary or damaged proteins and organelles, in order to maintain cell homeostasis. Under stress conditions autophagy is increased, and dysregulation of this highly conserved mechanism is linked to diseases including cancer [59,60]. Recent reports have demonstrated an important connection between EVs and autophagy during stress in the context of cancer [60].

Bhattacharya et al., [61] reported the interlinking role of autophagy and EV biogenesis in pancreatic cancer. G alpha interacting protein (GAIP) C-terminus (GIPC) was shown to be a master regulator of autophagy, with down regulation resulting in decreased glucose uptake in pancreatic cancer cells leading to metabolic stress. This stress in turn led to autophagy induction and an increase in the number of EVs released [61]. The chemical pesticide rotenone was also shown to induce chemical stress in cancer stem cells from both prostate and breast cancer cell line, resulting in extensive mitochondrial damage and induction of autophagy [62]. This in turn led to increased EV formation and release.

Interestingly, Dutta et al., [63] have shown that EVs released from breast cancer cells induced an increase in reactive oxygen species (ROS) in recipient human mammary epithelial cells (HMECs), leading to increased autophagy. These EV-treated HMECs also released growth factors promoting further tumorigenesis [63].

7. Heat Stress and Oxidative Stress

Heat stress has been found to influence EV release. Bewicke-Copley et al, [64] found that EVs released by heat shocked cells induced a bystander damage in unstressed populations. DNA damage and apoptosis was increased with cell viability reduced. While this could be a positive outcome in the cancer setting it was discovered that cells treated with EVs from heat shocked cells were more likely to survive a subsequent heat shock. This could mean that the EVs allowed the cells to develop an adaptive response and become resistant to hyperthermia therapies. Oxidative and heat stress have also been implicated in blood malignancies. The reason for relapse and poor prognosis in blood cancer is thought to be associated with inadequate NK-Cell function. Two lymphoma cell lines were found to release EVs bearing the NKG2D ligands MICA/B and ULBP1 and 2. These EVs were found to impair NK-Cell function by NKG2D receptor mediated cytotoxicity. When heat stress was applied EV secretion was enhanced, generating more soluble NKG2D ligands, aggravating impairment of the cytotoxic response. This is another consideration that should be taken into account when designing hyper-thermal anti-cancer therapies [65]. This study shows the role of oxidative and heat stressed EVs in this process [65], and supports the previously mentioned role of EVs from hypoxia treated tumour cells in NK cell cytotoxicity [51].

In contrast to these studies, heat stress was found to have a positive effect on the immunogenicity of EVs released by heat-treated ascites. The EVs were able to induce a tumour-specific cytotoxic T lymphocyte response and promote dendritic cell maturation [66]. Another study found that EVs derived from dendritic cells or tumour cells could induce specific anti-tumour immunity. Under hyperthermic conditions it was found that EVs derived from heat-stress treated tumour cells (HS-TEX) could activate dendritic and T cells, when compared to untreated cells. Intra-tumoural injection of the HS-TEX induced an anti-tumour immune response inhibiting tumour growth and increasing animal survival time [67]. Guo et al, [68] found similar results with HS-TEX in that they could induce an anti-tumour response. HS-TEX contained the ability to convert immunosuppressive regulatory T cells into T helper type 17 cells via interleukin-6 (IL-6). IL-6 promotes IL-17 expression which in turn causes rejection of established tumours. The ability of heat stress to aid in the release of EVs carrying drugs was investigated by Yang et al, [69]. The study examined whether heat stress facilitated in MCF-7 doxorubicin (DOX) cells releasing more EVs containing the drug. It was found that more EVs were released from the heat stress cells and that they also contained higher levels of DOX than non-heat stress cells of the same number. The EVs also inhibited MCF-7 proliferation and induced cell apoptosis in vitro along with inhibiting tumour growth in vivo.

8. Conclusions

Due to the recent interest in EVs in the cancer setting, the number of publications in this field has increased dramatically over the past decade. However, issues with interpretation of studies and reproducibility have arisen due to the different methods of isolation, characterisation and nomenclature employed. Table 1 highlights some of the EV isolation and characterisation methods routinely employed. Although novel approaches to improve EV samples purity and yield are constantly in development, ultracentrifugation remains the most frequently employed method, isolating EVs using speeds >100,000$\times$ g for >1 h [70,71]. This is often preceded by differential centrifugation at lower speeds to remove cell debris and proteins aggregates, and microfiltration. Many investigators couple a density based separation technique with a size exclusion isolation technique to improve the purity of samples. In recent years commercial EV isolation kits have also come on the market, however there are concerns that the isolated EVs may be accompanied by contaminating factors [71]. Refined isolation techniques are emerging e.g., tangential flow filtration [72], with the aim of supporting isolation of high yields of good manufacturing practice (GMP)-grade EVs for use in the clinical setting [73]. In the past, based on the information available at the time, researchers were ascribing functions to a specific EV type despite the fact that they were working with a heterogenous population, potentially containing vesicles of mixed origin (exosomes/microvesicles/apoptotic bodies) along with co-purified bound material in some cases [74]. This lead to the publication of the Minimal Information for Studies of Extracellular Vesicles (MISEV) 2014 [75]. This described a "set of biochemical, biophysical and functional standards that should be used to attribute any specific biological cargo or functions to EVs". As the field continued to rapidly evolve and expand, in 2017 EV-TRACK was published recommending that authors report specific experimental parameters for EV isolation and characterisation, to increase transparency and support interpretation of data presented and reproducibility of findings [76]. This included the "EV-METRIC", a checklist of information that should ideally be reported for EV experiments including: methods of isolation such as density gradient, EV density or Ultracentrifugation; protein analysis targeting $\geq$3 EV-enriched proteins, along with a negative control protein and the lysis buffer composition; and both quantitative and qualitative particle analysis should be performed including wide field and close up electron microscopy imaging [76].

Following consultation with investigators in the field of EV research, in 2018 updated MISEV guidelines were published [3]. This includes the reporting of EV isolation procedures in detail, with refinement of culture and harvesting conditions e.g., dead cell count in cultures should now be provided by investigators. For EVs from biological fluids and tissues, guidance on methods of collection and storage has been provided, along with the noting of clinicopathological details that could have an impact on the sample. If a study claims that an effect seen is caused by EVs, it should demonstrate that the function is EV specific, and not a result of cell-cell contact or other soluble mediators. Overall

MISEV2018 is aiming to standardise the quality of the research being conducted by allowing more reliable and reproducible results [3]. Continued progress in this exciting and rapidly expanding field of research, using EVs to unlock novel mechanisms of cancer progression and resistance to therapy will be dependent on widespread acceptance and implementation of these guidelines.

References

1. Yanez-Mo, M.; Siljander, P.R.; Andreu, Z.; Zavec, A.B.; Borras, F.E.; Buzas, E.I.; Buzas, K.; Casal, E.; Cappello, F.; Carvalho, J.; et al. Biological properties of extracellular vesicles and their physiological functions. *J. Extracell. Vesicles* **2015**, *4*, 27066. [CrossRef]
2. Abels, E.R.; Breakefield, X.O. Introduction to Extracellular Vesicles: Biogenesis, RNA Cargo Selection, Content, Release, and Uptake. *Cell. Mol. Neurobiol.* **2016**, *36*, 301–312. [CrossRef] [PubMed]
3. Théry, C.; Witwer, K.W.; Aikawa, E.; Alcaraz, M.J.; Anderson, J.D.; Andriantsitohaina, R.; Anna Antoniou, T.A.; Fabienne Archer, G.K.A.-S.; D Craig Ayre, J.-M.B.; Bachurski, D.; et al. Minimal information for studies of extracellular vesicles 2018 (MISEV2018): A position statement of the International Society for Extracellular Vesicles and update of the MISEV2014 guidelines. *J. Extracell. Vesicles* **2018**, *7*. [CrossRef] [PubMed]
4. Vader, P.; Mol, E.A.; Pasterkamp, G.; Schiffelers, R.M. Extracellular vesicles for drug delivery. *Adv. Drug Deliv. Rev.* **2016**, *106*, 148–156. [CrossRef] [PubMed]
5. Maas, S.L.N.; Breakefield, X.O.; Weaver, A.M. Extracellular Vesicles: Unique Intercellular Delivery Vehicles. *Trends Cell Biol.* **2017**, *27*, 172–188. [CrossRef] [PubMed]
6. O'Brien, K.; Rani, S.; Corcoran, C.; Wallace, R.; Hughes, L.; Friel, A.M.; McDonnell, S.; Crown, J.; Radomski, M.W.; O'Driscoll, L. Exosomes from triple-negative breast cancer cells can transfer phenotypic traits representing their cells of origin to secondary cells. *Eur. J. Cancer* **2013**, *49*, 1845–1859. [CrossRef]
7. Yokoi, A.; Yoshioka, Y.; Yamamoto, Y.; Ishikawa, M.; Ikeda, S.I.; Kato, T.; Kiyono, T.; Takeshita, F.; Kajiyama, H.; Kikkawa, F.; et al. Malignant extracellular vesicles carrying MMP1 mRNA facilitate peritoneal dissemination in ovarian cancer. *Nat. Commun.* **2017**, *8*, 14470. [CrossRef]
8. Patel, G.K.; Khan, M.A.; Bhardwaj, A.; Srivastava, S.K.; Zubair, H.; Patton, M.C.; Singh, S.; Khushman, M.; Singh, A.P. Exosomes confer chemoresistance to pancreatic cancer cells by promoting ROS detoxification and miR-155-mediated suppression of key gemcitabine-metabolising enzyme, DCK. *Br. J. Cancer* **2017**, *116*, 609–619. [CrossRef]
9. Corcoran, C.; Rani, S.; O'Brien, K.; O'Neill, A.; Prencipe, M.; Sheikh, R.; Webb, G.; McDermott, R.; Watson, W.; Crown, J.; et al. Docetaxel-resistance in prostate cancer: Evaluating associated phenotypic changes and potential for resistance transfer via exosomes. *PLoS ONE* **2012**, *7*, e50999. [CrossRef]
10. Ning, K.; Wang, T.; Sun, X.; Zhang, P.; Chen, Y.; Jin, J.; Hua, D. UCH-L1-containing exosomes mediate chemotherapeutic resistance transfer in breast cancer. *J. Surg. Oncol.* **2017**, *115*, 932–940. [CrossRef]
11. Wang, X.; Xu, C.; Hua, Y.; Sun, L.; Cheng, K.; Jia, Z.; Han, Y.; Dong, J.; Cui, Y.; Yang, Z. Exosomes play an important role in the process of psoralen reverse multidrug resistance of breast cancer. *J. Exp. Clin. Cancer Res.* **2016**, *35*, 186. [CrossRef] [PubMed]
12. Pokharel, D.; Padula, M.P.; Lu, J.F.; Jaiswal, R.; Djordjevic, S.P.; Bebawy, M. The Role of CD44 and ERM Proteins in Expression and Functionality of P-glycoprotein in Breast Cancer Cells. *Molecules* **2016**, *21*, 290. [CrossRef] [PubMed]
13. Crow, J.; Atay, S.; Banskota, S.; Artale, B.; Schmitt, S.; Godwin, A.K. Exosomes as mediators of platinum resistance in ovarian cancer. *Oncotarget* **2017**, *8*, 11917–11936. [CrossRef] [PubMed]
14. Torreggiani, E.; Roncuzzi, L.; Perut, F.; Zini, N.; Baldini, N. Multimodal transfer of MDR by exosomes in human osteosarcoma. *Int. J. Oncol.* **2016**, *49*, 189–196. [CrossRef] [PubMed]
15. Lopes-Rodrigues, V.; Di Luca, A.; Mleczko, J.; Meleady, P.; Henry, M.; Pesic, M.; Cabrera, D.; van Liempd, S.; Lima, R.T.; O'Connor, R.; et al. Identification of the metabolic alterations associated with the multidrug resistant phenotype in cancer and their intercellular transfer mediated by extracellular vesicles. *Sci. Rep.* **2017**, *7*, 44541. [CrossRef] [PubMed]
16. Cesi, G.; Philippidou, D.; Kozar, I.; Kim, Y.J.; Bernardin, F.; Van Niel, G.; Wienecke-Baldacchino, A.; Felten, P.; Letellier, E.; Dengler, S.; et al. A new ALK isoform transported by extracellular vesicles confers drug resistance to melanoma cells. *Mol. Cancer* **2018**, *17*, 145. [CrossRef] [PubMed]

17. Bouvy, C.; Wannez, A.; Laloy, J.; Chatelain, C.; Dogne, J.M. Transfer of multidrug resistance among acute myeloid leukemia cells via extracellular vesicles and their microRNA cargo. *Leuk. Res.* **2017**, *62*, 70–76. [CrossRef]
18. Wu, H.; Zeng, C.; Ye, Y.; Liu, J.; Mu, Z.; Xie, Y.; Chen, B.; Nong, Q.; Wu, D. Exosomes from Irradiated Nonsmall Cell Lung Cancer Cells Reduced Sensitivity of Recipient Cells to Anaplastic Lymphoma Kinase Inhibitors. *Mol. Pharm.* **2018**, *15*, 1892–1900. [CrossRef]
19. Pavlyukov, M.S.; Yu, H.; Bastola, S.; Minata, M.; Shender, V.O.; Lee, Y.; Zhang, S.; Wang, J.; Komarova, S.; Wang, J.; et al. Apoptotic Cell-Derived Extracellular Vesicles Promote Malignancy of Glioblastoma Via Intercellular Transfer of Splicing Factors. *Cancer Cell* **2018**, *34*, 119–135e110. [CrossRef]
20. Jelonek, K.; Wojakowska, A.; Marczak, L.; Muer, A.; Tinhofer-Keilholz, I.; Lysek-Gladysinska, M.; Widlak, P.; Pietrowska, M. Ionizing radiation affects protein composition of exosomes secreted in vitro from head and neck squamous cell carcinoma. *Acta Biochim. Pol.* **2015**, *62*, 265–272. [CrossRef]
21. Yu, X.; Harris, S.L.; Levine, A.J. The regulation of exosome secretion: A novel function of the p53 protein. *Cancer Res.* **2006**, *66*, 4795–4801. [CrossRef] [PubMed]
22. Lv, L.H.; Wan, Y.L.; Lin, Y.; Zhang, W.; Yang, M.; Li, G.L.; Lin, H.M.; Shang, C.Z.; Chen, Y.J.; Min, J. Anticancer drugs cause release of exosomes with heat shock proteins from human hepatocellular carcinoma cells that elicit effective natural killer cell antitumor responses in vitro. *J. Biol. Chem.* **2012**, *287*, 15874–15885. [CrossRef] [PubMed]
23. Takahashi, K.; Yan, I.K.; Kogure, T.; Haga, H.; Patel, T. Extracellular vesicle-mediated transfer of long non-coding RNA ROR modulates chemosensitivity in human hepatocellular cancer. *FEBS Open Bio* **2014**, *4*, 458–467. [CrossRef] [PubMed]
24. Takahashi, K.; Yan, I.K.; Wood, J.; Haga, H.; Patel, T. Involvement of extracellular vesicle long noncoding RNA (linc-VLDLR) in tumor cell responses to chemotherapy. *Mol. Cancer Res.* **2014**, *12*, 1377–1387. [CrossRef] [PubMed]
25. Kreger, B.T.; Johansen, E.R.; Cerione, R.A.; Antonyak, M.A. The Enrichment of Survivin in Exosomes from Breast Cancer Cells Treated with Paclitaxel Promotes Cell Survival and Chemoresistance. *Cancers* **2016**, *8*, 111. [CrossRef] [PubMed]
26. Keklikoglou, I.; Cianciaruso, C.; Güç, E.; Squadrito, M.L.; Spring, L.M.; Tazzyman, S.; Lambein, L.; Poissonnier, A.; Ferraro, G.B.; Baer, C.; et al. Chemotherapy elicits pro-metastatic extracellular vesicles in breast cancer models. *Nat. Cell Biol.* **2018**. [CrossRef]
27. Ji, R.; Zhang, B.; Zhang, X.; Xue, J.; Yuan, X.; Yan, Y.; Wang, M.; Zhu, W.; Qian, H.; Xu, W. Exosomes derived from human mesenchymal stem cells confer drug resistance in gastric cancer. *Cell Cycle* **2015**, *14*, 2473–2483. [CrossRef]
28. Crompot, E.; Van Damme, M.; Pieters, K.; Vermeersch, M.; Perez-Morga, D.; Mineur, P.; Maerevoet, M.; Meuleman, N.; Bron, D.; Lagneaux, L.; et al. Extracellular vesicles of bone marrow stromal cells rescue chronic lymphocytic leukemia B cells from apoptosis, enhance their migration and induce gene expression modifications. *Haematologica* **2017**, *102*, 1594–1604. [CrossRef]
29. Richards, K.E.; Zeleniak, A.E.; Fishel, M.L.; Wu, J.; Littlepage, L.E.; Hill, R. Cancer-associated fibroblast exosomes regulate survival and proliferation of pancreatic cancer cells. *Oncogene* **2017**, *36*, 1770–1778. [CrossRef]
30. Zheng, P.; Chen, L.; Yuan, X.; Luo, Q.; Liu, Y.; Xie, G.; Ma, Y.; Shen, L. Exosomal transfer of tumor-associated macrophage-derived miR-21 confers cisplatin resistance in gastric cancer cells. *J. Exp. Clin. Cancer Res.* **2017**, *36*, 53. [CrossRef] [PubMed]
31. Aung, T.; Chapuy, B.; Vogel, D.; Wenzel, D.; Oppermann, M.; Lahmann, M.; Weinhage, T.; Menck, K.; Hupfeld, T.; Koch, R.; et al. Exosomal evasion of humoral immunotherapy in aggressive B-cell lymphoma modulated by ATP-binding cassette transporter A3. *Proc. Natl. Acad. Sci. USA* **2011**, *108*, 15336–15341. [CrossRef] [PubMed]
32. Ma, X.; Chen, Z.; Hua, D.; He, D.; Wang, L.; Zhang, P.; Wang, J.; Cai, Y.; Gao, C.; Zhang, X.; et al. Essential role for TrpC5-containing extracellular vesicles in breast cancer with chemotherapeutic resistance. *Proc. Natl. Acad. Sci. USA* **2014**, *111*, 6389–6394. [CrossRef]
33. Wang, T.; Ning, K.; Lu, T.X.; Sun, X.; Jin, L.; Qi, X.; Jin, J.; Hua, D. Increasing circulating exosomes-carrying TRPC5 predicts chemoresistance in metastatic breast cancer patients. *Cancer Sci.* **2017**, *108*, 448–454. [CrossRef]

34. Chen, Y.; Wang, L.; Zhu, Y.; Chen, Z.; Qi, X.; Jin, L.; Jin, J.; Hua, D.; Ma, X. Breast cancer resistance protein (BCRP)-containing circulating microvesicles contribute to chemoresistance in breast cancer. *Oncol. Lett.* **2015**, *10*, 3742–3748. [CrossRef] [PubMed]
35. Yin, J.; Yan, X.; Yao, X.; Zhang, Y.; Shan, Y.; Mao, N.; Yang, Y.; Pan, L. Secretion of annexin A3 from ovarian cancer cells and its association with platinum resistance in ovarian cancer patients. *J. Cell Mol. Med.* **2012**, *16*, 337–348. [CrossRef] [PubMed]
36. Zhang, L.; Pan, L.; Xiang, B.; Zhu, H.; Wu, Y.; Chen, M.; Guan, P.; Zou, X.; Valencia, C.A.; Dong, B.; et al. Potential role of exosome-associated microRNA panels and in vivo environment to predict drug resistance for patients with multiple myeloma. *Oncotarget* **2016**, *7*, 30876–30891. [CrossRef] [PubMed]
37. Shiozawa, K.; Shuting, J.; Yoshioka, Y.; Ochiya, T.; Kondo, T. Extracellular vesicle-encapsulated microRNA-761 enhances pazopanib resistance in synovial sarcoma. *Biochem. Biophys. Res. Commun.* **2018**, *495*, 1322–1327. [CrossRef] [PubMed]
38. Montermini, L.; Meehan, B.; Garnier, D.; Lee, W.J.; Lee, T.H.; Guha, A.; Al-Nedawi, K.; Rak, J. Inhibition of oncogenic epidermal growth factor receptor kinase triggers release of exosome-like extracellular vesicles and impacts their phosphoprotein and DNA content. *J. Biol. Chem.* **2015**, *290*, 24534–24546. [CrossRef]
39. Shiau, J.-Y.; Chang, Y.-Q.; Nakagawa-Goto, K.; Lee, K.-H.; Shyur, L.-F. Phytoagent Deoxyelephantopin and Its Derivative Inhibit Triple Negative Breast Cancer Cell Activity through ROS-Mediated Exosomal Activity and Protein Functions. *Front. Pharmacol.* **2017**, *8*. [CrossRef]
40. Campanella, C.; D'Anneo, A.; Marino Gammazza, A.; Caruso Bavisotto, C.; Barone, R.; Emanuele, S.; Lo Cascio, F.; Mocciaro, E.; Fais, S.; Conway De Macario, E.; et al. The histone deacetylase inhibitor SAHA induces HSP60 nitration and its extracellular release by exosomal vesicles in human lung-derived carcinoma cells. *Oncotarget* **2016**, *7*, 28849–28867. [CrossRef]
41. Schlaepfer, I.R.; Nambiar, D.K.; Ramteke, A.; Kumar, R.; Dhar, D.; Agarwal, C.; Bergman, B.; Graner, M.; Maroni, P.; Singh, R.P.; et al. Hypoxia induces triglycerides accumulation in prostate cancer cells and extracellular vesicles supporting growth and invasiveness following reoxygenation. *Oncotarget* **2015**, *6*, 22836–22856. [CrossRef] [PubMed]
42. Lowry, M.C.; O'Driscoll, L. Can hi-jacking hypoxia inhibit extracellular vesicles in cancer? *Drug Discov. Today* **2018**, *23*, 1267–1273. [CrossRef] [PubMed]
43. Umezu, T.; Tadokoro, H.; Azuma, K.; Yoshizawa, S.; Ohyashiki, K.; Ohyashiki, J.H. Exosomal miR-135b shed from hypoxic multiple myeloma cells enhances angiogenesis by targeting factor-inhibiting HIF-1. *Blood* **2014**, *124*, 3748–3757. [CrossRef]
44. Wang, T.; Gilkes, D.M.; Takano, N.; Xiang, L.; Luo, W.; Bishop, C.J.; Chaturvedi, P.; Green, J.J.; Semenza, G.L. Hypoxia-inducible factors and RAB22A mediate formation of microvesicles that stimulate breast cancer invasion and metastasis. *Proc. Natl. Acad. Sci. USA* **2014**, *111*, E3234–E3242. [CrossRef] [PubMed]
45. King, H.W.; Michael, M.Z.; Gleadle, J.M. Hypoxic enhancement of exosome release by breast cancer cells. *BMC Cancer* **2012**, *12*, 421. [CrossRef] [PubMed]
46. Tadokoro, H.; Umezu, T.; Ohyashiki, K.; Hirano, T.; Ohyashiki, J.H. Exosomes Derived from Hypoxic Leukemia Cells Enhance Tube Formation in Endothelial Cells. *J. Biol. Chem.* **2013**, *288*, 34343–34351. [CrossRef] [PubMed]
47. Takahashi, K.; Yan, I.K.; Haga, H.; Patel, T. Modulation of hypoxia-signaling pathways by extracellular linc-RoR. *J. Cell Sci.* **2014**, *127*, 1585–1594. [CrossRef]
48. Kucharzewska, P.; Christianson, H.C.; Welch, J.E.; Svensson, K.J.; Fredlund, E.; Ringnér, M.; Mörgelin, M.; Bourseau-Guilmain, E.; Bengzon, J.; Belting, M. Exosomes reflect the hypoxic status of glioma cells and mediate hypoxia-dependent activation of vascular cells during tumor development. *Proc. Natl. Acad. Sci. USA* **2013**, *110*, 7312–7317. [CrossRef]
49. Li, L.; Li, C.; Wang, S.; Wang, Z.; Jiang, J.; Wang, W.; Li, X.; Chen, J.; Liu, K.; Li, C.; et al. Exosomes Derived from Hypoxic Oral Squamous Cell Carcinoma Cells Deliver miR-21 to Normoxic Cells to Elicit a Prometastatic Phenotype. *Cancer Res.* **2016**, *76*, 1770–1780. [CrossRef]
50. Ramteke, A.; Ting, H.; Agarwal, C.; Mateen, S.; Somasagara, R.; Hussain, A.; Graner, M.; Frederick, B.; Agarwal, R.; Deep, G. Exosomes secreted under hypoxia enhance invasiveness and stemness of prostate cancer cells by targeting adherens junction molecules. *Mol. Carcinog.* **2015**, *54*, 554–565. [CrossRef]
51. Berchem, G.; Noman, M.Z.; Bosseler, M.; Paggetti, J.; Baconnais, S.; Le Cam, E.; Nanbakhsh, A.; Moussay, E.; Mami-Chouaib, F.; Janji, B.; et al. Hypoxic tumor-derived microvesicles negatively regulate NK cell function by a mechanism involving TGF-β and miR23a transfer. *Oncoimmunology* **2016**, *5*, e1062968. [CrossRef]

52. Hsu, Y.-L.; Hung, J.-Y.; Chang, W.-A.; Jian, S.-F.; Lin, Y.-S.; Pan, Y.-C.; Wu, C.-Y.; Kuo, P.-L. Hypoxic Lung-Cancer-Derived Extracellular Vesicle MicroRNA-103a Increases the Oncogenic Effects of Macrophages by Targeting PTEN. *Mol. Ther.* **2018**, *26*, 568–581. [CrossRef] [PubMed]
53. Vallabhaneni, K.C.; Hassler, M.Y.; Abraham, A.; Whitt, J.; Mo, Y.Y.; Atfi, A.; Pochampally, R. Mesenchymal Stem/Stromal Cells under Stress Increase Osteosarcoma Migration and Apoptosis Resistance via Extracellular Vesicle Mediated Communication. *PLoS ONE* **2016**, *11*, e0166027. [CrossRef]
54. Vallabhaneni, K.C.; Penfornis, P.; Dhule, S.; Guillonneau, F.; Adams, K.V.; Mo, Y.Y.; Xu, R.; Liu, Y.; Watabe, K.; Vemuri, M.C.; et al. Extracellular vesicles from bone marrow mesenchymal stem/stromal cells transport tumor regulatory microRNA, proteins, and metabolites. *Oncotarget* **2015**, *6*, 4953–4967. [CrossRef]
55. Kanemoto, S.; Nitani, R.; Murakami, T.; Kaneko, M.; Asada, R.; Matsuhisa, K.; Saito, A.; Imaizumi, K. Multivesicular body formation enhancement and exosome release during endoplasmic reticulum stress. *Biochem. Biophys. Res. Commun.* **2016**, *480*, 166–172. [CrossRef] [PubMed]
56. Wang, N.; Tan, H.Y.; Li, S.; Feng, Y. Atg9b Deficiency Suppresses Autophagy and Potentiates Endoplasmic Reticulum Stress-Associated Hepatocyte Apoptosis in Hepatocarcinogenesis. *Theranostics* **2017**, *7*, 2325–2338. [CrossRef] [PubMed]
57. Javeed, N.; Sagar, G.; Dutta, S.K.; Smyrk, T.C.; Lau, J.S.; Bhattacharya, S.; Truty, M.; Petersen, G.M.; Kaufman, R.J.; Chari, S.T.; et al. Pancreatic Cancer-Derived Exosomes Cause Paraneoplastic β-cell Dysfunction. *Clin. Cancer Res.* **2015**, *21*, 1722–1733. [CrossRef]
58. Fulda, S.; Gorman, A.M.; Hori, O.; Samali, A. Cellular Stress Responses: Cell Survival and Cell Death. *Int. J. Cell Biol.* **2010**, *2010*, 214074. [CrossRef]
59. Gozuacik, D.; Akkoc, Y.; Ozturk, D.G.; Kocak, M. Autophagy-Regulating microRNAs and Cancer. *Front. Oncol.* **2017**, *7*, 65. [CrossRef]
60. Xu, J.; Camfield, R.; Gorski, S.M. The interplay between exosomes and autophagy—Partners in crime. *J. Cell Sci.* **2018**, *131*. [CrossRef]
61. Bhattacharya, S.; Pal, K.; Sharma, A.K.; Dutta, S.K.; Lau, J.S.; Yan, I.K.; Wang, E.; Elkhanany, A.; Alkharfy, K.M.; Sanyal, A.; et al. GAIP interacting protein C-terminus regulates autophagy and exosome biogenesis of pancreatic cancer through metabolic pathways. *PLoS ONE* **2014**, *9*, e114409. [CrossRef]
62. Kumar, D.; Gupta, D.; Shankar, S.; Srivastava, R.K. Biomolecular characterization of exosomes released from cancer stem cells: Possible implications for biomarker and treatment of cancer. *Oncotarget* **2014**, *6*, 3280–3291. [CrossRef] [PubMed]
63. Dutta, S.; Warshall, C.; Bandyopadhyay, C.; Dutta, D.; Chandran, B. Interactions between Exosomes from Breast Cancer Cells and Primary Mammary Epithelial Cells Leads to Generation of Reactive Oxygen Species Which Induce DNA Damage Response, Stabilization of p53 and Autophagy in Epithelial Cells. *PLoS ONE* **2014**, *9*, e97580. [CrossRef]
64. Bewicke-Copley, F.; Mulcahy, L.A.; Jacobs, L.A.; Samuel, P.; Akbar, N.; Pink, R.C.; Carter, D.R.F. Extracellular vesicles released following heat stress induce bystander effect in unstressed populations. *J. Extracell. Vesicles* **2017**, *6*, 1340746. [CrossRef]
65. Hedlund, M.; Nagaeva, O.; Kargl, D.; Baranov, V.; Mincheva-Nilsson, L. Thermal- and oxidative stress causes enhanced release of NKG2D ligand-bearing immunosuppressive exosomes in leukemia/lymphoma T and B cells. *PLoS ONE* **2011**, *6*, e16899. [CrossRef] [PubMed]
66. Zhong, H.; Yang, Y.; Ma, S.; Xiu, F.; Cai, Z.; Zhao, H.; Du, L. Induction of a tumour-specific CTL response by exosomes isolated from heat-treated malignant ascites of gastric cancer patients. *Int. J. Hyperth.* **2011**, *27*, 604–611. [CrossRef] [PubMed]
67. Chen, T.; Guo, J.; Yang, M.; Zhu, X.; Cao, X. Chemokine-containing exosomes are released from heat-stressed tumor cells via lipid raft-dependent pathway and act as efficient tumor vaccine. *J. Immunol.* **2011**, *186*, 2219–2228. [CrossRef]
68. Guo, D.; Chen, Y.; Wang, S.; Yu, L.; Shen, Y.; Zhong, H.; Yang, Y. Exosomes from heat-stressed tumour cells inhibit tumour growth by converting regulatory T cells to Th17 cells via IL-6. *Immunology* **2018**, *154*, 132–143. [CrossRef]
69. Yang, Y.; Chen, Y.; Zhang, F.; Zhao, Q.; Zhong, H. Increased anti-tumour activity by exosomes derived from doxorubicin-treated tumour cells via heat stress. *Int. J. Hyperther.* **2015**, *31*, 498–506. [CrossRef]
70. Wang, W.; Luo, J.; Wang, S. Recent Progress in Isolation and Detection of Extracellular Vesicles for Cancer Diagnostics. *Adv. Healthc. Mater.* **2018**, *7*. [CrossRef]

71. Van Deun, J.; Mestdagh, P.; Sormunen, R.; Cocquyt, V.; Vermaelen, K.; Vandesompele, J.; Bracke, M.; De Wever, O.; Hendrix, A. The impact of disparate isolation methods for extracellular vesicles on downstream RNA profiling. *J. Extracell. Vesicles* **2014**, *3*. [CrossRef] [PubMed]
72. Haraszti, R.A.; Miller, R.; Stoppato, M.; Sere, Y.Y.; Coles, A.; Didiot, M.C.; Wollacott, R.; Sapp, E.; Dubuke, M.L.; Li, X.; et al. Exosomes Produced from 3D Cultures of MSCs by Tangential Flow Filtration Show Higher Yield and Improved Activity. *Mol. Ther.* **2018**, *26*, 2838–2847. [CrossRef] [PubMed]
73. Andriolo, G.; Provasi, E.; Lo Cicero, V.; Brambilla, A.; Soncin, S.; Torre, T.; Milano, G.; Biemmi, V.; Vassalli, G.; Turchetto, L.; et al. Exosomes From Human Cardiac Progenitor Cells for Therapeutic Applications: Development of a GMP-Grade Manufacturing Method. *Front. Physiol.* **2018**, *9*, 1169. [CrossRef] [PubMed]
74. Thery, C.; Amigorena, S.; Raposo, G.; Clayton, A. Isolation and characterization of exosomes from cell culture supernatants and biological fluids. *Curr. Protoc. Cell Biol.* **2006**, *3*. [CrossRef] [PubMed]
75. Lotvall, J.; Hill, A.F.; Hochberg, F.; Buzas, E.I.; Di Vizio, D.; Gardiner, C.; Gho, Y.S.; Kurochkin, I.V.; Mathivanan, S.; Quesenberry, P.; et al. Minimal experimental requirements for definition of extracellular vesicles and their functions: A position statement from the International Society for Extracellular Vesicles. *J. Extracell. Vesicles* **2014**, *3*, 26913. [CrossRef] [PubMed]
76. Consortium, E.-T.; Van Deun, J.; Mestdagh, P.; Agostinis, P.; Akay, O.; Anand, S.; Anckaert, J.; Martinez, Z.A.; Baetens, T.; Beghein, E.; et al. EV-TRACK: Transparent reporting and centralizing knowledge in extracellular vesicle research. *Nat. Methods* **2017**, *14*, 228–232. [CrossRef]

4

Optimized Protocol for Isolation of Small Extracellular Vesicles from Human and Murine Lymphoid Tissues

Marie Bordas [1,2], Géraldine Genard [3], Sibylle Ohl [1], Michelle Nessling [4], Karsten Richter [4], Tobias Roider [5], Sascha Dietrich [5], Kendra K. Maaß [6,7,†] and Martina Seiffert [1,*,†]

[1] Division of Molecular Genetics, German Cancer Research Center (DKFZ), 69120 Heidelberg, Germany; m.bordas@dkfz.de (M.B.); s.ohl@dkfz.de (S.O.)
[2] Faculty of Biosciences, University of Heidelberg, 69120 Heidelberg, Germany
[3] Division of Biomedical Physics in Radiation Oncology, German Cancer Research Center (DKFZ), 69120 Heidelberg, Germany; g.genard@dkfz-heidelberg.de
[4] Central Unit Electron Microscopy, DKFZ, 69120 Heidelberg, Germany; m.nessling@dkfz.de (M.N.); k.richter@dkfz.de (K.R.)
[5] Department of Medicine V, Hematology, Oncology and Rheumatology, University of Heidelberg, 69120 Heidelberg, Germany; Tobias.Roider@med.uni-heidelberg.de (T.R.); Sascha.Dietrich@med.uni-heidelberg.de (S.D.)
[6] Hopp-Children's Cancer Center Heidelberg (KiTZ), 69120 Heidelberg, Germany; k.maass@kitz-heidelberg.de
[7] Division of Pediatric Neurooncology, German Cancer Research Center (DKFZ), 69120 Heidelberg, Germany
* Correspondence: m.seiffert@dkfz.de;
† These authors contributed equally to this work.

Abstract: Small extracellular vesicles (sEVs) are nanoparticles responsible for cell-to-cell communication released by healthy and cancer cells. Different roles have been described for sEVs in physiological and pathological contexts, including acceleration of tissue regeneration, modulation of tumor microenvironment, or premetastatic niche formation, and they are discussed as promising biomarkers for diagnosis and prognosis in body fluids. Although efforts have been made to standardize techniques for isolation and characterization of sEVs, current protocols often result in co-isolation of soluble protein or lipid complexes and of other extracellular vesicles. The risk of contaminated preparations is particularly high when isolating sEVs from tissues. As a consequence, the interpretation of data aiming at understanding the functional role of sEVs remains challenging and inconsistent. Here, we report an optimized protocol for isolation of sEVs from human and murine lymphoid tissues. sEVs from freshly resected human lymph nodes and murine spleens were isolated comparing two different approaches—(1) ultracentrifugation on a sucrose density cushion and (2) combined ultracentrifugation with size-exclusion chromatography. The purity of sEV preparations was analyzed using state-of-the-art techniques, including immunoblots, nanoparticle tracking analysis, and electron microscopy. Our results clearly demonstrate the superiority of size-exclusion chromatography, which resulted in a higher yield and purity of sEVs, and we show that their functionality alters significantly between the two isolation protocols.

Keywords: extracellular vesicles; exosomes; small extracellular vesicles; isolation; purification; size-exclusion chromatography; ultracentrifugation; sucrose density cushion; lymph node; spleen; solid tissue

---

1. Introduction

Extracellular vesicles (EVs) are lipid bilayer-enveloped nanovesicles secreted by both eukaryotic and prokaryotic cells and carrying cargos of proteins, lipids, and nucleic acids [1,2]. EVs contain both surface and luminal factors which can be used as markers for specific EV populations representing the different biogenesis pathways [3,4]. Although the definition of EVs is continuously being refined, currently three main subtypes of eukaryotic cell-derived EVs can be distinguished based on their size, composition, and cellular origin—small EVs (sEVs or exosomes, 30–150 nm), microvesicles (MVs, 100 nm^{-1} μm), and apoptotic bodies (1–5 μm) [5,6]. Unlike MVs, which originate from direct budding of the plasma membrane, sEVs stem from the endocytic compartment and are released after fusion of multivesicular bodies with the plasma membrane [5,7]. Due to the secretory release mechanism of MVs, it is well recognized that their cargo mirrors the cytoplasmic and surface composition of the parental cell. In contrast, several studies on sEV loading reported a specific enrichment or depletion of cellular proteins or RNAs in their cargoes, and several sorting mechanisms have been suggested [8–12].

sEVs have been shown to be taken up by various recipient cell types such as myeloid, stromal, and neuronal cells, among many others [13]. The delivery of sEV cargoes into recipient cells can lead to both transcriptional and proteomic changes as a result [1,14,15]. Depending on the origin of the sEVs and the recipient cells, sEV uptake can affect diverse biologic processes, e.g., inflammation, angiogenesis, immune response, or composition of the extracellular matrix [1]. Besides their functional properties, sEVs and their content, in particular microRNAs, are also discussed for their potential as diagnostic and prognostic biomarkers in pathological conditions [1,16]. More recently, researchers explored sEVs as a new therapeutic tool for targeted drug delivery [17,18].

Due to their large spectrum of action, the interest of the scientific community for sEVs has increased exponentially over the last few years. However, many technical limitations are encountered during isolation and purification of sEVs. In particular, the isolation of sEVs from solid tissues remains challenging, limiting studies with primary patient material and causing a biased use of cell line-derived sEVs. To overcome this limitation, we aimed to improve the isolation and purification of sEVs from lymphoid tissues of human and murine specimens by comparing two different isolation protocols. The first protocol is based on differential centrifugation combined with ultracentrifugation on a sucrose density cushion as previously described [19], whereas the second protocol combines differential centrifugation with size-exclusion chromatography (SEC) using the commercially available single qEV 35 nm columns from IZON (Izon, Christchurch, New Zealand) [20]. Previous studies have already compared the efficiency of qEV IZON columns with other accepted sEV isolation techniques and reported higher yields and quality of the final product, in particular for isolation of sEVs from plasma samples [21,22].

As starting material, we used three biopsies of lymph nodes (LNs) collected from patients with B-cell lymphoma and three spleens from a B-cell lymphoma mouse model [23,24]. By directly comparing the amount, purity, and functionality of sEVs obtained for both sample types with the two protocols, we demonstrate the superiority of the SEC-based isolation technique for lymphoid tissues.

2. Results

2.1. Isolation and Purification of sEVs from Human Lymph Nodes

Two protocols for sEV isolation from lymphoid tissues were performed in parallel on the same starting material to compare their efficiency in terms of (1) total amount of recovered sEVs, (2) purity of sEV preparation, and (3) reproducibility. After manual dissociation of LN biopsies of three B-cell lymphoma patients, the supernatants of the cell suspensions were collected and processed by differential centrifugation. The resulting pellets (100 K pellet) containing sEVs and soluble proteins and lipids was resuspended and split into two equal parts each, which were then combined either with SEC on IZON columns or differential centrifugation combined with ultracentrifugation on a 40% sucrose density cushion as illustrated in Figure 1. An identical volume of PBS (250 μL) was used for the final resuspension of sEVs isolated from IZON columns and the sucrose density cushions ("cushion"). Nanoparticle tracking analysis (NTA) revealed that the resuspended pellet from the "cushion" preparation as well as fractions 1 and 2 collected from the IZON column were enriched in the characteristic sEV size profile, with IZON fraction 2 accounting for the peak fraction (Figure 2A, left). SEV size profiles were also detected in the IZON fraction 3, although in lower concentrations. The absolute number of particles yielded from IZON peak fractions as assessed by NTA was 3.9- to 10.3-fold higher than the sEV particle number recovered using the sucrose density cushion (Figure 2A, right). We then performed protein quantification using a bicinchoninic acid (BCA) assay (Figure 2B, left). Due to their smaller size, protein complexes are able to enter the pores of the IZON column, and their elution is delayed, which can be observed as a second protein peak in the fraction F7 collected later [20,25].

The absolute amount of proteins recovered from IZON peak fractions was lower than the one obtained in the respective "cushion" preparation (1.4- to 2.2-fold lower; Figure 2B, right) which was less than the fold change detected by NTA for particle numbers. Calculation of the particle/protein ratios revealed lower values for "cushion" preparations compared to IZON fractions in two of the three samples (Figure 2D). Therefore, we hypothesized that the sucrose density cushion approach led to a larger amount of protein complexes co-isolated with the sEVs. The mean particle size and size distribution of sEVs were similar in IZON fractions 1 and 2 and the "cushion" preparation with 153, 157, and 148 nm for the peak IZON fractions, and 155, 163, and 152 nm for the corresponding "cushion" preparations (Figure 2C,E). In line with our hypothesis, immunoblot analysis revealed a lower signal for exosomal surface markers FLOTILLIN-1, CD81, CD9, and the luminal marker TSG101 in the "cushion" preparations compared to IZON fractions 1 and 2 for the same amount of protein loaded (Figure 2F and Figure S1).

As our study is one of the first to focus on solid tissues, we thoroughly validated the presence of contaminant proteins as recommended by the MISEV guidelines [26]. We neither detected the Golgi marker GM130 nor the mitochondrial marker CYTOCHROME C in both preparations. Surprisingly, we detected the endoplasmic reticulum (ER) protein CALNEXIN in sEVs isolated with both protocols. However, the amount of CALNEXIN in the sEV preparations was lower in comparison to the parental cell lysate (Figure 2F and Figure S1). Although partial contamination of the samples with cellular debris cannot be excluded, the presence of CALNEXIN but no markers from other cell organelles might be indicative for a specific sEV biogenesis pathway involving the ER in lymphoma cells. In addition, the IZON fraction F2 and "cushion" samples were analyzed by transmission electron microscopy (TEM).

The results illustrate that the SEC isolated samples allow a clear identification of sEVs for all of the three samples. However, we observed a higher heterogeneity in the "cushion" preparations, with sEVs barely detected in two out of three samples (Figure 2G). Employing immuno-electron microscopy, we confirmed the presence of the immune receptor MHC Class II (HLA-DR) on the surface of sEVs isolated by both approaches and thereby validate their immune cell origin (Figure 2H).

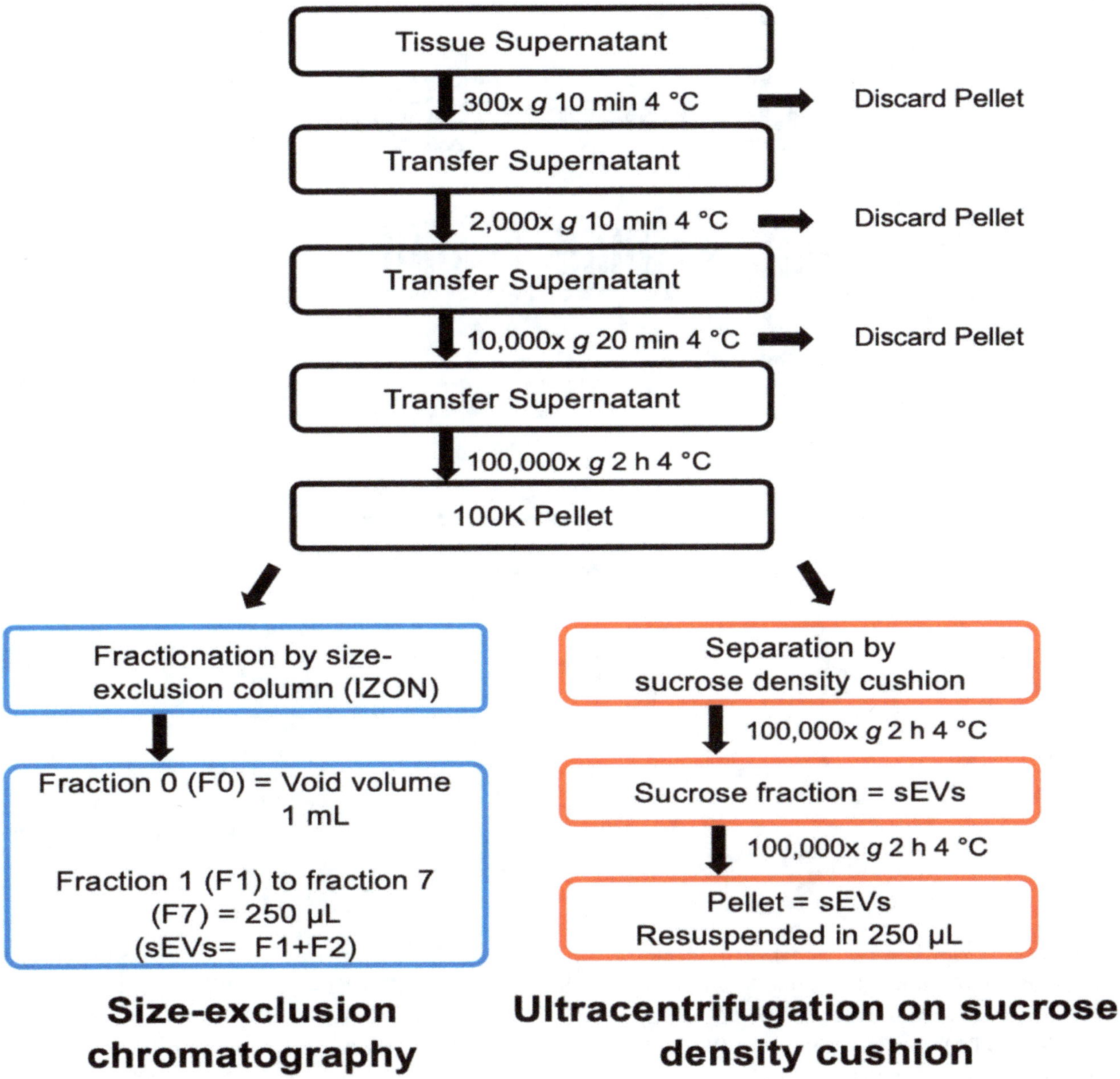

Figure 1. Experimental outline of the comparison of size-exclusion column-based (SEC) versus density-based small extracellular vesicle (sEV) isolations. Supernatant of dissociated lymphatic tissues was separated by differential ultracentrifugation and the resulting 100 K pellet was resuspended and split into two equal parts for direct method comparison. Equal volumes were loaded on either SEC columns or on a sucrose density cushion. Resulting sEV fractions were compared for yield, purity and functionality.

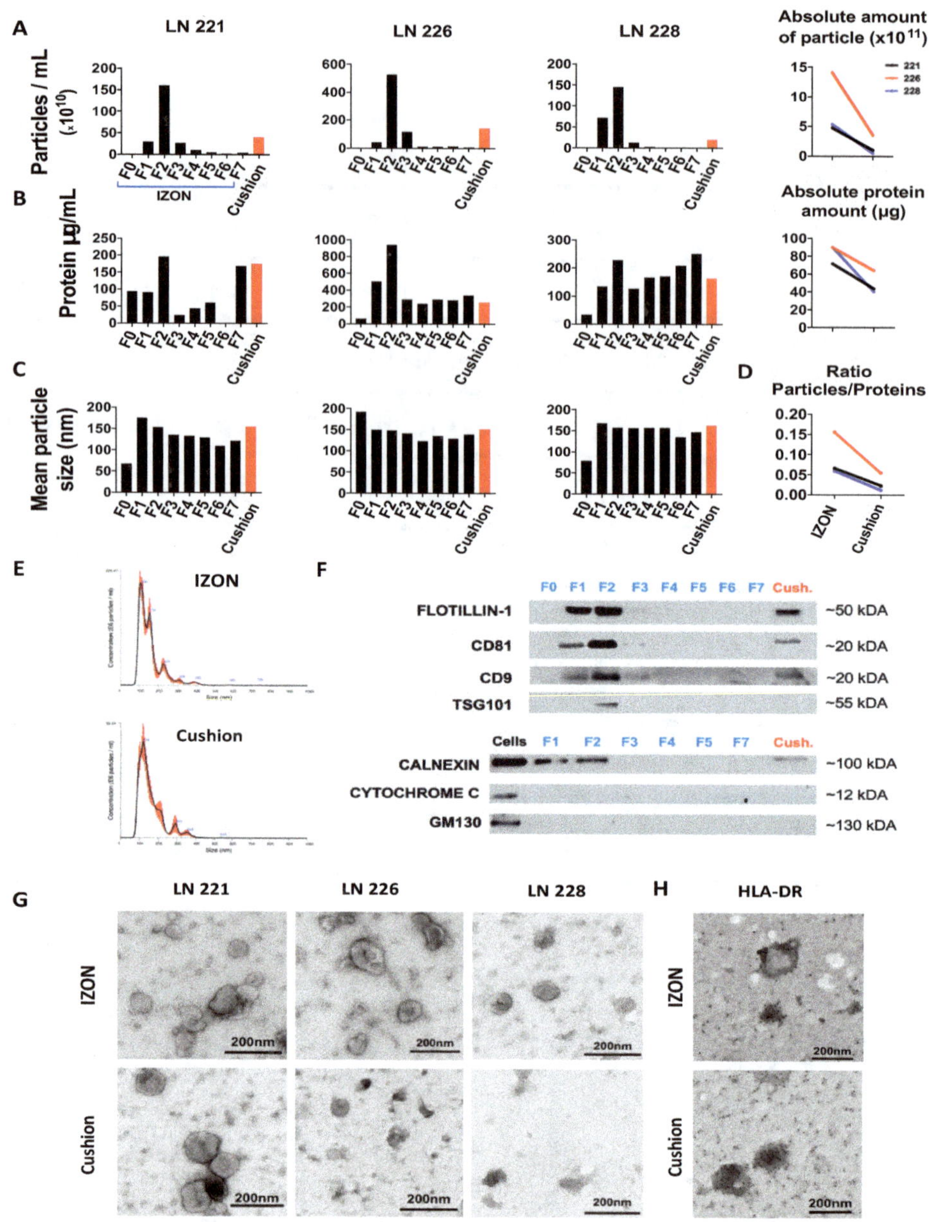

Figure 2. Comparative isolation and characterization of sEVs from human lymph nodes (LN). Size-exclusion column (SEC) fractions (F0 = void volume, 1 mL; F1, F2, F3, F4, F5, F6, F7 = serial fractions, 250 μL) and cushion fraction (pellet resuspended in 250 μL) were analyzed by nanosight tracking analysis (NTA), bicinchoninic acid (BCA) protein quantification, immunoblotting, and transmission electron microscopy (TEM). (**A**) Left: particle concentrations in IZON fractions F0–F7 and "cushion" fraction for three different human LN samples measured by NTA. Right: Absolute number of detected particles as sum of fraction 1 and fraction 2. For each sample, the particle concentration was normalized to the final volume of elution. (**B**) Left: BCA protein quantification for IZON fraction F0–F7 and the "cushion" fraction. Right: Absolute amount of protein in fraction 1 and fraction 2 (the protein concentration was normalized to the final volume of elution). (**C**) Mean particle size for IZON fraction F0–F7 and the "cushion" fraction analyzed by NTA. (**D**) Ratios of particles per protein amount are plotted for IZON and "cushion" fraction. (**E**) Representative particle distribution profile for IZON fraction 2 sample (left) and "cushion" sample analyzed by NTA (Sample LN221). (**F**) Immunoblotting analysis of FLOTILLIN-1, CD81, CD9, TSG101, CALNEXIN, CYTOCHROME C, and GM130 for indicated IZON fractions, the "cushion" fraction and parental cell lysates for one LN sample. (**G**) TEM images of IZON fraction 2 and the "cushion" fraction for the three indicated samples. (**H**) Immunogold electron microscopy for HLA-DR of one human LN sample (sample LN221). Scale bar: 200 μm.

2.2. Isolation and Purification of sEVs from Murine Spleen

Spleens from three mice with B-cell lymphoma were dissociated and processed as outlined in Figure 1. Similar to human LN samples, NTA results revealed an enriched particle concentration in IZON fractions 1 and 2, with fraction 2 being the peak fraction (Figure 3A, left). For one of the samples, sEVs were mainly detected in fractions 2 and 3, a difference we attribute to manual loading and elution of the IZON column. The absolute number of particles isolated was 4.8- to 27.7-fold higher in the IZON peak fractions in comparison to the respective cushion preparations (Figure 3A, right). The protein concentrations measured by BCA were more similar between IZON fractions and "cushion" and might be attributed to protein complexes co-isolated with the sEVs in the "cushion" preparation (Figure 3B, left).

The absolute amount of proteins recovered was 1.4- to 2.6-fold higher in the IZON preparations in comparison to the respective cushion preparations (Figure 3B, right). In line with these results, the particle/protein ratios were drastically reduced for "cushion" preparations in comparison to IZON fractions (Figure 3D). The mean particle sizes were 154, 142, and 157 nm in the IZON peak fractions, and 114, 143, and 155 nm for the corresponding "cushion" preparations (Figure 3C). Those results imply that, for one preparation at least, the obtained product was different when using the SEC or the sucrose density cushion approach. Additionally, we observed a difference in the size distribution profile depending on the isolation protocol used, which might be explained by different EV subpopulations isolated by the different approaches (Figure 3E).

Immunoblot results confirmed the exosomal identity of the particles in fractions 1, 2, and 3 and the cushion fraction by positive bands for the surface marker FLOTILLIN-1 but also the luminal markers ALIX and TSG101 in the IZON fractions 1–3 and in the "cushion" preparations (Figure 3F and Figure S2). FLOTILLIN-1 was only detected in IZON fractions, and TSG101 showed varying intensities being highly present in the "cushion" fraction while only weakly detected in the IZON fractions. Together with the variance in NTA size profiles, the immunoblotting results further suggested that different sEV subpopulations were isolated. Irrespective, we could exclude mitochondrial contaminations by ATP5A being absent from all sEV fractions (Figure 3F and Figure S2). In concordance with the human LN data, CALNEXIN could again be detected in sEV fractions from both protocols.

The quality and purity of the sEV isolations was further assessed by TEM (Figure 3G). We noted a high heterogeneity among the samples for the "cushion" preparations, with one sample highly enriched in lipidic structures (Figure 3G). Interestingly, we noticed the recurrent presence of small dark structures of an approximate size of 10 nm exclusively in "cushion" preparations (Black arrows, Figure 3G). A closer look at the particles revealed a specific geometrical shape, typical for ferritin (Figure 3H) [27,28]. We further observed a red color of the sEV pellets and suspensions which is typical for a contamination with erythrocyte-derived protein, strengthening our hypothesis (Figure 3I).

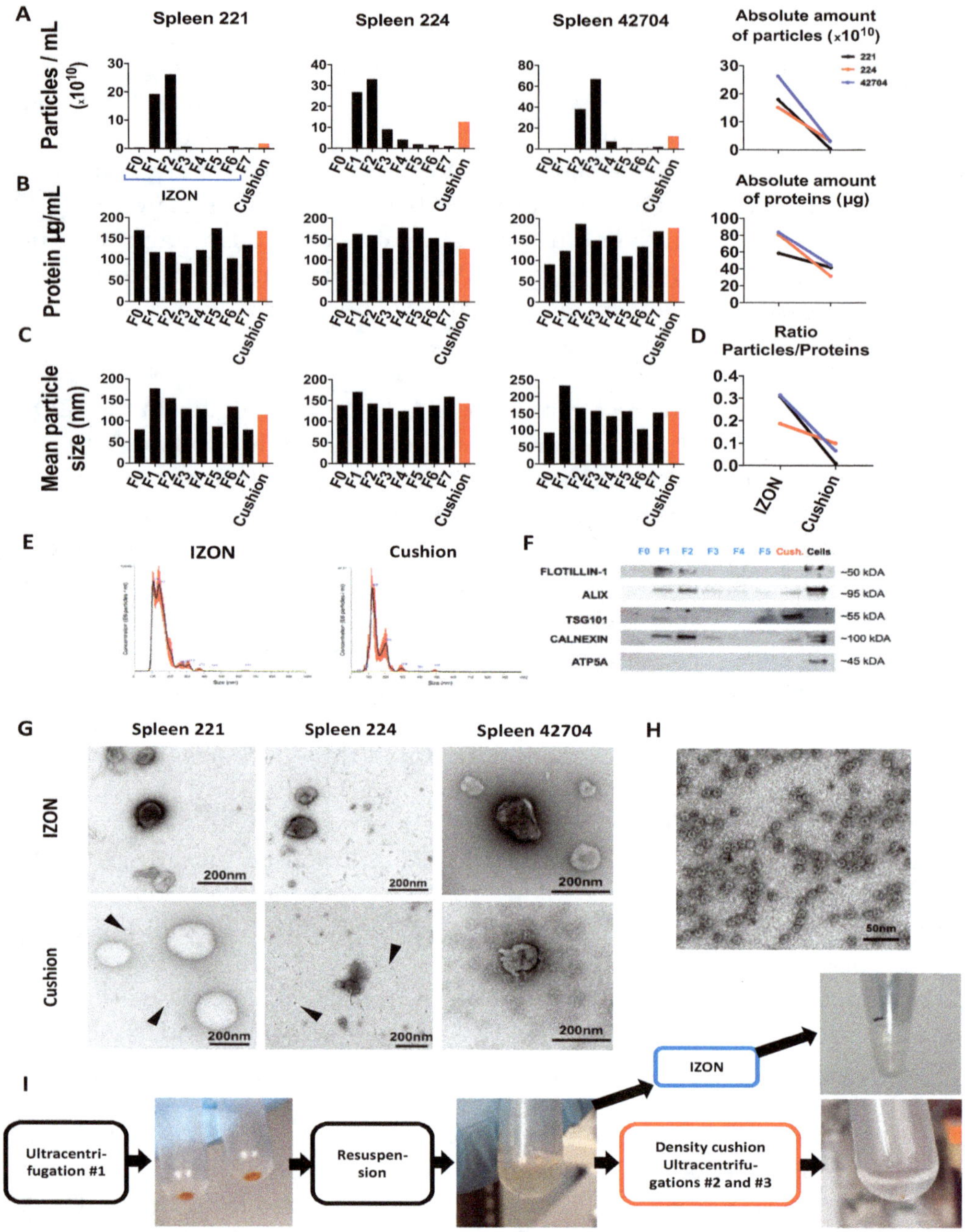

Figure 3. Isolation and characterization of murine spleen sEVs. (**A**) Left: SEV concentration in the different IZON Fractions and "cushion" preparations for three samples analyzed by NTA. Right: Absolute number of particles in indicated preparations. For each sample, the particle concentration in the two peak fractions or in the cushion product was normalized to the final volume of elution. (**B**) Left: protein quantification in the indicated preparations assessed by BCA assay. Right: absolute amount of protein in indicated preparations (the protein concentration was normalized to the final volume of elution). (**C**) Mean particle size of all fractions and the "cushion" preparations analyzed by NTA. (**D**) Ratios of particles per protein amount are plotted for IZON and "cushion" fraction. (**E**) One representative particle distribution profile for an IZON fraction 2 (left) and a "cushion" preparation analyzed by NTA (Spleen 42704). (**F**) Immunoblotting analysis of FLOTILLIN-1, ALIX, TSG101, CALNEXIN, and ATP5A for the different IZON fractions, the "cushion" preparation and parental cells for one spleen sample (spleen 224). (**G**) Transmission electron microscopy (TEM) images of IZON peak fraction and "cushion" preparation for the three indicated samples. (**H**) TEM image of ferritin-like structures found in "cushion" preparations. (**I**) Pictures of the sEV pellet, resuspended sEVs prior to application on the sucrose density cushion, and final pellet in PBS before resuspension.

2.3. Functional Analysis of sEVs Isolated by the Two Different Protocols

We and others have previously reported that tumor-derived sEVs (TEX) are able to induce an immunosuppressive phenotype in monocytes in vitro, with a typical upregulation of surface PD-L1 and HLA-DR [29–31]. We compared the potential of murine TEX isolated from three spleen samples using the two different approaches regarding their ability to induce such a phenotype. Both TEX preparations (IZON and "cushion") induced PD-L1 upregulation in monocytes, although to various degrees (Figure 4A and Figure S3 for gating strategy). However, for two of the three samples, "cushion" preparations did not induce an upregulation of HLA-DR (Figure 4B). These results indicate that both protocols resulted in sEV samples that induce a different immunosuppressive phenotype in monocytes. We also analyzed the expression of the activation marker ICAM-1 (CD54) in monocytes treated with TEX, which showed a much more drastic upregulation with the "cushion" preparations compared to the SEC-isolated TEX in all 3 analyzed samples (Figure 4C).

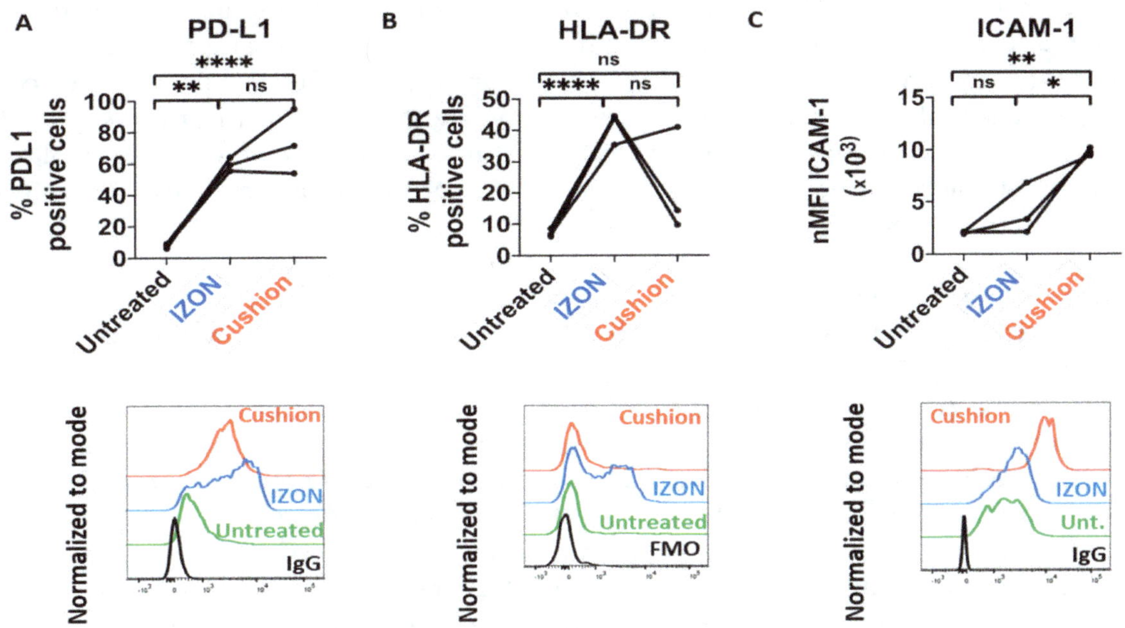

Figure 4. Response of murine monocytes upon tumor-derived sEVs (TEX) treatment. Bone marrow-derived monocytes were treated with 5 µg of the indicated sEV preparations for 8 h and analyzed by flow cytometry gating on CD11b+F4/80+$^{+}$CX3CR1+Ly6C+ cells ($n = 3$ mice per sEV preparation). (**A**) Top: percentage of PD-L1 positive cells among CD11b+F4/80+CX3CR1+Ly6C+ monocytes. Bottom: representative histogram including isotype antibody staining as negative control (IgG). (**B**) Percentage of MHC-II/HLA-DR positive cells among CD11b+F4/80+CX3CR1+Ly6C+ monocytes. Bottom: representative histogram including fluorescence-minus-one (FMO) staining as negative control. (**C**) Top: ICAM-1/CD54 expression presented as normalized mean fluorescence intensity (nMFI). Bottom: representative histogram. p-values were determined by one-way ANOVA with Tukey's multiple comparisons test. * $p < 0.05$; ** $p < 0.0021$; *** $p < 0.0002$; **** $p < 0.0001$.

3. Discussion

Multiple isolation approaches have been proposed for sEV preparations, including commercially available kits, ultrafiltration, polymer precipitation, immune-affinity capture, size-exclusion chromatography, ultracentrifugation, and ultracentrifugation combined with density cushion [32,33]. The selection of the isolation technique must consider the subsequent usage of the sEV preparations. Yield is generally prioritized when performing RNA or DNA sequencing. However, contamination with protein or lipid complexes must be avoided for proteomic analysis or functional assays. The sources and risk of protein contamination are even higher when isolation of sEVs is performed from solid tissues that require mechanical or enzymatic dissociation. Here, we compared two different protocols to isolate sEVs from solid lymphoid tissues: differential centrifugation combined to SEC using commercially available IZON columns and differential centrifugation combined to ultracentrifugation on sucrose density cushions. Although a total of three human LN and three mouse spleens are shown in our manuscript, our results are representative of larger cohorts of samples regularly analyzed in our laboratory. Both approaches led to efficient isolation of sEVs as shown by size characterization based on NTA analysis and the presence of exosomal markers by immunoblotting. However, further characterization of the preparations using BCA assay, TEM, and functional assay led us to conclude that the SEC approach is superior in terms of purity, quantity, and reproducibility.

In particular, our results strongly suggest that isolation of sEVs by the sucrose density cushion isolation protocol results in a more severe co-isolation of protein complexes with the sEVs. Using immunoblotting, we excluded contamination by mitochondrial and Golgi-derived proteins. We speculate that the presence of cellular debris in the supernatant of the dissociated tissues, which would lead to the sample contamination, was efficiently avoided by the rapid isolation of sEVs following organs' resection. However, the ER-derived protein CALNEXIN was found in sEV preparations using both isolation approaches. Such contamination likely results from the tissue dissociation. However, further investigations are required to verify that the presence of ER-proteins but not of proteins of other organelles could be the result of a specific packaging mechanism of tumor-derived sEVs. Furthermore, we also observed the presence of ferritin-like proteins in spleen sEVs isolated by the sucrose density cushion but not the SEC approach. The presence of ferritin seems to indicate an erythrocyte contamination. However, addition of erythrocyte lysis buffer to the supernatant would result in an increased release of hemoglobin. As erythrocytes are easier to separate from sEVs than hemoglobin, we do not recommend the usage of such buffer.

Previous studies focusing on plasma-derived sEVs reported lipoproteins as the main contaminants of sEV preparations [34–37]. Unfortunately, lipoproteins cannot be efficiently discriminated from sEVs when performing NTA analysis. However, contamination by lipoproteins of low and high density in sEV preparations seem less important when using the SEC-isolation approach [34–36]. In our study, we noticed the presence of large lipidic structures in one of the three murine samples isolated with the sucrose density cushion but not with SEC. Additional immunoblots are required to conclude on lipoprotein contamination in sEVs isolated with both approaches. A solution to limit lipoprotein contamination would be the combination of both SEC and sucrose density cushion. However, combination of isolation methods often results in a drastic loss of material. Other possible sources of contamination include secreted proteins, and extracellular matrix proteins. Investigations on such contaminants remain challenging, as these proteins could be considered as well as of exosomal origin.

We also would like to emphasize that ultracentrifugation combined with density cushion and differential centrifugation combined with SEC are isolation techniques that are based on the density or the size of EVs, respectively. Thus, it is possible that the use of a unique isolation protocol may impact on the distribution of sEV subpopulations in the preparations. In line with this hypothesis, different sEV marker proteins were enriched in sEVs isolated from spleens by the two different methods: sEV preparations obtained using the SEC approach were enriched in FLOTILIN-1 and ALIX but not TSG101, whereas the "cushion" preparations did show an enrichment in ALIX and TSG101 but not in FLOTILIN-1.

We previously reported that treatment of monocytes with TEX induces the upregulation of surface PD-L1 and major histocompatibility complex (MHC) II/HLA-DR. We compared the capability of TEX preparations of both isolation protocols to induce such a phenotype, using an identical amount of sEVs based on protein quantification. Treatment of monocytes with "cushion" preparations resulted in a more heterogeneous response of those two markers in comparison to SEC preparations, indicating different amounts of contaminant proteins from one "cushion" preparation to another. In particular, MHC II surface expression was increased when monocytes were treated with IZON preparations but not with "cushion" preparations, for two of three preparations. Yet, it is known that sEVs secreted by antigen-presenting cells are enriched in MHC II molecules, and that sEVs can promote the transfer of functional MHC II/antigen complexes to recipient cells [38,39]. On the contrary, we observed a higher upregulation of the monocyte activation marker ICAM-1 upon treatment with "cushion" preparations. These results highlight that contaminant proteins can interfere with biological results and lead to an incorrect conclusion of sEV-induced phenotypes. ICAM-1 expression on monocytes is a general activation marker and its upregulation can be induced by cytokines, lipoproteins, LPS etc. [40,41]. Given these results, we suspect that the PD-L1 upregulation observed in monocytes treated either with the IZON preparations or the "cushion" preparations is the consequence of monocytes' activation mainly by sEVs, whereas induced expression of ICAM-1 results from both sEVs and non-sEV contaminants. These results raise the hypothesis that soluble pro-inflammatory cytokines, secreted in B-cell lymphoma microenvironments, might contribute to the contamination in the "cushion" preparations, although further investigations would be required for firm conclusion.

As a conclusion, we strongly recommend the usage of SEC for sEV isolation from solid tissue represented here by lymphoid tissues. Multiple controls should be performed to validate the purity of the samples. Such controls include extensive immunoblotting of positive and negative exosomal markers. Reaching a complete purity of sEVs from biofluids or solid tissue seems unrealistic. Nevertheless, immunoblotting results in parallel to NTA analysis can provide a reliable estimation of preparations' contamination by protein complexes. TEM remains an indispensable tool to validate the presence and integrity of sEVs and to assess the amount of contamination by lipid complexes.

4. Materials and Methods

4.1. Animals

Eμ-TCL1 mice on C57BL/6 background were kindly provided by Carlo Croce (Ohio State University). C57BL/6 wild-type (WT) mice were purchased from Charles River Laboratories (Sulzfeld, Germany). Adoptive transfer of Eμ-TCL1 tumors was performed as previously described [42,43]. Briefly, $1–2 \times 10^7$ B-cells enriched from Eμ-TCL1 splenocytes were transplanted intraperitoneally (i.p.) into C57BL/6N WT animals. B-cell enrichment was performed using EasySep™ Mouse Pan-B Cell Isolation Kit (Stemcell Technologies, Vancouver, BC, Canada), yielding a purity above 95% of CD5+ CD19+ cells. Tumor load was assessed in the blood every week using flow cytometry as the proportion of CD5+ CD19+ cells among CD45+ cells. Animals with a tumor load >90% in peripheral blood were sacrificed; spleen was isolated and mechanically dissociated in PBS. All animal experiments were carried out according to institutional and governmental guidelines approved by the local authorities (Regierungspräsidium Karlsruhe, permit number G98/16, approved on 13 July 2016).

4.2. SEV-Free RPMI Medium

Fetal calf serum (FCS) (Gibco, Carlsbad, CA, USA) was ultra-centrifuged at 100,000× g for 18 h at 4 °C. FCS supernatant was filtered through a 0.22 μm filter. RPMI 1640 medium (Thermo Fisher Scientific Inc., Waltham, MA, USA) was supplemented with 10% sEV-free FCS and 1% penicillin/streptomycin (Gibco, Carlsbad, CA, USA). The medium was filtered through a 0.22 μm filter prior to use.

4.3. Isolation of Lymph Node Supernatants

Patient lymph node (LN) samples were obtained after the study protocols' approval by local ethics' committees from the Department of Medicine V of the University Clinic Heidelberg according to the declaration of Helsinki, and with patients' informed consent. LN samples were collected directly after biopsies from patients with diverse B-cell lymphomas. LNs were placed in 0.9% NaCl solution and processed immediately. Each LN was cut in small pieces with a maximum size of 2 mm. Cells were released in 50 mL of RPMI medium supplemented with sEV-free FCS (10%), Penicillin-Streptomycin (1%) and L-Glutamine (1%).

4.4. Isolation of Murine Spleen Supernatants

Entire spleens from three adoptively transferred Eμ-TCL1 mice were collected in 7 mL of 0.22 μm-filtered PBS each. Spleens were mechanically dissociated using MACS dissociator (Miltenyi Biotec, Bergisch Gladbach, Germany), using the program "m_spleen_01".

4.5. Differential Centrifugation

Collected supernatants of human LN and mouse spleen were centrifuged at 300× *g* for 10 min at 4 °C in a swing-out centrifuge to remove cellular debris. Resulting supernatants were transferred into new collection tubes and centrifuged at 2000× *g* for 20 min at 4 °C to remove larger apoptotic bodies. Resulting 2000× *g* supernatants were transferred into new collection tubes and centrifuged at 10,000× *g* for 40 min at 4 °C to remove MVs. Resulting 10,000× *g* supernatants were transferred into ultracentrifugation tubes (#5031, Seton Sci., Petaluma, CA, USA), and centrifuged at 100,000× *g* for 2 h at 4 °C on a Beckman Optima L-70 ultracentrifuge (Beckman Coulter GmbH, Krefeld, Germany) using a 40 Ti Swinging-Bucket Rotor. Resulting 100,000× *g* pellets were resuspended in 400 μL of 0.22-μm-filtered PBS and split in half for direct method comparison described below.

4.6. SEV Isolation on Sucrose Density Cushion

This protocol was adapted from a previous protocol established in our lab [19]. The half volume of the resuspended 100,000× *g* pellet was filled up with 0.22-μm-filtered PBS to 7 mL. The diluted pellet fraction was carefully applied onto 4 mL of a 40% sucrose cushion with a density of 1.12 g/mL without disturbing the cushion and centrifuged at 100,000× *g* for 2 h at 4 °C. The most upper PBS phase of around 6.5 mL was discarded. The following 3.5 mL high-density sucrose fraction containing the sEVs was recovered. The pellet was left untouched to avoid contaminating the sEV fraction with high molecular weight protein complexes. The sEVs were recovered by washing in 0.22-μm-filtered PBS by adding 7 mL of 0.22-μm-filtered PBS and centrifugation at 100,000× *g* for 2 h at 4 °C. The resulting sEV pellet was resuspended in 250 μL of 0.22-μm-filtered PBS.

4.7. SEV Isolation on Single qEV 35nm Columns, IZON

Single qEV 35 nm columns (Izon, Christchurch, New Zealand) were allowed to reach room temperature for 30 min. The resuspended pellet fraction (200 μL) was added onto the column. As soon as the sample volume was taken up by the column, 0.22-μm-filtered PBS was added to the top of the column tube. The following fractions were collected: F0 (1 mL = void volume of the column) and F1 to F7 (250 μL each), according to the manufacturer's instructions.

4.8. Bicinchonic Acid (BCA) Assay and Nanoparticle Tracking Analyzis (NTA)

Protein concentration of sEV samples was assessed employing Pierce™ BCA Protein Assay Kit (Thermo Fisher Scientific Inc., Waltham, MA, USA). 9 μL of each sEV sample was lysed with 1 μL of 10× RIPA buffer (Abcam, Cambridge, UK) and incubated for 30 min on a rotating wheel at 4 °C. Samples were then centrifuged at 17,000× *g* for 20 min at 4 °C. Resulting supernatants were subjected to the BCA assay according to the manufacturer's instructions. Absorbance was assessed with the use of a MITHRAS LB 940 plate reader (Berthold Technologies, Bad Wildbad, Germany). Particle quantification of sEV samples was performed via NTA using NanoSight LM10 equipped with a 405 nm laser (Malvern Instruments, Malvern, UK). For the NTA analysis, samples were diluted 1:500 to 1:1000 in 0.22-μm-filtered PBS. Camera level and detection threshold were set up at 13 and 5, respectively. The absence of background was verified using 0.2-μm-filtered PBS. For each sample, four videos of 60 s each were recorded and analyzed using the NTA 3.0 software version (Malvern Instruments, Malvern, UK).

4.9. *Immunoblotting*

SEVs and respective parental cells were lysed in RIPA buffer (Abcam, Cambridge, UK), and whole protein lysates were quantified via BCA™ Protein Assay Kit (Thermo Fisher Scientific Inc., Waltham, MA, USA). Per lane, 2.8 μg (human samples) or 3.4 μg (mouse samples) of protein were loaded onto 10% polyacrylamide gels. Following SDS-PAGE and protein transfer, membranes were blocked in 5% bovine serum albumin in Tris-buffered saline (TBS)-Tween 0.1%, and primary antibodies against FLOTILLIN-1 (1:1,000, Cell Signaling Technology, Danvers, MA, USA, #18634), CD81 (1:400, ProSci Inc., San Diego, CA, USA, #5195), CD9 (1:1000, Cell Signalling Technology, Danvers, MA, USA, #13174), TSG101 (1:1000, BD Bioscience, San Jose, CA, USA, #612697), ALIX (1:1000, Cell Signalling Technology, Danvers, MA, USA, #2171) CALNEXIN (1:500, GeneScript, Piscataway, NJ, USA, #A0124040), CYTOCHROME C (1:750, GeneScript, Piscataway, NJ, USA, #A0150740), GM130 (1:1000, Cell Signaling Technology, Danvers, MA, USA, #12480), ATP5A (1:1,000, Abcam, Cambridge, UK, #ab14748) were used in indicated dilutions in 5% bovine serum albumin in TBS-Tween 0.1%. Signals were visualized after secondary antibody hybridization by chemiluminescence detection reagent (Bio-Rad Lab, Hercules, CA, USA, #1705061) with GE Healthcare Amersham Imager 600 (GE Healthcare, Chicago, IL, USA).

4.10. *Electron Microscopy (EM)*

SEV fractions were adsorbed onto glow discharged carbon coated grids, washed in aqua bidest and negatively stained with 2% aqueous uranyl acetate. For immuno-EM, carbon-coated formvar grids were used and the immune reaction was performed after buffer wash including incubation with blocking agent (Aurion, Wageningen, The Netherlands), dilution series of primary antibody HLA-DR (Santa Cruz, Dallas, TX, USA #sc-51618) and Protein A-Au reporter (CMC, UMC Utrecht, The Netherlands). Micrographs were taken with a Zeiss EM 910 or EM 912 at 100 kV (Carl Zeiss, Oberkochen, Germany) using a slow scan CCD camera (TRS, Moorenweis, Germany).

4.11. *Functional Assay*

Murine monocytes were isolated from the bone marrow of C57BL/6 mice by magnetic depletion (EasySep™ Mouse Monocyte Isolation Kit, STEMCELL Technologies Inc., Vancouver, BC, Canada). 5×10^4 cells were cultured in 48-well plates in sEV-free RPMI medium and treated for 8 h with 5 μg of the respective sEV fractions referred above, as determined by BCA assay. Changes in PD-L1, HLA-DR and ICAM-1 expression were evaluated by flow cytometry (BD LSR Fortessa, BD Biosciences, San Jose, CA, USA). The following antibodies were used: PD-L1-PerCP (Biolegend, San Diego, CA, USA, #46-5982-82), HLA-DR-AlexaFluor700 (eBiosciences, San Diego, CA, USA, #56-5321-82), CD54-PE (Biolegend, San Diego, CA, USA, #116108), CX3CR1-BV711 (Biolegend, San Diego, CA, USA, #149031), Ly6C-APC-Cy7 (Biolegend, San Diego, CA, USA, #128015), CD11b-PeCy7 (Biolegend, San Diego, CA, USA, #101216), F4/80-FITC (Biolegend, San Diego, CA, USA, #123107), and the viability dye eFluorTM 506 (eBiosciences, San Diego, CA, USA, #65-0866).

4.12. *Statistical Analysis*

Results of the functional analysis were analyzed for statistical significance with GraphPad PRISM 8.0 software (GraphPad Software, San Diego, CA, USA), using one-way analysis of variance (ANOVA), followed by Tukey's multiple comparisons. The differences between means were considered significant if $p \leq 0.05$. The results are expressed as the means ± standard deviation.

4.13. *EV Track*

We have submitted all relevant data of our experiments to the EV-TRACK knowledgebase (EV-TRACK ID: EV200073) (Van Deun J, et al. EV-TRACK: transparent reporting and centralizing knowledge in extracellular vesicle research. Nature methods. 2017;14(3):228–32).

You may access and check the submission of experimental parameters to the EV-TRACK knowledgebase via the following URL: http://evtrack.org/review.php. Please use the EV-TRACK ID (EV200073) and the last name of the first author (Bordas) to access our submission.

Author Contributions: Conceptualization, M.B., G.G., K.K.M., and M.S.; methodology, M.B., G.G., S.O., M.N., K.R. and K.K.M.; resources, T.R. and S.D.; writing—original draft preparation, M.B. and G.G.; writing—review and editing, K.K.M. and M.S.; visualization, M.B., G.G., M.N. and K.R.; supervision, K.K.M. and M.S.; funding acquisition, M.S. All authors have read and agreed to the published version of the manuscript.

Acknowledgments: The authors thank Carolin Kolb, Mareike Knoll, and Angela Lenze for their help in collecting LN supernatants, and Franziska Haderk and Mariana Coelho for their critical review of the paper.

Abbreviations

BCA	Bicinchoninic acid
ER	Endoplasmatic reticulum
EVs	Extracellular vesicles
FCS	Fetal calf serum
i.p.	Intraperitoneal
LN	Lymph node
MHC	Major histocompatibility complex
MVs	Microvesicles
NTA	Nanoparticle Tracking Analysis
SEC	Size-exclusion chromatography
sEVs	Small extracellular vesicles
TEM	Transmission electron microscopy
TEX	Tumor-derived sEVs
WT	Wild-type

References

1. Kalluri, R.; LeBleu, V.S. The biology, function, and biomedical applications of exosomes. *Science* **2020**, *367*, eaau6977. [CrossRef] [PubMed]
2. Samanta, S.; Rajasingh, S.; Drosos, N.; Zhou, Z.; Dawn, B.; Rajasingh, J. Exosomes: New molecular targets of diseases. *Acta Pharmacol. Sin.* **2018**, *39*, 501–513. [CrossRef] [PubMed]
3. Campos-Silva, C.; Suárez, H.; Jara-Acevedo, R.; Linares-Espinós, E.; Martinez-Piñeiro, L.; Yáñez-Mó, M.; Valés-Gómez, M. High sensitivity detection of extracellular vesicles immune-captured from urine by conventional flow cytometry. *Sci. Rep.* **2019**, *9*, 1–12. [CrossRef] [PubMed]
4. Wu, D.; Yan, J.; Shen, X.; Sun, Y.; Thulin, M.; Cai, Y.; Wik, L.; Shen, Q.; Oelrich, J.; Qian, X.; et al. Profiling surface proteins on individual exosomes using a proximity barcoding assay. *Nat. Commun.* **2019**, *10*, 1–10. [CrossRef]
5. Maas, S.L.N.; Breakefield, X.O.; Weaver, A.M. Extracellular Vesicles: Unique Intercellular Delivery Vehicles. *Trends Cell Biol.* **2017**, *27*, 172–188. [CrossRef]
6. Raposo, G.; Stahl, P.D. Extracellular vesicles: A new communication paradigm? *Nat. Rev. Mol. Cell Biol.* **2019**, *20*, 509–510. [CrossRef]
7. Colombo, M.; Moita, C.; Van Niel, G.; Kowal, J.; Vigneron, J.; Benaroch, P.; Manel, N.; Moita, L.F.; Théry, C.; Raposo, G. Analysis of ESCRT functions in exosome biogenesis, composition and secretion highlights the heterogeneity of extracellular vesicles. *J. Cell Sci.* **2013**, *126*, 5553–5565. [CrossRef]
8. Palma, J.; Yaddanapudi, S.C.; Pigati, L.; Havens, M.A.; Jeong, S.; Weiner, G.A.; Weimer, K.M.E.; Stern, B.; Hastings, M.L.; Duelli, D.M. MicroRNAs are exported from malignant cells in customized particles. *Nucleic Acids Res.* **2012**, *40*, 9125–9138. [CrossRef]

9. Villarroya-Beltri, C.; Baixauli, F.; Gutiérrez-Vázquez, C.; Sánchez-Madrid, F.; Mittelbrunn, M. Sorting it out: Regulation of exosome loading. *Semin. Cancer Biol.* **2014**, *28*, 3–13. [CrossRef]
10. Leidal, A.M.; Huang, H.H.; Marsh, T.; Solvik, T.; Zhang, D.; Ye, J.; Kai, F.B.; Goldsmith, J.; Liu, J.Y.; Huang, Y.H.; et al. The LC3-conjugation machinery specifies the loading of RNA-binding proteins into extracellular vesicles. *Nat. Cell Biol.* **2020**, *22*, 187–199. [CrossRef]
11. Koppers-Lalic, D.; Hackenberg, M.; Bijnsdorp, I.V.; van Eijndhoven, M.A.J.; Sadek, P.; Sie, D.; Zini, N.; Middeldorp, J.M.; Ylstra, B.; de Menezes, R.X.; et al. Nontemplated nucleotide additions distinguish the small RNA composition in cells from exosomes. *Cell Rep.* **2014**, *8*, 1649–1658. [CrossRef] [PubMed]
12. Jeppesen, D.K.; Fenix, A.M.; Franklin, J.L.; Higginbotham, J.N.; Zhang, Q.; Zimmerman, L.J.; Liebler, D.C.; Ping, J.; Liu, Q.; Evans, R.; et al. Reassessment of Exosome Composition. *Cell* **2019**, *177*, 428–445.e18. [CrossRef] [PubMed]
13. Mathieu, M.; Martin-Jaular, L.; Lavieu, G.; Théry, C. Specificities of secretion and uptake of exosomes and other extracellular vesicles for cell-to-cell communication. *Nat. Cell Biol.* **2019**, *21*, 9–17. [CrossRef] [PubMed]
14. El Andaloussi, S.; Mäger, I.; Breakefield, X.O.; Wood, M.J.A. Extracellular vesicles: Biology and emerging therapeutic opportunities. *Nat. Rev. Drug Discov.* **2013**, *12*, 347–357. [CrossRef] [PubMed]
15. Zhang, X.; Yuan, X.; Shi, H.; Wu, L.; Qian, H.; Xu, W. Exosomes in cancer: Small particle, big player. *J. Hematol. Oncol.* **2015**, *8*, 83. [CrossRef]
16. Reclusa, P.; Taverna, S.; Pucci, M.; Durendez, E.; Calabuig, S.; Manca, P.; Serrano, M.J.; Sober, L.; Pauwels, P.; Russo, A.; et al. Exosomes as diagnostic and predictive biomarkers in lung cancer. *J. Thorac. Dis.* **2017**, *9*, S1373–S1382. [CrossRef]
17. Liu, Y.; Li, D.; Liu, Z.; Zhou, Y.; Chu, D.; Li, X.; Jiang, X.; Hou, D.; Chen, X.; Chen, Y.; et al. Targeted exosome-mediated delivery of opioid receptor Mu siRNA for the treatment of morphine relapse. *Sci. Rep.* **2015**, *5*, 17543. [CrossRef]
18. Yong, T.; Zhang, X.; Bie, N.; Zhang, H.; Zhang, X.; Li, F.; Hakeem, A.; Hu, J.; Gan, L.; Santos, H.A.; et al. Tumor exosome-based nanoparticles are efficient drug carriers for chemotherapy. *Nat. Commun.* **2019**, *10*, 3838. [CrossRef]
19. Haderk, F.; Hanna, B.; Richter, K.; Schnölzer, M.; Zenz, T.; Stilgenbauer, S.; Lichter, P.; Seiffert, M. Extracellular vesicles in chronic lymphocytic leukemia. *Leuk. Lymphoma* **2013**, *54*, 1826–1830. [CrossRef]
20. Böing, A.N.; van der Pol, E.; Grootemaat, A.E.; Coumans, F.A.W.; Sturk, A.; Nieuwland, R. Single-step isolation of extracellular vesicles by size-exclusion chromatography. *J. Extracell. Vesicles* **2014**, *3*. [CrossRef]
21. Lobb, R.J.; Becker, M.; Wen, S.W.; Wong, C.S.F.; Wiegmans, A.P.; Leimgruber, A.; Möller, A. Optimized exosome isolation protocol for cell culture supernatant and human plasma. *J. Extracell. Vesicles* **2015**, *4*, 27031. [CrossRef] [PubMed]
22. Stranska, R.; Gysbrechts, L.; Wouters, J.; Vermeersch, P.; Bloch, K.; Dierickx, D.; Andrei, G.; Snoeck, R. Comparison of membrane affinity-based method with size-exclusion chromatography for isolation of exosome-like vesicles from human plasma. *J. Transl. Med.* **2018**, *16*, 1–9. [CrossRef] [PubMed]
23. Bichi, R.; Shinton, S.A.; Martin, E.S.; Koval, A.; Calin, G.A.; Cesari, R.; Russo, G.; Hardy, R.R.; Croce, C.M. Human chronic lymphocytic leukemia modeled in mouse by targeted TCL1 expression. *Proc. Natl. Acad. Sci. USA* **2002**, *99*, 6955–6960. [CrossRef] [PubMed]
24. Gorgun, G.; Ramsay, A.G.; Holderried, T.A.W.; Zahrieh, D.; Dieu, R.L.; Liu, F.; Quackenbush, J.; Croce, C.M.; Gribben, J.G. Eμ-TCL1 mice represent a model for immunotherapeutic reversal of chronic lymphocytic leukemia-induced T-cell dysfunction. *Proc. Natl. Acad. Sci. USA* **2009**, *106*, 6250–6255. [CrossRef] [PubMed]
25. Gámez-Valero, A.; Monguió-Tortajada, M.; Carreras-Planella, L.; Franquesa, M.; Beyer, K.; Borràs, F.E. Size-Exclusion Chromatography-based isolation minimally alters Extracellular Vesicles' characteristics compared to precipitating agents. *Sci. Rep.* **2016**, *6*, 33641. [CrossRef] [PubMed]
26. Théry, C.; Witwer, K.W.; Aikawa, E.; Alcaraz, M.J.; Anderson, J.D.; Andriantsitohaina, R.; Antoniou, A.; Arab, T.; Archer, F.; Atkin-Smith, G.K.; et al. Minimal information for studies of extracellular vesicles 2018 (MISEV2018): A position statement of the International Society for Extracellular Vesicles and update of the MISEV2014 guidelines. *J. Extracell. Vesicles* **2018**, *7*, 1535750. [CrossRef]
27. Sana, B.; Johnson, E.; Sheah, K.; Poh, C.L.; Lim, S. Iron-based ferritin nanocore as a contrast agent. *Biointerphases* **2010**, *5*, FA48–FA52. [CrossRef]

28. Falvo, E.; Tremante, E.; Arcovito, A.; Papi, M.; Elad, N.; Boffi, A.; Morea, V.; Conti, G.; Toffoli, G.; Fracasso, G.; et al. Improved Doxorubicin Encapsulation and Pharmacokinetics of Ferritin–Fusion Protein Nanocarriers Bearing Proline, Serine, and Alanine Elements. *Biomacromolecules* **2015**, *17*, 514–522. [CrossRef]
29. Fleming, V.; Hu, X.; Weller, C.; Weber, R.; Groth, C.; Riester, Z.; Hüser, L.; Sun, Q.; Nagibin, V.; Kirschning, C.; et al. Melanoma extracellular vesicles generate immunosuppressive myeloid cells by upregulating PD-L1 via TLR4 signaling. *Cancer Res.* **2019**, *79*, 4715–4728. [CrossRef]
30. Gabrusiewicz, K.; Li, X.; Wei, J.; Hashimoto, Y.; Marisetty, A.L.; Ott, M.; Wang, F.; Hawke, D.; Yu, J.; Healy, L.M.; et al. Glioblastoma stem cell-derived exosomes induce M2 macrophages and PD-L1 expression on human monocytes. *OncoImmunology* **2018**, *7*, e1412909. [CrossRef]
31. Haderk, F.; Schulz, R.; Iskar, M.; Cid, L.L.; Worst, T.; Willmund, K.V.; Schulz, A.; Warnken, U.; Seiler, J.; Benner, A.; et al. Tumor-derived exosomes modulate PD-L1 expression in monocytes. *Sci. Immunol.* **2017**, *2*, 1–12. [CrossRef] [PubMed]
32. Gurunathan, S.; Kang, M.H.; Jeyaraj, M.; Qasim, M.; Kim, J.H. Review of the Isolation, Characterization, Biological Function, and Multifarious Therapeutic Approaches of Exosomes. *Cells* **2019**, *8*, 307. [CrossRef] [PubMed]
33. Gardiner, C.; Di Vizio, D.; Sahoo, S.; Théry, C.; Witwer, K.M.; Wauben, M.; Hill, A.F. Techniques used for the isolation and characterization of extracellular vesicles: Results of a worldwide survey. *J. Extracell. Vesicles* **2016**, *5*, 32945. [CrossRef] [PubMed]
34. Yuana, Y.; Levels, J.; Grootemaat, A.; Sturk, A.; Nieuwland, R. Co-isolation of extracellular vesicles and high-density lipoproteins using density gradient ultracentrifugation. *J. Extracell. Vesicles* **2014**, *3*. [CrossRef] [PubMed]
35. Karimi, N.; Cvjetkovic, A.; Jang, C.J.; Crescitelli, S.; Feizi, M.A.H.; Nieuwland, R.; Lötvall, J.; Lässer, C. Detailed analysis of the plasma extracellular vesicle proteome after separation from lipoproteins. *Cell. Mol. Life Sci.* **2018**, *75*, 2873–2886. [CrossRef]
36. Brennan, K.; Martin, K.; FitzGerald, S.P.; O'Sullivan, J.; Wu, Y.; Blanco, A.; Richardson, C.; Mc Gee, M.M. A comparison of methods for the isolation and separation of extracellular vesicles from protein and lipid particles in human serum. *Sci. Rep.* **2020**, *10*. [CrossRef]
37. Takov, K.; Yellon, D.M.; Davidson, S. Comparison of small extracellular vesicles isolated from plasma by ultracentrifugation or size-exclusion chromatography: Yield, purity and functional potential. *J. Extracell. Vesicles* **2018**, *8*, 1560809. [CrossRef]
38. Raposo, G.; Nijman, H.W.; Stoorvogel, W.; Liejendekker, R.; Harding, C.V.; Melief, C.K.; Geuze, H.J. B Lymphocytes Secrete Antigen-Presenting Vesicles. *J. Exp. Med.* **1996**, *1*, 1161–1172. [CrossRef]
39. André, F.; Chaput, N.; Schartz, N.E.C.; Flament, C.; Aubert, N.; Bernard, J.; Lemonnier, F.; Raposo, G.; Escudier, B.; Hsu, D.-H.; et al. Exosomes as Potent Cell-Free Peptide-Based Vaccine. I. Dendritic Cell-Derived Exosomes Transfer Functional MHC Class I/Peptide Complexes to Dendritic Cells. *J. Immunol.* **2004**, *172*, 2126–2136.
40. Möst, J.; Schwaeble, W.; Drach, J.; Sommerauer, A.; Dierich, M.P. Regulation of the Expression of ICAM-1 on Human Monocytes and Monocytic Tumor Cell Lines. *J. Immunol.* **1992**, *148*, 1635–1642.
41. Fujihara, M.; Ikebuchi, K.; Yamaguchi, M.; Abe, H.; Niwa, K.; Sekiguchi, S. Effects of Liposome-Encapsulated Hemoglobin on Phorbol Ester-Induced Superoxide Production and Expression of Costimulatory Molecules by Monocytes in Vitro. *Artif. Cells Blood Substit. Immobil. Biotechnol.* **1998**, *26*, 487–495. [CrossRef] [PubMed]
42. McClanahan, F.; Riches, J.C.; Miller, S.; Day, W.P.; Kotsiou, E.; Neuberg, D.; Croce, C.M.; Capasso, M.; Gribben, J.G. Mechanisms of PD-L1/PD-1–mediated CD8 T-cell dysfunction in the context of aging-related immune defects in the Eμ-TCL1 CLL mouse model. *Blood* **2015**, *126*, 212–221. [CrossRef] [PubMed]
43. Hanna, B.; McClanahan, F.; Yazdanparast, H.; Zaborsky, N.; Kalter, V.; Rössner, P.M.; Benner, A.; Dürr, C.; Egle, A.; Gribben, J.G.; et al. Depletion of CLL-associated patrolling monocytes and macrophages controls disease development and repairs immune dysfunction in vivo. *Leukemia* **2016**, *30*, 570–579. [CrossRef] [PubMed]

5

Exploiting Manipulated Small Extracellular Vesicles to Subvert Immunosuppression at the Tumor Microenvironment through Mannose Receptor/CD206 Targeting

Maria Luisa Fiani *, Valeria Barreca, Massimo Sargiacomo, Flavia Ferrantelli, Francesco Manfredi and Maurizio Federico *

National Center for Global Health, Istituto Superiore di Sanità, 00161 Rome, Italy; barrecavaleria@gmail.com (V.B.); massimo.sargiacomo@iss.it (M.S.); flavia.ferrantelli@iss.it (F.F.); francesco.manfredi@iss.it (F.M.)

* Correspondence: maria.fiani@iss.it (M.L.F.); maurizio.federico@iss.it (M.F.);

Abstract: Immunosuppression at tumor microenvironment (TME) is one of the major obstacles to be overcome for an effective therapeutic intervention against solid tumors. Tumor-associated macrophages (TAMs) comprise a sub-population that plays multiple pro-tumoral roles in tumor development including general immunosuppression, which can be identified in terms of high expression of mannose receptor (MR or CD206). Immunosuppressive TAMs, like other macrophage sub-populations, display functional plasticity that allows them to be re-programmed to inflammatory macrophages. In order to mitigate immunosuppression at the TME, several efforts are ongoing to effectively re-educate pro-tumoral TAMs. Extracellular vesicles (EVs), released by both normal and tumor cells types, are emerging as key mediators of the cell to cell communication and have been shown to have a role in the modulation of immune responses in the TME. Recent studies demonstrated the enrichment of high mannose glycans on the surface of small EVs (sEVs), a subtype of EVs of endosomal origin of 30–150 nm in diameter. This characteristic renders sEVs an ideal tool for the delivery of therapeutic molecules into MR/CD206-expressing TAMs. In this review, we report the most recent literature data highlighting the critical role of TAMs in tumor development, as well as the experimental evidences that has emerged from the biochemical characterization of sEV membranes. In addition, we propose an original way to target immunosuppressive TAMs at the TME by endogenously engineered sEVs for a new therapeutic approach against solid tumors.

Keywords: tumor-associated macrophages; tumor microenvironment; macrophage polarization; mannose receptor; exosomes; extracellular vesicles; HIV-1 Nef

---

1. Introduction

Both immunosuppression and genetic escape are formidable weapons through which tumors can elude host immune surveillance. Solid tumors develop in a quite complex context, referred to as tumor microenvironment (TME) [1,2], which is composed of both cellular and non-cellular elements, usually resulting in an immunosuppressive behavior. Counteracting such a general effect would favor both spontaneous and therapeutic anti-tumor immunity, hence critically contributing to control tumor cell growth. Therefore, subverting TME immunosuppression represents a major goal for anticancer immunotherapies.

Both normal and tumor cells constitutively release membrane-bilayered vesicles, commonly referred to as extracellular vesicles (EVs) [3,4]. They differ in the mechanisms of biogenesis and secretion, giving rise to the generation of a heterogeneous population of vesicles with different sizes and contents [5,6], which include small EVs (sEVs) or exosomes and microvesicles or ectosomes. Exosomes are vesicles of 30–150 nm diameter generated by inward budding of endosomal membranes to form intraluminal vesicles that accumulate in intracellular organelles called multivesicular bodies (MVBs). MVBs ultimately fuse with the plasma membrane, thereby releasing intraluminal vesicles into the extracellular environment (Figure 1). On the contrary, ectosomes are 100–500 nm vesicles shed by direct budding from the plasma membrane [7,8]. Different types of EVs often show overlapping features that make difficult to obtain relatively pure preparations when purified from cell-conditioned media or biological fluids. In this review, we will use the term sEV to refer to EV types co-isolated by typical purification methods and exosomes to distinguish EV whose subcellular biogenesis strictly derives from multivesicular bodies/endosomes [8].

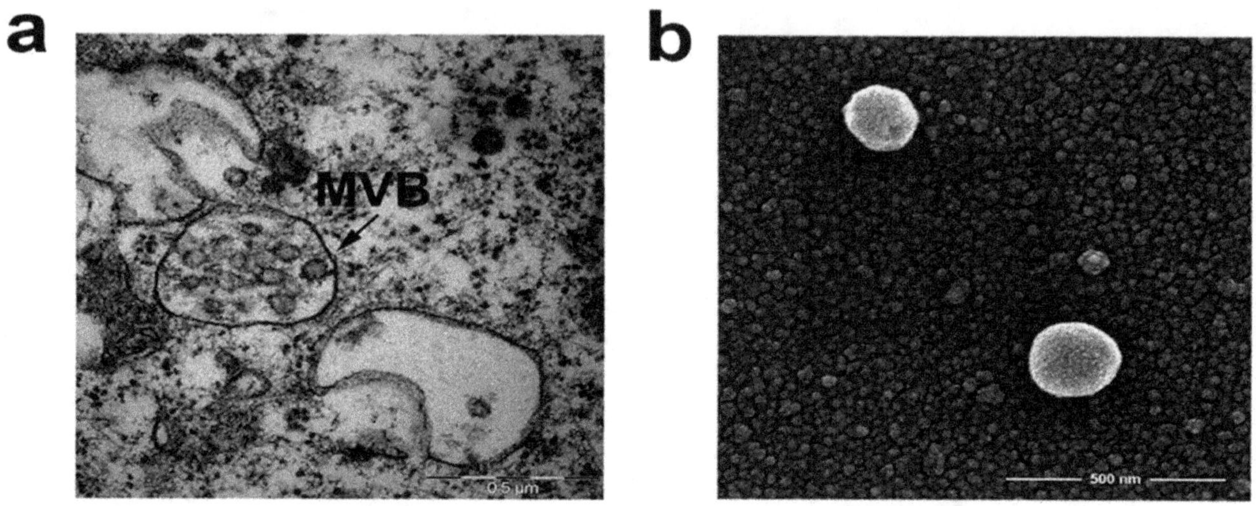

Figure 1. Electron microscopy of multivesicular bodies (MVB) and small extracellular vesicles (sEVs) (**a**) TEM micrograph of multivesicular bodies with intraluminal vesicles in Mel501, a melanoma cell line (**b**) SEM (Scanning Electron Microscope). Micrograph of sEVs purified from conditioned medium of Mel501 cells by differential centrifugations. Courtesy of Francesca Iosi and the Microscopy Area of the ISS Core Facilities.

SEVs carry a complex cargo of nucleic acids, proteins, and lipids that largely reflects the characteristics and the functional state of the cells they originate from, and that will be delivered to neighboring or distant cells [9,10]. As a result, the functions of those recipient cells will be modulated by sEVs in a manner that is strictly dependent on the nature of producer cells, making sEVs central players in intercellular communication and reprogramming of target cells [11]. Ectosomes generation is a much less known process that requires the accumulation of their cargo at the cytosolic surface of specific plasma membrane microdomains [7,12].

SEVs-mediated transfer of molecular and genetic material from one cell to another, either locally or at long distance, is a key contributor to the mechanisms of intercellular communication involved in various physiological and pathological conditions [13–15]. Moreover, for these reasons, sEVs are now considered powerful tools for clinical applications, including advanced diagnostics, therapeutics, and regenerative medicine [16–19].

The molecular composition of sEVs is determined by the cell type of origin as well as by the intracellular pathway followed en route to their release into the extracellular space [8,20,21]. This heterogeneity confers to sEVs distinct properties, such as tropism to certain organs, and uptake by specific cell types. In the case of tumor-derived sEVs, these events often lead to the impairment of immune responses at TME [22], also favoring pre-metastatic niche formation and metastasis [23,24].

In tumor cells, sEV biogenesis and ultimately sEV composition is a complex and regulated process, which involves many different molecules associated with the sEV biogenesis pathway [4,25].

Whatever the cell type of origin, sEVs can be characterized, although not exclusively, by the presence of different types of cell surface proteins, such as tetraspanins, (i.e., CD9, CD81, CD63), ESCRT (endosomal sorting complex required for transport) proteins (Alix and TSG101), integrins, RNA, DNA, lipids such as ceramide and the atypical phospholipid, lysobisphosphatidic acid (LBPA) [26], and oligosaccharides [27].

TME exerts a key influence on tumor cells, and the resulting sEVs, responsible for proteins and genetic material transfer from primary tumor cells, play a crucial role in metastatic colonization and in the formation of the pre-metastatic niche, driving recipient cells to acquire a pro-tumorigenic phenotype [23,28]. The selective conditions present in TME, such as the generation of a hypoxic [29] and acidic environment [30], strongly influence sEV secretion by tumor cells, thus contributing to the malignant tumor phenotype. Furthermore, sEV membrane composition reflects TME changes and conceivably influence and control the different mechanisms of entry or interaction of sEVs with target cells supporting tumor growth [31,32].

Interestingly, it has been described that major players of immunosuppression at the TME, i.e., immunosuppressive tumor-associated macrophages (TAMs), express on their surface high levels of mannose receptor (MR, CD206) [33]. The MR is an endocytic receptor with a high affinity for high mannose oligosaccharides, glycans highly enriched on the surface of sEVs [34]. In this review, literature data regarding both TAM functions and the molecular structure of sEVs are reviewed. In addition, we propose an original way to exploit typical molecular signatures of both TAMs and sEVs to counteract the immunosuppression at the TME.

2. The Tumor Microenvironment

In solid tumors, cancer cells are embedded within a milieu that favors their proliferation and comprises both cellular and non-cellular components. Fibroblasts, endothelial cells, and essentially all types of immune cells are part of the TME [35,36]. Among non-cellular components, tumor-derived sEVs play a key role in immune suppression. TME composition can vary among different tumors, and between primary and metastatic neo-formations in the same patient, and is tightly associated with the clinical outcome of cancer patients.

TMEs can be categorized based on different criteria. In terms of abundance of tumor-infiltrating cytotoxic CD8+ T lymphocytes (CTLs), TMEs can be distinguished in either hot/inflamed, with the highest content of CTLs, or cold/desert, with a virtual absence of infiltrated CTLs [37]. TME core infiltrated by CTLs represents a favorable condition for an effective anticancer immune response, both spontaneous and induced by immunotherapeutic interventions.

TME is populated by different kinds of immune cells having immune suppressive actions. Among these are myeloid-derived suppressor cells, neutrophils, CD4+ Treg lymphocytes, and immunosuppressive M2-like TAMs [38]. These latter cells can represent up to 50% of the tumor mass, and play a key role in the immune evasion at TME by secreting proteases, angiogenic factors, and pro-tumoral products. The functional plasticity of TAMs modifies their phenotype and activity in response to a great number of microenvironmental stimuli, although the mechanisms that determine the different polarization states are still to be elucidated [39]. These different functional states often coexist and can significantly vary between different tumors [40,41]. TAMs can also dispose at the tumor margin, where they can interact with CTLs, thus inhibiting their infiltration towards tumor cells [42]. For all these reasons, immunosuppressive TAMs have been identified as a major cell target for novel designs of cancer immunotherapies focused on improving the overall anti-tumor immune response. A schematic representation of cells populating TEM is illustrated in Figure 2.

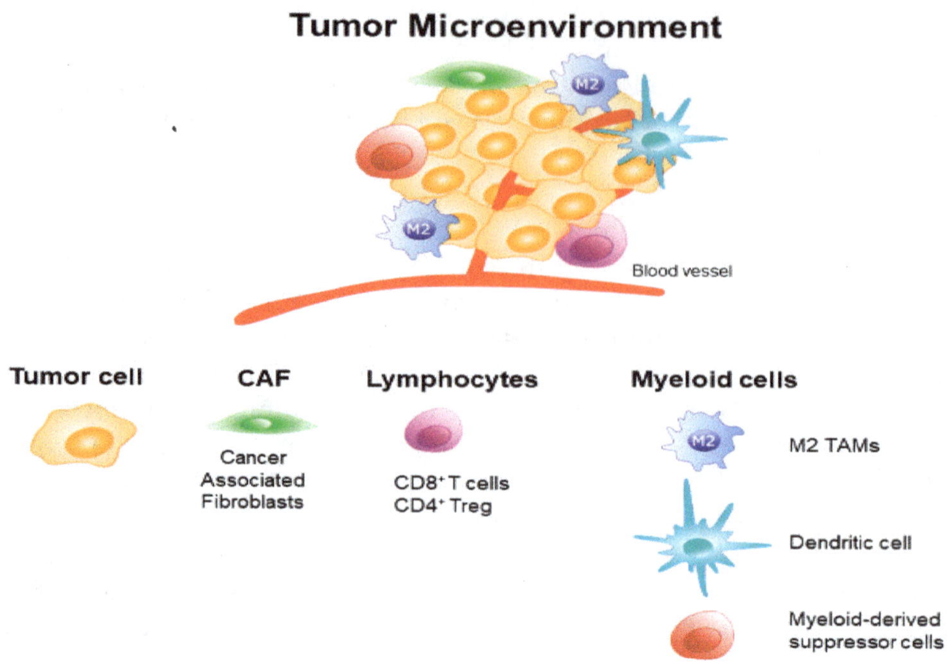

Figure 2. Schematic representation of cells populating the tumor microenvironment (TME).

3. The TAM-Mediated Immunosuppression at the TME

A large number of macrophages infiltrate solid tumors, thereby influencing several aspects of tumor development [43]. The most relevant effects include suppression of anticancer immunity, angiogenesis promotion, and support for metastasis. Macrophages are recruited at TME, in response to the secretion by tumor cells and other TME cell types, of a number of chemoattractant soluble factors, including vascular-endothelial growth factor A (VEGF-A) [44], chemokine ligand 2 (CCL2) [45], and colony-stimulating factor 1 (CSF-1) [46].

TME-populating macrophages can be schematically distinguished in M1- and M2-like macrophages. M1-like macrophages show both pro-inflammatory and immune-stimulatory properties, thus exerting an anti-tumor function. On the other hand, M2-like macrophages favor tumor angiogenesis and immunosuppression. Such a distinction, although useful from both therapeutic and diagnostic points of view, is now outdated, due to the identification of a large number of intermediate subclasses, i.e., up to 19 [47]. They have been identified through most recent transcriptomic techniques, e.g., single-cell mass cytometry and single-cell RNA sequencing [48–50], and in vivo represent a continuum of functional phenotypes with intermediates showing overlapping features.

TAMs can be characterized by the expression of different surface markers [47,51], distinct metabolic changes [52,53], and a broad transcriptional repertoire with the involvement of key transcription factors, which can be activated by the environmental signals received. In particular, members of the signal transducer and activator of transcription (STAT), peroxisome proliferator-activated receptors (PPARs), interferon regulatory factor (IRF), and nuclear transcription factor-κB (NF-κB) families are essential for macrophage polarization toward the M1 profile [41,54,55].

M2-like TAMs contribute to tumor angiogenesis by secreting soluble factors inducing endothelial cell proliferation, including VEGF-A, interleukin (IL)-1β, IL-6, tumor necrosis factor (TNF)α, CXCL8, and fibroblast growth factor (FGF)-2 [56]. In particular, the secretion of VEGF-A by perivascular TAMs can increase vascular permeability and access of tumor cells to peripheral blood circulation [57]. On the other hand, the production of proteases, e.g., matrix-metalloproteases, induces degradation of extracellular matrix and the consequent liberation of embedded soluble factors released by both cancer and stromal cells having pro-tumoral effects and favoring metastasization.

TAM-mediated immunosuppression at TME is essentially mediated by three concurrent mechanisms: (i) Release of soluble immunosuppressive factors, e.g., IL-10, CCL22, and transforming growth factor (TGF)-α as well as factors recruiting regulatory T cells (Treg) [58]; (ii) expression of ligands for lymphocyte suppressor factors PD-1 and CTLA-4, i.e., PDL-1 and CD80, as well as other checkpoint inhibitors with similar functions, including B7-H4, V-domain Ig suppressor of T cell activation (VISTA) [59], and vascular endothelial receptor (CLEVER) [60], and (iii) starving the TME of L-arginine, i.e., an essential factor for T-cell activity, through the release of arginase-1 [61]. Figure 3 illustrates the principal mechanisms of TAM mediated immunosuppression at the TME.

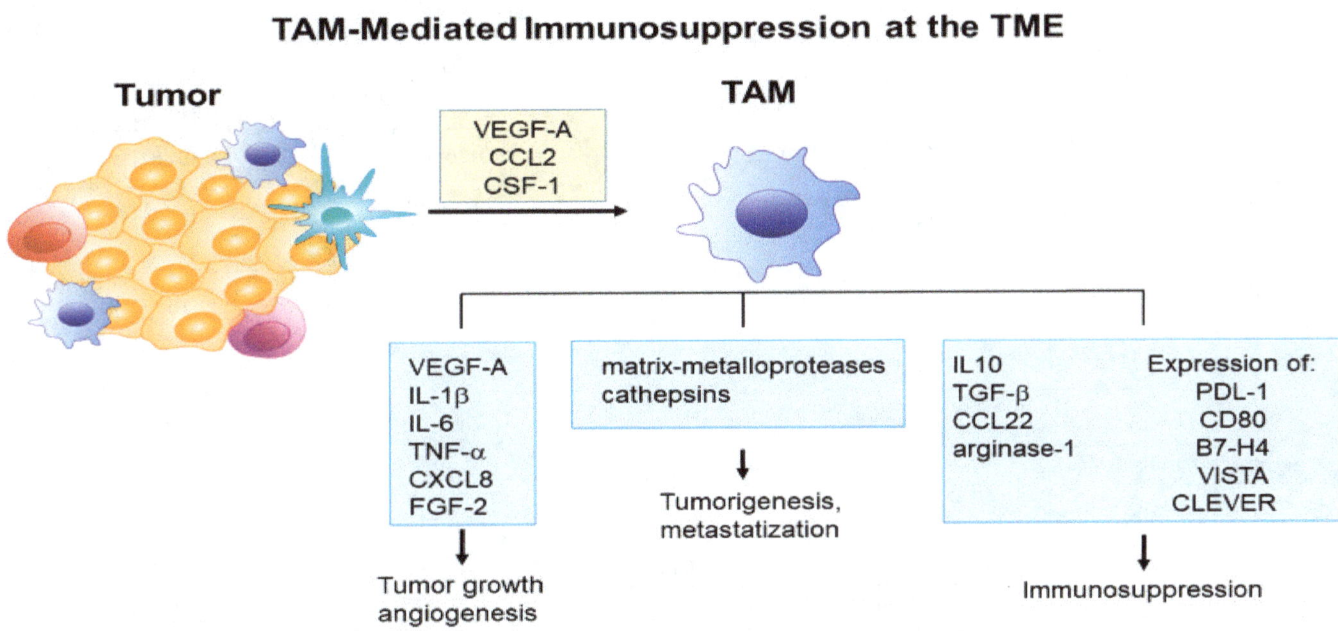

Figure 3. Principal mechanisms of tumor-associated macrophage (TAM) mediated immunosuppression at the tumor microenvironment (TME).

The multiple immunosuppressive signals at play within the TME greatly reduce the efficacy of current immunotherapies. Therefore, new strategies to effectively reprogram the various immunosuppressive cell types at the TME are urgently needed.

4. Re-Programming of TAMs

Immunosuppressive TAMs represent a privileged therapeutic target for the treatment of solid tumors, especially in the case immune checkpoint blockers (ICBs) are used. Given the enormous therapeutic value of TAMs re-education towards M1-like macrophages to promote tumor regression, much attention has focused on effective strategies aimed at targeting TAMs, including the blockade of the M2 phenotype, enhanced activation of M1 macrophages and reprogramming of TAMs toward M1-like phenotype [58,62–65]. Many of these different approaches against immunosuppressive TAMs have been summarized in Table 1 and strategies directed at TAM reprogramming illustrated in Figure 4.

Table 1. Selected strategies to target tumor-associated macrophages (TAMs).

Mechanism of Action	Active Agent	Vehicle Carrier	Target	References
Depletion of M2 TAMs	Shiga toxins	Shigella Flexneri attenuated strain	TAMs	[66]
	Immunotoxins		TAMs Receptors	[67,68]
	Bisphosphonates (e.g., clodrolip, zoledronic acid)	Liposomes	TAMs, Kupffer cells	[69,70]
	Trabectedin		TAMs	[71]
	Tyrosine Kinase Inhibitors (e.g., Dasatinib, Bosutinib)		endothelial and myeloid cells in TEM, TAMs	[72,73]
Inhibition of circulating monocyte recruitment into tumor	CCR2 inhibitors; anti-CCR2/CCL2 blocking antibodies		TAMs CCR2	[45,74–76]
	Antagonists of CXCL12/CXCR4 axis		TAMs CXCR4	[77,78]
	anti-CSF-1R antibody		TAMs CSF-1R	[79,80]
	neutralizing CD11b antibody		CD11b on Myeloid Cells	[81,82]
Blockade of M2 Phenotype	Tyrosine kinase inhibitors or drugs blocking STAT3		TAMs STAT3	[83,84]
	drugs blocking STAT6		TAMs STAT6	[85]
Enhanced Activation of M1 Macrophages	Th1 cytokines like IFN-γ		TAMs STAT1 stimulation	[86,87]
	metformin		TAMs AMPKα1 stimulation	[88]
	toll-like receptor agonists, CpG-ODNs; PI3Kγ deletion		TAMs NF-κB stimulation	[89–91]
Reprogramming TAMs Toward M1-Like Phenotype	mRNAs; miRNA	Targeted Nanocarriers	TAMs	[92,93]
	siRNA	Different types of Nanoparticles	TAMs	[94–96]
	anti-CD40 antibody		TAMs CD40	[97–99]
	anti-MARCO antibody		TAMs MARCO	[100]
	gefitinib/vorinostat	Trastuzumab-modified Mannosylated Liposomes	TAMs MR	[101]
	Drug free	Mannosylated Liposomes	TAMs MR	[102]
	RP-182 Peptide		TAMs MR	[103]

AMPKα1, AMP-activated protein kinase; CCL2, C–C chemokine ligand 2; CCR2, C–C chemokine receptor type 2; CSF-1, Colony-Stimulating Factor 1; CSF-1R, colony-stimulating factor 1 receptor; CXCL12, C–X–C motif chemokine 12; CXCR4, C-X-C chemokine receptor type 4; CpG-ODN, unmethylated cytosineguanine (CpG) oligodeoxynucleotides; IFN-γ, interferon gamma; MARCO, macrophage receptor with collagenous structure; MR, mannose receptor/CD206; NF-κB, nuclear factor kappa B; PI3Kγ, phosphoinositide 3-kinase; STAT, signal transducer and activator of transcription.

Promising results have been obtained with direct activation of M1-like macrophages by Th1 cytokines like IFN-γ [87], and by targeting toll-like receptors (TLR) and/or CD40 with agonists and monoclonal antibodies [97–99]. However, the onset of systemic inflammation limited the therapeutic efficacy of these approaches in vivo, and additional investigations are ongoing to circumvent this hurdle. In any case, considering the functional plasticity of macrophages, re-educating M2- versus an M1-like macrophage phenotype currently appears the most attractive therapeutic option.

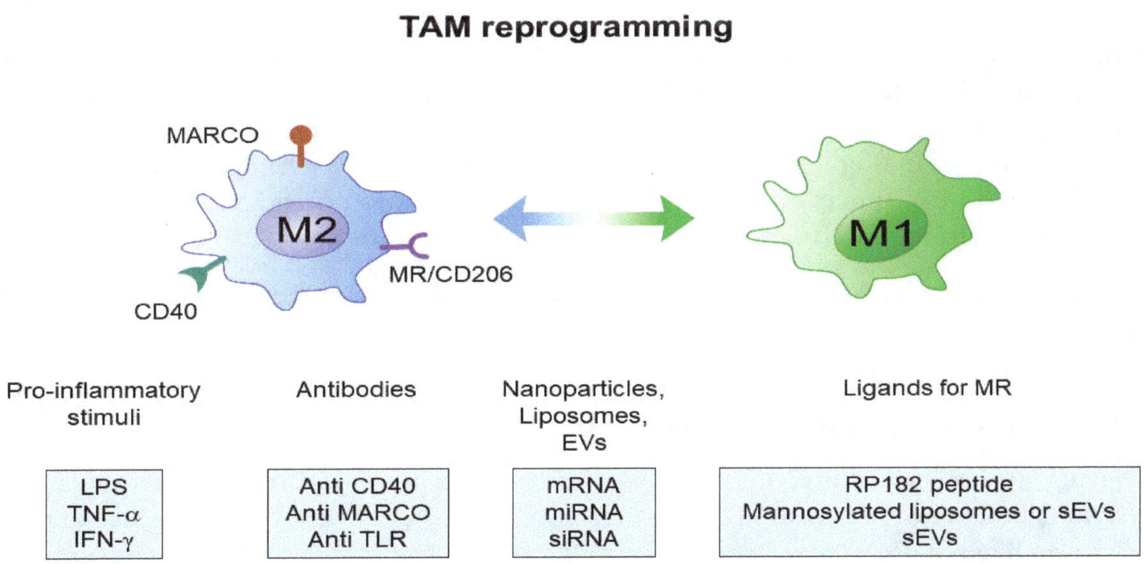

Figure 4. Scheme of current strategies for tumor-associated macrophage (TAM) reprogramming.

A major hindrance to effective targeting of immunosuppressive TAMs is represented by the scarcity of specific protein markers expressed on M2 macrophages. Some potential targets, whose expression also correlates with poor prognosis, have been investigated. Among them, the selective targeting of MARCO (macrophage receptor with collagenous structure) with monoclonal antibodies has been recently used to promote a switch to an M1-activated phenotype [100].

Increasing evidence suggests that a valuable alternative is represented by targeting the MR, which is highly expressed on M2, but not M1 macrophages [92,103–105].

5. The Mannose Receptor in M1 Polarization

MR is expressed on TAMs where is a prototypical marker of M2-type activation. It is also expressed on the surface of immature dendritic cells (DCs), liver sinusoidal endothelial cells, and other tissue macrophages. Earlier studies have demonstrated that MR expression is strongly down-regulated by IFN-γ [106], and upregulated by interleukin-4 (IL-4) [107]. The MR is a 175 kDa Type I integral membrane protein that belongs to the family of C-type lectin receptors and binds glycoconjugates terminated in mannose, fucose, or *N*-acetil-β-D-glucosamine (GlcNAc) in a calcium-dependent manner [108–110]. The receptor contains three distinct extracellular domains, i.e., an N-terminal cysteine-rich domain (CR) that binds sulfated carbohydrates, a fibronectin type II domain (FNII) that binds collagen, and eight tandem C-type lectin carbohydrate-recognition domains (CRDs) [111,112]. CRDs have only weak affinity affinities for single sugars, and several CRDs need to be clustered to achieve high-affinity binding to oligosaccharides. This clustering allows for the internalization of mannosylated proteins and other exogenous molecules, including allergens and microbial products.

MR is a highly effective clathrin-dependent endocytic receptor that constantly recycles between the plasma membrane and the early endosomal compartment [111]. Most part of MRs is intracellular, while only ~15% of the cellular pool can be found on the cell surface. Like other members of the C-type lectin receptor family, the MR undergoes conformational changes upon ligand binding or as pH decreases in intracellular compartments [111,113]. Once acidification takes place in the endosomal compartment, the MR dissociates from its ligands, and the empty receptor recycles back to the plasma membrane.

Several approaches have been adopted to target the MR and selectively deliver therapeutic nanoparticles. Among these, drug-free mannosylated liposomes have been shown to induce effective anti-tumor activity by enhancing the expression ratio of CD86/MR [102]. In another study, mannosylated nanoparticles suitable for intracellular delivery of drug carriers have been shown to selectively target with high specificity MR expressing macrophages [104].

MR conformational changes that occur upon ligand binding have been recently exploited to target M2 macrophages and induce reprogramming toward M1 phenotypes. For instance, precision targeting with short peptides showed some potential for the intracellular delivery of therapeutically relevant molecules [114]. In addition, a very recent report showed that direct binding of MR with a synthetic peptide (RP-182), i.e., an analogue of naturally occurring antimicrobial peptides, activates phagocytosis and autophagy in M2-like macrophages, reverting these cells into an anti-tumor M1-like phenotype with increased M1 cytokine production and phagocytosis of cancer cells [103].

On the other hand, also cell-secreted sEVs can be considered attractive candidates to specifically target M2-like macrophages via the MR, since they expose high mannose and other classes of N-linked oligosaccharides on their surface [115,116].

6. Extracellular Vesicles for Anti-Tumor Therapy

Pioneering studies have shown that sEVs secreted by DCs pulsed with cancer peptides successfully eradicate established tumors in mice [117]. Furthermore, tumor-derived EVs are a source of neoantigens that, once internalized by DCs, could cross-prime CD8+ T cells and lead to tumor rejection [118]. Since these early studies, the field of sEVs-based cancer therapeutics has attracted many efforts, and sEVs have emerged as promising tools for targeted drug delivery. Despite a growing interest in these nanovesicles as natural carriers, there are still many open questions that need further investigation. For example, specific recognition by target cells is of fundamental importance for an effective delivery of bioactive molecules. EVs uptake may occur via receptor-mediated endocytosis or phagocytosis, or direct fusion with the plasma membrane. Some studies have pointed at integrins [24] and scavenging receptors [119] as mediators of EVs targeting, but current knowledge on this matter is rapidly evolving and has been recently comprehensively reviewed [5,7,120–122].

The different modes of sEV uptake may result in distinct localization and functional effects of the sEVs components, but it is still unknown whether a specific route of entry is to be preferred for a successful transfer of EVs cargo. Thus, understanding through which mechanisms sEVs deliver their content into target cells is a central point that needs to be further elucidated. To the best of current knowledge, while either non-selective uptake or direct fusion with the plasma membrane of target cells seem to be the preferential mechanisms of bulk sEV incorporation (Figure 5) [123], alternative and more specific routes of uptake may depend on the characteristics of surface components of both sEV and target cells.

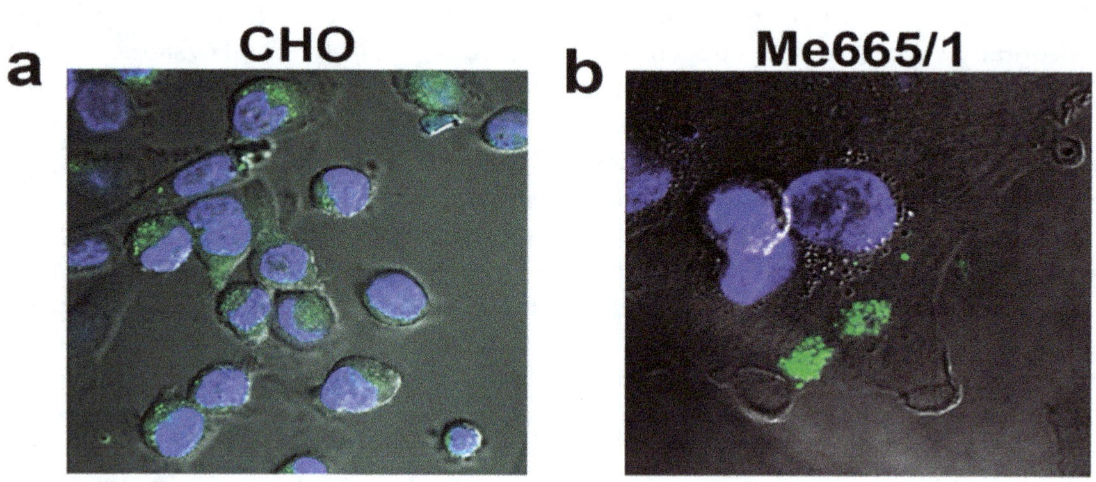

Figure 5. Fluorescent sEVs uptake by different cell lines. Confocal fluorescence microscopy images of green fluorescent sEVs derived from melanoma Me665/1 cells transferred on (**a**) CHO cells and (**b**) Me665/1 cells in nonspecific conditions [66]. Blue-fluorescent nuclei are stained with DAPI.

Tumor-derived EVs can be efficiently taken up by DCs for antigen processing and cross-presentation to tumor-specific CTLs [124]. Immature DCs (iDCs) internalize EVs more efficiently than mature DCs, whereas mature DCs retain more EVs on the cell surface [125]. The surface of iDCs harbors sugar-binding C-type lectin receptors (CLRs) [126,127], which is a characteristic shared with M2-like macrophages.

Glycomic studies conducted to date demonstrate that surface glycoprofiles of sEVs contain high amounts of mannose and other classes of N-linked oligosaccharides [27,34,128,129] making them suitable ligands for the MR. Recently, it has also been shown that mannose-modified serum sEVs display elevated uptake by murine DCs [116]. This evidence, together with a number of studies showing that mannosylation of both liposomes [102] and synthetic nanocarriers [130,131] enhance cellular uptake by M2 macrophages, inducing stimulation and polarization of macrophages toward the M1 phenotype, point to sEVs as powerful ligands of M2 macrophages. Furthermore, the affinity of EVs membrane components (i.e., proteins, lipids, and glycans) for certain tissues greatly affects biodistribution of EVs in vivo, thus, encouraging studies aimed at altering the surface of EVs to improve targeting to selected organs. Many different strategies have been adopted, but there is now accumulating evidence that carbohydrates on the vesicle surface participate in the recognition and uptake of EVs by phagocytes. Interestingly, manipulation of surface glycans on EVs either by removal of sialic acid [132] or by treatment with *N*-glycosidases [133,134] alters the uptake capacity of different cells. In the first case, the change also affects the in vivo biodistribution of sEVs showing accumulation of desialylated sEVs in the lungs. Altogether, these results point to the importance of *N*-glycosylation in cellular uptake, but since different cell types respond differently to glycosylation changes appears evident that the cell to be targeted, with its endowment of specific protein receptors, represent the cornerstone of receptor ligand recognition.

7. Molecular Basis of TAM Re-Programming by Engineered sEVs

The subversion of TAM-mediated immunosuppression at TME through macrophage transcriptional reprogramming represents a quite attractive option for anticancer combined immunotherapy. Like normal cells, tumor cells also constitutively release sEVs, which interact primarily with TME cell constituents. SEVs, as naturally occurring vesicles, have a low intrinsic immunogenic profile, are able to avoid, at least in part, the degradative pathway and possess the ability to overcome the blood-brain barrier. Consequently, they have emerged as an important means to deliver therapeutic agents. In this context, engineering tumor-derived sEVs with molecules inducing an inflammatory-like macrophage phenotype would be instrumental in alleviating the immunosuppression at TME. Over the past few years, several engineering strategies have been devised to manipulate tumor-derived sEVs in order to induce cellular and innate immunity. SEV engineering can be carried out either at the level of producer cells or directly on purified sEVs. The different approaches used mostly depend on cargo properties, such as hydrophilicity, hydrophobicity, and molecular weight, as each method has a different loading capacity. For example, chemicophysical methods like electroporation are widely used for loading relatively large molecules. such as siRNA or miRNA into sEVs [19]. Particularly for cancer research, the observation that exosomal miRNAs effectively engage target mRNA and suppress gene expression in recipient cells has been heavily favored. Due to the availability of various cellular engineering methods, different types of RNAs that are released via sEVs have also been exploited and recently reviewed [11].

During the past decade, many efforts have been devoted to transfect sEV-producing cells with plasmids encoding protein sequences that, once uploaded in the nanovesicles, are able to alter the phenotype of target cells [135–138]. However, a method to target M2 macrophages with sEVs capable of inducing their reprogramming to M1 anti-tumor phenotype is not yet available. One major hurdle in EVs research for an effective therapeutic application is represented by the lack of optimized isolation, characterization and quantification procedures. Various purification techniques, such as differential ultracentrifugation, density gradients, precipitation, filtration, size exclusion chromatography, and immunoisolation are currently used to obtain less heterogeneous sEVs preparation; however, quantification often relies on the total protein content of EVs. Fluorescent labeling techniques are more accurate and have the main advantage that track EVs in vivo [139,140]. The different approaches for TAMs reprogramming by engineered EVs are schematically illustrated in Figure 6.

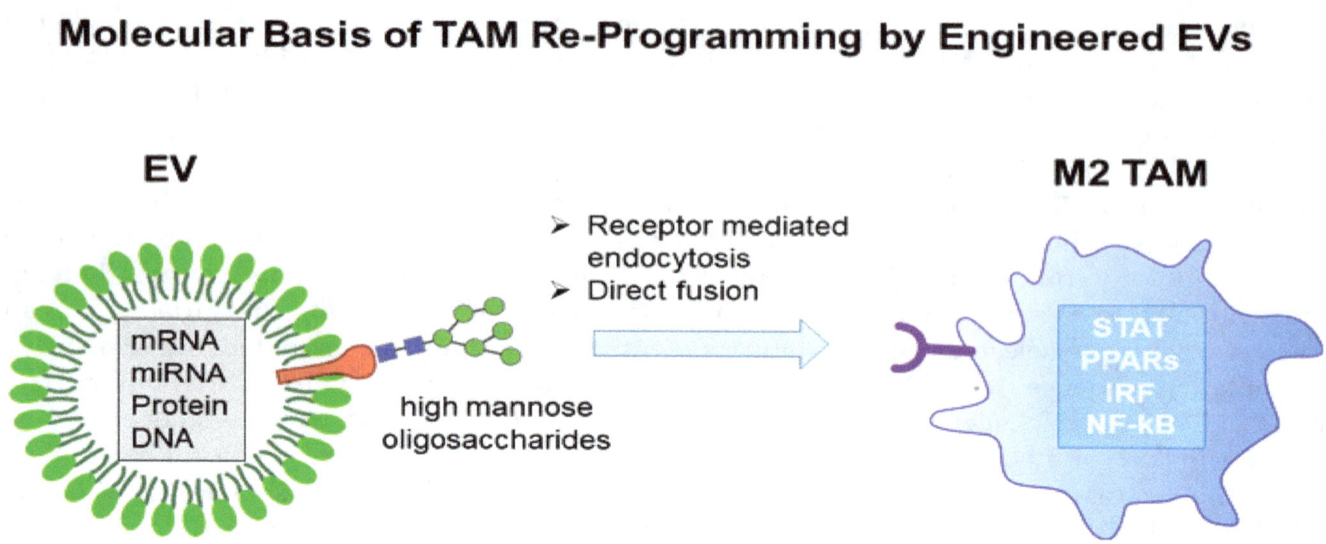

Figure 6. TAMs reprogramming by engineered EVs.

8. HIV-1 Nef Protein as Effector of TAM Reprogramming

A plausible candidate for reprogramming M2 macrophages is the Human immunodeficiency virus (HIV)-1 Negative Regulatory Factor (Nef) protein [141], a 27 kilodalton (kDa) scaffold protein, which lacks enzymatic activities. After synthesis at free ribosomes, Nef reaches both intracellular and plasma membranes with which it tightly interacts through its N-terminal myristoylation. Nef acts as a scaffold/adaptor element in triggering activation of signal transducing molecules like p21 PAK-2, NF-κB, STATs, ERK1/2, Vav, and Src family kinases. In most cases, signal activation occurs upon Nef association with lipid raft microdomains at cell membranes [142–144]. The fact that also sEV membranes are enriched in lipid raft microdomains explains why Nef can be found in EVs [145–149].

Cumulate literature data demonstrate that the presence of Nef inside macrophages induces a strong pro-inflammatory response. In particular, Nef switches on the transcription of many inflammatory genes, as well as the release of inflammatory factors like CCL3, CCL4, IL-1β, IL-6, TNF-α [150,151], and interferon gamma (IFN)-γ [152]. This potent pro-inflammatory response is mediated by the activation of several signal transduction molecules, including STAT-1, 2, and 3, NF-κB, JNK, ERK1/2, and MAPK [153,154]. The inflammatory effects of Nef on macrophages depends on four glutamate-acidic cluster domain located at 62–65 amino acid position [155].

Data from many independent investigation groups strongly support the idea that Nef associates with EVs at low levels [156,157]. Conversely, we identified a Nef mutant incorporating in sEVs/EVs at quite high levels [158]. This Nef mutant (referred to as Nef^{mut}) is defective for the most part of Nef functions, including down-regulation of cell membrane receptors, Nef-associated kinase (NAK) activation, an increase of HIV expression [159]. Nevertheless, it maintains an unaltered acidic cluster domain that correlates with the induction of cell activation in antigen-presenting cells when it is delivered by nanovesicles [160].

Considering this evidence, one may hypothesize that the delivery of Nef^{mut}-engineered EVs inside M2-TAMs would be instrumental in re-educating macrophages at TME from an M2-like to an M1-like phenotype. In the case of solid tumors, this design could be applied through a quite simple strategy, i.e., tumor cell engineering for Nef^{mut} expression through retroviral vector-mediated transduction. In this way, only actively replicating cells at the TME, hence preferably tumor cells, are expected to be transduced. Nef-engineered tumor-derived sEVs may diffuse within TME, thereby preferentially entering MR positive TAMs in view of the mannose expressed on the sEV surface. Once internalized by macrophages, Nef^{mut} might switch on intracellular signals—ultimately leading to the release of pro-inflammatory factors. These factors might act in an autocrine/paracrine loop to induce the reprogramming of macrophage transcriptional profile toward the M1-like phenotype (Figure 7).

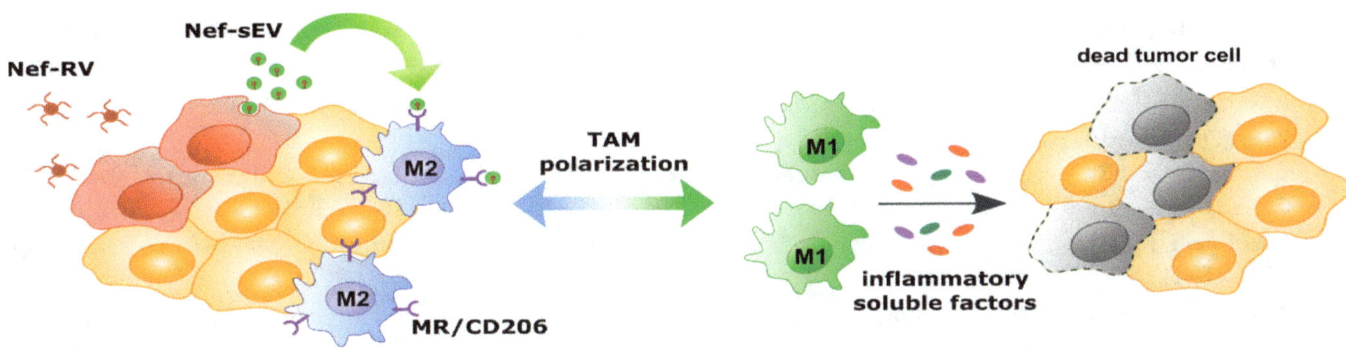

Figure 7. Scheme of the proposed mechanism for TAM reprogramming. Tumor cells are transfected with retroviral vectors expressing Nef^{mut} (Nef-RV). Nef-engineered sEVs (Nef-sEV) are then released into the TME infiltrated with M2 like macrophages expressing the mannose receptor (MR/CD206). MR mediated uptake of Nef-Sev might induce polarization of M2 into M1 like macrophages—ultimately leading to the release of pro-inflammatory factors.

Eventually, this mechanism, which essentially hijacks the sEV-mediated intercellular communication at the TME, is expected to alleviate the immune suppression at the TME, thereby favoring the action of anticancer adaptive immune responses.

Hopefully, once supported by experimental confirmation, this design would have a therapeutic utility in the battle against solid tumors.

9. Conclusions

Immunosuppression at TME protects cancer cells from both spontaneous and artificially generated host immune responses. Hence, subverting immunosuppression should be considered a priority for any anticancer immunotherapeutic strategy. Even if M2-like TAMs are major players, other cell types contribute to the general immunosuppression at TME, including CD4+ Tregs lymphocytes, myeloid-derived suppressor cells, and neutrophils [42]. Similar to macrophages, neutrophils can polarize in pro- and anti-tumor phenotypes, depending on the stimuli they receive at TME. Interestingly, it has been reported that IFN-γ can polarize neutrophils towards an anti-tumor phenotype [161]. Considering the quite high levels of IFN-γ transcripts induced by Nef in macrophages, the delivery of Nef^{mut}-engineered, tumor-derived sEVs to M2-like TAMs is expected to have paracrine anti-tumor effects also on neutrophils that populate the TME.

The strategy we propose is certainly only one of many potential new anti-tumor therapeutic approaches that the manipulation of sEVs/EVs can offer. The increase in knowledge of the sEV/EVs biology, mainly regarding the mechanisms of cell entry, will favor the implementation of new and more efficient therapeutic approaches against tumors and infectious diseases.

References

1. Fridman, W.H.; Pages, F.; Sautes-Fridman, C.; Galon, J. The immune contexture in human tumours: Impact on clinical outcome. *Nat. Rev. Cancer* **2012**, *12*, 298–306. [CrossRef] [PubMed]
2. Quail, D.F.; Joyce, J.A. Microenvironmental regulation of tumor progression and metastasis. *Nat. Med.* **2013**, *19*, 1423–1437. [CrossRef] [PubMed]
3. van Niel, G.; D'Angelo, G.; Raposo, G. Shedding light on the cell biology of extracellular vesicles. *Nat. Rev. Mol. Cell Biol.* **2018**, *19*, 213–228. [CrossRef] [PubMed]
4. Kalluri, R.; LeBleu, V.S. The biology, function, and biomedical applications of exosomes. *Science* **2020**, *367*, eaau6977. [CrossRef]
5. Raposo, G.; Stahl, P.D. Extracellular vesicles: A new communication paradigm? *Nat. Rev. Mol. Cell Biol.* **2019**, *20*, 509–510. [CrossRef] [PubMed]
6. Colombo, M.; Raposo, G.; Théry, C. Biogenesis, Secretion, and Intercellular Interactions of Exosomes and Other Extracellular Vesicles. *Annu. Rev. Cell Dev. Biol.* **2014**, *30*, 255–289. [CrossRef]
7. Meldolesi, J. Exosomes and Ectosomes in Intercellular Communication. *Curr. Biol. CB* **2018**, *28*, R435–R444. [CrossRef]
8. Thery, C.; Witwer, K.W.; Aikawa, E.; Alcaraz, M.J.; Anderson, J.D.; Andriantsitohaina, R.; Antoniou, A.; Arab, T.; Archer, F.; Atkin-Smith, G.K.; et al. Minimal information for studies of extracellular vesicles 2018 (MISEV2018): A position statement of the International Society for Extracellular Vesicles and update of the MISEV2014 guidelines. *J. Extracell Vesicles* **2018**, *7*, 1535750. [CrossRef]
9. Jeppesen, D.K.; Fenix, A.M.; Franklin, J.L.; Higginbotham, J.N.; Zhang, Q.; Zimmerman, L.J.; Liebler, D.C.; Ping, J.; Liu, Q.; Evans, R.; et al. Reassessment of Exosome Composition. *Cell* **2019**, *177*, 428–445. [CrossRef]
10. Kowal, J.; Arras, G.; Colombo, M.; Jouve, M.; Morath, J.P.; Primdal-Bengtson, B.; Dingli, F.; Loew, D.; Tkach, M.; Théry, C. Proteomic comparison defines novel markers to characterize heterogeneous populations of extracellular vesicle subtypes. *Proc. Natl. Acad. Sci. USA* **2016**, *113*, E968–E977. [CrossRef]
11. O'Brien, K.; Breyne, K.; Ughetto, S.; Laurent, L.C.; Breakefield, X.O. RNA delivery by extracellular vesicles in mammalian cells and its applications. *Nat. Rev. Mol. Cell Biol.* **2020**. [CrossRef] [PubMed]
12. Cocucci, E.; Meldolesi, J. Ectosomes and exosomes: Shedding the confusion between extracellular vesicles. *Trends Cell Biol.* **2015**, *25*, 364–372. [CrossRef] [PubMed]
13. Mathivanan, S.; Ji, H.; Simpson, R.J. Exosomes: Extracellular organelles important in intercellular communication. *J. Proteom.* **2010**, *73*, 1907–1920. [CrossRef]
14. Maia, J.; Caja, S.; Strano Moraes, M.C.; Couto, N.; Costa-Silva, B. Exosome-Based Cell-Cell Communication in the Tumor Microenvironment. *Front. Cell Dev. Biol.* **2018**, *6*, 18. [CrossRef]
15. Andaloussi, S.E.L.; Mager, I.; Breakefield, X.O.; Wood, M.J. Extracellular vesicles: Biology and emerging therapeutic opportunities. *Nat. Rev. Drug Discov.* **2013**, *12*, 347–357. [CrossRef]
16. Tkach, M.; Thery, C. Communication by Extracellular Vesicles: Where We are and Where We Need to Go. *Cell* **2016**, *164*, 1226–1232. [CrossRef] [PubMed]
17. Xu, R.; Greening, D.W.; Zhu, H.J.; Takahashi, N.; Simpson, R.J. Extracellular vesicle isolation and characterization: Toward clinical application. *J. Clin. Investig.* **2016**, *126*, 1152–1162. [CrossRef]
18. Ratajczak, M.Z.; Kucia, M.; Jadczyk, T.; Greco, N.J.; Wojakowski, W.; Tendera, M.; Ratajczak, J. Pivotal role of paracrine effects in stem cell therapies in regenerative medicine: Can we translate stem cell-secreted paracrine factors and microvesicles into better therapeutic strategies? *Leukemia* **2012**, *26*, 1166–1173. [CrossRef]
19. Luan, X.; Sansanaphongpricha, K.; Myers, I.; Chen, H.; Yuan, H.; Sun, D. Engineering exosomes as refined biological nanoplatforms for drug delivery. *Acta Pharmacol. Sin.* **2017**, *38*, 754–763. [CrossRef]
20. Nogues, L.; Benito-Martin, A.; Hergueta-Redondo, M.; Peinado, H. The influence of tumour-derived extracellular vesicles on local and distal metastatic dissemination. *Mol. Asp. Med.* **2017**. [CrossRef]
21. Maas, S.L.N.; Breakefield, X.O.; Weaver, A.M. Extracellular Vesicles: Unique Intercellular Delivery Vehicles. *Trends Cell Biol.* **2017**, *27*, 172–188. [CrossRef] [PubMed]
22. Clayton, A.; Mitchell, J.P.; Court, J.; Mason, M.D.; Tabi, Z. Human tumor-derived exosomes selectively impair lymphocyte responses to interleukin-2. *Cancer Res.* **2007**, *67*, 7458–7466. [CrossRef]

23. Peinado, H.; Aleckovic, M.; Lavotshkin, S.; Matei, I.; Costa-Silva, B.; Moreno-Bueno, G.; Hergueta-Redondo, M.; Williams, C.; Garcia-Santos, G.; Ghajar, C.; et al. Melanoma exosomes educate bone marrow progenitor cells toward a pro-metastatic phenotype through MET. *Nat. Med.* **2012**, *18*, 883–891. [CrossRef] [PubMed]
24. Hoshino, A.; Costa-Silva, B.; Shen, T.L.; Rodrigues, G.; Hashimoto, A.; Tesic Mark, M.; Molina, H.; Kohsaka, S.; Di Giannatale, A.; Ceder, S.; et al. Tumour exosome integrins determine organotropic metastasis. *Nature* **2015**. [CrossRef] [PubMed]
25. Hessvik, N.P.; Llorente, A. Current knowledge on exosome biogenesis and release. *Cell. Mol. Life Sci.* **2017**. [CrossRef] [PubMed]
26. Gruenberg, J. Life in the lumen: The multivesicular endosome. *Traffic* **2020**, *21*, 76–93. [CrossRef]
27. Williams, C.; Royo, F.; Aizpurua-Olaizola, O.; Pazos, R.; Boons, G.J.; Reichardt, N.C.; Falcon-Perez, J.M. Glycosylation of extracellular vesicles: Current knowledge, tools and clinical perspectives. *J. Extracell Vesicles* **2018**, *7*, 1442985. [CrossRef]
28. Costa-Silva, B.; Aiello, N.M.; Ocean, A.J.; Singh, S.; Zhang, H.; Thakur, B.K.; Becker, A.; Hoshino, A.; Mark, M.T.; Molina, H.; et al. Pancreatic cancer exosomes initiate pre-metastatic niche formation in the liver. *Nat. Cell Biol.* **2015**, *17*, 816–826. [CrossRef]
29. Laoui, D.; Van Overmeire, E.; Di Conza, G.; Aldeni, C.; Keirsse, J.; Morias, Y.; Movahedi, K.; Houbracken, I.; Schouppe, E.; Elkrim, Y.; et al. Tumor hypoxia does not drive differentiation of tumor-associated macrophages but rather fine-tunes the M2-like macrophage population. *Cancer Res.* **2014**, *74*, 24–30. [CrossRef]
30. Colegio, O.R.; Chu, N.Q.; Szabo, A.L.; Chu, T.; Rhebergen, A.M.; Jairam, V.; Cyrus, N.; Brokowski, C.E.; Eisenbarth, S.C.; Phillips, G.M.; et al. Functional polarization of tumour-associated macrophages by tumour-derived lactic acid. *Nature* **2014**, *513*, 559–563. [CrossRef]
31. Whiteside, T.L. Tumor-Derived Exosomes and Their Role in Cancer Progression. *Adv. Clin. Chem.* **2016**, *74*, 103–141. [CrossRef] [PubMed]
32. Han, L.; Lam, E.W.; Sun, Y. Extracellular vesicles in the tumor microenvironment: Old stories, but new tales. *Mol. Cancer* **2019**, *18*, 59. [CrossRef] [PubMed]
33. Murray, P.J.; Allen, J.E.; Biswas, S.K.; Fisher, E.A.; Gilroy, D.W.; Goerdt, S.; Gordon, S.; Hamilton, J.A.; Ivashkiv, L.B.; Lawrence, T.; et al. Macrophage activation and polarization: Nomenclature and experimental guidelines. *Immunity* **2014**, *41*, 14–20. [CrossRef] [PubMed]
34. Gerlach, J.Q.; Griffin, M.D. Getting to know the extracellular vesicle glycome. *Mol. Biosyst.* **2016**, *12*, 1071–1081. [CrossRef]
35. Nistico, P.; Ciliberto, G. Biological mechanisms linked to inflammation in cancer: Discovery of tumor microenvironment-related biomarkers and their clinical application in solid tumors. *Int. J. Biol. Markers* **2020**, *35*, 8–11. [CrossRef]
36. Hinshaw, D.C.; Shevde, L.A. The Tumor Microenvironment Innately Modulates Cancer Progression. *Cancer Res.* **2019**, *79*, 4557–4566. [CrossRef]
37. Binnewies, M.; Roberts, E.W.; Kersten, K.; Chan, V.; Fearon, D.F.; Merad, M.; Coussens, L.M.; Gabrilovich, D.I.; Ostrand-Rosenberg, S.; Hedrick, C.C.; et al. Understanding the tumor immune microenvironment (TIME) for effective therapy. *Nat. Med.* **2018**, *24*, 541–550. [CrossRef]
38. Beatty, G.L.; Winograd, R.; Evans, R.A.; Long, K.B.; Luque, S.L.; Lee, J.W.; Clendenin, C.; Gladney, W.L.; Knoblock, D.M.; Guirnalda, P.D.; et al. Exclusion of T Cells From Pancreatic Carcinomas in Mice Is Regulated by Ly6Clow F4/80+ Extratumoral Macrophages. *Gastroenterology* **2015**, *149*, 201–210. [CrossRef]
39. Mosser, D.M.; Edwards, J.P. Exploring the full spectrum of macrophage activation. *Nat. Rev. Immunol.* **2008**, *8*, 958–969. [CrossRef]
40. Yang, M.; McKay, D.; Pollard, J.W.; Lewis, C.E. Diverse Functions of Macrophages in Different Tumor Microenvironments. *Cancer Res.* **2018**, *78*, 5492–5503. [CrossRef]
41. Martinez, F.O.; Gordon, S. The M1 and M2 paradigm of macrophage activation: Time for reassessment. *F1000prime Rep.* **2014**, *6*, 13. [CrossRef] [PubMed]
42. Galdiero, M.R.; Marone, G.; Mantovani, A. Cancer Inflammation and Cytokines. *Cold Spring Harb. Perspect. Biol.* **2018**, *10*. [CrossRef] [PubMed]
43. Hanahan, D.; Coussens, L.M. Accessories to the crime: Functions of cells recruited to the tumor microenvironment. *Cancer Cell* **2012**, *21*, 309–322. [CrossRef]

44. Pollard, J.W. Tumour-educated macrophages promote tumour progression and metastasis. *Nat. Rev. Cancer* **2004**, *4*, 71–78. [CrossRef]
45. Qian, B.Z.; Li, J.; Zhang, H.; Kitamura, T.; Zhang, J.; Campion, L.R.; Kaiser, E.A.; Snyder, L.A.; Pollard, J.W. CCL2 recruits inflammatory monocytes to facilitate breast-tumour metastasis. *Nature* **2011**, *475*, 222–225. [CrossRef] [PubMed]
46. Lin, E.Y.; Nguyen, A.V.; Russell, R.G.; Pollard, J.W. Colony-stimulating factor 1 promotes progression of mammary tumors to malignancy. *J. Exp. Med.* **2001**, *193*, 727–740. [CrossRef] [PubMed]
47. Aras, S.; Zaidi, M.R. TAMeless traitors: Macrophages in cancer progression and metastasis. *Br. J. Cancer* **2017**, *117*, 1583–1591. [CrossRef]
48. Cuccarese, M.F.; Dubach, J.M.; Pfirschke, C.; Engblom, C.; Garris, C.; Miller, M.A.; Pittet, M.J.; Weissleder, R. Heterogeneity of macrophage infiltration and therapeutic response in lung carcinoma revealed by 3D organ imaging. *Nat. Commun.* **2017**, *8*, 14293. [CrossRef]
49. Huang, Y.K.; Wang, M.; Sun, Y.; Di Costanzo, N.; Mitchell, C.; Achuthan, A.; Hamilton, J.A.; Busuttil, R.A.; Boussioutas, A. Macrophage spatial heterogeneity in gastric cancer defined by multiplex immunohistochemistry. *Nat. Commun.* **2019**, *10*, 3928. [CrossRef]
50. Azizi, E.; Carr, A.J.; Plitas, G.; Cornish, A.E.; Konopacki, C.; Prabhakaran, S.; Nainys, J.; Wu, K.; Kiseliovas, V.; Setty, M.; et al. Single-Cell Map of Diverse Immune Phenotypes in the Breast Tumor Microenvironment. *Cell* **2018**, *174*, 1293–1308. [CrossRef]
51. Gordon, S.; Pluddemann, A.; Mukhopadhyay, S. Plasma membrane receptors of tissue macrophages: Functions and role in pathology. *J. Pathol.* **2020**, *250*, 656–666. [CrossRef]
52. Langston, P.K.; Shibata, M.; Horng, T. Metabolism Supports Macrophage Activation. *Front. Immunol.* **2017**, *8*, 61. [CrossRef]
53. O'Neill, L.A.; Pearce, E.J. Immunometabolism governs dendritic cell and macrophage function. *J. Exp. Med.* **2016**, *213*, 15–23. [CrossRef]
54. Xue, J.; Schmidt, S.V.; Sander, J.; Draffehn, A.; Krebs, W.; Quester, I.; De Nardo, D.; Gohel, T.D.; Emde, M.; Schmidleithner, L.; et al. Transcriptome-based network analysis reveals a spectrum model of human macrophage activation. *Immunity* **2014**, *40*, 274–288. [CrossRef]
55. Lawrence, T.; Natoli, G. Transcriptional regulation of macrophage polarization: Enabling diversity with identity. *Nat. Rev. Immunol.* **2011**, *11*, 750–761. [CrossRef]
56. Chanmee, T.; Ontong, P.; Konno, K.; Itano, N. Tumor-associated macrophages as major players in the tumor microenvironment. *Cancers* **2014**, *6*, 1670–1690. [CrossRef]
57. De Palma, M.; Biziato, D.; Petrova, T.V. Microenvironmental regulation of tumour angiogenesis. *Nat. Rev. Cancer* **2017**, *17*, 457–474. [CrossRef]
58. Beltraminelli, T.; De Palma, M. Biology and therapeutic targeting of tumour-associated macrophages. *J. Pathol.* **2020**, *250*, 573–592. [CrossRef]
59. Blando, J.; Sharma, A.; Higa, M.G.; Zhao, H.; Vence, L.; Yadav, S.S.; Kim, J.; Sepulveda, A.M.; Sharp, M.; Maitra, A.; et al. Comparison of immune infiltrates in melanoma and pancreatic cancer highlights VISTA as a potential target in pancreatic cancer. *Proc. Natl. Acad. Sci. USA* **2019**, *116*, 1692–1697. [CrossRef]
60. Mantovani, A.; Bonecchi, R. One Clever Macrophage Checkpoint. *Clin. Cancer Res. Off. J. Am. Assoc. Cancer Res.* **2019**, *25*, 3202–3204. [CrossRef]
61. Rodriguez, P.C.; Quiceno, D.G.; Zabaleta, J.; Ortiz, B.; Zea, A.H.; Piazuelo, M.B.; Delgado, A.; Correa, P.; Brayer, J.; Sotomayor, E.M.; et al. Arginase I production in the tumor microenvironment by mature myeloid cells inhibits T-cell receptor expression and antigen-specific T-cell responses. *Cancer Res.* **2004**, *64*, 5839–5849. [CrossRef]
62. Cassetta, L.; Pollard, J.W. Targeting macrophages: Therapeutic approaches in cancer. *Nat. Rev. Drug Discov.* **2018**, *17*, 887–904. [CrossRef]
63. Mantovani, A.; Marchesi, F.; Malesci, A.; Laghi, L.; Allavena, P. Tumour-associated macrophages as treatment targets in oncology. *Nat. Rev. Clin. Oncol.* **2017**, *14*, 399–416. [CrossRef]
64. Genard, G.; Lucas, S.; Michiels, C. Reprogramming of Tumor-Associated Macrophages with Anticancer Therapies: Radiotherapy versus Chemo- and Immunotherapies. *Front. Immunol.* **2017**, *8*, 828. [CrossRef]
65. Anfray, C.; Ummarino, A.; Andon, F.T.; Allavena, P. Current Strategies to Target Tumor-Associated-Macrophages to Improve Anti-Tumor Immune Responses. *Cells* **2019**, *9*, 46. [CrossRef]

66. Galmbacher, K.; Heisig, M.; Hotz, C.; Wischhusen, J.; Galmiche, A.; Bergmann, B.; Gentschev, I.; Goebel, W.; Rapp, U.R.; Fensterle, J. Shigella mediated depletion of macrophages in a murine breast cancer model is associated with tumor regression. *PLoS ONE* **2010**, *5*, e9572. [CrossRef]
67. Bak, S.P.; Walters, J.J.; Takeya, M.; Conejo-Garcia, J.R.; Berwin, B.L. Scavenger Receptor-A–Targeted Leukocyte Depletion Inhibits Peritoneal Ovarian Tumor Progression. *Cancer Res.* **2007**, *67*, 4783. [CrossRef]
68. Nagai, T.; Tanaka, M.; Tsuneyoshi, Y.; Xu, B.; Michie, S.A.; Hasui, K.; Hirano, H.; Arita, K.; Matsuyama, T. Targeting tumor-associated macrophages in an experimental glioma model with a recombinant immunotoxin to folate receptor beta. *Cancer Immunol. Immunother.* **2009**, *58*, 1577–1586. [CrossRef]
69. Lehenkari, P.P.; Kellinsalmi, M.; Näpänkangas, J.P.; Ylitalo, K.V.; Mönkkönen, J.; Rogers, M.J.; Azhayev, A.; Väänänen, H.K.; Hassinen, I.E. Further insight into mechanism of action of clodronate: Inhibition of mitochondrial ADP/ATP translocase by a nonhydrolyzable, adenine-containing metabolite. *Mol. Pharmacol.* **2002**, *61*, 1255–1262. [CrossRef]
70. Zhang, W.; Zhu, X.D.; Sun, H.C.; Xiong, Y.Q.; Zhuang, P.Y.; Xu, H.X.; Kong, L.Q.; Wang, L.; Wu, W.Z.; Tang, Z.Y. Depletion of tumor-associated macrophages enhances the effect of sorafenib in metastatic liver cancer models by antimetastatic and antiangiogenic effects. *Clin. Cancer Res. Off. J. Am. Assoc. Cancer Res.* **2010**, *16*, 3420–3430. [CrossRef]
71. Allavena, P.; Signorelli, M.; Chieppa, M.; Erba, E.; Bianchi, G.; Marchesi, F.; Olimpio, C.O.; Bonardi, C.; Garbi, A.; Lissoni, A.; et al. Anti-inflammatory properties of the novel antitumor agent yondelis (trabectedin): Inhibition of macrophage differentiation and cytokine production. *Cancer Res.* **2005**, *65*, 2964–2971. [CrossRef] [PubMed]
72. Liang, W.; Kujawski, M.; Wu, J.; Lu, J.; Herrmann, A.; Loera, S.; Yen, Y.; Lee, F.; Yu, H.; Wen, W.; et al. Antitumor activity of targeting SRC kinases in endothelial and myeloid cell compartments of the tumor microenvironment. *Clin. Cancer Res. Off. J. Am. Assoc. Cancer Res.* **2010**, *16*, 924–935. [CrossRef]
73. Ozanne, J.; Prescott, A.R.; Clark, K. The clinically approved drugs dasatinib and bosutinib induce anti-inflammatory macrophages by inhibiting the salt-inducible kinases. *Biochem. J.* **2015**, *465*, 271–279. [CrossRef]
74. Brana, I.; Calles, A.; LoRusso, P.M.; Yee, L.K.; Puchalski, T.A.; Seetharam, S.; Zhong, B.; de Boer, C.J.; Tabernero, J.; Calvo, E. Carlumab, an anti-C-C chemokine ligand 2 monoclonal antibody, in combination with four chemotherapy regimens for the treatment of patients with solid tumors: An open-label, multicenter phase 1b study. *Target. Oncol.* **2015**, *10*, 111–123. [CrossRef] [PubMed]
75. Kitamura, T.; Qian, B.-Z.; Soong, D.; Cassetta, L.; Noy, R.; Sugano, G.; Kato, Y.; Li, J.; Pollard, J.W. CCL2-induced chemokine cascade promotes breast cancer metastasis by enhancing retention of metastasis-associated macrophages. *J. Exp. Med.* **2015**, *212*, 1043–1059. [CrossRef] [PubMed]
76. Bonapace, L.; Coissieux, M.M.; Wyckoff, J.; Mertz, K.D.; Varga, Z.; Junt, T.; Bentires-Alj, M. Cessation of CCL2 inhibition accelerates breast cancer metastasis by promoting angiogenesis. *Nature* **2014**, *515*, 130–133. [CrossRef] [PubMed]
77. Sánchez-Martín, L.; Estecha, A.; Samaniego, R.; Sánchez-Ramón, S.; Vega, M.Á.; Sánchez-Mateos, P. The chemokine CXCL12 regulates monocyte-macrophage differentiation and RUNX3 expression. *Blood* **2011**, *117*, 88–97. [CrossRef] [PubMed]
78. Hughes, R.; Qian, B.Z.; Rowan, C.; Muthana, M.; Keklikoglou, I.; Olson, O.C.; Tazzyman, S.; Danson, S.; Addison, C.; Clemons, M.; et al. Perivascular M2 Macrophages Stimulate Tumor Relapse after Chemotherapy. *Cancer Res.* **2015**, *75*, 3479–3491. [CrossRef]
79. Ao, J.Y.; Zhu, X.D.; Chai, Z.T.; Cai, H.; Zhang, Y.Y.; Zhang, K.Z.; Kong, L.Q.; Zhang, N.; Ye, B.G.; Ma, D.N.; et al. Colony-Stimulating Factor 1 Receptor Blockade Inhibits Tumor Growth by Altering the Polarization of Tumor-Associated Macrophages in Hepatocellular Carcinoma. *Mol. Cancer Ther.* **2017**, *16*, 1544–1554. [CrossRef]
80. Zhu, Y.; Knolhoff, B.L.; Meyer, M.A.; Nywening, T.M.; West, B.L.; Luo, J.; Wang-Gillam, A.; Goedegebuure, S.P.; Linehan, D.C.; De Nardo, D.G. CSF1/CSF1R Blockade Reprograms Tumor-Infiltrating Macrophages and Improves Response to T-cell Checkpoint Immunotherapy in Pancreatic Cancer Models. *Cancer Res.* **2014**, *74*, 5057. [CrossRef]
81. Zhang, Q.-Q.; Hu, X.-W.; Liu, Y.-L.; Ye, Z.-J.; Gui, Y.-H.; Zhou, D.-L.; Qi, C.-L.; He, X.-D.; Wang, H.; Wang, L.-J. CD11b deficiency suppresses intestinal tumor growth by reducing myeloid cell recruitment. *Sci. Rep.* **2015**, *5*, 15948. [CrossRef]

82. Ahn, G.O.; Tseng, D.; Liao, C.H.; Dorie, M.J.; Czechowicz, A.; Brown, J.M. Inhibition of Mac-1 (CD11b/CD18) enhances tumor response to radiation by reducing myeloid cell recruitment. *Proc. Natl. Acad. Sci. USA* **2010**, *107*, 8363–8368. [CrossRef]
83. Sun, L.; Chen, B.; Jiang, R.; Li, J.; Wang, B. Resveratrol inhibits lung cancer growth by suppressing M2-like polarization of tumor associated macrophages. *Cell. Immunol.* **2017**, *311*, 86–93. [CrossRef]
84. Edwards, J.P.; Emens, L.A. The multikinase inhibitor sorafenib reverses the suppression of IL-12 and enhancement of IL-10 by PGE_2 in murine macrophages. *Int. Immunopharmacol.* **2010**, *10*, 1220–1228. [CrossRef]
85. Dong, R.; Gong, Y.; Meng, W.; Yuan, M.; Zhu, H.; Ying, M.; He, Q.; Cao, J.; Yang, B. The involvement of M2 macrophage polarization inhibition in fenretinide-mediated chemopreventive effects on colon cancer. *Cancer Lett.* **2017**, *388*, 43–53. [CrossRef]
86. Dunn, G.P.; Koebel, C.M.; Schreiber, R.D. Interferons, immunity and cancer immunoediting. *Nat. Rev. Immunol.* **2006**, *6*, 836–848. [CrossRef]
87. Parker, B.S.; Rautela, J.; Hertzog, P.J. Antitumour actions of interferons: Implications for cancer therapy. *Nat. Rev. Cancer* **2016**, *16*, 131–144. [CrossRef]
88. Ding, L.; Liang, G.; Yao, Z.; Zhang, J.; Liu, R.; Chen, H.; Zhou, Y.; Wu, H.; Yang, B.; He, Q. Metformin prevents cancer metastasis by inhibiting M2-like polarization of tumor associated macrophages. *Oncotarget* **2015**, *6*, 36441–36455. [CrossRef]
89. Huang, L.; Xu, H.; Peng, G. TLR-mediated metabolic reprogramming in the tumor microenvironment: Potential novel strategies for cancer immunotherapy. *Cell. Mol. Immunol.* **2018**, *15*, 428–437. [CrossRef]
90. Chen, N.; Wei, M.; Sun, Y.; Li, F.; Pei, H.; Li, X.; Su, S.; He, Y.; Wang, L.; Shi, J.; et al. Self-assembly of poly-adenine-tailed CpG oligonucleotide-gold nanoparticle nanoconjugates with immunostimulatory activity. *Small* **2014**, *10*, 368–375. [CrossRef]
91. Kaneda, M.M.; Messer, K.S.; Ralainirina, N.; Li, H.; Leem, C.J.; Gorjestani, S.; Woo, G.; Nguyen, A.V.; Figueiredo, C.C.; Foubert, P.; et al. PI3Kγ is a molecular switch that controls immune suppression. *Nature* **2016**, *539*, 437–442. [CrossRef] [PubMed]
92. Zhang, F.; Parayath, N.N.; Ene, C.I.; Stephan, S.B.; Koehne, A.L.; Coon, M.E.; Holland, E.C.; Stephan, M.T. Genetic programming of macrophages to perform anti-tumor functions using targeted mRNA nanocarriers. *Nat. Commun.* **2019**, *10*, 3974. [CrossRef]
93. Cai, X.; Yin, Y.; Li, N.; Zhu, D.; Zhang, J.; Zhang, C.-Y.; Zen, K. Re-polarization of tumor-associated macrophages to pro-inflammatory M1 macrophages by microRNA-155. *J. Mol. Cell Biol.* **2012**, *4*, 341–343. [CrossRef]
94. Zhang, M.; Gao, Y.; Caja, K.; Zhao, B.; Kim, J.A. Non-viral nanoparticle delivers small interfering RNA to macrophages in vitro and in vivo. *PLoS ONE* **2015**, *10*, e0118472. [CrossRef] [PubMed]
95. Jia, N.; Wu, H.; Duan, J.; Wei, C.; Wang, K.; Zhang, Y.; Mao, X. Polyethyleneimine-coated Iron Oxide Nanoparticles as a Vehicle for the Delivery of Small Interfering RNA to Macrophages In Vitro and In Vivo. *JoVE* **2019**, e58660. [CrossRef]
96. Liang, S.; Zheng, J.; Wu, W.; Li, Q.; Saw, P.E.; Chen, J.; Xu, X.; Yao, H.; Yao, Y. A Robust Nanoparticle Platform for RNA Interference in Macrophages to Suppress Tumor Cell Migration. *Front. Pharm.* **2018**, *9*, 1465. [CrossRef]
97. Zippelius, A.; Schreiner, J.; Herzig, P.; Muller, P. Induced PD-L1 expression mediates acquired resistance to agonistic anti-CD40 treatment. *Cancer Immunol. Res.* **2015**, *3*, 236–244. [CrossRef] [PubMed]
98. Rodell, C.B.; Arlauckas, S.P.; Cuccarese, M.F.; Garris, C.S.; Li, R.; Ahmed, M.S.; Kohler, R.H.; Pittet, M.J.; Weissleder, R. TLR7/8-agonist-loaded nanoparticles promote the polarization of tumour-associated macrophages to enhance cancer immunotherapy. *Nat. Biomed. Eng.* **2018**, *2*, 578–588. [CrossRef]
99. Vogel, D.Y.; Glim, J.E.; Stavenuiter, A.W.; Breur, M.; Heijnen, P.; Amor, S.; Dijkstra, C.D.; Beelen, R.H. Human macrophage polarization in vitro: Maturation and activation methods compared. *Immunobiology* **2014**, *219*, 695–703. [CrossRef]
100. Georgoudaki, A.M.; Prokopec, K.E.; Boura, V.F.; Hellqvist, E.; Sohn, S.; Ostling, J.; Dahan, R.; Harris, R.A.; Rantalainen, M.; Klevebring, D.; et al. Reprogramming Tumor-Associated Macrophages by Antibody Targeting Inhibits Cancer Progression and Metastasis. *Cell Rep.* **2016**, *15*, 2000–2011. [CrossRef]

101. Peng, H.; Chen, B.; Huang, W.; Tang, Y.; Jiang, Y.; Zhang, W.; Huang, Y. Reprogramming Tumor-Associated Macrophages To Reverse EGFRT790M Resistance by Dual-Targeting Codelivery of Gefitinib/Vorinostat. *Nano Lett.* **2017**, *17*, 7684–7690. [CrossRef] [PubMed]
102. Ye, J.; Yang, Y.; Dong, W.; Gao, Y.; Meng, Y.; Wang, H.; Li, L.; Jin, J.; Ji, M.; Xia, X.; et al. Drug-free mannosylated liposomes inhibit tumor growth by promoting the polarization of tumor-associated macrophages. *Int. J. Nanomed.* **2019**, *14*, 3203–3220. [CrossRef]
103. Jaynes, J.M.; Sable, R.; Ronzetti, M.; Bautista, W.; Knotts, Z.; Abisoye-Ogunniyan, A.; Li, D.; Calvo, R.; Dashnyam, M.; Singh, A.; et al. Mannose receptor (CD206) activation in tumor-associated macrophages enhances adaptive and innate antitumor immune responses. *Sci. Transl. Med.* **2020**, *12*. [CrossRef] [PubMed]
104. Chen, P.; Zhang, X.; Venosa, A.; Lee, I.H.; Myers, D.; Holloway, J.A.; Prud'homme, R.K.; Gao, D.; Szekely, Z.; Laskin, J.D.; et al. A Novel Bivalent Mannosylated Targeting Ligand Displayed on Nanoparticles Selectively Targets Anti-Inflammatory M2 Macrophages. *Pharmaceutics* **2020**, *12*, 243. [CrossRef] [PubMed]
105. Irache, J.M.; Salman, H.H.; Gamazo, C.; Espuelas, S. Mannose-targeted systems for the delivery of therapeutics. *Expert Opin. Drug Deliv.* **2008**, *5*, 703–724. [CrossRef]
106. Harris, N.; Super, M.; Rits, M.; Chang, G.; Ezekowitz, R.A. Characterization of the murine macrophage mannose receptor: Demonstration that the downregulation of receptor expression mediated by interferon-gamma occurs at the level of transcription. *Blood* **1992**, *80*, 2363–2373. [CrossRef]
107. Stein, M.; Keshav, S.; Harris, N.; Gordon, S. Interleukin 4 potently enhances murine macrophage mannose receptor activity: A marker of alternative immunologic macrophage activation. *J. Exp. Med.* **1992**, *176*, 287–292. [CrossRef]
108. Taylor, M.E.; Drickamer, K. Mammalian sugar-binding receptors: Known functions and unexplored roles. *FEBS J.* **2019**, *286*, 1800–1814. [CrossRef]
109. Pontow, S.E.; Kery, V.; Stahl, P.D. Mannose receptor. *Int. Rev. Cytol.* **1992**, *137b*, 221–244. [CrossRef]
110. Blum, J.S.; Stahl, P.D.; Diaz, R.; Fiani, M.L. Purification and characterization of the D-mannose receptor from J774 mouse macrophage cells. *Carbohydr. Res.* **1991**, *213*, 145–153. [CrossRef]
111. Martinez-Pomares, L. The mannose receptor. *J. Leukoc. Biol.* **2012**, *92*, 1177–1186. [CrossRef]
112. Stahl, P.D.; Ezekowitz, R.A. The mannose receptor is a pattern recognition receptor involved in host defense. *Curr. Opin. Immunol.* **1998**, *10*, 50–55. [CrossRef]
113. Hu, Z.; Shi, X.; Yu, B.; Li, N.; Huang, Y.; He, Y. Structural Insights into the pH-Dependent Conformational Change and Collagen Recognition of the Human Mannose Receptor. *Structure* **2018**, *26*, 60–71. [CrossRef]
114. Scodeller, P.; Simon-Gracia, L.; Kopanchuk, S.; Tobi, A.; Kilk, K.; Saalik, P.; Kurm, K.; Squadrito, M.L.; Kotamraju, V.R.; Rinken, A.; et al. Precision Targeting of Tumor Macrophages with a CD206 Binding Peptide. *Sci. Rep.* **2017**, *7*, 14655. [CrossRef] [PubMed]
115. Costa, J. Glycoconjugates from extracellular vesicles: Structures, functions and emerging potential as cancer biomarkers. *Biochim. Biophys. Acta* **2017**, *1868*, 157–166. [CrossRef] [PubMed]
116. Choi, E.S.; Song, J.; Kang, Y.Y.; Mok, H. Mannose-Modified Serum Exosomes for the Elevated Uptake to Murine Dendritic Cells and Lymphatic Accumulation. *Macromol. Biosci.* **2019**, *19*, e1900042. [CrossRef] [PubMed]
117. Zitvogel, L.; Regnault, A.; Lozier, A.; Wolfers, J.; Flament, C.; Tenza, D.; Ricciardi-Castagnoli, P.; Raposo, G.; Amigorena, S. Eradication of established murine tumors using a novel cell-free vaccine: Dendritic cell derived exosomes. *Nat. Med.* **1998**, *4*, 594–600. [CrossRef]
118. Wolfers, J.; Lozier, A.; Raposo, G.; Regnault, A.; Théry, C.; Masurier, C.; Flament, C.; Pouzieux, S.; Faure, F.; Tursz, T.; et al. Tumor-derived exosomes are a source of shared tumor rejection antigens for CTL cross-priming. *Nat. Med.* **2001**, *7*, 297–303. [CrossRef]
119. Plebanek, M.P.; Mutharasan, R.K.; Volpert, O.; Matov, A.; Gatlin, J.C.; Thaxton, C.S. Nanoparticle Targeting and Cholesterol Flux Through Scavenger Receptor Type B-1 Inhibits Cellular Exosome Uptake. *Sci. Rep.* **2015**, *5*, 15724. [CrossRef]
120. Mulcahy, L.A.; Pink, R.C.; Carter, D.R. Routes and mechanisms of extracellular vesicle uptake. *J. Extracell Vesicles* **2014**, *3*. [CrossRef]
121. Mathieu, M.; Martin-Jaular, L.; Lavieu, G.; Thery, C. Specificities of secretion and uptake of exosomes and other extracellular vesicles for cell-to-cell communication. *Nat. Cell Biol.* **2019**, *21*, 9–17. [CrossRef]
122. Gonda, A.; Kabagwira, J.; Senthil, G.N.; Wall, N.R. Internalization of Exosomes through Receptor-Mediated Endocytosis. *Mol. Cancer Res.* **2019**, *17*, 337–347. [CrossRef]

123. Zanetti, C.; Gallina, A.; Fabbri, A.; Parisi, S.; Palermo, A.; Fecchi, K.; Boussadia, Z.; Carollo, M.; Falchi, M.; Pasquini, L.; et al. Cell Propagation of Cholera Toxin CTA ADP-Ribosylating Factor by Exosome Mediated Transfer. *Int. J. Mol. Sci.* **2018**, *19*, 1521. [CrossRef] [PubMed]
124. Andre, F.; Schartz, N.E.C.; Movassagh, M.; Flament, C.; Pautier, P.; Morice, P.; Pomel, C.; Lhomme, C.; Escudier, B.; Le Chevalier, T.; et al. Malignant effusions and immunogenic tumour-derived exosomes. *Lancet* **2002**, *360*, 295–305. [CrossRef]
125. Montecalvo, A.; Shufesky, W.J.; Beer Stolz, D.; Sullivan, M.G.; Wang, Z.; Divito, S.J.; Papworth, G.D.; Watkins, S.C.; Robbins, P.D.; Larregina, A.T.; et al. Exosomes As a Short-Range Mechanism to Spread Alloantigen between Dendritic Cells during T Cell Allorecognition. *J. Immunol.* **2008**, *180*, 3081. [CrossRef] [PubMed]
126. Lee, R.T.; Hsu, T.L.; Huang, S.K.; Hsieh, S.L.; Wong, C.H.; Lee, Y.C. Survey of immune-related, mannose/fucose-binding C-type lectin receptors reveals widely divergent sugar-binding specificities. *Glycobiology* **2011**, *21*, 512–520. [CrossRef]
127. Robinson, M.J.; Sancho, D.; Slack, E.C.; LeibundGut-Landmann, S.; Reis e Sousa, C. Myeloid C-type lectins in innate immunity. *Nat. Immunol.* **2006**, *7*, 1258–1265. [CrossRef]
128. Krishnamoorthy, L.; Bess, J.W., Jr.; Preston, A.B.; Nagashima, K.; Mahal, L.K. HIV-1 and microvesicles from T cells share a common glycome, arguing for a common origin. *Nat. Chem. Biol.* **2009**, *5*, 244–250. [CrossRef]
129. Shimoda, A.; Sawada, S.I.; Sasaki, Y.; Akiyoshi, K. Exosome surface glycans reflect osteogenic differentiation of mesenchymal stem cells: Profiling by an evanescent field fluorescence-assisted lectin array system. *Sci. Rep.* **2019**, *9*, 11497. [CrossRef]
130. Chen, P.; Zhang, X.; Jia, L.; Prud'homme, R.K.; Szekely, Z.; Sinko, P.J. Optimal structural design of mannosylated nanocarriers for macrophage targeting. *J. Control. Release Off. J. Control. Release Soc.* **2014**, *194*, 341–349. [CrossRef]
131. Ahowesso, C.; Black, P.N.; Saini, N.; Montefusco, D.; Chekal, J.; Malosh, C.; Lindsley, C.W.; Stauffer, S.R.; Di Russo, C.C. Chemical inhibition of fatty acid absorption and cellular uptake limits lipotoxic cell death. *Biochem. Pharmacol.* **2015**, *98*, 167–181. [CrossRef]
132. Royo, F.; Cossío, U.; de Angulo, A.R.; Llop, J.; Falcon-Perez, J.M. Modification of the glycosylation of extracellular vesicles alters their biodistribution in mice. *Nanoscale* **2019**, *11*, 1531–1537. [CrossRef]
133. Williams, C.; Pazos, R.; Royo, F.; Gonzalez, E.; Roura-Ferrer, M.; Martinez, A.; Gamiz, J.; Reichardt, N.C.; Falcon-Perez, J.M. Assessing the role of surface glycans of extracellular vesicles on cellular uptake. *Sci. Rep.* **2019**, *9*, 11920. [CrossRef] [PubMed]
134. Lucchetti, D.; Colella, F.; Perelli, L.; Ricciardi-Tenore, C.; Calapa, F.; Fiori, M.E.; Carbone, F.; De Maria, R.; Sgambato, A. CD147 Promotes Cell Small Extracellular Vesicles Release during Colon Cancer Stem Cells Differentiation and Triggers Cellular Changes in Recipient Cells. *Cancers* **2020**, *12*, 260. [CrossRef] [PubMed]
135. Yuan, Z.; Kolluri, K.K.; Gowers, K.H.; Janes, S.M. TRAIL delivery by MSC-derived extracellular vesicles is an effective anticancer therapy. *J. Extracell Vesicles* **2017**, *6*, 1265291. [CrossRef] [PubMed]
136. Sterzenbach, U.; Putz, U.; Low, L.-H.; Silke, J.; Tan, S.-S.; Howitt, J. Engineered Exosomes as Vehicles for Biologically Active Proteins. *Mol. Ther. J. Am. Soc. Gene Ther.* **2017**, *25*, 1269–1278. [CrossRef] [PubMed]
137. Cooks, T.; Pateras, I.S.; Jenkins, L.M.; Patel, K.M.; Robles, A.I.; Morris, J.; Forshew, T.; Appella, E.; Gorgoulis, V.G.; Harris, C.C. Mutant p53 cancers reprogram macrophages to tumor supporting macrophages via exosomal miR-1246. *Nat. Commun.* **2018**, *9*, 771. [CrossRef] [PubMed]
138. Zhang, Z.; Dombroski, J.A.; King, M.R. Engineering of Exosomes to Target Cancer Metastasis. *Cell Mol. Bioeng.* **2020**, *13*, 1–16. [CrossRef]
139. Maas, S.L.N.; de Vrij, J.; van der Vlist, E.J.; Geragousian, B.; van Bloois, L.; Mastrobattista, E.; Schiffelers, R.M.; Wauben, M.H.M.; Broekman, M.L.D.; Nolte-'t Hoen, E.N.M. Possibilities and limitations of current technologies for quantification of biological extracellular vesicles and synthetic mimics. *J. Control. Release Off. J. Control. Release Soc.* **2015**, *200*, 87–96. [CrossRef]
140. Coscia, C.; Parolini, I.; Sanchez, M.; Biffoni, M.; Boussadia, Z.; Zanetti, C.; Fiani, M.L.; Sargiacomo, M. Generation, quantification, and tracing of metabolically labeled fluorescent exosomes. In *Lentiviral Vectors and Exosomes as Gene and Protein Delivery Tools*; Federico, M., Ed.; Springer: New York, NY, USA, 2016; pp. 217–235.
141. Pereira, E.A.; daSilva, L.L. HIV-1 Nef: Taking Control of Protein Trafficking. *Traffic* **2016**, *17*, 976–996. [CrossRef]

142. Mukhamedova, N.; Hoang, A.; Dragoljevic, D.; Dubrovsky, L.; Pushkarsky, T.; Low, H.; Ditiatkovski, M.; Fu, Y.; Ohkawa, R.; Meikle, P.J.; et al. Exosomes containing HIV protein Nef reorganize lipid rafts potentiating inflammatory response in bystander cells. *PLoS Pathog.* **2019**, *15*, e1007907. [CrossRef]
143. Zheng, Y.H.; Plemenitas, A.; Linnemann, T.; Fackler, O.T.; Peterlin, B.M. Nef increases infectivity of HIV via lipid rafts. *Curr. Biol. CB* **2001**, *11*, 875–879. [CrossRef]
144. Zheng, Y.H.; Plemenitas, A.; Fielding, C.J.; Peterlin, B.M. Nef increases the synthesis of and transports cholesterol to lipid rafts and HIV-1 progeny virions. *Proc. Natl. Acad. Sci. USA* **2003**, *100*, 8460–8465. [CrossRef] [PubMed]
145. Sami Saribas, A.; Cicalese, S.; Ahooyi, T.M.; Khalili, K.; Amini, S.; Sariyer, I.K. HIV-1 Nef is released in extracellular vesicles derived from astrocytes: Evidence for Nef-mediated neurotoxicity. *Cell Death Dis.* **2017**, *8*, e2542. [CrossRef]
146. Pužar Dominkuš, P.; Ferdin, J.; Plemenitaš, A.; Peterlin, B.M.; Lenassi, M. Nef is secreted in exosomes from Nef.GFP-expressing and HIV-1-infected human astrocytes. *J. Neurovirology* **2017**, *23*, 713–724. [CrossRef]
147. McNamara, R.P.; Costantini, L.M.; Myers, T.A.; Schouest, B.; Maness, N.J.; Griffith, J.D.; Damania, B.A.; MacLean, A.G.; Dittmer, D.P. Nef Secretion into Extracellular Vesicles or Exosomes Is Conserved across Human and Simian Immunodeficiency Viruses. *mBio* **2018**, *9*. [CrossRef] [PubMed]
148. Lee, J.H.; Wittki, S.; Brau, T.; Dreyer, F.S.; Kratzel, K.; Dindorf, J.; Johnston, I.C.; Gross, S.; Kremmer, E.; Zeidler, R.; et al. HIV Nef, paxillin, and Pak1/2 regulate activation and secretion of TACE/ADAM10 proteases. *Mol. Cell* **2013**, *49*, 668–679. [CrossRef] [PubMed]
149. Arenaccio, C.; Chiozzini, C.; Columba-Cabezas, S.; Manfredi, F.; Affabris, E.; Baur, A.; Federico, M. Exosomes from human immunodeficiency virus type 1 (HIV-1)-infected cells license quiescent CD4+ T lymphocytes to replicate HIV-1 through a Nef- and ADAM17-dependent mechanism. *J. Virol.* **2014**, *88*, 11529–11539. [CrossRef]
150. Swingler, S.; Brichacek, B.; Jacque, J.M.; Ulich, C.; Zhou, J.; Stevenson, M. HIV-1 Nef intersects the macrophage CD40L signalling pathway to promote resting-cell infection. *Nature* **2003**, *424*, 213–219. [CrossRef]
151. Olivetta, E.; Percario, Z.; Fiorucci, G.; Mattia, G.; Schiavoni, I.; Dennis, C.; Jager, J.; Harris, M.; Romeo, G.; Affabris, E.; et al. HIV-1 Nef induces the release of inflammatory factors from human monocyte/macrophages: Involvement of Nef endocytotic signals and NF-kappa B activation. *J. Immunol.* **2003**, *170*, 1716–1727. [CrossRef] [PubMed]
152. Mangino, G.; Percario, Z.A.; Fiorucci, G.; Vaccari, G.; Manrique, S.; Romeo, G.; Federico, M.; Geyer, M.; Affabris, E. In vitro treatment of human monocytes/macrophages with myristoylated recombinant Nef of human immunodeficiency virus type 1 leads to the activation of mitogen-activated protein kinases, IkappaB kinases, and interferon regulatory factor 3 and to the release of beta interferon. *J. Virol.* **2007**, *81*, 2777–2791. [CrossRef]
153. Federico, M.; Percario, Z.; Olivetta, E.; Fiorucci, G.; Muratori, C.; Micheli, A.; Romeo, G.; Affabris, E. HIV-1 Nef activates STAT1 in human monocytes/macrophages through the release of soluble factors. *Blood* **2001**, *98*, 2752–2761. [CrossRef]
154. Percario, Z.; Olivetta, E.; Fiorucci, G.; Mangino, G.; Peretti, S.; Romeo, G.; Affabris, E.; Federico, M. Human immunodeficiency virus type 1 (HIV-1) Nef activates STAT3 in primary human monocyte/macrophages through the release of soluble factors: Involvement of Nef domains interacting with the cell endocytotic machinery. *J. Leukoc. Biol.* **2003**, *74*, 821–832. [CrossRef]
155. Mangino, G.; Percario, Z.A.; Fiorucci, G.; Vaccari, G.; Acconcia, F.; Chiarabelli, C.; Leone, S.; Noto, A.; Horenkamp, F.A.; Manrique, S.; et al. HIV-1 Nef induces proinflammatory state in macrophages through its acidic cluster domain: Involvement of TNF alpha receptor associated factor 2. *PLoS ONE* **2011**, *6*, e22982. [CrossRef] [PubMed]
156. Luo, X.; Fan, Y.; Park, I.W.; He, J.J. Exosomes are unlikely involved in intercellular nef transfer. *PLoS ONE* **2015**, *10*, e0124436. [CrossRef]
157. Olivetta, E.; Arenaccio, C.; Manfredi, F.; Anticoli, S.; Federico, M. The Contribution of Extracellular Nef to HIV-Induced Pathogenesis. *Curr. Drug Targets* **2016**, *17*, 46–53. [CrossRef] [PubMed]
158. Lattanzi, L.; Federico, M. A strategy of antigen incorporation into exosomes: Comparing cross-presentation levels of antigens delivered by engineered exosomes and by lentiviral virus-like particles. *Vaccine* **2012**, *30*, 7229–7237. [CrossRef]

159. D'Aloja, P.; Santarcangelo, A.C.; Arold, S.; Baur, A.; Federico, M. Genetic and functional analysis of the human immunodeficiency virus (HIV) type 1-inhibiting F12-HIVnef allele. *J. Gen. Virol.* **2001**, *82*, 2735–2745. [CrossRef] [PubMed]
160. Sistigu, A.; Bracci, L.; Valentini, M.; Proietti, E.; Bona, R.; Negri, D.R.; Ciccaglione, A.R.; Tritarelli, E.; Nisini, R.; Equestre, M.; et al. Strong CD8+ T cell antigenicity and immunogenicity of large foreign proteins incorporated in HIV-1 VLPs able to induce a Nef-dependent activation/maturation of dendritic cells. *Vaccine* **2011**, *29*, 3465–3475. [CrossRef]
161. Andzinski, L.; Kasnitz, N.; Stahnke, S.; Wu, C.F.; Gereke, M.; von Kockritz-Blickwede, M.; Schilling, B.; Brandau, S.; Weiss, S.; Jablonska, J. Type I IFNs induce anti-tumor polarization of tumor associated neutrophils in mice and human. *Int. J. Cancer.* **2016**, *138*, 1982–1993. [CrossRef]

6

Potential use of Extracellular Vesicles Generated by Microbubble-Assisted Ultrasound as Drug Nanocarriers for Cancer Treatment

Yuana Yuana [1,2,*], Banuja Balachandran [1], Kim M. G. van der Wurff-Jacobs [3], Raymond M. Schiffelers [4] and Chrit T. Moonen [1]

1 Imaging Division, University Medical Center Utrecht, 3584 CX Utrecht, The Netherlands
2 Department of Biomedical Engineering, TU Eindhoven, 5600 MB Eindhoven, The Netherlands
3 Department of Pharmaceutical Sciences, University of Utrecht, 3584 CG Utrecht, The Netherlands
4 Laboratory of Clinical Chemistry and Haematology, University Medical Center Utrecht, 3584 CX Utrecht, The Netherlands
* Correspondence: Y.yuana@tue.nl;

Abstract: Extracellular vesicles (EVs)-carrying biomolecules derived from parental cells have achieved substantial scientific interest for their potential use as drug nanocarriers. Ultrasound (US) in combination with microbubbles (MB) have been shown to trigger the release of EVs from cancer cells. In the current study, the use of microbubbles-assisted ultrasound (USMB) to generate EVs containing drug cargo was investigated. The model drug, CellTracker™ green fluorescent dye (CTG) or bovine serum albumin conjugated with fluorescein isothiocyanate (BSA FITC) was loaded into primary human endothelial cells in vitro using USMB. We found that USMB loaded CTG and BSA FITC into human endothelial cells (HUVECs) and triggered the release of EVs containing these compounds in the cell supernatant within 2 h after treatment. The amount of EV released seemed to be correlated with the increase of US acoustic pressure. Co-culturing these EVs resulted in uptake by the recipient tumour cells within 4 h. In conclusion, USMB was able to load the model drugs into endothelial cells and simultaneously trigger the release of EVs-carrying model drugs, highlighting the potential of EVs as drug nanocarriers for future drug delivery in cancer.

Keywords: drug delivery; extracellular vesicles; lysosome; nanocarriers; ultrasound

---

1. Introduction

A major problem of drug delivery in cancer is that drugs cannot always efficiently bypass the multiple biological barriers. In most of the tissues, the endothelial lining is dense and does not permit the rapid exchange of materials such as drugs between the blood and the interstitial tumour space. Improvements in the drug delivery system to treat cancer including the delivery vehicles may improve the efficacy and safety of drugs [1,2].

Nanosized-lipid bilayer vesicles, known as extracellular vesicles (EVs), carry biomolecules such as proteins, lipids and genetic information that originated from the host cells [3]. The nanometre size and targeting modalities enable EVs to penetrate the interstitial tumour space. When EVs are loaded with a drug, this drug cargo can be delivered and transferred to dedicated cancer cells [1,4]. In general, drugs can be loaded into host cells allowing their natural mechanism to package drugs into EVs, also known as endogenous loading [5]. Drugs also can be exogenously loaded into EVs after isolation/purification steps using physical or chemical methods [6]. The feasibility and safety of EVs also have been shown in human clinical trials [7]. Thus, EVs hold potentials as drug nanocarriers.

Ultrasound (US) in combination with gaseous microbubbles (MB), beyond their common use for diagnosis, represents an emerging approach for localized drug delivery [8,9]. Under the action of US waves, MBs transiently affect the permeability of biological barriers (e.g., cell membrane, endothelial lining), thus leading to the uptake and enhanced accumulation of drugs in the region where the US-induced pressure surpasses a certain threshold [10,11]. Despite the potential benefits of USMB-guided drug delivery, the micron size of MB used in this system limits their action exclusively to the vascular bed [12,13]. Thus, future refinements in USMB-guided drug delivery to enhance drug uptake to the targeted tumour cells are warranted.

Recently, we showed that EVs with a diameter of less than 200 nm were released from head-and-neck cancer cells (FaDu) in vitro for up to 4-fold at 2–4 h after USMB [14]. The fact that USMB triggered a rapid release of EVs shows the significance of this technology to generate EVs and opens opportunities for us to study whether the drug-loaded EVs can be generated following the loading of the drug into the host cells using USMB.

In this study, we used human endothelial cells (HUVECs) as an in vitro model. CellTracker™ green fluorescent dye (CTG) and bovine serum albumin (BSA) with fluorescein isothiocyanate (FITC) conjugate were used as model drugs. These compounds have distinct characteristics: CTG could freely pass through cell membranes and be transformed into a cell-impermeant fluorescent product in the cytosol, whereas BSA FITC conjugate could pass through cell membranes via endocytosis. Different US acoustic pressures (0.6, 0.7 and 0.8 MPa) were applied to load CTG and BSA FITC into HUVECs. Cells and cell supernatants were harvested and measured to investigate if model drugs were present in the cells and EVs. Subsequently, we highlighted the potential use of EVs as drug nanocarriers by demonstrating the uptake of drug-loaded EVs generated by USMB after 4 h being co-cultured with the recipient cancer cells, FaDu and MDA-MB-231 human breast carcinoma cells. Finally, we utilized the Cre-LoxP system to confirm these findings [15].

2. Results

2.1. Loading Model Drugs into Cells

Using the bright-field microscopy, we observed that directly after US application at 0.6, 0.7 and 0.8 MPa, most cells remained attached to the culture chamber (Figure 1A). Increasing the pressure to 1 MPa detached almost 50% of the cells from the surface of the culture chamber, and therefore this condition was not used further in this study (Figure 1A). Untreated and USMB-treated HUVECs (0.6, 0.7 and 0.8 MPa) were harvested and stained using propidium iodide (PI) to determine the percentage of live and dead cells using flow cytometry. The percentages of live and dead cells in treated samples were comparable with the untreated samples (Figure 1B). These results indicated that the US acoustic pressures 0.6, 0.7 and 0.8 MPa could be used safely for loading the model drug into cells in vitro. We observed that HUVECs could be loaded directly with CTG by incubating CTG for 30 min (Figure 1C). However, USMB did not enhance the loading of CTG in HUVECs (Figure 1C).

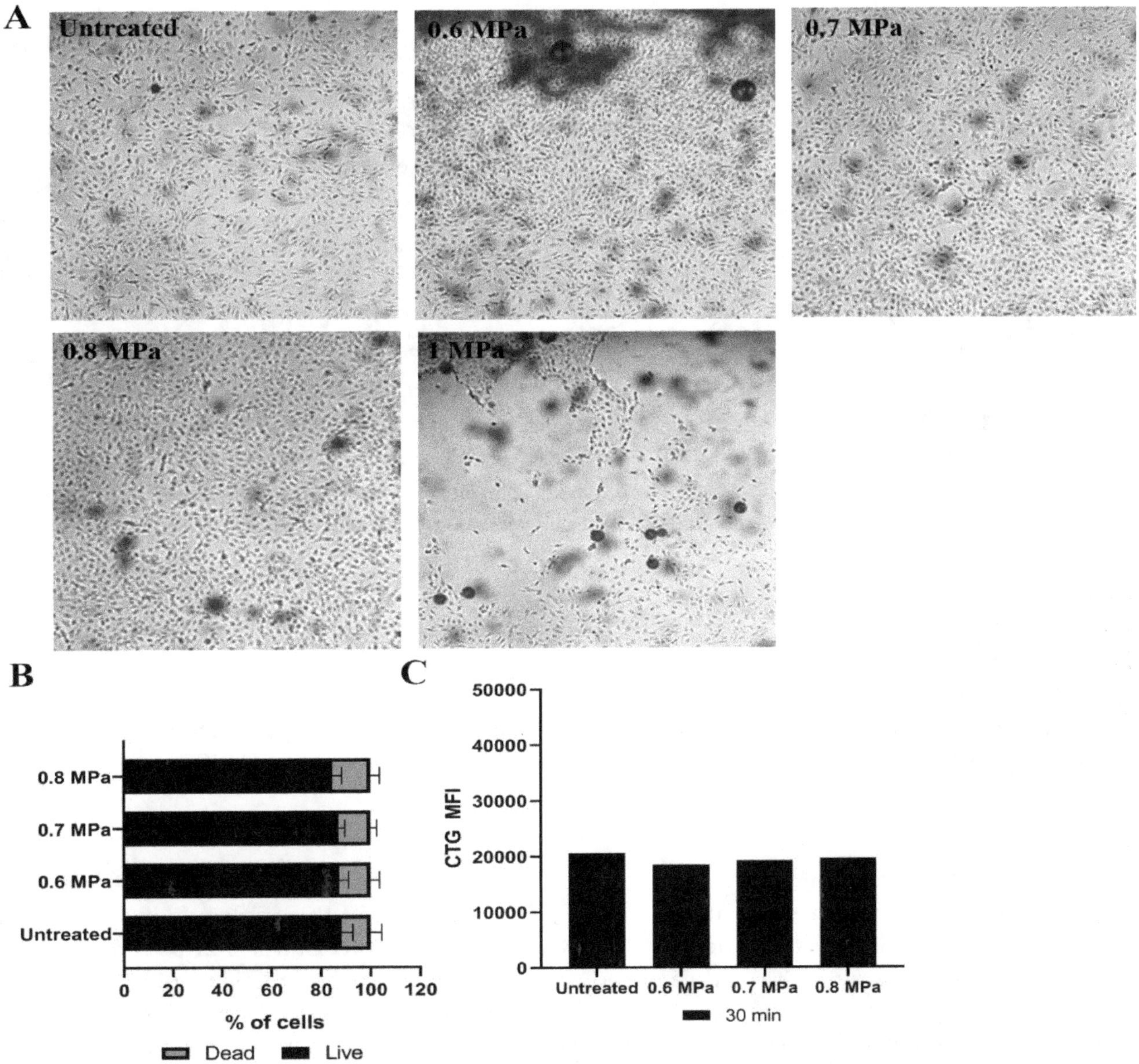

Figure 1. Human endothelial cells (HUVECs) stained with CellTracker™ green fluorescent dye (CTG) and treated at different ultrasound (US) acoustic pressures. Microscopy pictures of HUVECs captured before (untreated) and directly after US treatment (0.6, 0.7, 0.8 and 1 MPa) are presented at 10× magnification (**A**). Afterwards, cells were harvested and stained with PI to determine live and dead cells. The mean percentages of live and dead cells are presented (**B**). US acoustic pressures of 0.6, 0.7 and 0.8 MPa were tested to increase CTG signal intensities in HUVECs after incubation of CTG for 30 min (**C**). Untreated condition was HUVEs after incubation of CTG for 30 min. After 2 h of post-USMB, HUVECs were harvested and measured using flow cytometry (**C**). The mean fluorescent intensities (MFI) of CTG were presented (**C**). As a control, HUVECs without CTG staining were used. MFI results presented are the mean of two independent experiments.

By using USMB, our objective was to increase the delivery of BSA FITC into the cells. We found that loading BSA FITC using different US acoustic pressures was not correlated with the increase of BSA FITC signal in the cells. At 0.6 MPa, BSA FITC signal intensities in HUVECs measured by flow cytometry were the highest compared to untreated and treated HUVECs at 0.7 and 0.8 MPa (Figure 2A). Using a fluorescence microscope, we observed that BSA FITC was present in the HUVECs after treatment with different US acoustic pressures (0.6, 0.7 and 0.8 MPa) and also in the untreated HUVECs (Figure 2B).

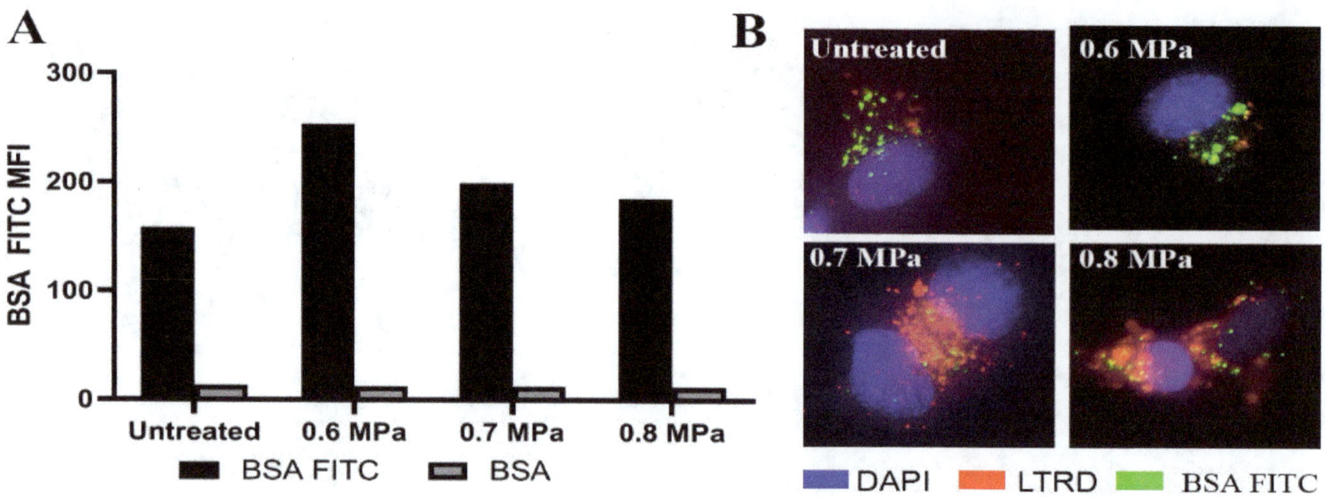

Figure 2. HUVECs loaded with bovine serum albumin conjugated with fluorescein isothiocyanate (BSA FITC) before and after USMB application at different acoustic pressures. After administering BSA FITC with USMB, HUVECs were incubated for 1 h and then, the medium was refreshed. After 2 h, HUVECs were harvested and measured with flow cytometry to quantify the BSA FITC signal, depicted as MFI (**A**). The results presented were the mean of two independent experiments. These cells were also stained with a nuclear stain, DAPI (blue) and a lysosomal marker, LTRD (**B**). Co-localisation of red and green colours indicate the presence of BSA FITC in the lysosomal compartment (**B**). These images were representative images taken at 60× magnification using a fluorescence microscope (Keyence BZ9000, Mechelen, Belgium).

This BSA FITC signal was co-localised with LysoTraker™ Red DND-99 (LTRD) staining in HUVECs, especially after US treatments (Figure 2B). The presence of this molecule in the lysosomal compartment of these cells possibly indicates the degradation of BSA in this compartment. We also co-administered HUVECs with DQ-Red BSA and MB before application of US at 0.6, 0.7 and 0.8 MPa to check if lysosomes were involved in the degradation of BSA. The bright red spots which were indicative for de-quenched fluorescence molecules of DQ-Red BSA when reaching the degradative endo-lysosomes, were detected in HUVECs after US treatment (Figure S1). These red spots were absent in untreated HUVECs (Figure S1). Thus, these results confirmed the presence of BSA in lysosomal compartments after US treatment.

2.2. Triggering EV Release Containing Model Drug Cargo

Using anti-CD9 and anti-CD63-coated magnetic beads, we captured EVs from the cell supernatants of HUVECs which have been incubated with CTG for 30 min and treated with USMB. USMB triggered an increased release of EVs from HUVEC shown by an increased level of mean fluorescent intensity (MFI) of EVs positive for CD9 and CD63 after USMB treatment especially at 0.8 MPa compared to the untreated, 0.6 and 0.7 MPa conditions (Figure 3A,B). EVs captured by anti-CD9 and anti-CD63 beads carried CTG and the MFI levels of CTG seemed to increase when US acoustic pressures were increased (Figure 3A,B). The measurement of the same cell supernatants using a micro flow cytometer showed that the concentrations of EVs exposing CD9 and CD63 increased 2- and 3-fold subsequently after USMB compared to untreated counterparts, especially at 0.7 and 0.8 MPa (Figure 3C). These EVs also carried CTG and their concentration was significantly higher at 0.8 MPa ($p < 0.05$, Figure 3C). The percentages of EVs positive for CD9 or CD63 which carried CTG generated by USMB at 0.8 MPa are about 20% and 37% respectively.

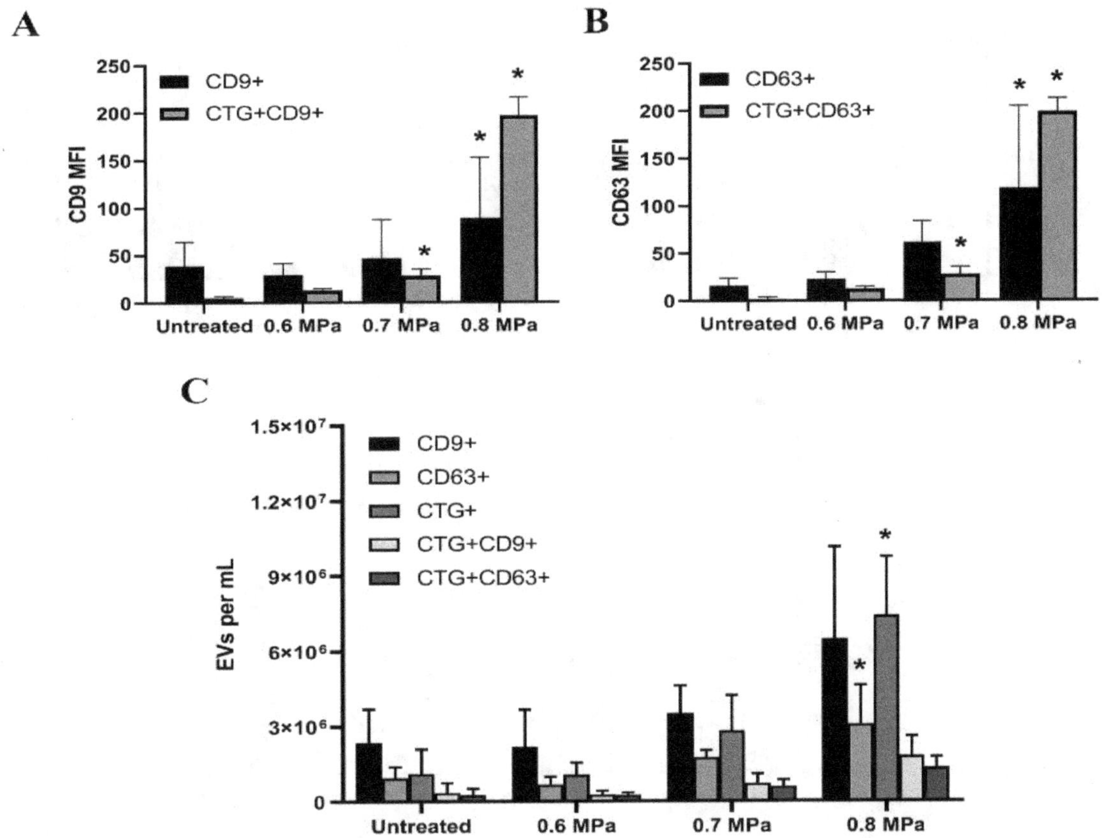

Figure 3. Extracellular vesicles (EVs)-containing supernatants of HUVECs after CTG and USMB treatments measured using flow cytometry. EVs were captured by anti-CD9 and anti-CD63 beads in the supernatants of CTG-stained HUVEC followed by USMB treatment. These EVs were stained with anti-CD9 and anti-CD63 detection antibodies. The MFI of CD9 and CD63 levels in untreated and USMB treated samples (0.6, 0.7 and 0.8 MPa) were quantified using flow cytometry (**A**,**B**). These captured CD9 EVs and CD63 EVs carried CTG (**A**,**B**). Samples were measured in triplicate and these were data from two independent experiments. The same samples were measured using a micro flow cytometer. Anti-CD9 and -CD63 were used to detect EVs bearing CD9 and CD63 antigens present in the cell supernatants (**C**). Total EVs carrying CTG and EVs exposing CD9 or CD63 which carried CTG were also quantified in the same samples (**C**). Results are reported in EVs per mL sample. These were data from two independent experiments. As controls, EVs without CTG were used and all results presented have been corrected from these controls (**A**–**C**). Statistical analysis was performed using one-way ANOVA followed by Tukey's test. The p value < 0.05 was considered significant (*).

HUVECs loaded with BSA FITC also released EVs containing BSA FITC (Figures 4 and 5). These EVs from the cell supernatants were captured by anti-CD9 and anti-CD63 beads (Figure 4A,B). As controls, BSA was used and EVs present in the supernatant of HUVECs were measured. EVs exposing CD9 and CD63 generated by USMB showed higher BSA FITC fluorescent signal (BSA FITC MFI) than the BSA FITC signal detected from EVs generated by untreated counterparts (Figure 4A,B). However, the increased MFI levels of BSA FITC were not correlated with increasing US pressures. Using the micro flow cytometer, we could detect the concentration of EVs carrying BSA FITC increased significantly at the conditions of 0.7 and 0.8 MPa (2.4×10^6/mL and 2.2×10^6/mL; Figure 4C) compared to the untreated conditions. The percentages of EVs positive for CD9 or CD63 which carried BSA FITC generated by USMB at 0.7 MPA are about 12% and 25% respectively, whereas these generated by USMB at 0.8 MPA are about 10% and 18% respectively. Because of the sensitivity of the micro flow cytometer in measuring single EVs, differences in EV concentrations in condition media of HUVEC treated using different US pressures were detectable.

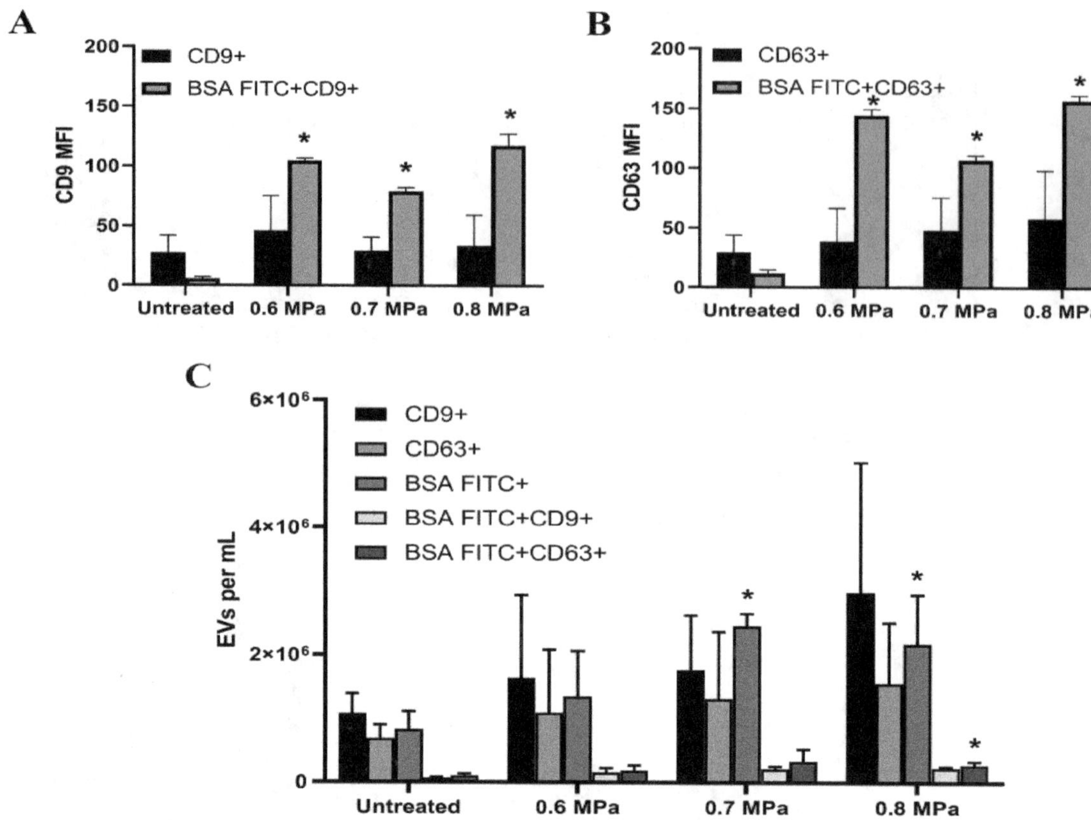

Figure 4. EVs-containing supernatants of HUVECs after co-administration of BSA FITC and MB followed by US treatments measured using flow cytometry. Anti-CD9 and anti-CD63 capture assays coupled to flow cytometry measurement were used to detect CD9 and CD63 EVs present in the conditioned media of HUVEC co-administered with BSA FITC together with MB. Anti-CD9 captured EVs with CD9 on the surface, whereas anti-CD63 captured EVs with CD63 on the surface. Anti-CD9 and anti-CD63 were used as the detection antibodies (**A**,**B**). The MFI of CD9 and CD63 levels in untreated and USMB treated samples (0.6, 0.7 and 0.8 MPa) and also the BSA FITC signals in these captured CD9 EVs and CD63 were quantified (**A**,**B**). Samples were measured in triplicate and these were data from two independent experiments. The same samples were measured by the micro flow cytometer. Anti-CD9 and -CD63 were used to detect EVs bearing CD9 and CD63 antigens present in the conditioned media (**C**). Total EVs carrying BSA-FITC, indicated by EVs exposing CD9 and CD63, which carried BSA-FITC were quantified in the same samples (**C**). Results are reported in EVs per mL sample. These were data from two independent experiments. EVs with BSA were used as controls and all results presented have been corrected from these controls (**A**–**C**). Statistical analysis was performed using one-way ANOVA followed by Tukey's test. The p value < 0.05 was considered significant (*).

To visualise EVs carrying BSA FITC, we also performed immunogold electron microscopy. We used anti-CD9 and anti-CD63 to detect EVs exposing CD9 and CD63 on their surface. Anti-mouse IgG secondary antibody labelled with 6-nm gold particles was used to trace EVs exposing CD9 or CD63. To investigate whether these EVs were carrying BSA FITC, we stained the samples with anti FITC secondary antibody labelled with 10-nm gold particles. We found that CD9-exposing EV carried BSA FITC (Figure 5A). Some CD63-exposing EVs carried BSA FITC (Figure 5B). In Figure 5C, EV carrying BSA FITC was clearly detected. Negative controls were samples containing EVs with BSA

(Figure 5D–F). EVs imaged by electron microscopy mostly have a diameter of around 100 nm. This is also confirmed by NTA measurements (Figure S2). These results show that HUVECs which were loaded with BSA FITC released EVs carrying BSA FITC after USMB treatment.

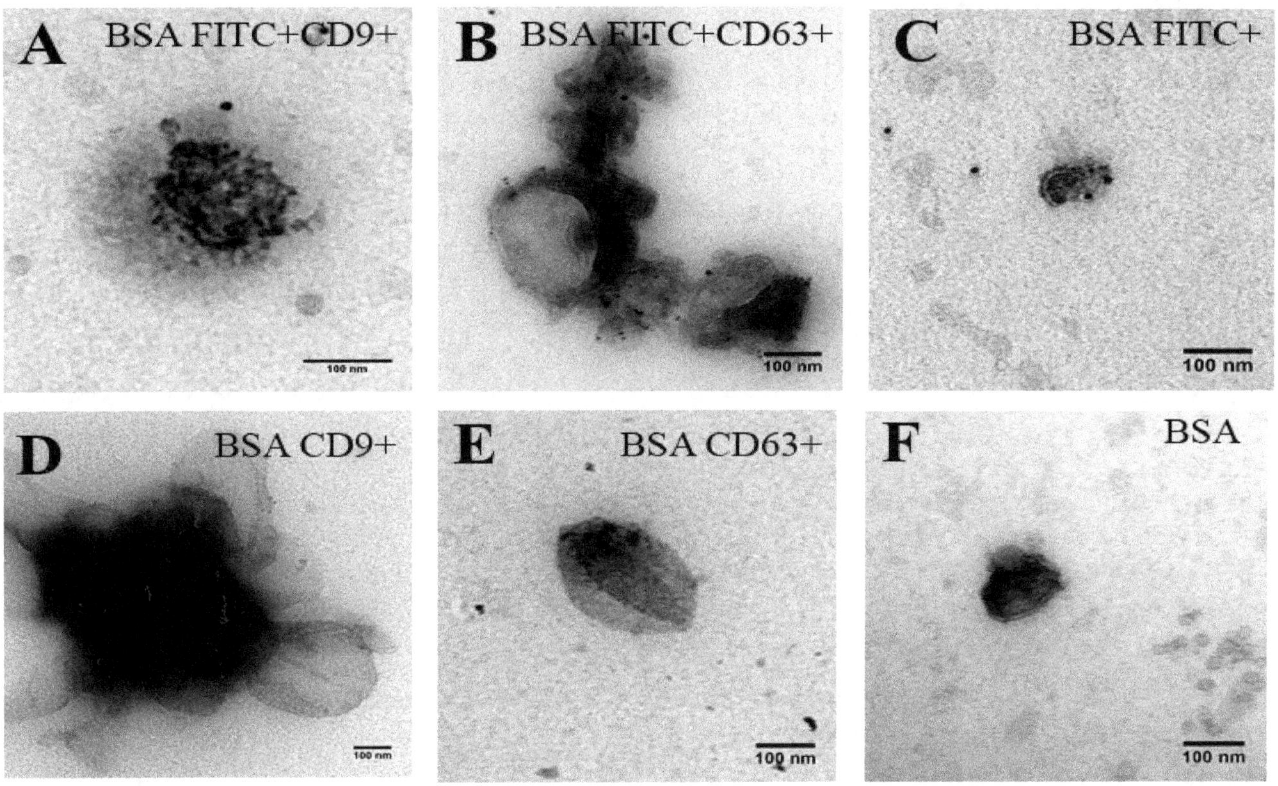

Figure 5. Immunogold electron microscopy of cell supernatant containing EVs derived from HUVEC treated with USMB (0.7 MPa) in the presence of BSA FITC as a model drug. EVs were stained with anti-CD9 or anti-CD63 and the detection antibody was anti mouse IgG secondary antibody labelled with 6-nm gold particles (**A**,**B**). IgG1 isotype control was used as a control for anti-CD9 and anti-CD63 staining (**C**). Anti FITC secondary antibody labelled with 10-nm gold particles stained BSA FITC present on EVS (**A**–**C**). As controls, cell supernatant containing EVs with non-conjugated BSA processed and stained in the same manner (**D**–**F**).

2.3. Uptake of EVs Carrying CTG and BSA FITC

We performed flow cytometry measurement to quantify the EV uptake by FaDu and MDA-MB-231 after 4 h of co-culturing. The uptake of EV carrying CTG or BSA FITC is shown by the percentages of positive cells for CTG or BSA FITC (Figure 6A,B). After co-culturing with undiluted EVs carrying CTG, there were 0.3% FaDU cells and 0.5% MDA-MB-231 cells positive for CTG (Figure 6A). We also monitored the uptake of EVs containing CTG by MDA-MB-231 cells at 0, 2 and 4 h using a fluorescent microscope (Figure S3). After co-culturing for 4 h, only 2–3 cells per 1.7 cm^2 were positive for CTG. Thus, the overall results show that the uptake of EVs containing CTG by these cancer cells was low. The same cancer cells took up more EVs carrying BSA FITC (Figure 6B). Particularly, 9.7% of the FaDu cells and 30.5% of the MDA-MB-231 cells were positive after co-culturing with undiluted EVs carrying BSA FITC as detected by flow cytometry (Figure 6B).

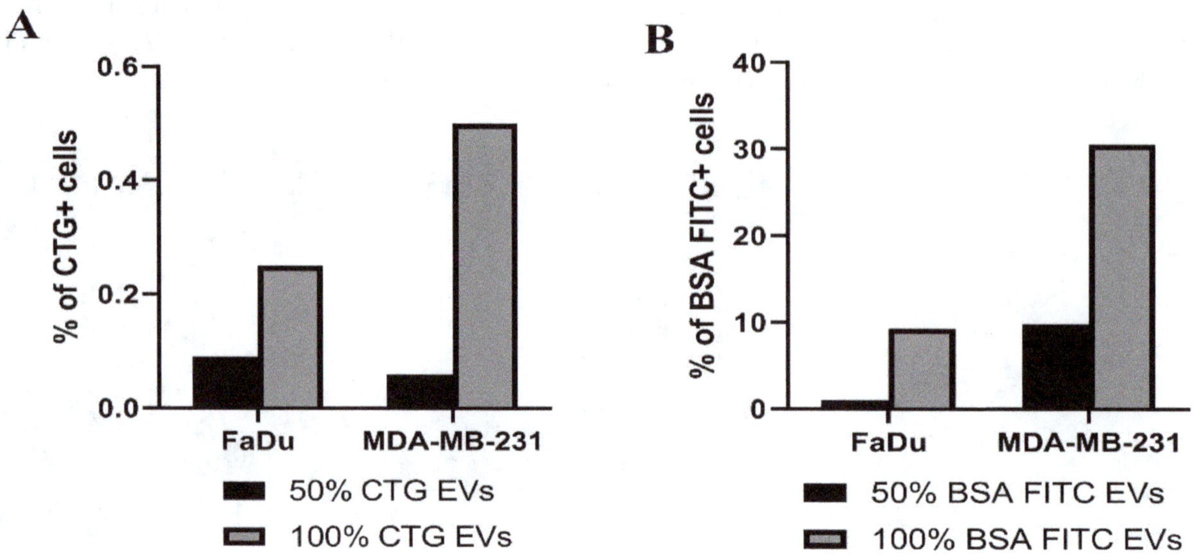

Figure 6. Uptake of EVs containing CTG or BSA-FITC by FaDu and MDA-MB-231 cells. EVs containing CTG (**A**) or BSA-FITC (**B**) at a concentration of 50% and 100% were co-cultured with FaDu or MDA-MB-231 cells for 4 h. The uptake of these EVs was measured by using flow cytometry. Percentages of FaDu and MDA-MB-231 cells exposing CTG or BSA FITC after EV uptake at 4 h-time point are presented (**A**,**B** accordingly). Control conditions were cells co-cultured with EVs which contained no CTG or BSA FITC. Percentages of the positive cells have been corrected for their controls and the uptake of EVs was normalized to 1×10^9 particles/mL.

We also stained the cells with the lysosomal tracker, LTRD, and nucleic acid stain DAPI, to localize the presence of EVs containing BSA FITC in the cells. Using fluorescent and confocal microscopes, we observed that EVs containing BSA FITC were taken up by cells and they were not present in the lysosomal compartment of these cells (Figure 7A,B). The FITC fluorescent signal was observed in the cell cytoplasm or in the vicinity of the cell nucleus (Figure 7A,B accordingly, and Supplementary Video S1).

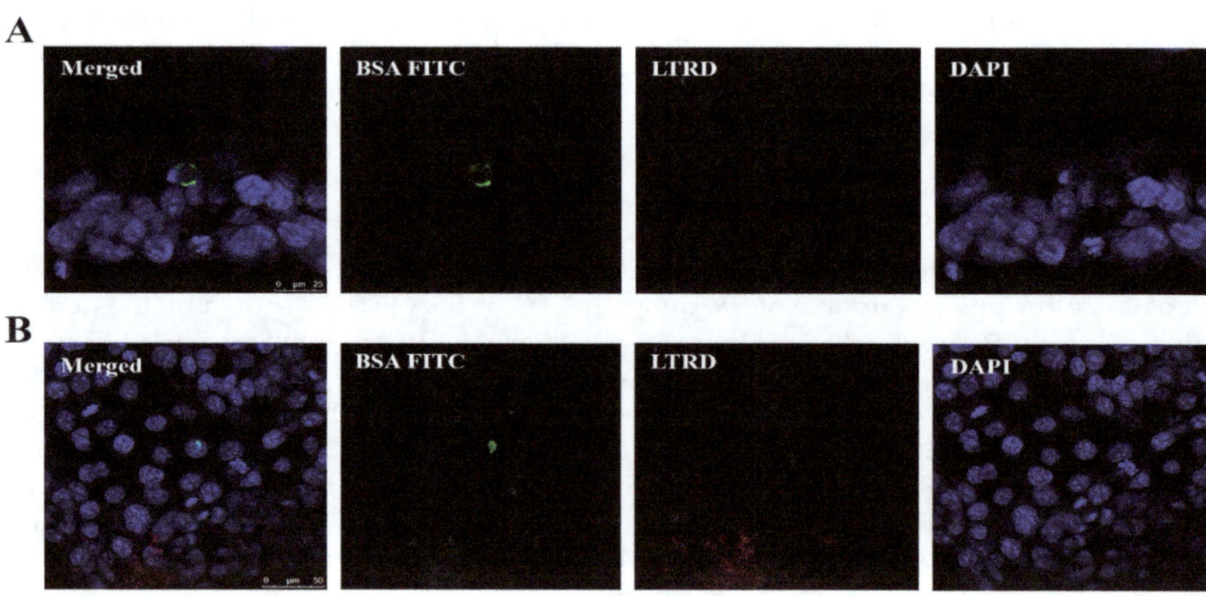

Figure 7. Uptake of EVs containing BSA FITC by FaDu cells. FaDu cells were co-cultured with EVs containing BSA FITC (undiluted). After 4 h, cells were washed with acid wash buffer followed by PBS and fixed prior to staining. LTRD probe (red) was used to stain lysosome and DAPI (blue) to stain the nucleus. Merged images show the uptake of BSA FITC by FaDu cells. BSA FITC signal (green) seems to be present in the cell cytoplasm (**A**, **merged**) or in the vicinity of the cell nucleus (**B**, **merged**).

To confirm the uptake of EVs by recipient tumour cells and the release of their cargo in these cells, we co-cultured EVs containing Cre recombinase with T47D receptor cells. The expression of Cre recombinase in EVs has been confirmed by PCR (Figure S4). The colour switch from red to green when receptor cells took up EVs containing Cre recombinase was monitored using a fluorescence microscope and flow cytometer (Figure 8A,B). Using a fluorescent microscope, we monitored the uptake of EVs containing Cre recombinase (100% concentration) at 0, 2 and 4 h (Figure 8A). At 2 h, we could not detect the green Cre-LoxP cells. At 4 h, we located green Cre-LoxP cells (2 cells/0.7 cm^2 chamber slide, Figure 8A white arrow). Flow cytometry analysis also shows that Cre-LoxP cells were present after co-culturing T47D recipient cells with EVs containing Cre recombinase at a concentration of 50% and 100% (Figure 8Bi,ii). Cells co-cultured with EVs without Cre recombinase served as control (Figure 8Biii). The percentage of green Cre-LoxP cells increased with increasing concentration of EV containing Cre recombinase (7.33% to 10.5%; Figure 8Bi–ii). A red-to-green colour switch observed in reporter-expressing T47D human mammary cells after co-culturing USMB-generated EV carrying active cre recombines confirms the uptake of EVs and transfer of their active cargo into the recipient cells.

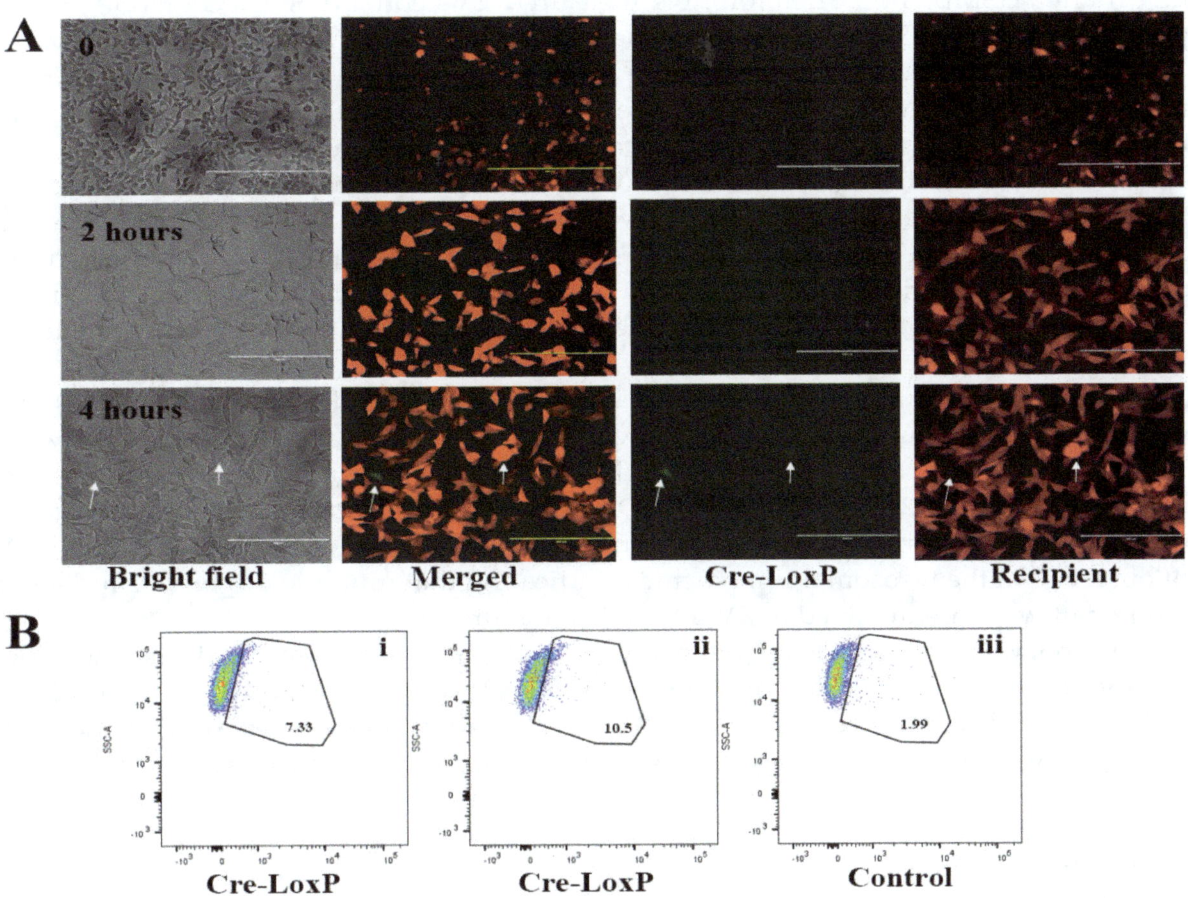

Figure 8. Uptake of EVs carrying Cre recombinase by T47D recipient tumour cells. Undiluted (100%) EVs carrying Cre-recombinase were added to T47D recipient cells (red). At 0, 2 and 4 h, bright field and fluorescent images were taken (**A**). Images were taken at 10× magnification at time point 0. Other images were taken at 20× magnification. Cre-LoxP cells which have a green fluorescent colour were located and shown with a white arrow, both in the bright field and fluorescent images. This uptake was also monitored using flow cytometry. Flow cytometry measurements were performed to assess the presence of Cre-LoxP cells which exposed green fluorescent signal after the co-culturing of EVs carrying Cre recombinase at a concentration of 50% and 100% for 4 h (**Bi–ii**). Controls were cells that were co-cultured with EVs without Cre recombinase (**Biii**).

3. Discussion

In the current study, we observed that increased US acoustic pressure is not always favourable for drug loading. By increasing US acoustic pressure to 1 MPa, cells were already detached from the culture chamber. Lammertink et al. [16] reported that HUVEC detachment was already observed after the application of 0.5 MPa. However, in that study HUVEC was grown on collagen-coated OptiCells™ which may explain this discrepancy. We also noticed that as CTG could freely pass through cell membranes, incubation for 30 min was already sufficient to stain/load HUVECs with CTG and therefore, applying USMB had no additional effect. In the case of BSA FITC loading, applying USMB at 0.6 MPa enhanced the loading. However, increasing US acoustic pressure to 0.7 or 0.8 MPa did not further enhance the loading of this molecule into cells. Based on the degradation of DQ-Red BSA and the de-quenching of the fluorescence after USMB loading, it was likely that our model drug was trapped and degraded in the lysosomal compartment. In conclusion, the type of the model drug is one of parameters which may influence the loading using USMB. It is also clear that this loading would not be optimal if the model drug would be trapped in the lysosomal compartment.

Our previous findings indicated that levels of EVs exposing CD9 or CD63 significantly increased at 2 and 4 h after USMB [14]. Therefore, we measured EVs containing CTG or BSA FITC in the cell supernatant 2 h after USMB treatment. The concentration of EVs containing CTG was 3-fold higher than the concentration of EVs containing BSA FITC, especially when the highest US acoustic pressure was applied (0.8 MPa). This difference may be caused by the mechanism through which these compounds are taken up and sorted by the cells [17]. As CTG could freely enter cell membranes and reside in the cytosol, this compound may avoid endosomal entrapment and perhaps easily be packed into EVs. In contrast, BSA FITC which was internalised by cells via endocytosis may lead to the delivery of endocytosed cargo into the endosomal-lysosomal degradative pathway, influencing the packing of this cargo into EVs [18,19]. It may be necessary to reduce the lysosomal function to prevent the drug compound degradation after internalization and eventually increase the release of EV containing drug [19,20].

Multiple different cellular entry routes are available for EVs and other nanoparticles to cross a cell's plasma membrane during in vitro and in vivo cell exposure. This cell uptake likely depends on the type of recipient cells and the EV cargo [17,21,22]. In our study, EVs prepared in a similar manner were internalised 2–3-fold more by MDA-MB-231 than FaDu cells. We observed that EVs containing CTG were taken up less than EVs containing BSA FITC. As EVs containing BSA FITC were not observed in the lysosomal compartment indicating that multiple or perhaps independently contributing pathways are involved in EV internalisation processes.

In conclusion, we have demonstrated the feasibility of USMB to load model drugs into endothelial cells allowing the cell's natural mechanism to generate EVs and package this compound into EVs. Co-culturing these EVs with FaDu and MDA-MB-231 cells resulted in the uptake of EVs and the release of model drug cargo in these cells. These results highlight the potential of EVs as drug nanocarriers for drug delivery.

4. Materials and Methods

4.1. Cell Culture

HUVECs (pooled donors, Lonza, Verviers, Belgium) were cultured in endothelial cell growth medium 2 with supplement mix (PromoCell GmbH, Heidelberg, Germany). MDA-MB-231 (ATCC® HTB-26™, LGC Standards GmbH, Wesel, Germany) and FaDu (ATCC® HTB-43™) cells were cultured in high glucose Dulbecco's modified Eagle medium (DMEM) (Sigma-Aldrich, Zwijndrecht, The Netherlands) and 10% foetal bovine serum (FBS) (Sigma-Aldrich). For FaDu cells, this medium was also supplemented with 1% non-essential amino acid (Sigma-Aldrich, Zwijndrecht, The Netherlands). MDA-MB-231 expressing Cre recombinase and T47D receptor cells were a kind gift from Prof. Raymond Schiffelers and these stable cell lines have been generated as described by Zomer et al. [15]. MDA-MB231 cells expressing Cre recombinase were cultured in high glucose DMEM, whereas T47D receptor cells were cultured in DMEM: F12 medium (Lonza, Breda, The Netherlands). In these culture media, 10% FBS, 1% penicillin-streptomycin (Fisher Scientific, Landsmeer, The Netherlands) and 5 µg/mL puromycin (Fisher scientific) were added. All cells were maintained in standard cell culture flasks in a humidified incubator at 37 °C and 5% CO_2 and passaged twice a week to maintain 80% confluency.

4.2. Cell Preparation for USMB

For USMB experiments, cells were prepared according to our previously reported method [14]. Briefly, cells were seeded in cell culture chambers (treated CLINIcell® 175 µm thick polycarbonate walls, 25 cm^2) (Mabio, Tourcoing, France) 1 day prior to USMB. Media used for maintaining cells in the culture chambers are the same media for maintaining cells in the culture flasks. Except for HUVECs, medium M199 containing L-glutamine (Sigma-Aldrich) and 20% FBS was used. Prior to USMB treatment, these culture media were filtered using 100 KDa centrifugal filter units (Amicon ultra-15, Merck Millipore, Amsterdam, The Netherlands) to deplete EVs related to bovine serum.

4.3. Loading Model Drug Using USMB

SonoVue™ (Bracco, Milan, Italy) lipid shelled MB, encapsulating sulphur hexafluoride gas (SF6) were used in the USMB experiments [23]. Microbubbles (MB) were prepared according to the manufacturer's guidelines.

For co-administrating CTG, cells were first washed with phosphate-buffered saline (PBS) and incubated with 10 nM CTG in serum-free medium for 30 min. As a negative control, cells without CTG were used. Prior to USMB, MB (700 µL) were mixed with 9.5 mL EV-depleted serum-containing medium. This MB-containing medium (final concentration of 0.7–3.5 × 10^8 MB) was then injected and the cell culture chamber was placed upside down, allowing the MB to rise by floatation towards the cells, ensuring close contact between cells and MB. For co-administrating BSA FITC, cells were first starved for 4 h in serum-free medium before BSA FITC (50 µg/mL) and MB were administered. As a negative control, non-conjugated BSA was administered at the same concentration. We also used DQ-Red BSA (Fisher scientific) to observe the loading of BSA by USMB. Once DQ-Red BSA has reached the degradative endo-lysosomes, it is broken down into smaller fragments which lead to fluorescence de-quenching shown as bright red spots in endo-lysosomes [24], and therefore tracking the location of this molecule in the cell.

Cells and MB were exposed to pulsed ultrasound (10% duty cycle, 1 kHz pulse repetition frequency and 100 µs pulse duration) with varying acoustic pressures at 0.6, 0.7 and 0.8 MPa peak negative pressure (PNP) as was calibrated using a 125 µm glass fibre hydrophone (Precision Acoustics, Dorchester, UK). The piezoelectric unfocused single element transducer with a diameter of 20 mm (Precision Acoustics) operating at 1.5 MHz was placed at the bottom of a water tank, 8 cm below a cell culture chamber frame. The water surface was another 12 cm above the frame and heated to 37 °C. When exposing cells to US, the cell culture chamber was mounted in the frame and moved over the transducer for 80 s to expose the whole cell culture chamber surface, which resulted in a total exposure of approximately 5 s for each area of the cell culture chamber [14,25]. For USMB treatment control, untreated cells were used.

After US exposure, CTG-treated cells were washed carefully with PBS and incubated for another 2 h in EV-depleted serum-containing medium. For BSA FITC-treated cells, we first incubated these cells for 1 h before they were washed carefully with PBS and incubated for another 2 h. After incubation, the conditioned medium was harvested and centrifuged at 300× *g* for 10 min at room temperature (RT) to remove detached cells and large cellular debris. Conditioned medium containing EVs was collected, snap-frozen in liquid nitrogen and stored at −80 °C for EV measurements. Attached cells in the cell culture chamber were detached using trypsin-EDTA solution (Sigma-Aldrich). After serum-containing medium was added, cells were pelleted by centrifuging at 250× *g* for 3 min at RT. To wash cells, cold PBS was added and cells were pelleted again. Finally, cells were re-suspended in 300 µL PBS containing 0.5% BSA and used for flow cytometry measurement.

4.4. Flow Cytometry Measurement

The cell suspension was mixed with 5 µL of 10 µg/mL propidium iodide (PI) (Fisher Scientific) and incubated for 5 min. Up to 10,000 events were measured using BD FACScanto II (Becton Dickinson, CA, USA) and the fluorescent intensity of CTG or BSA FITC together with PI was monitored.

For EV measurement using flow cytometry, conditioned medium containing EVs was thawed quickly at 37 °C. We captured EVs using immuno-magnetic beads and measured these EV-bead complexes using BD FACScanto II according to our previously reported method [14]. Using 96-well round-bottom plate 70 µL conditioned medium containing EVs was mixed with 10 µL ExoCap magnetic capture beads CD9 or CD63 (JSR Life Sciences, Leuven, Belgium) to obtain end concentration of 6×10^4 beads per sample. This mixture was incubated overnight at 4 °C with gentle agitation. On the next day, the magnetic plate separator was used to hold EV-bead complexes during washing steps. The supernatant was removed and 100 µL 0.5% BSA-containing PBS (0.45 µm filtered) was added followed by gentle agitation to wash EV-bead complexes for 5 min at RT. After placing the plate on top of the magnetic plate separator, the supernatant was removed. Hundred microliter of diluted mouse anti-human CD9, mouse anti-CD63, or the corresponding mouse IgG1 isotype control coupled to Alexa Fluor 647 were added to the respective wells (see Supplementary Table S1 for the dilutions and manufacturer). This mixture was incubated at RT in the dark for 30 min with gentle agitation. Lastly, the plate was again placed on top of the magnetic plate separator and the antibody/isotype control solution was removed. The plate was washed once with 100 µL 0.5% BSA-containing PBS per well. Samples containing EVs attached to the beads were re-suspended in 250 µL 0.5% BSA/PBS and measured using FACS canto-II. Two independent experiments were performed and from each experiment, samples were measured in triplicate to assess the levels of CD9 and CD63.

The concentration of EVs exposing CD9 or CD63, and EVs carrying CTG or BSA FITC were measured using the A60-Micro flow cytometer (Apogee Flow Systems, Hertfordshire, UK). This flow cytometer enables the measurement of biological particles down to about 100 nm diameter by light scatter [26]. Twenty microliter of conditioned medium containing EVs was incubated with 2.5 µL of mouse-anti human CD9 coupled to PE and 2.5 µL of mouse-anti human CD63 coupled to Alexa Fluor 647. As negative controls, mouse IgG1 isotype controls coupled to PE and to Alexa Fluor 647 were used (please see Supplementary Table S1 for the dilutions of antibodies/isotype controls and their manufacturer). The mixture of samples with antibodies or IgG1 isotype controls were incubated for 15 min and diluted to 225 µL before being measured with flowrate 3.01 µL/ min for 120 s.

4.5. Immunogold Electron Microscopy

To investigate whether BSA FITC was carried by EVs after USMB, immunogold electron microscopy was performed as described previously [14,27]. EVs from conditioned medium collected from USMB-treated HUVEC in the presence of BSA FITC or non-conjugated BSA were measured. A thin layer of carbon on the 100 mesh formvar-carbon-coated copper grids (Electron Microscopy Sciences, Hatfield, PA, USA) was evaporated using a carbon-vacuum evaporator according to the manufacturer's instructions (Edwards Auto 306, West Sussex, UK) just before sample application. Grids were floated directly on top of 10 µL of cell supernatant containing EVs and incubated for 7 min at RT. The grids were washed three times using PBS (0.22 µm filtered). Next, the grids were incubated in a blocking solution (Aurion, Wageningen, The Netherlands) for 30 min at RT followed by washing three times with 0.1% BSA-c (Aurion). For double immuno-labelling, the grids were incubated overnight at 4 °C in a mixture of 20 µL of mouse anti-human primary antibody (anti-CD9 or anti-CD63 antibody). For negative controls, mouse IgG1 isotype control was used. Unbound antibodies were removed from the grids in six washing steps by placing the grids on top of 0.1% BSA-c solution. Afterwards, the grids were incubated with 10 µL goat anti mouse IgG secondary antibody labelled with 6-nm gold particles

(Aurion) for 1 h at RT. Grids were washed six times with 0.1% BSA-c before incubation with 10 μL mouse anti-FITC labelled with 10-nm gold particles for 1 h at RT. Dilutions of the antibodies, isotype control and immunogold particles can be found in Supplementary Table S1. Grids were washed again six times with 0.1% BSA-c, followed by three times washing with PBS. To fix the labelled sample, the grid was incubated for 5 min with 2% glutaraldehyde (Sigma-Aldrich) and washed six times with milli-Q water. Next, the grids were transferred to 0.4% uranyl acetate in 2% methyl cellulose and incubated for 10 min. Excess solution on the grids was removed by blotting the grid at a 45° angle once from the side of the grid with filter paper. Grids were imaged using a Tecnai 12 electron microscope (FEI, Fisher Scientific) operated at 80 kV. Images were recorded at 60,000x magnification and processed using ImageJ [28]. The presence of a 6-nm gold particle on the EV surface indicated the presence of CD9 or CD63 on EV surface, whereas the presence of a 10-nm gold particle indicated the presence of BSA FITC in EV.

4.6. EV Isolation and Concentration

Before the EV isolation, the Amicon® ultra centrifugal filter 100 kDa (Merck Millipore) was washed with EV-depleted serum-containing medium by centrifuging at 4000× *g* for 30 min. The eluate was thrown. Next, 4 mL of supernatant containing EV with CTG, BSA FITC, or Cre recombinase which was collected after USMB treatment on HUVEC cells was added to the filtration device and centrifuged at 4000× *g* for 30 min. The eluate was thrown and the filter was washed once with EV-depleted serum-containing medium followed by centrifuging at 4000× *g* for 30 min. Finally, the EV concentrate was collected and the same medium was added up to get 750 μL EV suspension. As negative controls, the supernatant containing EVs without model drug that was generated at the same USMB condition was used.

4.7. Uptake Assay

Prior to the uptake assay, 70,000 cells (FaDu, MDA-MB231, or T47D reporter cells) were plated in the 8-chambers Millicell® EZ slides (Merck Millipore) and incubated overnight. From 750 μL EV suspension, 500 μL was added to the cells (100% concentration). The rest of this EV suspension was added to cells in another chamber (50% concentration) and 250 μL EV-depleted serum-containing medium was added to fill up the chamber. We also performed a nanoparticle tracking analysis (NTA) to estimate the concentration of EVs added for co-culturing [29] (Supplementary Procedure S1). Cells were co-cultured with EVs and measured at time points 0, 2 and 4 h using a fluorescent microscope (Nikon Ti2-U from Nikon Instruments Europe BV, Amsterdam, The Netherlands; EVOS FL from Fisher scientific). At 4 h-time point cells were washed with PBS and followed by acid wash buffer (0.5 M NaCl, 0.2 M acetic acid) to remove EVs which were on the cell surface and not taken up. Cells were washed once more with PBS before fixing with 4% paraformaldehyde (Sigma-Aldrich) or harvesting them for further analysis. For fluorescence microscopy measurement, fixed cells were stained with 50 nM LysoTraker™ Red DND-99 (LTRD) (Fisher Scientific) to stain the cell lysosomal compartment and one drop of Fluoroshield with DAPI (Sigma-Aldrich) to stain the cell nuclei; and imaged using a confocal microscope (TCS SP5 X, Leica Microsystems B.V., Amsterdam, The Netherlands). For flow cytometry measurement, cells were detached using trypsin-EDTA solution. After serum containing-medium was added, cells were pelleted by centrifuging at 250× *g* for 3 min at RT. The supernatant was carefully removed and cells were finally re-suspended in 300 μL PBS containing 0.5% BSA for flow cytometry measurement using BD FACScanto II. The uptake of CTG, BSA FITC and Cre recombinase by cells was measured at wavelength 530/30 nm.

4.8. RT-PCR

To detect Cre recombinase gene expression in T47D reporter cells, RT-PCR was performed. T47D receptor cells after 4 h of co-culturing with EV containing Cre recombinase were harvested. The cell pellet was dissolved in 1 mL Trizol (Fisher Scientific) per 1 million cells. The cell lysate was incubated at RT for 10–20 min. This lysate could be frozen at −20 °C prior to RNA isolation. For RNA isolation, chloroform (Sigma) was added to the cell lysate at a dilution of 1:5. The mixture was shaken vigorously for 15 s and then incubated for 2 min at RT. Next, this mixture was centrifuged for 15 min at 12,000× *g* at 4 °C. The upper aqueous phase was collected and 1 µL of GlycoBlue (Fisher Scientific) was added. Isopropanol (Merck Millipore) at a dilution of 1:2 was added to each sample and vortexed thoroughly. The sample was incubated for 10 min at RT and then centrifuged for 10 min at 12,000× *g* at 4 °C. The supernatant was carefully removed and the pellet was resuspended in 80% ethanol (Merck Millipore) at the same volume as the cell lysate. Next, the sample was centrifuged for 5 min at 7500× *g* at 4 °C. The supernatant was carefully removed and the pellet was air-dried for 10–15 min at RT. RNA was reconstituted in 20 µL nuclease-free water (Fisher Scientific) and quantified by Nanodrop (DeNovix Inc., Wilmington, DE, USA). The cDNA was prepared using Superscript 4 (Fisher Scientific) according to the manufacturer's instructions. The cDNA was amplified using a PCR with Cre-specific primers (forward primer: 5′ GCCTGCATTACCGGTCGATGC 3′; reverse primer: 5′ GTGGCAGATGGC GCGGCAACA 3′) [15]. Thermal cycle conditions used for all reactions were as follows: 5 min at 95 °C, followed by 40 cycles consisting of denaturation for 15 s at 95 °C, annealing for 30 s at 58 °C and extension for 1 min at 72 °C. PCR reactions were concluded with incubation for 7 min at 72 °C to complete the extension of all synthesised products. PCR products were then visualised on a 1.5% TAE agarose gel (Roche Diagnostics, Manneheim, Germany).

4.9. Statistical Analysis

Statistical analysis was performed using GraphPad Prism 8 (La Jolla, CA, USA). ANOVA (one-way) followed by Tukey's multiple comparison tests were used to analyse and define significant differences between samples (untreated, 0.6, 0.7 and 0.8 MPa). Differences were considered significant when p-values were <0.05.

Author Contributions: Conceptualisation, Y.Y., R.M.S. and C.T.M.; methodology, Y.Y., R.M.S. and C.T.M.; formal analysis, Y.Y., B.B. and K.M.G.v.d.W.-J.; investigation, Y.Y., B.B. and K.M.G.v.d.W.-J.; writing—original draft preparation, Y.Y.; writing—review and editing, Y.Y., B.B., R.M.S. and C.T.M.; supervision, Y.Y. and C.T.M.; project administration, Y.Y.; funding acquisition, Y.Y. All authors have read and agreed to the published version of the manuscript.

Acknowledgments: Authors acknowledge technical support given by Roel Deckers (ultrasound setup) from Imaging Division, University Medical Center Utrecht, Utrecht, the Netherlands and George Posthuma (transmission electron microscopy) from Cell Microscopy Center, Cell Biology Department, University Medical Center Utrecht, The Netherlands. This work was funded by Focused Ultrasound Foundation (High Risk Track grant, YY).

References

1. Balachandran, B.; Yuana, Y. Extracellular vesicles-based drug delivery system for cancer treatment. *Cogent Med.* **2019**, *6*, 1–23. [CrossRef]
2. Liu, D.; Yang, F.; Xiong, F.; Gu, N. The Smart Drug Delivery System and Its Clinical Potential. *Theranostics* **2016**, *6*, 1306–1323. [CrossRef] [PubMed]
3. Yuana, Y.; Sturk, A.; Nieuwland, R. Extracellular vesicles in physiological and pathological conditions. *Blood Rev.* **2013**, *27*, 31–39. [CrossRef] [PubMed]
4. Walker, S.; Busatto, S.; Pham, A.; Tian, M.; Suh, A.; Carson, K.; Quintero, A.; Lafrence, M.; Malik, H.; Santana, M.X.; et al. Extracellular vesicle-based drug delivery systems for cancer treatment. *Theranostics* **2019**, *9*, 8001–8017. [CrossRef] [PubMed]
5. Tang, K.; Zhang, Y.; Zhang, H.; Xu, P.; Liu, J.; Ma, J.; Lv, M.; Li, D.; Katirai, F.; Shen, G.X.; et al. Delivery of chemotherapeutic drugs in tumour cell-derived microparticles. *Nat. Commun.* **2012**, *3*, 1282. [CrossRef]

6. Van der Meel, R.; Fens, M.H.; Vader, P.; van Solinge, W.W.; Eniola-Adefeso, O.; Schiffelers, R.M. Extracellular vesicles as drug delivery systems: Lessons from the liposome field. *J. Control Release* **2014**, *195*, 72–85. [CrossRef]
7. Fais, S.; O'Driscoll, L.; Borras, F.E.; Buzas, E.; Camussi, G.; Cappello, F.; Carvalho, J.; Cordeiro da Silva, A.; Del Portillo, H.; El Andaloussi, S.; et al. Evidence-Based Clinical Use of Nanoscale Extracellular Vesicles in Nanomedicine. *ACS Nano.* **2016**, *10*, 3886–3899. [CrossRef]
8. Mitragotri, S. Healing sound: The use of ultrasound in drug delivery and other therapeutic applications. *Nat. Rev. Drug Discov.* **2005**, *4*, 255–260. [CrossRef]
9. Sennoga, C.A.; Kanbar, E.; Auboire, L.; Dujardin, P.A.; Fouan, D.; Escoffre, J.M.; Bouakaz, A. Microbubble-mediated ultrasound drug-delivery and therapeutic monitoring. *Expert Opin. Drug Deliv.* **2016**, 1–13. [CrossRef]
10. Bouakaz, A.; Zeghimi, A.; Doinikov, A.A. Sonoporation: Concept and Mechanisms. In *Therapeutic Ultrasound*; Escoffre, J.-M., Bouakaz, A., Eds.; Springer International Publishing: Cham, Switzerland, 2016; pp. 175–189. ISBN 978-3-319-22536-4.
11. Lentacker, I.; De Cock, I.; Deckers, R.; De Smedt, S.C.; Moonen, C.T. Understanding ultrasound induced sonoporation: Definitions and underlying mechanisms. *Adv. Drug Deliv. Rev.* **2014**, *72*, 49–64. [CrossRef]
12. Suzuki, R.; Klibanov, A.L. Co-administration of Microbubbles and Drugs in Ultrasound-Assisted Drug Delivery: Comparison with Drug-Carrying Particles. *Adv. Exp. Med. Biol.* **2016**, *880*, 205–220. [PubMed]
13. Zhao, Y.Z.; Du, L.N.; Lu, C.T.; Jin, Y.G.; Ge, S.P. Potential and problems in ultrasound-responsive drug delivery systems. *Int. J. Nanomedicine* **2013**, *8*, 1621–1633. [PubMed]
14. Yuana, Y.; Jiang, L.; Lammertink, B.H.A.; Vader, P.; Deckers, R.; Bos, C.; Schiffelers, R.M.; Moonen, C.T. Microbubbles-Assisted Ultrasound Triggers the Release of Extracellular Vesicles. *Int. J. Mol. Sci.* **2017**, *18*, 1610. [CrossRef] [PubMed]
15. Zomer, A.; Maynard, C.; Verweij, F.J.; Kamermans, A.; Schafer, R.; Beerling, E.; Schiffelers, R.M.; de Wit, E.; Berenguer, J.; Ellenbroek, S.I.; et al. In Vivo imaging reveals extracellular vesicle-mediated phenocopying of metastatic behavior. *Cell* **2015**, *161*, 1046–1057. [CrossRef] [PubMed]
16. Lammertink, B.; Deckers, R.; Storm, G.; Moonen, C.; Bos, C. Duration of ultrasound-mediated enhanced plasma membrane permeability. *Int. J. Pharm.* **2015**, *482*, 92–98. [CrossRef] [PubMed]
17. Donahue, N.D.; Acar, H.; Wilhelm, S. Concepts of nanoparticle cellular uptake, intracellular trafficking, and kinetics in nanomedicine. *Adv. Drug Deliv. Rev.* **2019**, *143*, 68–96. [CrossRef]
18. Hessvik, N.P.; Llorente, A. Current knowledge on exosome biogenesis and release. *Cell. Mol. Life Sci.* **2018**, *75*, 193–208. [CrossRef]
19. Ortega, F.G.; Roefs, M.T.; de Miguel Perez, D.; Kooijmans, S.A.; de Jong, O.G.; Sluijter, J.P.; Schiffelers, R.M.; Vader, P. Interfering with endolysosomal trafficking enhances release of bioactive exosomes. *Nanomed. Nanotechnol. Biol. Med.* **2019**, *20*, 1–12. [CrossRef]
20. Safaei, R.; Larson, B.J.; Cheng, T.C.; Gibson, M.A.; Otani, S.; Naerdemann, W.; Howell, S.B. Abnormal lysosomal trafficking and enhanced exosomal export of cisplatin in drug-resistant human ovarian carcinoma cells. *Mol. Cancer Ther.* **2005**, *4*, 1595–1604. [CrossRef]
21. Van Niel, G.; D'Angelo, G.; Raposo, G. Shedding light on the cell biology of extracellular vesicles. *Nat. Rev. Mol. Cell Biol.* **2018**, *19*, 213–228. [CrossRef]
22. Lázaro-Ibáñez, E.; Neuvonen, M.; Takatalo, M.; Thanigai Arasu, U.; Capasso, C.; Cerullo, V.; Rhim, J.S.; Rilla, K.; Yliperttula, M.; Siljander, P.R.M. Metastatic state of parent cells influences the uptake and functionality of prostate cancer cell-derived extracellular vesicles. *J. Extracell. Vesicles* **2017**, *6*, 1–12. [CrossRef] [PubMed]
23. Schneider, M. Characteristics of SonoVuetrade mark. *Echocardiography* **1999**, *16*, 743–746. [CrossRef]
24. Marwaha, R.; Sharma, M. DQ-Red BSA Trafficking Assay in Cultured Cells to Assess Cargo Delivery to Lysosomes. *Bio-Protocol* **2017**, *7*, 1–12. [CrossRef] [PubMed]
25. Lammertink, B.H.A.; Bos, C.; van der Wurff-Jacobs, K.M.; Storm, G.; Moonen, C.T.; Deckers, R. Increase of intracellular cisplatin levels and radiosensitization by ultrasound in combination with microbubbles. *J. Control. Release* **2016**, *238*, 157–165. [CrossRef] [PubMed]

26. Van der Pol, E.; Coumans, F.A.; Grootemaat, A.E.; Gardiner, C.; Sargent, I.L.; Harrison, P.; Sturk, A.; van Leeuwen, T.G.; Nieuwland, R. Particle size distribution of exosomes and microvesicles determined by transmission electron microscopy, flow cytometry, nanoparticle tracking analysis, and resistive pulse sensing. *J. Thromb. Haemost* **2014**, *12*, 1182–1192. [CrossRef] [PubMed]
27. Cizmar, P.; Yuana, Y. Detection and Characterization of Extracellular Vesicles by Transmission and Cryo-Transmission Electron Microscopy. *Methods Mol. Biol.* **2017**, *1660*, 221–232.
28. Schneider, C.A.; Rasband, W.S.; Eliceiri, K.W. NIH Image to ImageJ: 25 years of image analysis. *Nat. Methods* **2012**, *9*, 671–675. [CrossRef]
29. Van der Pol, E.; Coumans, F.; Varga, Z.; Krumrey, M.; Nieuwland, R. Innovation in detection of microparticles and exosomes. *J. Thromb. Haemost.* **2013**, *11*, 36–45. [CrossRef] [PubMed]

7

Tiny Actors in the Big Cellular World: Extracellular Vesicles Playing Critical Roles in Cancer

Ancuta Jurj [1,†], Cecilia Pop-Bica [1,†], Ondrej Slaby [2,3], Cristina D. Ştefan [4], William C. Cho [5], Schuyler S. Korban [6] and Ioana Berindan-Neagoe [1,7,*]

1 Research Center for Functional Genomics, Biomedicine and Translational Medicine, "Iuliu Haţieganu" University of Medicine and Pharmacy, 400337 Cluj-Napoca, Romania; ancajurj15@gmail.com (A.J.); cecilia.bica8@gmail.com (C.P.-B.)
2 Central European Institute of Technology, Masaryk University, 625 00 Brno, Czech Republic; on.slaby@gmail.com
3 Department of Pathology, Faculty Hospital Brno and Faculty of Medicine, Masaryk University, 625 00 Brno, Czech Republic
4 SingHealth Duke-NUS Global Health Institute, Singapore 169857, Singapore; cristinastefan10@gmail.com
5 Department of Clinical Oncology, Queen Elizabeth Hospital, Hong Kong, China; chocs@ha.org.hk
6 Department of Natural Resources and Environmental Sciences, University of Illinois at Urbana-Champaign, Urbana, IL 61801, USA; korban@illinois.edu
7 Department of Functional Genomics and Experimental Pathology, "Prof. Dr. Ion Chiricuta" Oncology Institute, 400015 Cluj-Napoca, Romania
* Correspondence: ioananeagoe29@gmail.com
† These authors contribution equally.

Abstract: Communications among cells can be achieved either via direct interactions or via secretion of soluble factors. The emergence of extracellular vesicles (EVs) as entities that play key roles in cell-to-cell communication offer opportunities in exploring their features for use in therapeutics; i.e., management and treatment of various pathologies, such as those used for cancer. The potential use of EVs as therapeutic agents is attributed not only for their cell membrane-bound components, but also for their cargos, mostly bioactive molecules, wherein the former regulate interactions with a recipient cell while the latter trigger cellular functions/molecular mechanisms of a recipient cell. In this article, we highlight the involvement of EVs in hallmarks of a cancer cell, particularly focusing on those molecular processes that are influenced by EV cargos. Moreover, we explored the roles of RNA species and proteins carried by EVs in eliciting drug resistance phenotypes. Interestingly, engineered EVs have been investigated and proposed as therapeutic agents in various in vivo and in vitro studies, as well as in several clinical trials.

Keywords: extracellular vesicles; cancer; therapeutic agents; cell-to-cell communication

---

1. Introduction

Both solid and hematological malignant tumors are not isolated entities. In fact, they involve complex systemic networks involving cell-to-cell communications between tumor cells and accompanying modified cells. Moreover, both tumor progression and invasion are sustained by a complex microenvironment. This is comprised of networks of components, including cancer-associated fibroblasts, endothelial cells, lymphocytes, and macrophages, as well as secreted factors and elements of the extracellular matrix. Interactions among neighboring cells through a direct cell–cell contact is essential for tumor growth and development, while intercellular communication provides a complex system of secreted factors [1].

To manage all components present in multicellular organisms, cellular communication is critical. McCrea et al. wrote an inspirational quote on the intercellular communication "the music that the nucleus hears" [2]. Communication involves sharing of information through several signaling mechanisms that are either direct (intracrine/autocrine and juxtacrine) and/or indirect (endocrine, paracrine, and synaptic) communications [3]. In this regard, all types of cells have been shown to release and receive both soluble factors and membrane-derived vesicles, the latter receiving increasing attention in the past decades [4]. The first instance of the presence of membrane-derived vesicles is observed in reticulocytes, wherein released vesicles would remove transferrin receptors from the cell, an important step in their maturation to erythrocytes [5]. Early on, these membrane-derived vesicles have been initially deemed as cellular "garbage bags". Subsequently, numerous studies have been undertaken to investigate membrane-derived vesicles detected on primary cells [6]; i.e., primary cells of the immune and nervous systems, and cancer cell lines [7]. It has been reported that extracellular vesicles (EVs) can be isolated from various bodily fluids, as they play important roles in the management of various normal physiological processes, including stem cell maintenance [8], immune surveillance [9], tissue repair [10], and blood coagulation [4].

It is reported that physical and molecular characteristics of EVs have impacts on various biological processes, including cancer development, progression, and metastasis [11]. Moreover, small sizes of EVs offer critical properties, including immune system escape, biocompatibility, and biodegradability, as well as transfer of their contents into both neighboring and distant cells. During biogenesis, EVs acquire important bioactive molecules that regulate several biological processes. Thus, cancer-derived EVs have been largely described as possessing both pro- and antitumor functions. For example, tumor-derived EVs interact with immune cells by delivering negative signals and interfering with their antitumor functions. By suppressing immune cell functions, EVs promote cancer progression and facilitate tumor escape. Moreover, EVs carry important molecules and factors that either directly or indirectly influence several processes, including development and maturation, as well as antitumor activities in immune cells [11]. Conversely, antitumor effects of EVs have been observed in dendritic cell-derived EVs, and these are capable of being used in immunotherapy [12].

It has been observed that EVs are tightly linked to tumorigenesis [13], spread of pathogenic agents and viruses (e.g., the Human Immunodeficiency Virus-1 [HIV-1]), amyloid-β-derived peptides [14], and α-synuclein [15] (linked to Alzheimer's and Parkinson's diseases). Due to varied compositions of EVs, they have been deemed useful in the fields of both diagnostics and therapeutics [16]. Moreover, EVs can be potentially useful in serving as drug delivery vehicles by transporting several molecular species as part of normal cell-to-cell communication.

In this review, we will discuss the potential and role(s) of EVs in modulating both physiological and pathological processes, as well as how these entities can be used as therapeutic agents [17].

2. The War Waged Inside the Cell

EVs are described based on their size, cellular origin (endosome- or plasma membrane-derived), biological function, and biogenesis process. Moreover, when described based on their biogenesis, EVs are cataloged into apoptotic bodies, microvesicles, and exosomes [18]. These major classes are cell-based vesicles having diameters ranging between 30 and 2000 nm (Table 1). Furthermore, these entities exhibit different properties that help distinguish them among all main classes of EVs. Differences among different EV classes are based on the content, size, route of biogenesis, and surface markers [19].

One of the largest cell-based vesicles is those of apoptotic bodies that are released by any type of cell once apoptotic processes are activated. Specifically, apoptotic EVs are generated during plasma membrane blebbing during apoptosis, as these are phagocytosed by macrophages and then fused with lysosomes [20]. Generally, these EVs are known to carry nuclear fragments and cellular organelles as a result of cell fragmentation [20]. Furthermore, these EVs are characterized by a flip of phosphatidylserine along an external layer, a permeable membrane, and expression of phagocytosis-promoting signals (calreticulin [21] and calnexin [22]), as well as chemokines and adhesion molecules, including ICAM3 and CX3CL1/fractalkine, and MHC class II molecules [23]. These are all important for direct antigen presentation CD4+ T cells and immunological memory activation [23].

Microvesicles, also known as ectosomes, are usually larger than 0.2 μm in size, and they are released outward from the plasma membrane via budding or shedding into the extracellular matrix. The process of microvesicle formation is mediated through a complex process involving cytoskeletal protein contraction and phospholipid redistribution [24]. During biogenesis, microvesicles are mainly composed of a plasma membrane and of cytosolic-associated proteins [19]. Microvesicles are involved in several key functions, including intercellular communication, signal transduction, and immune regulation. In particular, these entities mediate tumor invasion, inflammation, metastases, stem-cell renewal, and expansion [25]. During biogenesis, microvesicles receive important structural components, including Flotillin-2, Annexin V, integrins, selectin, CD40, and metalloproteinase [26].

In contrast, exosomes are between 30–100 nm in size, and are generated using the endosomal pathway [25]. Exosome biogenesis begins with the formation of early endosomes that undergo inward (or reverse) budding and then subsequent formation of intraluminal vesicles (ILVs), and referred to as multivesicular bodies (MVBs) or late endosomes. As a final step, late endosomes may either directly fuse with lysosomes, wherein the endocytosed cargo is degraded, or they may fuse with the plasmalemma releasing its ILVs (exosomes) to the extracellular space [25,27]. ESCRT (endosomal sorting complexes required for transport) is a molecular complex that plays an important role in MVB formation and regulation (Figure 1). Specifically, ESCRT is formed from the other four molecular complexes, including ESCRT-0, -I, -II, and -III. These multi-protein complexes are responsible for different functions, depending on their components. ESCRT-0 is dependent on ubiquitin and determines clustering of the cargo, ESCRT-I and ESCRT-II play important roles in bud formation, and ESCRT-III determines scission of vesicles. In addition, accessory proteins (VPS4 ATPase) are implicated in the final steps of ESCRT functions, namely of dissociation and recycling. In many studies, other ESCRT-independent pathways of MVB formation have been observed [28]. Some classes of molecules implicated in ESCRT-independent mechanisms of exosome biogenesis are represented by proteolipid proteins, tetraspanins, and heat shock proteins [29].

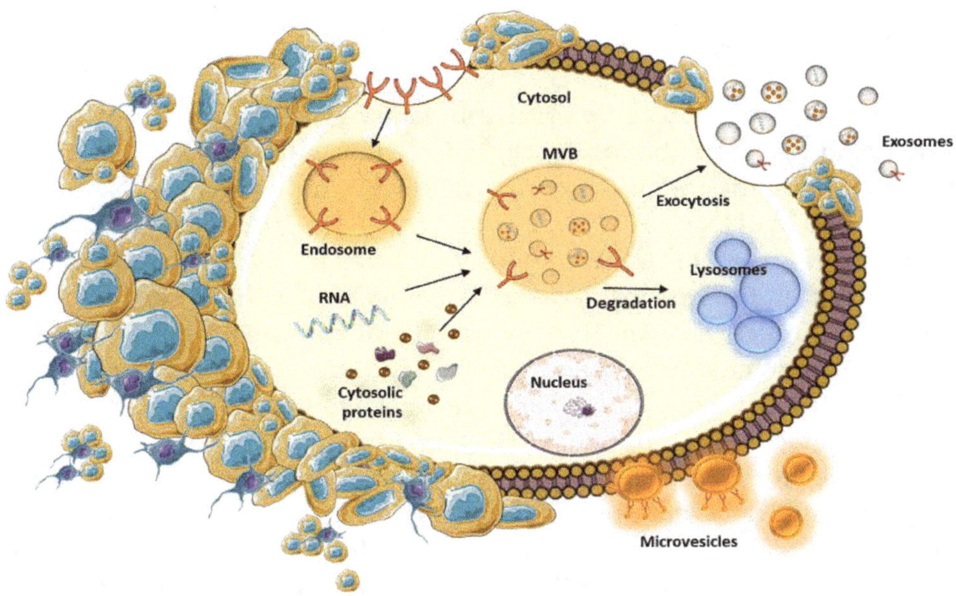

Figure 1. Biogenesis mechanisms of EVs, exosomes, and microvesicles. Endocytosis, an active process, begins with the generation of endosomes after cells are internalized within the extracellular fluid material to form internal vesicles and early and late endosomes. Furthermore, multivesicular bodies (MVBs) are formed via inward budding of a late endosomal membrane. Moreover, MVBs can fuse with either the plasmalemma, releasing their cargo into extracellular space, or with lysosomes, wherein their contents are degraded.

In general, following MVB fusion with the plasmalemma, exosomes are secreted from cells. This mechanism is regulated via two mechanisms, constitutive and inducible. The constitutive mechanism is managed by a plethora of molecules, including heterotrimeric G-proteins, flotillins, and glycosphingolipids [30], while inducible secretion is determined by stress stimuli, including thrombin, DNA damage, hypoxia, heat shock, and lipopolysaccharide (LPS) stimulation [27].

Table 1. Major characteristics of EVs.

Characteristics	Exosome	Multivesicular Body	Apoptotic Body	References
Size	Homologous 30–100 nm	Heterogenous 100–1000 nm	Heterogenous 1–5 μm	[31–33]
Origin	Multivesicular bodies fusion with cellular membrane	Direct outward budding or blebbing from the cellular membrane	Cellular membrane blebbing during cell death, cellular debris	[33,34]
Density	1.13–1.19 g/mL	1.25–1.30 g/mL	1.16–1.28 g/mL	[35]
Contents	Nucleic acids (DNA, mRNAs, miRs), lipids, specific proteins	Nucleic acids (DNA, mRNAs, miRs), lipids, specific proteins	Cellular organelles, cytosolic content (RNAs, fragmented DNA, proteins)	[33]
Protein components	Multivesicular body biogenesis (ALIX, TSG101), tetraspanins (CD9, CD63, CD81, CD82)	Death receptors (CD40 ligands), Cell adhesion (selectins, integrins)	Transcription and protein synthesis (histones)	[25,36]
Lipids	Lipidic molecules from the donor cells (include BMP)	Lipids from plasma membrane and resemble the donor cells (without BMP)	Characterized by phosphatidylserine externalization	[36,37]
Mechanism of release	Constitutive and/or cellular activation, depends on the cell type of origin	Cytoskeleton rearrangements, generation of membrane curvature, vesicle release, relocation of phospholipids to the outer membrane	Rho-associated kinase I and myosin ATPase activity	[37–39]
Determinant of controlled contents	The cellular origin and physiological state of the cell	No direct correlation	The cellular origin and stimuli	[35]
Markers	Membrane impermeable (PI negative), CD63, TSG101, Alix, flotillin, tetraspanins, HSP70, HSP90	Membrane impermeable (PI negative), selectin, integrin, flotillin-2, Annexin A1	Membrane permeable (PI positive), histone, DNA, Annexin V	[25,32]

MV, microvesicle; BMP, bone morphogenetic protein; PI, propidium iodide.

During biogenesis, exosomes receive critical bioactive molecules from donor cells, including nucleic acids, lipids, and proteins, that are specific for each cell type [40]. For composition of both exosomes and microvesicles, the following components are important: mRNAs, microRNAs (miR), non-coding RNAs, DNAs (mtDNA, ssDNA, and dsDNA), mRNA cytoplasmic proteins, and lipid raft-interacting proteins (Figure 2) [41]. Recent attention has focused on understanding how DNAs are packaged within EVs. In this regard, several research groups have reported on the presence of DNAs (mtDNA, ssDNA, and dsDNA) in EVs secreted from various types of malignancies, including melanoma, breast, lung, pancreas, and prostate cancer [42]. However, there is little knowledge of the origin, biological significance, and mechanism of DNA packaging in EVs. Conversely, few studies have reported that DNA is located along the outer surface and not within EVs [43,44]. Thus, it is proposed that outer surfaces of EVs are capable of interacting with proteins, nucleic acids, and other molecules regulating motility, aggregation, and various other important processes for EVs [45]. Furthermore, cargos within these vesicles can influence recipient cells [46], thus suggesting that exchanges of EV cargos between either normal or cancer cells may represent an effective and efficient intercellular communication when cells have particular physiological behaviors, but these are dramatically altered in cancer cells. Alongside nucleic acids, exosomal proteins are specific, and they are present in endocytic compartments of donor cell membranes, as well as in cellular membranes, the nucleus, the cytosol, and the Golgi apparatus, as well as in the endoplasmic reticulum and mitochondria, but at lower frequencies for these latter two organelles [47]. Tetraspanins (CD9, CD63, CD81, and CD82) are among some of the most typical proteins present in exosomes, alongside GPI-anchored proteins and receptors. Moreover, within interiors of exosomes, several molecular species of a parent cell are encased, and these are represented by structural components, heat shock proteins, chaperones, and enzymes involved in metabolic processes, among many others (Figure 2) [17,27].

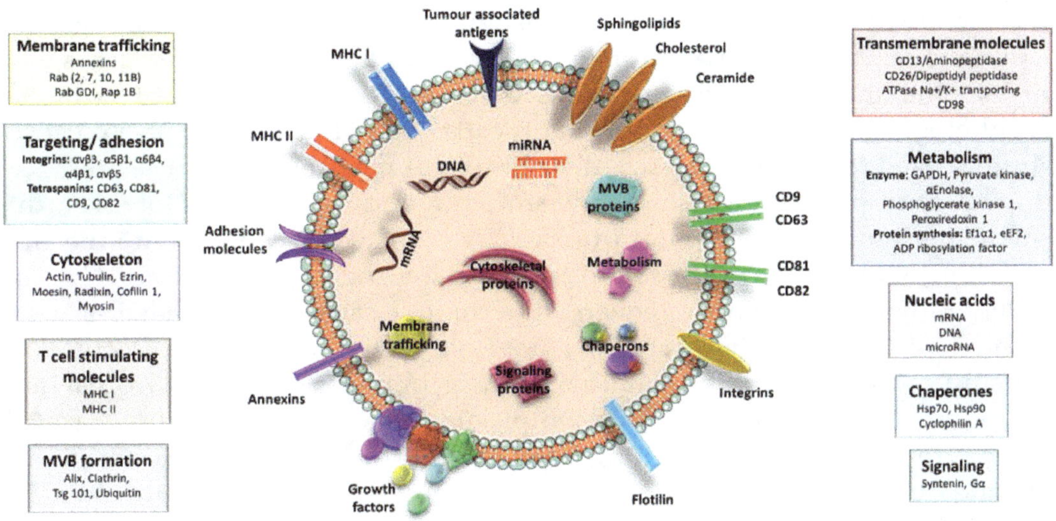

Figure 2. EVs' cargo profile. EVs are secreted by a wide range of cells, normal and tumor, having the capacity to deliver various bioactive molecules including nucleic acids, specific proteins, and lipids from the donor cells to recipient cells

Interestingly, EVs are carriers of essential soluble immune mediators, including cytokines and chemokines. Several cytokines, such as IL-1α, IL-1β, IL-6, IL18, and IL-32, are engulfed within EVs. In endothelial cell-derived apoptotic bodies, IL-1α is present; whereas, IL-18 is associated with EVs shed from surfaces of macrophages. Additionally, IL-6 and IL-32 are secreted by mast cells upon IL-1 stimulation [48]. Moreover, heat-stressed tumor cells have been shown to release EVs with different CCL compositions compared to their nonstressed counterpart [49].

3. EVs Isolation and Characterization

EVs can be isolated from different biological fluids (plasma, serum, saliva, milk, and urine, among others), as well as from cell culture supernatants. There are several available methodologies to remove undesirable particles from samples of interest. In cell cultures, EVs are separated from other components of cell media using differential centrifugation. This technique utilizes centrifugal force to separate contaminants from EVs, along with several necessary steps to remove cells, cell debris, and large microvesicles in order to obtain purified EVs [50]. Another isolation technique, density gradient centrifugation, separates EVs into specific layers in different solutions (sucrose, iohexol, and iodixanol) depending on their buoyant densities [51]. In this method, subcellular components, including mitochondria, endosomes, and peroxisomes, are successfully separated into distinct layers within the density gradient solution [52]. In yet another method, size-exclusion chromatography utilizes porous beads to separate biomolecules based on their hydrodynamic radii [53]; thus, biological samples are filtered through a column of porous beads of radii smaller than those of EVs [54]. Similarly, filter-based enrichment methods also depend on the sizes of EVs for separation, but instead of porous beads, sieves are used. Further, antibody enrichment methods are based on selecting for markers specific for EVs, such as CD9, CD63, and CD81, thus serving as complementary to size-based methods, thereby capable of specific selection of EVs [55]. Recently, acoustics and/or microfluidics methods have been developed that will isolate EVs in label-free and contact-free manners [56,57]. In addition, EVs can be separated from biological samples via precipitation using different chemicals, such as polyethylene glycol (PEG), sodium acetate, or protamine. It has been reported that using PEG, both EVs and proteins are precipitated into a pellet that can be further analyzed [58]. Similarly, magnetic beads coated with antibodies for common EV surface proteins (CD9, CD63, and CD81) are used [59]; whereas, a fluidic technique, ExoTIC (exosome total isolation kit), utilizes step-wise nanoporous membranes to trap molecules or particles of specific sizes, thereby allowing for smaller molecules and particles to flow through a membrane filter [60]. This latter method may be deemed as the most accurate size-based method used to isolate EVs from biological samples with a high yield of intact EV structures.

As EVs, of nano-sizes, must be quantified and evaluated for purity, there are several methods that can determine the numbers of vesicles released and cell type (detection of surface antigens), as well as EV morphological traits [61]. Dynamic light scattering (DLS) is based on a particle's Brownian motion in solution, used to measure the size distribution of particles, as well as their zeta potentials, measuring diameters of particles ranging between 1 nm and 6 μm [62]. However, this technique does not provide any biochemical data of purified EVs [62]. In another technique similar to DLS, nanoparticle tracking analysis (NTA) is used to measure concentration, count, and size distribution of EVs based on their Brownian motion; moreover, this technique can measure smaller-sized EVs, ranging from 1 to 1000 nm [63]. In yet another technique, flow cytometry is used to indirectly quantify EVs as it is based on using specific antibodies that accurately recognize EV markers from a liquid medium. However, flow cytometry cannot evaluate the complex profiles of subsets of EVs. Similar to DLS and NTA, flow cytometry is capable of providing data on EV size, count, and distribution [64]. Finally, both EV purity and quality can be determined using transmission electron microscopy (TEM) wherein standard traits, such as cup-like structures and lipid bilayers, can be determined [65]; whereas, EV purity can be assessed based on presence or absence of protein markers [50].

4. Biological Roles of EVs

EVs, particularly exosomes, play important roles in cells by influencing several biological processes. Their effects on receptor cells can be exerted via various mechanisms, such as phagocytosis, direct receptor binding, and receptor-dependent internalization. Thus, EVs can deliver information through a wide range of mechanisms, thereby playing important roles in tissue repair [10], stem cell maintenance [8], and immune surveillance [9]. Due to their pleiotropic actions, EVs have been, time and time again, deemed as signalomes.

It has been reported that EVs can influence activities of immune cells present both in the tumor microenvironment and in the circulatory system [66]. Once EVs are internalized into targeted cells, they release their cargo and exert their role by activating different biological mechanisms. EVs can mediate the activation of immune cells by promoting proliferation and survival of hematopoietic stem cells, as well as activation of monocytes [67], B lymphocytes [66], and NK cells [68]. EVs can also inhibit immune responses via regulation of NK and $CD8^+$ cell activities [69] and activation of Treg cells, as well as inhibition of dendritic cell (DC) maturation [70] and formation [71]. For those EVs derived from stem cells, they have been demonstrated to regulate stem cell maintenance with implications in tissue regeneration [72]. In addition, it has also been shown that EVs can modify stem cells to develop into either a liver cell phenotype [73] or a lung phenotype [74].

5. Pathological Roles of EVs

It is important to point out that EVs can be secreted by malignant or deregulated cells. During biogenesis processes, EVs are loaded with important bioactive molecules from malignant cells that influence the phenotype(s) of target cells. It has been reported that EVs are implicated in the formation of a premetastatic milieu throughout the body [75]. Moreover, EVs are also involved in other critical biological processes and have the capability of stimulating tumor progression [13]. This process is sustained by EVs via delivery and release of their targets into a target cell(s). Alongside tumor progression, EVs have the capability of carrying out other critical processes, including cell proliferation, tumor growth [76], angiogenesis [77–85], matrix remodeling, metastasis [75,86–96], immune escape [69,97–109], resistance to apoptosis [110–113], deregulation of energetic metabolism [114–117], sustaining proliferative signaling [94,118–120], evading growth suppression [121–123], deregulating and tumor-promoting inflammation [100,124,125] (Figure 3).

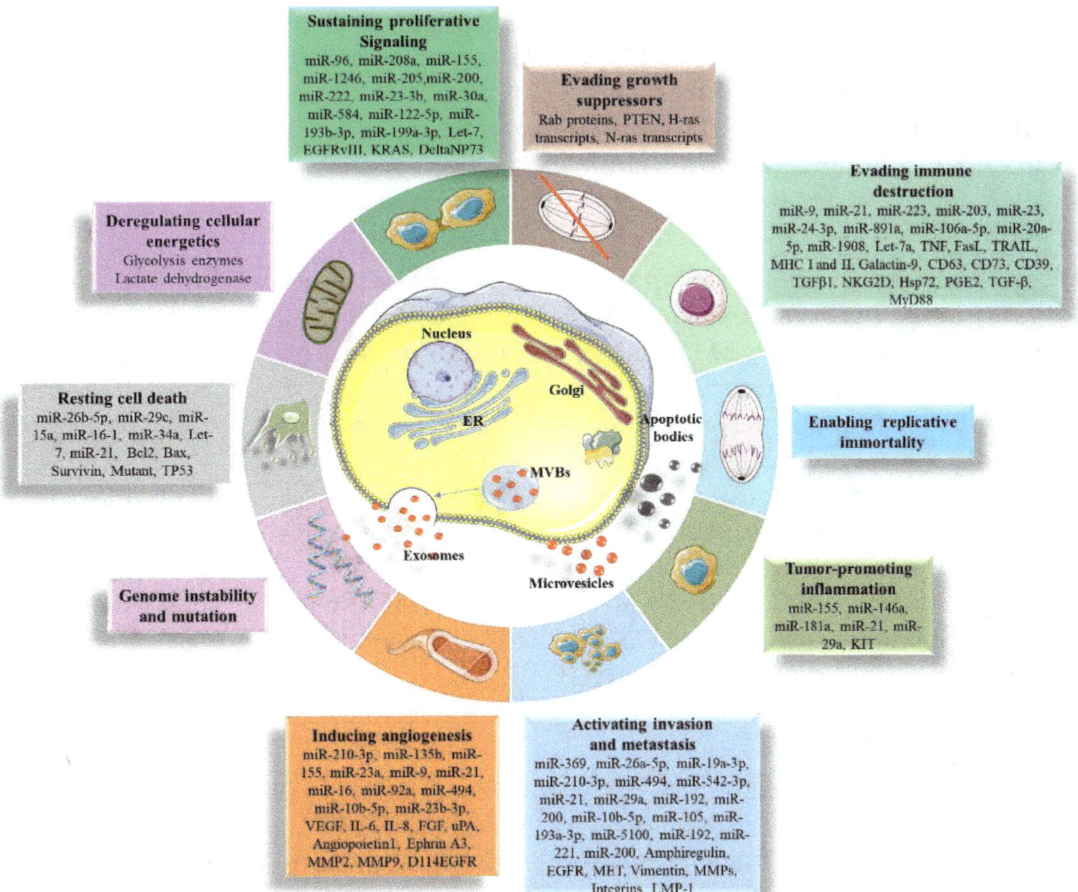

Figure 3. A schematic representation of the impact of tumor-derived EVs on the hallmarks of cancer. Pro-oncogenic molecules can be transported through the cellular membrane by EVs and microvesicles. Molecules transported via EVs have been reported to contribute to each of the hallmarks of cancer. Abbreviations: ER, endoplasmic reticulum; MVBs, multivesicular bodies

5.1. Promoting Cell Proliferation and Resistance to Apoptosis

EV transfer can modify particular signaling pathways in the target cell, modifying proliferation and resistance to apoptosis, among other processes. For example, it has been reported that in gastric cancer, cell proliferation can be enhanced through exosomal transfer of CD97 that activates the Mitogen-Activated Protein Kinase (MAPK) pathway [126]. In chronic myeloid leukemia, it has been observed that cellular proliferation is promoted via induction of phosphatidylinositol 3-kinase (PI3K)/protein kinase B (AKT) and MAPK pathways [127]. For instance, melanoma-derived EVs transfer PDGFR-8, which in turn activates the PI3K/AKT pathway in target cells [128]. Moreover, PI3K/AKT and MAPK pathways are reported to be activated in both gastric and bladder carcinomas by EVs [129]. In addition, EVs derived from glioblastoma are reported to promote cell proliferation in a CLIC1-dependent manner [130]. Soekmadji et al. have demonstrated that EVs derived from prostate cancer cells cultured in the presence of androgens are enriched in CD9, which promotes proliferation of androgen-deprived cells [131]; whereas, Matsumoto et al. have reported that mice injected with melanoma-derived EVs result in accelerated in vivo growth of murine melanomas [132].

EVs can also alter target cell(s) via their miR content as it has been shown that miR-93-5 from esophageal cancer-derived EVs inhibits phosphatase and tensin protein (PTEN) expression stimulating cell proliferation [133]. Other important examples of EVs' role in stimulating cell proliferation has been reported in colon cancer wherein EVs carry higher levels of miR-200b and miR-193a [134] and of pancreatic cancer-derived EVs loaded with miR-23b-3p and of papillary thyroid cancer-derived EVs loaded with miR-222 [135]. Furthermore, it has been observed that tumor-derived EVs actively transfer miR-106a-5p, miR-891a, miR-24-3p, and miR-20a-5p that promote cell proliferation via alteration of Microtubule Affinity-Regulating Kinase1 (MARK1) signaling in human nasopharynx cancer [136]. Moreover, EV miR-302b is delivered from lung carcinoma cell lines to target cells, leading to cell growth inhibition via the TGFβRII/ERK signaling pathway [137], while EV miR-584 accelerates cell proliferation in hepatocellular cancer cells [138].

5.2. Promoting Cell Migration

In addition to their effects on cell proliferation, EVs secreted by tumor cells can also alter the migratory status of malignant cells. EVs derived from nasopharyngeal carcinoma carrying epithelial-mesenchymal transition (EMT)-inducing signals, including HIF1α, TGFβ [98], and matrix metalloproteinases (MMPs), were reported to improve the migratory capacity of tumor cells [139]. Interestingly, EVs from a hypoxic prostate cell line have been shown to lead to increased mobility and invasiveness in a naïve human prostate cancer cells [140]. Moreover, EVs secreted from muscle-invasive bladder cancer contributed to decreased levels of E-cadherin as well as to enhanced migration and invasion in uroepithelial cells [141,142]. In another study, EV miR-105 was reported to stimulate invasion in both the respiratory and central nervous systems by inhibiting ZO-1 in endothelial cells, leading to enhanced cell migration [143]. Furthermore, it has been observed that EV miR-21 stimulates invasion of esophageal tumor cells by activating the programmed cell death 4 (PDCD4)/ c-Jun NH_2-terminal kinase (JNK) axis [90].

5.3. Sustaining Angiogenesis

It has been reported that induction of a mutated epidermal growth factor receptor variant III (EGFRvIII) in glioma cells would lead to increased vesiculation and transfer of the mutated EGFRvIII to other cells and to increased vascular endothelial growth factor (VEGF) production [94]. In addition, it has been observed that EVs from primary glioblastoma cells are loaded with miRs that influence angiogenesis [76]. Recently, it has been demonstrated that EGFR can be transferred to endothelial cells wherein expression of VEGF is induced along with subsequent autocrine activation of VEGF-R2 [77]. Thus, these findings suggest that EVs can result in tumor growth by stimulating cancer cell proliferation and activating angiogenesis in adjacent endothelial cells [77]. Kim et al. have reported that sphingomyelin expressed on tumor cells-derived EVs stimulate processes, such as migration and angiogenesis, in endothelial cells [144]. It has been observed that such EVs secreted by tumor cells are enriched in MMPs as well as in CD147. These components have been proposed to play roles in both hydrolysis of the extracellular matrix and initiation of angiogenesis [145]. Interestingly, it has also been observed that pSTAT5 can be transferred to endothelial cells via EVs, and that it is capable of activating ERK1/2 along with subsequent angiogenesis stimulation [146]. Moreover, miR-214 is also responsible for promoting angiogenesis by suppressing Ataxia Telangiectasia Mutated (ATM) expression and preventing senescence [147]. In fact, mesenchymal stem cells-derived EVs have also been shown to stimulate the angiogenesis process, as demonstrated in vivo in an ischemic heart model [148].

Colon cancer cells have been shown to transfer miR-25-3p to endothelial cells not only by stimulating angiogenesis but also by increasing vascular permeability [149]. EVs secreted by hepatocellular carcinoma cells have been shown to transfer miR-103 to endothelial cells, leading to a reduction in the integrity of endothelial junctions, and thereby increasing vascular permeability [150]. The angiogenesis process has also been shown to be stimulated by miR-145-5p and miR-14-3p from lung cancer-derived EVs [151]. Moreover, in lung cancer cells, release of EV miR-21 stimulates angiogenesis in nontumor lung cells [90]. In another study, miR-9 exhibits proangiogenic activity by reducing expression levels of the *SOCS5* gene and by promoting Janus kinase/signal transducers and activators of transcription (JAK-STAT) signaling, thereby supporting migration of endothelial cells and tumor angiogenesis [152]. Furthermore, increased expression levels of EV miR-9 can differentiate an osteoblast precursor cell line into osteoblast cells and upregulate angiogenesis via an AMPK-dependent pathway [153].

From a therapeutic perspective, it has been observed that EVs can be used to shed bevacizumab, an anti-VEGF antibody, thus leading to decreased efficacy in glioblastoma [154]. Additionally, some cancers are capable of secreting VEGF isoforms with reduced affinities for bevacizumab, leading to another therapy escape mechanism [155]. Another antiangiogenic agent commonly used throughout the field of oncology is sorafenib. Hepatocellular carcinoma-derived EVs have been shown to activate the HGF/MET/AKT pathway in sensitive hepatocellular carcinoma cells, thereby inducing sorafenib resistance. Moreover, it has been observed that more invasive cell lines are capable of better inducing sorafenib resistance compared to less invasive cell lines, thus demonstrating that different malignant subclones are capable of sharing their acquired resistance [156].

It has been reported that sorafenib induces increased expression of linc-ROR in EVs secreted by hepatocellular carcinoma cells [157]. EVs have also been shown to transfer resistance to sunitinib, a similar compound to sorafenib, to hepatocellular carcinoma subclones [157], as well as to different subclones of renal cell carcinoma [158].

5.4. *Immune System Evasion*

One of the important functions of the immune system is to recognize and to destroy particular cells that present alterations when compared to self-antigens of unaltered (normal) cells. However, this function can be evaded by malignant cells either by changing surface antigens of malignant cells or by influencing the immune system. The role(s) of EVs in this process has been reported in various studies [80]. It has been demonstrated that EVs secreted from tumor-derived macrophages are enriched with particular miRs that enhance the local invasion of breast cancer cells [103]. In fact, the effects induced by EVs are related to modulation of the immune response. Furthermore, it has been demonstrated that EVs of tumor cells are capable of promoting immune escape by determining regulatory T cell expansion [159] and by shedding FAS ligand (FASL), as well as by inducing $CD8^{+}$T cell apoptosis and increasing expression of the *MMP9* gene in melanoma cells [79,160].

Recently, it has been reported that EVs can express PD-L1, thus suppressing activities of antitumor T-cells [161]. Moreover, it has been observed that EV PD-L1 expression is inversely correlated with nivolumab and pembrolizumab response [162]. These findings are of particular importance in checkpoint blockade therapy as this reveals that EVs can act as decoys for therapeutic agents. As checkpoint blockers, this would allow for adjustment of the dosage of therapy by taking into consideration EV expression of particular markers, such as PD-L1. In other cancers, such as head and neck squamous cell carcinoma, it has been observed that there are differences between EV cargos in patients experiencing relapse compared to those who remain in remission at two years following ipilimumab therapy [163]. More specifically, it has been observed that for patients in remission, at two years, have lower numbers of EVs positive for both CD3 and CTLA4. Conversely, it has been shown that patients who relapsed after two years have increased numbers of EVs derived from Treg cells, thus demonstrating the importance of EVs in mirroring the T-cell response to tumor cells [163].

Immunomodulatory effects of EVs have also been reported in gastric cancer [164]. It has been observed that EVs isolated from gastroepiploic veins have shown increased levels of TGF-β1 expression for patients presenting either lymph nodes or distant metastasis. This finding has demonstrated the role of EVs in preparing an immunosuppressive premetastatic niche for engraftment of circulating tumor cells [164]. Although not explored in the abovementioned study, it is likely that checkpoint inhibitors could reverse these observed generated immunosuppressive premetastatic niches along with reduced probability of gastric cancer reaching advanced stages.

In other studies, it has been observed that EV miR-212-3p from pancreatic cancer cells have degraded RFXAP mRNAs in dendritic cells (DCs), leading to immune tolerance by minimizing expression of MHC II [165]. Furthermore, hypoxic tumor cells-derived EVs influence functions of natural killer (NK) cells by delivering miR-23a and TGFβ [166], while miR-214 secreted from human embryonic kidney cells induces immunological tolerance responses in CD4+ T-cells [167].

5.5. Transferring Mutations

Tumor-derived EVs have DNA fragments that can be transferred to recipient cells [45]. It has been reported that resistant melanoma cells can activate the MAPK pathway in sensitive melanoma cells through an EV-mediated truncated ALK transfer [168]. Moreover, EVs positive for EGFRvIII have been shown to activate both MAPK and PI3K/AKT pathways [94]; whereas, β-catenin-mutated colon cancer cells are reported to transfer their mutation to β-catenin wild-type cells along with subsequent activation of the β-catenin/WNT pathway [169]. In addition, a mutated SMAD4 is observed to be transferred from resistant to sensitive ovarian cancer cells, leading to an increased platinum resistance [80].

6. EVs in Cancer Stem Cells

As EVs play important roles in cancer cells, it is known that particular subpopulation(s) within a malignant mass, cancer stem cells (CSCs), present significant chemoresistance and are generally deemed as seeds for relapse [170]. EVs derived from CSCs are reported to transfer particular information to other cells. For example, EVs derived from renal cell carcinoma stem cells have been shown to carry a specific miR signature that influences levels of PTEN in target cells. This change is functionally translated into increased EMT followed by a subsequent increase in frequency of metastasis [88,171].

EVs derived from glioblastoma stem cells contain miR-21, which can be transferred to endothelial cells, leading to upregulation of angiogenesis via the miR-21/VEGF pathway [172]. In another study, macrophages treated with glioblastoma cancer stem cell-derived EVs can skew macrophages to an anti-inflammatory phenotype (M2), associated with increased expression of PD-L1 on surfaces of these cells, thus demonstrating immunosuppressive roles of these EVs [173]. On the other hand, EVs from thyroid CSC spheroids can induce a stem cell-like phenotype in recipient cells by increasing levels of SOX2. Moreover, it has been shown that EVs derived from these cells also increase the EMT through SLUG upregulation [174].

EVs from CSCs have also been shown to influence the immune system, as EVs derived from colorectal CSC are reported to increase IL-1β in neutrophils, thereby inducing a pro-inflammatory environment [175].

7. EVs in Drug Resistance

One of the most heavily investigated characteristics of EVs is their ability to transfer resistance to particular therapeutic compounds. This is due to their capability of transferring specific molecular traits, such as efflux pumps or pathway regulation, thus rendering a phenotype better adapted to a particular selected therapeutic strategy [80]. Often, efflux pumps are transferred from resistant to sensitive cells [176–180]. These efflux pumps induce tumor resistance, corresponding to the transfer of ATP-binding cassette (ABC) family members, of which the multidrug resistance 1 (MDR1) and multidrug resistance-associated protein 1 (MRP1) have attracted attention in oncology [176–180]. More specifically, MRP1 can be transferred from resistant acute promyelocytic leukemia to sensitive cells [176]. Additionally, in breast cancer, MDR1 can be induced by EVs through the activation of NFATc3 [181]. On the other hand, it has been demonstrated that p-STAT3 can be transferred to 5-fluorouracil-sensitive colorectal cancer cells to increase their resistance to 5-fluorouracil [182]. Furthermore, it has been observed that CLIC1 can be transferred to gastric cancer cells, thereby increasing levels of MDR1 and BCL2 and leading both to increased drug efflux and decreased apoptosis [183].

It is important to point out that other important molecular species, including both coding and non-coding RNAs, can also be transferred in EVs, which can also contribute enhanced cell resistance to various drug/compound treatments.

As platinum compounds are important components of the oncology arsenal, studies have been undertaken to assess transfer of resistance to these compounds. Often, it has been demonstrated that miRs influence resistance to platinum. For example, miR-19b influences resistance to platinum in colon cancer [184], while both miR-425-3p and miR-96 influence resistance to platinum in lung cancer cells [185,186]. Moreover, transfer of lncRNA HOTTIP increases resistance to platinum in gastric cancer cells, while increased serum HOTTIP lncRNA is associated with poor response to platinum [187]. Furthermore, coding RNAs are reported to influence sensitivity to platinum. For example, transfer of DNMT1 mRNA increases the resistance of ovarian cancer to platinum compounds [188].

Several other compounds are reported to be transferred through EVs as well For example, resistance to 5-fluorouracil in colon cancer cells is induced by both miR-145 and miR-34a [189], while the resistance of breast cancer cells to both adriamycin and tamoxifen are mediated by miR-222 transfer [190,191], and resistance of pancreatic cancer cells to gemcitabine is mediated by miR-155 transfer, leading to TP53INP1 modulation [192].

Interestingly, some pathways are more frequently targeted by some of the miRs, it has been reported that the PI3K/AKT pathway can be targeted by miR-21 in breast cancer cells [193] and by miR-1238 in glioblastoma cells [194].

8. EVs Used as Diagnostic Markers

EVs have been deemed as useful diagnostic markers in detecting the presence of a disease once the characteristics of malignancy are known. However, current methodologies for isolation and characterization of EVs are costly and not sufficiently standardized for de novo diagnostic protocols.

Nevertheless, one set of markers useful for diagnostics consists of fusion genes present in an assessed disease. These fusions occur more or less frequently depending on various malignancies, with hematologic malignancies, sarcomas, and prostate cancer presenting the most frequent fusion events [195]. For example, presence of *BCR-ABL* fusion genes in EVs, secreted by chronic myelogenous leukemia (CML) [18], in a patient's plasma correlate with remission status in CML patients [196]. Although this approach cannot be directly transferred to a clinical diagnosis, as CML can be easily assessed in a patient's blood, this can serve as an example for use in solid tumors, such as prostate cancer. The prostate cancer malignancy presents gene fusions in ~50% of cases, particularly of the *TMPRSS2–ERG* fusion gene as it is highly frequent [197]. Such an approach requires use of urine samples as isolated EVs present alterations in RNA signature(s) compared to those of control samples, including presence of the *TMPRSS2–ERG* fusion gene [198].

However, several common cancers do not present high frequencies of fusion genes. As a result, alternative strategies must be explored. For example, HER2-HER3 dimers from EVs have been assessed in HER2-positive breast cancer patients participating in a clinical trial (NCT04288141). Although the primary objective of this study was to identify a marker for resistance to anti-HER2 therapy, assessment of HER2-HER3 dimers from EVs may aid in identifying the tumor load in HER2-positive breast cancer patients (NCT04288141).

One of the most common alternative approaches under consideration for use of EVs as biomarkers is that of the dosage of the RNA species, particularly of miRs, determined by qRT-PCR followed by protein assessment, using either ELISA or mass spectrometry [199]. However, a major problem that may arise, particularly in assessing RNAs content in EVs, is that of sensitivity of RNA species to particular transport and storage conditions. Moreover, it has been observed that RNA assessment has rarely made it to a clinical setting, as these assessments have been generally constrained to viral loads, particularly of RNA viruses.

Thus, future studies should focus on either genetic or proteomic markers present in EVs, as these are more likely to be amenable for clinical implementation.

9. EVs Used in Anti-Cancer Therapy

In recent years, accumulated knowledge of characteristics and cargos of EVs has suggested that these structures could serve as valuable biomarkers for diagnostic/prognosis, as well as therapeutic agents for treatment of various pathologies [200]. The emergence of EVs in cancer therapy serves as a valuable nanotechnology to overcome major worldwide cancer management problems [201]. Currently, there are many studies recommending use of EVs as delivery vectors for treatment of various cancer, following manipulation and engineering of these EVs to carry various molecules useful as therapeutic agents (Figure 4) [61,202,203].

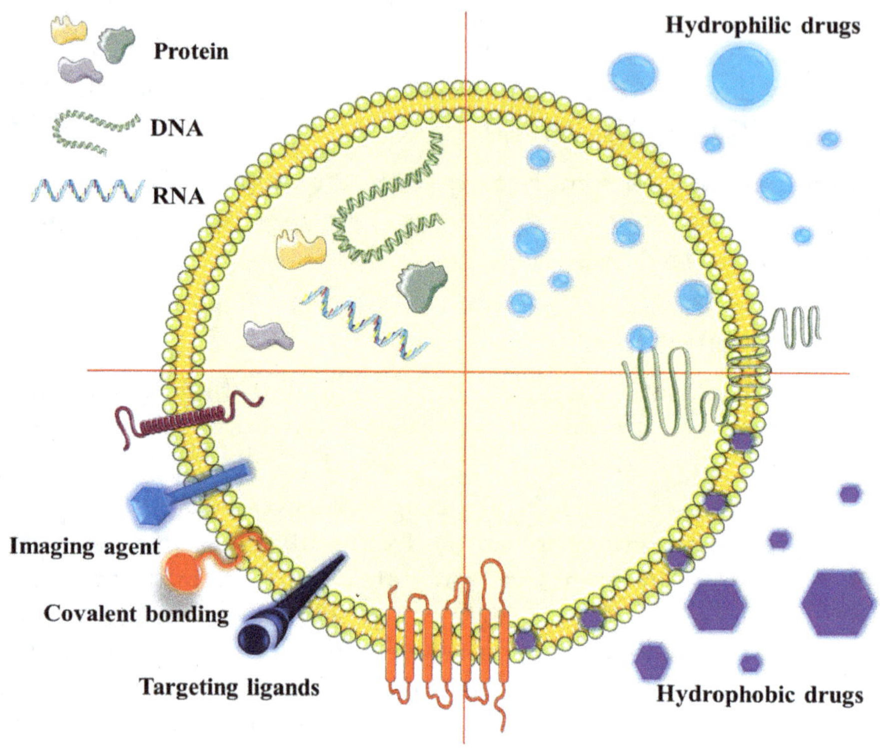

Figure 4. Properties of EVs useful in serving as drug delivery systems. These EVs consist of a lipid bilayer and an aqueous core, as they can incorporate hydrophilic drugs, hydrophobic drugs, nucleic acids (DNA, RNA), and proteins, as well as compounds (targeting ligands, covalent bonding, and imaging agents) that can be specifically attached to surfaces of EVs.

Overall, use of EVs as delivery agents will aid in the transport of internal cargo via enhanced endocytosis, thus protecting the contents from degradation. In contrast to liposomes or other nanoparticles used as carriers, EVs can serve as ideal bioparticles for targeted therapies [204,205]. Interestingly, it is suggested that biodistribution of EVs is influenced by cell origin and characteristics, with cell-specific tropism, thereby highlighting their potential use in the field of precision medicine [206]. In this arena, studies have reported on the efficiency of EVs as biocompatible drug vectors, as well as exhibiting low cytotoxicity and immunogenicity, and demonstrating their internalizing capabilities within a cell, as well as crossing the blood–brain barrier [207,208]. These EVs are capable of encapsulating various molecules, such as siRNA, miR, and various chemotherapeutics [207]. For example, Ma et al. have demonstrated that EVs carrying anti-cancer compounds can be absorbed by regenerated tumor cells, thus offering opportunities for their use in overcoming acquired drug resistance during cancer therapy [208]. Furthermore, it is reported that EVs are more likely to be internalized under acidic conditions; therefore, tumor cells are preferentially targeted by EVs rather than cells from surrounding healthy tissues [209]. Moreover, paclitaxel-loaded EVs have been used to improve the efficiency of treatment in multidrug-resistant tumor cells [210]. Recently, it has been demonstrated that tumor-derived EVs exhibit tropism toward their parental tumor cells [211], wherein engineered EVs, derived from fibrosarcoma and cervical cancer cell lines encapsulating the drug Doxil, are monitored both in vivo and in vitro using either HT1080 or HeLa tumors/cell lines. As expected, mice treated with Doxil-encapsulated EVs have higher levels of Doxil at the tumor site than those treated with Doxil alone, thereby reducing nonspecific cytotoxic effects of this drug [211].

In another study on small cell lung cancer (SCLC), sFlt-1-enriched EVs (soluble fms-like tyrosine kinase-1) are reported to act as tumor suppressors in mice via suppression of angiogenesis and induction of apoptosis in SCLC tumor cells [212].

Furthermore, in vitro and in vivo experiments of colorectal cancer cells revealed that EVs carrying miR-128-3p enhanced sensitivities to oxaliplatin by targeting *Bmi1* and *MRP5* genes [213]. In another study, the inhibitory effects on cell proliferation and EMT of miR-34c were evaluated using EVs derived from mesenchymal stem cells for delivery of miR-34c into nasopharyngeal carcinoma cell lines, and increased sensitivity to radiotherapy was observed [214]. Moreover, EVs delivering miR-199a-3p successfully suppressed both invasion and proliferation of ovarian cancer cell lines [215].

Currently, numerous clinical trials are investigating potential uses of EVs for either diagnostic/prognostic purposes or for therapeutic treatments of cancer (Table 2). These clinical trials assessing the use of microvesicles underlines their critical roles in malignancies. For example, some of these ongoing studies are evaluating engineered EVs for use as therapeutics for the treatment of pancreatic cancer. While in a completed phase II clinical trial, a vaccine developed with tumor antigen-loaded dendritic cell-derived EVs for NSCLC patients responsive to induction chemotherapy have yielded promising results [216]. It is reported dendritic cell-derived EVs manufactured with IFN-γ serve as a viable immunotherapeutic for NSCLC patients [216]. Moreover, this construct boosts NKp30-dependent NK cell functions, but without adverse consequences on antigen-specific T cell responses when used as maintenance immunotherapy for these NSCLC patients [216].

Table 2. Clinical studies exploring the use of EVs in cancer research studies.

Clinical Trial Identifier/Phase Status	Malignancy Investigated	EVs Use
NCT03236675/active, not recruiting	NSCLC	Detection of *EML4-ALK* fusion transcripts and T790M EGFR mutation
NCT03108677/recruiting	Osteosarcoma	Biomarkers for lung metastases, based on the RNS profile
NCT03985696/recruiting	Non-Hodgkin B-cell Lymphomas	Investigate EVs roles in immunotherapy, as carriers of therapeutic targets (CD20, PDL-1)
NCT03217266/recruiting	Soft tissue sarcoma	Detection of cell-free circulating tumor DNA mutations.
NCT02310451/unknown	Melanoma	Investigation of the effect of EVs produced by senescent melanoma cells
NCT03800121/recruiting	Sarcoma	Biomarkers for recurrence.
NCT03102268/unknown	Cholangiocarcinoma	Characterization of the ncRNAs in tumor derived EVs
NCT03911999/recruiting	Prostate cancer	Investigation of the relationship of urinary EVs and the aggressiveness of prostate cancer
NCT03711890/recruiting	Pancreatic cancer	Diagnostic biomarkers
NCT02869685/unknown	NSCLC	Detection of PD-L1 mRNA in plasma EVs
NCT03488134/active, not recruiting	Thyroid cancer	Urine exosomal proteins as biomarkers
NCT04258735/recruiting	Breast cancer	Diagnostic makers in a genomic panel
NCT02862470/active, not recruiting	Thyroid cancer	Urine EVs for the use as prognostic biomarkers
NCT01159288/completed	NSCLC	Treatment as tumor antigen-loaded dendritic cell-derived EVs
NCT04227886/recruiting	Rectal cancer	Biomarkers for toxicities and response to neoadjuvant therapy
NCT03608631/not yet recruiting	Pancreatic cancer	Treatment - mesenchymal stromal cells-derived EVs with KRAS G12D siRNA
NCT01779583/unknown	Gastric cancer	Prognostic and predictive biomarkers
NCT03874559/recruiting	Rectal cancer	Diagnostic biomarkers

Abbreviations: EVs- Extracellular vesicles; ncRNA–non-coding RNA; NSCLC–non-small cell lung cancer; PDL-1-programmed cell death ligand 1; siRNA–silence interfering RNA.

All the abovementioned features of EVs render them as suitable candidates for targeted therapies, especially for cancer. However, there are some challenges in attempts for use in broad applications for cancer therapy, such as lack of standardized methods of isolation and purification of EVs, and challenges in identifying optimized methods for loading EVs with therapeutic compounds [217–219]. As of now, there are several studies on the use of engineered EVs loaded with different molecules/drugs for in vitro and/or in vivo experiments in cancer research, and these are summarized in Table 3.

Table 3. Studies focused on investigating the effect of EVs-based therapy in in vivo and in vitro.

Pathology	EVs/Extracellular Vesicles Derived From	Cargo	Method of Engineering	In Vitro/In Vivo	Effect	Reference
Ovarian cancer	Fibroblasts from normal omentum	miR-199a-3p	Electroporation	SKOV3ip1, OVCAR3, CaOV3 and SKOV3-13	Inhibition of ovarian cancer cell proliferation, invasiveness, and c-Met expression.	[215]
				BALB/c nude mice	Inhibition of ovarian cancer peritoneal dissemination.	
Cancer	M1 macrophages	aCD47 and SIRPα	Polarization and conjugation	4T1tumor-bearing BALB/c mice	enhanced the phagocytosis of macrophages	[220]
Cancer	Bel7402 cell line	Doxorubicin-loaded PSiNPs (porous silicone nanoparticles)	Incubation	BALB/c mice and C57BL/6 mice bearing H22 tumors	Enhanced tumor accumulation of doxorubicin	[221]
Small cell lung cancer	BEAS-2B and NCI-H69 cell lines	sFlt-1	Cloning sFlt-1 into a lentivirus and obtaining engineered cell lines overexpressing sFlt-1	Nude mice with NCI-H69 xenografts	Induction of tumor apoptosis and inhibition tumor cell proliferation.	[212]
Glioma	RAW264.7 cells	Doxorubicin	Incubation	GL261 cells and RAW264.7 cells	Uptake of loaded EVs is higher in cancer cells than in normal cells	[222]
				C57BL/6 mice	Increased blood circulation time	
Breast cancer	Artificial chimeric EVs (ACEs)	Doxorubicin	Integration of RBCs and MCF-7 cell membrane proteins into synthetic phospholipid bilayers.	MCF-7 cells	Inhibition of cellular growth	[223]
				BALB/c nude mice and ICR mice	Doxorubicin accumulation in tumor improving anti-tumor efficacy	
Hepatocellular carcinoma	Plasma of healthy blood donors	miR-31 and miR-451	Electroporation	HepG2 cells	Increased cancer cell apoptosis.	[224]
Breast cancer	MSC	Doxorubicin	Electroporation	BT-474 and MDA-MB231 cells	Reduced cell viability, but with no significant differences between free DOX and EVs encapsulate DOX	[225]
Her2+ Breast Cancer	HEK 293T cells	siRNA	pLEX-LAMP-DARPin lentiviral transduction in HEK 293T cells	SKBR3 cells	Increased suppression of target gene (*TPD52*) compared to untreated cells and negative control (unloaded EVs)	[226]
Breast cancer	MSC	miR-379	lentiviral transduction of MSCs	BALB/c nude mice	Reduction in tumor size compared to the negative control (NTC extracellular vesicles)	[227]
NSCLC	RAW 264.7 cells	Paclitaxel	Sonication and incubation (including vectorization of EVs-AA-PEG-exoPTX)	C57BL/6 mice with established mCherry-3LL-M27 metastases	Stronger suppression of metastases growth and greater survival time as compared to Taxol, or non-vectorized exoPTX formulation	[228]
Pancreatic cancer	Normal fibroblast-like mesenchymal cells	siRNA or shRNA targeting $Kras^{G12D}$	Electroporation	Panc-1 cells	Enhanced apoptosis and decreased proliferation	[229]
				Nu/nu mice with orthotopic Panc-1 tumors	Controlled growth of tumors	
Chronic myeloid leukemia	HEK293T cells	Imatinib (IL3 EVs)	Incubation	LAMA84 and K562R cells	Reduction in cell viability compared to empty imatinib loaded EVs	[230]
				NOD/SCID mice	Reduction in tumor size	
Melanoma	B16BL6 cells	CpG-DNA (SAV-LA EVs)	Incubation	C57BL/6J mice and BALB/c nu/nu mice	Inhibition of tumor growth.	[231]
Breast cancer	immature mouse dendritic cell line (imDC)	Doxorubicin (iRGD-positive EVs)	Electroporation	MDA-MB-231	Inhibition of cell proliferation	[232]
				MDA-MB-231 tumor-bearing BALB/c nude mice	Inhibition of tumor growth due to effective accumulation of Dox at tumor sites	
Breast cancer	HEK293	let-7 (GE11-positive EVs)	lipofection	RAG2−/− mice	Suppression of tumor growth	[233]

Abbreviations: EVs, extracellular vesicles; MSC, mesenchymal stem cells; NSCLC, non-small cell lung cancer; siRNA, small interfering RNA; shRNA, short hairpin RNA; DC; dendritic cells.

10. Conclusions

EVs represent particles released from both normal and malignant cells that have important biological roles in ensuring cell-to-cell communication, not only for neighboring cells but also for distant cells. EVs are classified as EVs, multivesicular bodies and apoptotic bodies, of different sizes, origin, and protein and lipid compositions. These EVs play critical roles in pathological states of cells, regulating all hallmarks of cancer cells and resistance to drug treatments, thus highlighting the potential of these entities in the management of cancer. EV capabilities in carrying different active biomolecules, such as different RNA species, DNA, and proteins for targeting recipient cells without triggering immune responses, have rendered them as valuable biological entities for use as therapeutic agents that can overcome the shortcomings of complex diseases, such as cancer.

Author Contributions: Conceptualization, A.J. and C.P.-B.; methodology, W.C.C.; validation, O.S. and C.D.Ş.; formal analysis, C.D.Ş.; investigation, A.J., C.P-B., W.C.C., and I.B.-N.; writing—original draft preparation, A.J., C.P-B., O.S., C.D.Ş., W.C.C., and I.B.-N.; writing—review and editing, W.C.C., S.S.K., and I.B.-N.; visualization, A.J., C.P.-B., O.S., C.D.Ş., W.C.C., and I.B.-N.; supervision, I.B.-N.; project administration, I.B.-N.; funding acquisition, I.B.-N. All authors have read and agreed to the published version of the manuscript.

References

1. Maia, J.; Caja, S.; Strano Moraes, M.C.; Couto, N.; Costa-Silva, B. Exosome-Based Cell-Cell Communication in the Tumor Microenvironment. *Front. Cell Dev. Biol.* **2018**, *6*, 18. [CrossRef] [PubMed]
2. McCrea, P.D.; Gu, D.; Balda, M.S. Junctional Music that the Nucleus Hears: Cell-Cell Contact Signaling and the Modulation of Gene Activity. *Cold Spring Harb. Perspect. Biol.* **2009**, *1*, a002923. [CrossRef] [PubMed]
3. Brücher, B.L.D.M.; Jamall, I.S. Cell-Cell Communication in the Tumor Microenvironment, Carcinogenesis, and Anticancer Treatment. *Cell Physiol. Biochem.* **2014**, *34*, 213–243. [CrossRef] [PubMed]
4. Lee, Y.; EL Andaloussi, S.; Wood, M.J.A. Exosomes and microvesicles: Extracellular vesicles for genetic information transfer and gene therapy. *Hum. Mol. Genet.* **2012**, *21*, R125–R134. [CrossRef] [PubMed]
5. Harding, C.; Heuser, J.; Stahl, P. Endocytosis and intracellular processing of transferrin and colloidal gold-transferrin in rat reticulocytes: Demonstration of a pathway for receptor shedding. *Eur. J. Cell Biol.* **1984**, *35*, 256–263.
6. Lai, R.C.; Chen, T.S.; Lim, S.K. Mesenchymal stem cell exosome: A novel stem cell-based therapy for cardiovascular disease. *Regen. Med.* **2011**, *6*, 481–492. [CrossRef]
7. Guescini, M.; Genedani, S.; Stocchi, V.; Agnati, L.F. Astrocytes and Glioblastoma cells release exosomes carrying mtDNA. *J. Neural Transm.* **2010**, *117*, 1–4. [CrossRef]
8. Ratajczak, J.; Miekus, K.; Kucia, M.; Zhang, J.; Reca, R.; Dvorak, P.; Ratajczak, M.Z. Embryonic stem cell-derived microvesicles reprogram hematopoietic progenitors: Evidence for horizontal transfer of mRNA and protein delivery. *Leukemia* **2006**, *20*, 847–856. [CrossRef]
9. Raposo, G.; Nijman, H.W.; Stoorvogel, W.; Liejendekker, R.; Harding, C.V.; Melief, C.J.; Geuze, H.J. B lymphocytes secrete antigen-presenting vesicles. *J. Exp. Med.* **1996**, *183*, 1161–1172. [CrossRef]
10. Gatti, S.; Bruno, S.; Deregibus, M.C.; Sordi, A.; Cantaluppi, V.; Tetta, C.; Camussi, G. Microvesicles derived from human adult mesenchymal stem cells protect against ischaemia-reperfusion-induced acute and chronic kidney injury. *Nephrol. Dial. Transplant.* **2011**, *26*, 1474–1483. [CrossRef]
11. Olejarz, W.; Dominiak, A.; Żołnierzak, A.; Kubiak-Tomaszewska, G.; Lorenc, T. Tumor-Derived Exosomes in Immunosuppression and Immunotherapy. *J. Immunol. Res.* **2020**, *2020*, 6272498. [CrossRef] [PubMed]
12. Besse, B.; Charrier, M.; Lapierre, V.; Dansin, E.; Lantz, O.; Planchard, D.; Le Chevalier, T.; Livartoski, A.; Barlesi, F.; Laplanche, A.; et al. Dendritic cell-derived exosomes as maintenance immunotherapy after first line chemotherapy in NSCLC. *OncoImmunology* **2016**, *5*, e1071008. [CrossRef] [PubMed]

13. Rak, J.; Guha, A. Extracellular vesicles—Vehicles that spread cancer genes. *Bioessays* **2012**, *34*, 489–497. [CrossRef] [PubMed]
14. Bellingham, S.A.; Guo, B.B.; Coleman, B.M.; Hill, A.F. Exosomes: Vehicles for the Transfer of Toxic Proteins Associated with Neurodegenerative Diseases? *Front. Physiol.* **2012**, *3*, 124. [CrossRef] [PubMed]
15. Emmanouilidou, E.; Melachroinou, K.; Roumeliotis, T.; Garbis, S.D.; Ntzouni, M.; Margaritis, L.H.; Stefanis, L.; Vekrellis, K. Cell-Produced -Synuclein Is Secreted in a Calcium-Dependent Manner by Exosomes and Impacts Neuronal Survival. *J. Neurosci.* **2010**, *30*, 6838–6851. [CrossRef] [PubMed]
16. Tai, Y.-L.; Chu, P.-Y.; Lee, B.-H.; Chen, K.-C.; Yang, C.-Y.; Kuo, W.-H.; Shen, T.-L. Basics and applications of tumor-derived extracellular vesicles. *J. Biomed. Sci.* **2019**, *26*, 35. [CrossRef]
17. EL Andaloussi, S.; Mäger, I.; Breakefield, X.O.; Wood, M.J.A. Extracellular vesicles: Biology and emerging therapeutic opportunities. *Nat. Rev. Drug Discov.* **2013**, *12*, 347–357. [CrossRef]
18. Jurj, A.; Pasca, S.; Teodorescu, P.; Tomuleasa, C.; Berindan-Neagoe, I. Basic knowledge on BCR-ABL1-positive extracellular vesicles. *Biomark. Med.* **2020**, *14*, 451–458. [CrossRef]
19. Gebara, N.; Rossi, A.; Skovronova, R.; Aziz, J.M.; Asthana, A.; Bussolati, B. Extracellular Vesicles, Apoptotic Bodies and Mitochondria: Stem Cell Bioproducts for Organ Regeneration. *Curr. Transpl. Rep.* **2020**, *7*, 105–113. [CrossRef]
20. Kakarla, R.; Hur, J.; Kim, Y.J.; Kim, J.; Chwae, Y.-J. Apoptotic cell-derived exosomes: Messages from dying cells. *Exp. Mol. Med.* **2020**, *52*, 1–6. [CrossRef]
21. Gardai, S.J.; McPhillips, K.A.; Frasch, S.C.; Janssen, W.J.; Starefeldt, A.; Murphy-Ullrich, J.E.; Bratton, D.L.; Oldenborg, P.-A.; Michalak, M.; Henson, P.M. Cell-Surface Calreticulin Initiates Clearance of Viable or Apoptotic Cells through trans-Activation of LRP on the Phagocyte. *Cell* **2005**, *123*, 321–334. [CrossRef] [PubMed]
22. Lunavat, T.R.; Cheng, L.; Kim, D.-K.; Bhadury, J.; Jang, S.C.; Lässer, C.; Sharples, R.A.; López, M.D.; Nilsson, J.; Gho, Y.S.; et al. Small RNA deep sequencing discriminates subsets of extracellular vesicles released by melanoma cells—Evidence of unique microRNA cargos. *RNA Biol.* **2015**, *12*, 810–823. [CrossRef] [PubMed]
23. Caruso, S.; Poon, I.K.H. Apoptotic Cell-Derived Extracellular Vesicles: More Than Just Debris. *Front. Immunol.* **2018**, *9*, 1486. [CrossRef]
24. Akers, J.C.; Gonda, D.; Kim, R.; Carter, B.S.; Chen, C.C. Biogenesis of extracellular vesicles (EV): Exosomes, microvesicles, retrovirus-like vesicles, and apoptotic bodies. *J. Neurooncol.* **2013**, *113*, 1–11. [CrossRef]
25. Jurj, A.; Zanoaga, O.; Braicu, C.; Lazar, V.; Tomuleasa, C.; Irimie, A.; Berindan-Neagoe, I. A Comprehensive Picture of Extracellular Vesicles and Their Contents. Molecular Transfer to Cancer Cells. *Cancers* **2020**, *12*, 298. [CrossRef] [PubMed]
26. Borges, F.T.; Reis, L.A.; Schor, N. Extracellular vesicles: Structure, function, and potential clinical uses in renal diseases. *Braz. J. Med. Biol. Res.* **2013**, *46*, 824–830. [CrossRef]
27. Ailawadi, S.; Wang, X.; Gu, H.; Fan, G.-C. Pathologic function and therapeutic potential of exosomes in cardiovascular disease. *Biochim. Biophys. Acta (BBA) Mol. Basis Dis.* **2015**, *1852*, 1–11. [CrossRef]
28. Kajimoto, T.; Okada, T.; Miya, S.; Zhang, L.; Nakamura, S. Ongoing activation of sphingosine 1-phosphate receptors mediates maturation of exosomal multivesicular endosomes. *Nat. Commun.* **2013**, *4*, 2712. [CrossRef]
29. Kowal, J.; Tkach, M.; Théry, C. Biogenesis and secretion of exosomes. *Curr. Opin. Cell Biol.* **2014**, *29*, 116–125. [CrossRef]
30. Ekström, E.J.; Bergenfelz, C.; von Bülow, V.; Serifler, F.; Carlemalm, E.; Jönsson, G.; Andersson, T.; Leandersson, K. WNT5A induces release of exosomes containing pro-angiogenic and immunosuppressive factors from malignant melanoma cells. *Mol. Cancer* **2014**, *13*, 88. [CrossRef]
31. Chuo, S.T.-Y.; Chien, J.C.-Y.; Lai, C.P.-K. Imaging extracellular vesicles: Current and emerging methods. *J. Biomed. Sci.* **2018**, *25*, 91. [CrossRef] [PubMed]
32. Willms, E.; Cabañas, C.; Mäger, I.; Wood, M.J.A.; Vader, P. Extracellular Vesicle Heterogeneity: Subpopulations, Isolation Techniques, and Diverse Functions in Cancer Progression. *Front. Immunol.* **2018**, *9*, 738. [CrossRef]
33. Doyle, L.; Wang, M. Overview of Extracellular Vesicles, Their Origin, Composition, Purpose, and Methods for Exosome Isolation and Analysis. *Cells* **2019**, *8*, 727. [CrossRef] [PubMed]
34. Xie, C.; Ji, N.; Tang, Z.; Li, J.; Chen, Q. The role of extracellular vesicles from different origin in the microenvironment of head and neck cancers. *Mol. Cancer* **2019**, *18*, 83. [CrossRef] [PubMed]

35. Zhang, Y.; Liu, Y.; Liu, H.; Tang, W.H. Exosomes: Biogenesis, biologic function and clinical potential. *Cell Biosci.* **2019**, *9*, 19. [CrossRef] [PubMed]
36. Zaborowski, M.P.; Balaj, L.; Breakefield, X.O.; Lai, C.P. Extracellular Vesicles: Composition, Biological Relevance, and Methods of Study. *BioScience* **2015**, *65*, 783–797. [CrossRef]
37. Simeone, P.; Bologna, G.; Lanuti, P.; Pierdomenico, L.; Guagnano, M.T.; Pieragostino, D.; Del Boccio, P.; Vergara, D.; Marchisio, M.; Miscia, S.; et al. Extracellular Vesicles as Signaling Mediators and Disease Biomarkers across Biological Barriers. *Int. J. Mol. Sci.* **2020**, *21*, 2514. [CrossRef]
38. Catalano, M.; O'Driscoll, L. Inhibiting extracellular vesicles formation and release: A review of EV inhibitors. *J. Extracell. Vesicles* **2020**, *9*, 1703244. [CrossRef]
39. Joshi, B.S.; de Beer, M.A.; Giepmans, B.N.G.; Zuhorn, I.S. Endocytosis of Extracellular Vesicles and Release of Their Cargo from Endosomes. *ACS Nano* **2020**, *14*, 4444–4455. [CrossRef]
40. Théry, C.; Ostrowski, M.; Segura, E. Membrane vesicles as conveyors of immune responses. *Nat. Rev. Immunol.* **2009**, *9*, 581–593. [CrossRef]
41. Villarroya-Beltri, C.; Baixauli, F.; Gutiérrez-Vázquez, C.; Sánchez-Madrid, F.; Mittelbrunn, M. Sorting it out: Regulation of exosome loading. *Semin. Cancer Biol.* **2014**, *28*, 3–13. [CrossRef]
42. Thakur, B.K.; Zhang, H.; Becker, A.; Matei, I.; Huang, Y.; Costa-Silva, B.; Zheng, Y.; Hoshino, A.; Brazier, H.; Xiang, J.; et al. Double-stranded DNA in exosomes: A novel biomarker in cancer detection. *Cell Res.* **2014**, *24*, 766–769. [CrossRef] [PubMed]
43. Fischer, S.; Cornils, K.; Speiseder, T.; Badbaran, A.; Reimer, R.; Indenbirken, D.; Grundhoff, A.; Brunswig-Spickenheier, B.; Alawi, M.; Lange, C. Indication of Horizontal DNA Gene Transfer by Extracellular Vesicles. *PLoS ONE* **2016**, *11*, e0163665. [CrossRef] [PubMed]
44. Shelke, G.; Jang, S.C.; Yin, Y.; Lässer, C.; Lötvall, J. Human mast cells release extracellular vesicle-associated DNA. *Matters* **2016**, *2*, e201602000034. [CrossRef]
45. Kawamura, Y.; Yamamoto, Y.; Sato, T.-A.; Ochiya, T. Extracellular vesicles as trans-genomic agents: Emerging roles in disease and evolution. *Cancer Sci.* **2017**, *108*, 824–830. [CrossRef]
46. Record, M.; Subra, C.; Silvente-Poirot, S.; Poirot, M. Exosomes as intercellular signalosomes and pharmacological effectors. *Biochem. Pharmacol.* **2011**, *81*, 1171–1182. [CrossRef] [PubMed]
47. Choi, D.-S.; Kim, D.-K.; Kim, Y.-K.; Gho, Y.S. Proteomics, transcriptomics and lipidomics of exosomes and ectosomes. *Proteomics* **2013**, *13*, 1554–1571. [CrossRef]
48. Yáñez-Mó, M.; Siljander, P.R.-M.; Andreu, Z.; Bedina Zavec, A.; Borràs, F.E.; Buzas, E.I.; Buzas, K.; Casal, E.; Cappello, F.; Carvalho, J.; et al. Biological properties of extracellular vesicles and their physiological functions. *J. Extracell. Vesicles* **2015**, *4*, 27066. [CrossRef] [PubMed]
49. Chen, T.; Guo, J.; Yang, M.; Zhu, X.; Cao, X. Chemokine-Containing Exosomes Are Released from Heat-Stressed Tumor Cells via Lipid Raft-Dependent Pathway and Act as Efficient Tumor Vaccine. *J. Immunol.* **2011**, *186*, 2219–2228. [CrossRef]
50. Li, I.; Nabet, B.Y. Exosomes in the tumor microenvironment as mediators of cancer therapy resistance. *Mol. Cancer* **2019**, *18*, 32. [CrossRef]
51. Cvjetkovic, A.; Lötvall, J.; Lässer, C. The influence of rotor type and centrifugation time on the yield and purity of extracellular vesicles. *J. Extracell. Vesicles* **2014**, *3*, 23111. [CrossRef] [PubMed]
52. Konoshenko, M.Y.; Lekchnov, E.A.; Vlassov, A.V.; Laktionov, P.P. Isolation of Extracellular Vesicles: General Methodologies and Latest Trends. *Biomed. Res. Int.* **2018**, *2018*, 8545347. [CrossRef] [PubMed]
53. Böing, A.N.; van der Pol, E.; Grootemaat, A.E.; Coumans, F.A.W.; Sturk, A.; Nieuwland, R. Single-step isolation of extracellular vesicles by size-exclusion chromatography. *J. Extracell. Vesicles* **2014**, *3*, 23430. [CrossRef] [PubMed]
54. Yamamoto, K.R.; Alberts, B.M.; Benzinger, R.; Lawhorne, L.; Treiber, G. Rapid bacteriophage sedimentation in the presence of polyethylene glycol and its application to large-scale virus purification. *Virology* **1970**, *40*, 734–744. [CrossRef]
55. Li, P.; Kaslan, M.; Lee, S.H.; Yao, J.; Gao, Z. Progress in Exosome Isolation Techniques. *Theranostics* **2017**, *7*, 789–804. [CrossRef]
56. Liu, F.; Vermesh, O.; Mani, V.; Ge, T.J.; Madsen, S.J.; Sabour, A.; Hsu, E.-C.; Gowrishankar, G.; Kanada, M.; Jokerst, J.V.; et al. The Exosome Total Isolation Chip. *ACS Nano* **2017**, *11*, 10712–10723. [CrossRef]

57. Wu, M.; Ouyang, Y.; Wang, Z.; Zhang, R.; Huang, P.-H.; Chen, C.; Li, H.; Li, P.; Quinn, D.; Dao, M.; et al. Isolation of exosomes from whole blood by integrating acoustics and microfluidics. *Proc. Natl. Acad. Sci. USA* **2017**, *114*, 10584–10589. [CrossRef]
58. Gallart-Palau, X.; Serra, A.; Wong, A.S.W.; Sandin, S.; Lai, M.K.P.; Chen, C.P.; Kon, O.L.; Sze, S.K. Extracellular vesicles are rapidly purified from human plasma by PRotein Organic Solvent PRecipitation (PROSPR). *Sci. Rep.* **2015**, *5*, 14664. [CrossRef]
59. Heath, N.; Grant, L.; De Oliveira, T.M.; Rowlinson, R.; Osteikoetxea, X.; Dekker, N.; Overman, R. Rapid isolation and enrichment of extracellular vesicle preparations using anion exchange chromatography. *Sci. Rep.* **2018**, *8*, 5730. [CrossRef]
60. Merchant, M.L.; Powell, D.W.; Wilkey, D.W.; Cummins, T.D.; Deegens, J.K.; Rood, I.M.; McAfee, K.J.; Fleischer, C.; Klein, E.; Klein, J.B. Microfiltration isolation of human urinary exosomes for characterization by MS. *Proteom. Clin. Appl.* **2010**, *4*, 84–96. [CrossRef]
61. Usman, W.M.; Pham, T.C.; Kwok, Y.Y.; Vu, L.T.; Ma, V.; Peng, B.; Chan, Y.S.; Wei, L.; Chin, S.M.; Azad, A.; et al. Efficient RNA drug delivery using red blood cell extracellular vesicles. *Nat. Commun.* **2018**, *9*, 2359. [CrossRef]
62. Gercel-Taylor, C.; Atay, S.; Tullis, R.H.; Kesimer, M.; Taylor, D.D. Nanoparticle analysis of circulating cell-derived vesicles in ovarian cancer patients. *Anal. Biochem.* **2012**, *428*, 44–53. [CrossRef] [PubMed]
63. Vestad, B.; Llorente, A.; Neurauter, A.; Phuyal, S.; Kierulf, B.; Kierulf, P.; Skotland, T.; Sandvig, K.; Haug, K.B.F.; Øvstebø, R. Size and concentration analyses of extracellular vesicles by nanoparticle tracking analysis: A variation study. *J. Extracell. Vesicles* **2017**, *6*, 1344087. [CrossRef] [PubMed]
64. Headland, S.E.; Jones, H.R.; D'Sa, A.S.V.; Perretti, M.; Norling, L.V. Cutting-edge analysis of extracellular microparticles using ImageStream(X) imaging flow cytometry. *Sci. Rep.* **2014**, *4*, 5237. [CrossRef] [PubMed]
65. Linares, R.; Tan, S.; Gounou, C.; Brisson, A.R. Imaging and Quantification of Extracellular Vesicles by Transmission Electron Microscopy. *Methods Mol. Biol.* **2017**, *1545*, 43–54. [CrossRef]
66. Sprague, D.L.; Elzey, B.D.; Crist, S.A.; Waldschmidt, T.J.; Jensen, R.J.; Ratliff, T.L. Platelet-mediated modulation of adaptive immunity: Unique delivery of CD154 signal by platelet-derived membrane vesicles. *Blood* **2008**, *111*, 5028–5036. [CrossRef]
67. Baj-Krzyworzeka, M.; Mytar, B.; Szatanek, R.; Surmiak, M.; Węglarczyk, K.; Baran, J.; Siedlar, M. Colorectal cancer-derived microvesicles modulate differentiation of human monocytes to macrophages. *J. Transl. Med.* **2016**, *14*, 36. [CrossRef]
68. Simhadri, V.R.; Reiners, K.S.; Hansen, H.P.; Topolar, D.; Simhadri, V.L.; Nohroudi, K.; Kufer, T.A.; Engert, A.; Pogge von Strandmann, E. Dendritic Cells Release HLA-B-Associated Transcript-3 Positive Exosomes to Regulate Natural Killer Function. *PLoS ONE* **2008**, *3*, e3377. [CrossRef]
69. Clayton, A.; Mitchell, J.P.; Court, J.; Linnane, S.; Mason, M.D.; Tabi, Z. Human Tumor-Derived Exosomes Down-Modulate NKG2D Expression. *J. Immunol.* **2008**, *180*, 7249–7258. [CrossRef]
70. Eken, C.; Gasser, O.; Zenhaeusern, G.; Oehri, I.; Hess, C.; Schifferli, J.A. Polymorphonuclear Neutrophil-Derived Ectosomes Interfere with the Maturation of Monocyte-Derived Dendritic Cells. *J. Immunol.* **2008**, *180*, 817–824. [CrossRef]
71. Yu, S.; Liu, C.; Su, K.; Wang, J.; Liu, Y.; Zhang, L.; Li, C.; Cong, Y.; Kimberly, R.; Grizzle, W.E.; et al. Tumor Exosomes Inhibit Differentiation of Bone Marrow Dendritic Cells. *J. Immunol.* **2007**, *178*, 6867–6875. [CrossRef] [PubMed]
72. Camussi, G.; Deregibus, M.-C.; Bruno, S.; Grange, C.; Fonsato, V.; Tetta, C. Exosome/microvesicle-mediated epigenetic reprogramming of cells. *Am. J. Cancer Res.* **2011**, *1*, 98–110.
73. Jang, Y.-Y.; Collector, M.I.; Baylin, S.B.; Diehl, A.M.; Sharkis, S.J. Hematopoietic stem cells convert into liver cells within days without fusion. *Nat. Cell Biol.* **2004**, *6*, 532–539. [CrossRef]
74. Quesenberry, P.J.; Aliotta, J.M. Cellular phenotype switching and microvesicles. *Adv. Drug Deliv. Rev.* **2010**, *62*, 1141–1148. [CrossRef] [PubMed]
75. Peinado, H.; Alečković, M.; Lavotshkin, S.; Matei, I.; Costa-Silva, B.; Moreno-Bueno, G.; Hergueta-Redondo, M.; Williams, C.; García-Santos, G.; Ghajar, C.; et al. Melanoma exosomes educate bone marrow progenitor cells toward a pro-metastatic phenotype through MET. *Nat. Med.* **2012**, *18*, 883–891. [CrossRef] [PubMed]

76. Skog, J.; Würdinger, T.; van Rijn, S.; Meijer, D.H.; Gainche, L.; Curry, W.T.; Carter, B.S.; Krichevsky, A.M.; Breakefield, X.O. Glioblastoma microvesicles transport RNA and proteins that promote tumour growth and provide diagnostic biomarkers. *Nat. Cell Biol.* **2008**, *10*, 1470–1476. [CrossRef]
77. Al-Nedawi, K.; Meehan, B.; Kerbel, R.S.; Allison, A.C.; Rak, J. Endothelial expression of autocrine VEGF upon the uptake of tumor-derived microvesicles containing oncogenic EGFR. *Proc. Natl. Acad. Sci. USA* **2009**, *106*, 3794–3799. [CrossRef]
78. Hu, C.; Meiners, S.; Lukas, C.; Stathopoulos, G.T.; Chen, J. Role of exosomal microRNAs in lung cancer biology and clinical applications. *Cell Prolif.* **2020**, *53*. [CrossRef]
79. Kumar, A.; Deep, G. Exosomes in hypoxia-induced remodeling of the tumor microenvironment. *Cancer Lett.* **2020**, *488*, 1–8. [CrossRef]
80. Lee, J.-K.; Park, S.-R.; Jung, B.-K.; Jeon, Y.-K.; Lee, Y.-S.; Kim, M.-K.; Kim, Y.-G.; Jang, J.-Y.; Kim, C.-W. Exosomes Derived from Mesenchymal Stem Cells Suppress Angiogenesis by Down-Regulating VEGF Expression in Breast Cancer Cells. *PLoS ONE* **2013**, *8*, e84256. [CrossRef]
81. Umezu, T.; Ohyashiki, K.; Kuroda, M.; Ohyashiki, J.H. Leukemia cell to endothelial cell communication via exosomal miRNAs. *Oncogene* **2013**, *32*, 2747–2755. [CrossRef] [PubMed]
82. Kalinina, N.; Klink, G.; Glukhanyuk, E.; Lopatina, T.; Efimenko, A.; Akopyan, Z.; Tkachuk, V. miR-92a regulates angiogenic activity of adipose-derived mesenchymal stromal cells. *Exp. Cell Res.* **2015**, *339*, 61–66. [CrossRef]
83. Liu, Y.; Luo, F.; Wang, B.; Li, H.; Xu, Y.; Liu, X.; Shi, L.; Lu, X.; Xu, W.; Lu, L.; et al. STAT3-regulated exosomal miR-21 promotes angiogenesis and is involved in neoplastic processes of transformed human bronchial epithelial cells. *Cancer Lett.* **2016**, *370*, 125–135. [CrossRef] [PubMed]
84. Mao, G.; Liu, Y.; Fang, X.; Liu, Y.; Fang, L.; Lin, L.; Liu, X.; Wang, N. Tumor-derived microRNA-494 promotes angiogenesis in non-small cell lung cancer. *Angiogenesis* **2015**, *18*, 373–382. [CrossRef]
85. Grange, C.; Tapparo, M.; Collino, F.; Vitillo, L.; Damasco, C.; Deregibus, M.C.; Tetta, C.; Bussolati, B.; Camussi, G. Microvesicles Released from Human Renal Cancer Stem Cells Stimulate Angiogenesis and Formation of Lung Premetastatic Niche. *Cancer Res.* **2011**, *71*, 5346–5356. [CrossRef] [PubMed]
86. Sidhu, S.S.; Mengistab, A.T.; Tauscher, A.N.; LaVail, J.; Basbaum, C. The microvesicle as a vehicle for EMMPRIN in tumor–stromal interactions. *Oncogene* **2004**, *23*, 956–963. [CrossRef] [PubMed]
87. Le, M.T.N.; Hamar, P.; Guo, C.; Basar, E.; Perdigão-Henriques, R.; Balaj, L.; Lieberman, J. miR-200–containing extracellular vesicles promote breast cancer cell metastasis. *J. Clin. Invest.* **2014**, *124*, 5109–5128. [CrossRef] [PubMed]
88. Liao, J.; Liu, R.; Shi, Y.-J.; Yin, L.-H.; Pu, Y.-P. Exosome-shuttling microRNA-21 promotes cell migration and invasion-targeting PDCD4 in esophageal cancer. *Int. J. Oncol.* **2016**, *48*, 2567–2579. [CrossRef]
89. Zhou, W.; Fong, M.Y.; Min, Y.; Somlo, G.; Liu, L.; Palomares, M.R.; Yu, Y.; Chow, A.; O'Connor, S.T.F.; Chin, A.R.; et al. Cancer-secreted miR-105 destroys vascular endothelial barriers to promote metastasis. *Cancer Cell* **2014**, *25*, 501–515. [CrossRef]
90. Wang, L.; He, J.; Hu, H.; Tu, L.; Sun, Z.; Liu, Y.; Luo, F. Lung CSC-derived exosomal miR-210-3p contributes to a pro-metastatic phenotype in lung cancer by targeting FGFRL1. *J. Cell Mol. Med.* **2020**, *24*, 6324–6339. [CrossRef]
91. Higginbotham, J.N.; Demory Beckler, M.; Gephart, J.D.; Franklin, J.L.; Bogatcheva, G.; Kremers, G.-J.; Piston, D.W.; Ayers, G.D.; McConnell, R.E.; Tyska, M.J.; et al. Amphiregulin Exosomes Increase Cancer Cell Invasion. *Curr. Biol.* **2011**, *21*, 779–786. [CrossRef]
92. Al-Nedawi, K.; Meehan, B.; Micallef, J.; Lhotak, V.; May, L.; Guha, A.; Rak, J. Intercellular transfer of the oncogenic receptor EGFRvIII by microvesicles derived from tumour cells. *Nat. Cell Biol.* **2008**, *10*, 619–624. [CrossRef]
93. Fong, M.Y.; Zhou, W.; Liu, L.; Alontaga, A.Y.; Chandra, M.; Ashby, J.; Chow, A.; O'Connor, S.T.F.; Li, S.; Chin, A.R.; et al. Breast-cancer-secreted miR-122 reprograms glucose metabolism in premetastatic niche to promote metastasis. *Nat. Cell Biol.* **2015**, *17*, 183–194. [CrossRef] [PubMed]
94. Garnier, D.; Magnus, N.; Meehan, B.; Kislinger, T.; Rak, J. Qualitative changes in the proteome of extracellular vesicles accompanying cancer cell transition to mesenchymal state. *Exp. Cell Res.* **2013**, *319*, 2747–2757. [CrossRef] [PubMed]

95. Tauro, B.J.; Mathias, R.A.; Greening, D.W.; Gopal, S.K.; Ji, H.; Kapp, E.A.; Coleman, B.M.; Hill, A.F.; Kusebauch, U.; Hallows, J.L.; et al. Oncogenic H-Ras Reprograms Madin-Darby Canine Kidney (MDCK) Cell-derived Exosomal Proteins Following Epithelial-Mesenchymal Transition. *Mol. Cell Proteom.* **2013**, *12*, 2148–2159. [CrossRef]
96. Aga, M.; Bentz, G.L.; Raffa, S.; Torrisi, M.R.; Kondo, S.; Wakisaka, N.; Yoshizaki, T.; Pagano, J.S.; Shackelford, J. Exosomal HIF1α supports invasive potential of nasopharyngeal carcinoma-associated LMP1-positive exosomes. *Oncogene* **2014**, *33*, 4613–4622. [CrossRef] [PubMed]
97. Cai, Z.; Yang, F.; Yu, L.; Yu, Z.; Jiang, L.; Wang, Q.; Yang, Y.; Wang, L.; Cao, X.; Wang, J. Activated T Cell Exosomes Promote Tumor Invasion via Fas Signaling Pathway. *J. Immunol.* **2012**, *188*, 5954–5961. [CrossRef]
98. Gao, F.; Zhao, Z.-L.; Zhao, W.-T.; Fan, Q.-R.; Wang, S.-C.; Li, J.; Zhang, Y.-Q.; Shi, J.-W.; Lin, X.-L.; Yang, S.; et al. miR-9 modulates the expression of interferon-regulated genes and MHC class I molecules in human nasopharyngeal carcinoma cells. *Biochem. Biophys. Res. Commun.* **2013**, *431*, 610–616. [CrossRef]
99. Fabbri, M.; Paone, A.; Calore, F.; Galli, R.; Gaudio, E.; Santhanam, R.; Lovat, F.; Fadda, P.; Mao, C.; Nuovo, G.J.; et al. MicroRNAs bind to Toll-like receptors to induce prometastatic inflammatory response. *Proc. Natl. Acad. Sci. USA* **2012**, *109*, E2110–E2116. [CrossRef]
100. Ma, X.; Chen, Z.; Hua, D.; He, D.; Wang, L.; Zhang, P.; Wang, J.; Cai, Y.; Gao, C.; Zhang, X.; et al. Essential role for TrpC5-containing extracellular vesicles in breast cancer with chemotherapeutic resistance. *Proc. Natl. Acad. Sci. USA* **2014**, *111*, 6389–6394. [CrossRef]
101. Klibi, J.; Niki, T.; Riedel, A.; Pioche-Durieu, C.; Souquere, S.; Rubinstein, E.; Le Moulec, S.; Guigay, J.; Hirashima, M.; Guemira, F.; et al. Blood diffusion and Th1-suppressive effects of galectin-9–containing exosomes released by Epstein-Barr virus–infected nasopharyngeal carcinoma cells. *Blood* **2009**, *113*, 1957–1966. [CrossRef]
102. Yang, M.; Chen, J.; Su, F.; Yu, B.; Su, F.; Lin, L.; Liu, Y.; Huang, J.-D.; Song, E. Microvesicles secreted by macrophages shuttle invasion-potentiating microRNAs into breast cancer cells. *Mol. Cancer* **2011**, *10*, 117. [CrossRef] [PubMed]
103. Zhou, M.; Chen, J.; Zhou, L.; Chen, W.; Ding, G.; Cao, L. Pancreatic cancer derived exosomes regulate the expression of TLR4 in dendritic cells via miR-203. *Cell. Immunol.* **2014**, *292*, 65–69. [CrossRef] [PubMed]
104. Berchem, G.; Noman, M.Z.; Bosseler, M.; Paggetti, J.; Baconnais, S.; Le cam, E.; Nanbakhsh, A.; Moussay, E.; Mami-Chouaib, F.; Janji, B.; et al. Hypoxic tumor-derived microvesicles negatively regulate NK cell function by a mechanism involving TGF-β and miR23a transfer. *OncoImmunology* **2016**, *5*, e1062968. [CrossRef]
105. Huber, V.; Fais, S.; Iero, M.; Lugini, L.; Canese, P.; Squarcina, P.; Zaccheddu, A.; Colone, M.; Arancia, G.; Gentile, M.; et al. Human Colorectal Cancer Cells Induce T-Cell Death Through Release of Proapoptotic Microvesicles: Role in Immune Escape. *Gastroenterology* **2005**, *128*, 1796–1804. [CrossRef] [PubMed]
106. Clayton, A.; Mitchell, J.P.; Court, J.; Mason, M.D.; Tabi, Z. Human Tumor-Derived Exosomes Selectively Impair Lymphocyte Responses to Interleukin-2. *Cancer Res.* **2007**, *67*, 7458–7466. [CrossRef] [PubMed]
107. Ashiru, O.; Boutet, P.; Fernandez-Messina, L.; Aguera-Gonzalez, S.; Skepper, J.N.; Vales-Gomez, M.; Reyburn, H.T. Natural Killer Cell Cytotoxicity Is Suppressed by Exposure to the Human NKG2D Ligand MICA*008 That Is Shed by Tumor Cells in Exosomes. *Cancer Res.* **2010**, *70*, 481–489. [CrossRef]
108. Condamine, T.; Ramachandran, I.; Youn, J.-I.; Gabrilovich, D.I. Regulation of Tumor Metastasis by Myeloid-Derived Suppressor Cells. *Annu. Rev. Med.* **2015**, *66*, 97–110. [CrossRef] [PubMed]
109. Xiang, X.; Poliakov, A.; Liu, C.; Liu, Y.; Deng, Z.; Wang, J.; Cheng, Z.; Shah, S.V.; Wang, G.-J.; Zhang, L.; et al. Induction of myeloid-derived suppressor cells by tumor exosomes. *Int. J. Cancer* **2009**, *124*, 2621–2633. [CrossRef]
110. Khan, S.; Jutzy, J.M.S.; Aspe, J.R.; McGregor, D.W.; Neidigh, J.W.; Wall, N.R. Survivin is released from cancer cells via exosomes. *Apoptosis* **2011**, *16*, 1–12. [CrossRef]
111. Cappellesso, R.; Tinazzi, A.; Giurici, T.; Simonato, F.; Guzzardo, V.; Ventura, L.; Crescenzi, M.; Chiarelli, S.; Fassina, A. Programmed cell death 4 and microRNA 21 inverse expression is maintained in cells and exosomes from ovarian serous carcinoma effusions: PDCD4 and miR-21 Expression in OSC and Exosomes. *Cancer Cytopathol.* **2014**, *122*, 685–693. [CrossRef] [PubMed]
112. Taylor, D.D.; Gercel-Taylor, C. MicroRNA signatures of tumor-derived exosomes as diagnostic biomarkers of ovarian cancer. *Gynecol. Oncol.* **2008**, *110*, 13–21. [CrossRef]
113. Yu, X.; Harris, S.L.; Levine, A.J. The Regulation of Exosome Secretion: A Novel Function of the p53 Protein. *Cancer Res.* **2006**, *66*, 4795–4801. [CrossRef] [PubMed]

114. Xavier, C.P.R.; Caires, H.R.; Barbosa, M.A.G.; Bergantim, R.; Guimarães, J.E.; Vasconcelos, M.H. The Role of Extracellular Vesicles in the Hallmarks of Cancer and Drug Resistance. *Cells* **2020**, *9*, 1141. [CrossRef]
115. Graner, M.W.; Alzate, O.; Dechkovskaia, A.M.; Keene, J.D.; Sampson, J.H.; Mitchell, D.A.; Bigner, D.D. Proteomic and immunologic analyses of brain tumor exosomes. *FASEB J.* **2009**, *23*, 1541–1557. [CrossRef] [PubMed]
116. Webber, J.; Stone, T.C.; Katilius, E.; Smith, B.C.; Gordon, B.; Mason, M.D.; Tabi, Z.; Brewis, I.A.; Clayton, A. Proteomics Analysis of Cancer Exosomes Using a Novel Modified Aptamer-based Array (SOMAscan^{TM}) Platform. *Mol. Cell Proteom.* **2014**, *13*, 1050–1064. [CrossRef] [PubMed]
117. Welton, J.L.; Khanna, S.; Giles, P.J.; Brennan, P.; Brewis, I.A.; Staffurth, J.; Mason, M.D.; Clayton, A. Proteomics Analysis of Bladder Cancer Exosomes. *Mol. Cell Proteom.* **2010**, *9*, 1324–1338. [CrossRef] [PubMed]
118. Demory Beckler, M.; Higginbotham, J.N.; Franklin, J.L.; Ham, A.-J.; Halvey, P.J.; Imasuen, I.E.; Whitwell, C.; Li, M.; Liebler, D.C.; Coffey, R.J. Proteomic Analysis of Exosomes from Mutant KRAS Colon Cancer Cells Identifies Intercellular Transfer of Mutant KRAS. *Mol. Cell Proteom.* **2013**, *12*, 343–355. [CrossRef]
119. Soldevilla, B.; Rodríguez, M.; San Millán, C.; García, V.; Fernández-Periañez, R.; Gil-Calderón, B.; Martín, P.; García-Grande, A.; Silva, J.; Bonilla, F.; et al. Tumor-derived exosomes are enriched in ΔNp73, which promotes oncogenic potential in acceptor cells and correlates with patient survival. *Hum. Mol. Genet.* **2014**, *23*, 467–478. [CrossRef]
120. Ohshima, K.; Inoue, K.; Fujiwara, A.; Hatakeyama, K.; Kanto, K.; Watanabe, Y.; Muramatsu, K.; Fukuda, Y.; Ogura, S.; Yamaguchi, K.; et al. Let-7 MicroRNA Family Is Selectively Secreted into the Extracellular Environment via Exosomes in a Metastatic Gastric Cancer Cell Line. *PLoS ONE* **2010**, *5*, e13247. [CrossRef]
121. Abd Elmageed, Z.Y.; Yang, Y.; Thomas, R.; Ranjan, M.; Mondal, D.; Moroz, K.; Fang, Z.; Rezk, B.M.; Moparty, K.; Sikka, S.C.; et al. Neoplastic Reprogramming of Patient-Derived Adipose Stem Cells by Prostate Cancer Cell-Associated Exosomes: Tumor Exosomes Trigger Stem Cell Transformation. *Stem Cells* **2014**, *32*, 983–997. [CrossRef]
122. Ostenfeld, M.S.; Jeppesen, D.K.; Laurberg, J.R.; Boysen, A.T.; Bramsen, J.B.; Primdal-Bengtson, B.; Hendrix, A.; Lamy, P.; Dagnaes-Hansen, F.; Rasmussen, M.H.; et al. Cellular Disposal of miR23b by RAB27-Dependent Exosome Release Is Linked to Acquisition of Metastatic Properties. *Cancer Res.* **2014**, *74*, 5758–5771. [CrossRef] [PubMed]
123. Putz, U.; Howitt, J.; Doan, A.; Goh, C.-P.; Low, L.-H.; Silke, J.; Tan, S.-S. The Tumor Suppressor PTEN Is Exported in Exosomes and Has Phosphatase Activity in Recipient Cells. *Sci. Signal.* **2012**, *5*, ra70. [CrossRef]
124. Meehan, K.; Vella, L.J. The contribution of tumour-derived exosomes to the hallmarks of cancer. *Crit. Rev. Clin. Lab. Sci.* **2016**, *53*, 121–131. [CrossRef]
125. Xiao, H.; Lässer, C.; Shelke, G.V.; Wang, J.; Rådinger, M.; Lunavat, T.R.; Malmhäll, C.; Lin, L.H.; Li, J.; Li, L.; et al. Mast cell exosomes promote lung adenocarcinoma cell proliferation—Role of KIT-stem cell factor signaling. *Cell Commun. Signal.* **2014**, *12*, 64. [CrossRef]
126. Li, C. CD97 promotes gastric cancer cell proliferation and invasion through exosome-mediated MAPK signaling pathway. *World J. Gastroenterol. WJG* **2015**, *21*, 6215. [CrossRef] [PubMed]
127. Qu, J.-L.; Qu, X.-J.; Zhao, M.-F.; Teng, Y.-E.; Zhang, Y.; Hou, K.-Z.; Jiang, Y.-H.; Yang, X.-H.; Liu, Y.-P. Gastric cancer exosomes promote tumour cell proliferation through PI3K/Akt and MAPK/ERK activation. *Dig. Liver Dis.* **2009**, *41*, 875–880. [CrossRef] [PubMed]
128. Vella, L.J.; Behren, A.; Coleman, B.; Greening, D.W.; Hill, A.F.; Cebon, J. Intercellular Resistance to BRAF Inhibition Can Be Mediated by Extracellular Vesicle–Associated PDGFRβ. *Neoplasia* **2017**, *19*, 932–940. [CrossRef]
129. Yang, L.; Wu, X.-H.; Wang, D.; Luo, C.-L.; Chen, L.-X. Bladder cancer cell-derived exosomes inhibit tumor cell apoptosis and induce cell proliferation in vitro. *Mol. Med. Rep.* **2013**, *8*, 1272–1278. [CrossRef] [PubMed]
130. Setti, M.; Osti, D.; Richichi, C.; Ortensi, B.; Del Bene, M.; Fornasari, L.; Beznoussenko, G.; Mironov, A.; Rappa, G.; Cuomo, A.; et al. Extracellular vesicle-mediated transfer of CLIC1 protein is a novel mechanism for the regulation of glioblastoma growth. *Oncotarget* **2015**, *6*, 31413–31427. [CrossRef]
131. The Australian Prostate Cancer Collaboration BioResource; Soekmadji, C.; Riches, J.D.; Russell, P.J.; Ruelcke, J.E.; McPherson, S.; Wang, C.; Hovens, C.M.; Corcoran, N.M.; Hill, M.M.; et al. Modulation of paracrine signaling by CD9 positive small extracellular vesicles mediates cellular growth of androgen deprived prostate cancer. *Oncotarget* **2017**, *8*, 52237–52255. [CrossRef] [PubMed]

132. Matsumoto, A.; Takahashi, Y.; Nishikawa, M.; Sano, K.; Morishita, M.; Charoenviriyakul, C.; Saji, H.; Takakura, Y. Accelerated growth of B16 BL 6 tumor in mice through efficient uptake of their own exosomes by B16 BL 6 cells. *Cancer Sci.* **2017**, *108*, 1803–1810. [CrossRef]
133. Liu, M.X.; Liao, J.; Xie, M.; Gao, Z.K.; Wang, X.H.; Zhang, Y.; Shang, M.H.; Yin, L.H.; Pu, Y.P.; Liu, R. miR-93-5p Transferred by Exosomes Promotes the Proliferation of Esophageal Cancer Cells via Intercellular Communication by Targeting PTEN. *Biomed. Environ. Sci.* **2018**, *31*, 171–185. [CrossRef]
134. Teng, Y.; Ren, Y.; Hu, X.; Mu, J.; Samykutty, A.; Zhuang, X.; Deng, Z.; Kumar, A.; Zhang, L.; Merchant, M.L.; et al. MVP-mediated exosomal sorting of miR-193a promotes colon cancer progression. *Nat. Commun.* **2017**, *8*, 14448. [CrossRef]
135. Lee, J.C.; Zhao, J.-T.; Gundara, J.; Serpell, J.; Bach, L.A.; Sidhu, S. Papillary thyroid cancer–derived exosomes contain miRNA-146b and miRNA-222. *J. Surg. Res.* **2015**, *196*, 39–48. [CrossRef] [PubMed]
136. Graner, M.W.; Schnell, S.; Olin, M.R. Tumor-derived exosomes, microRNAs, and cancer immune suppression. *Semin. Immunopathol.* **2018**, *40*, 505–515. [CrossRef] [PubMed]
137. Lässer, C.; Théry, C.; Buzás, E.I.; Mathivanan, S.; Zhao, W.; Gho, Y.S.; Lötvall, J. The International Society for Extracellular Vesicles launches the first massive open online course on extracellular vesicles. *J. Extracell. Vesicles* **2016**, *5*, 34299. [CrossRef] [PubMed]
138. Kogure, T.; Lin, W.-L.; Yan, I.K.; Braconi, C.; Patel, T. Intercellular nanovesicle-mediated microRNA transfer: A mechanism of environmental modulation of hepatocellular cancer cell growth. *Hepatology* **2011**, *54*, 1237–1248. [CrossRef]
139. You, Y.; Shan, Y.; Chen, J.; Yue, H.; You, B.; Shi, S.; Li, X.; Cao, X. Matrix metalloproteinase 13-containing exosomes promote nasopharyngeal carcinoma metastasis. *Cancer Sci.* **2015**, *106*, 1669–1677. [CrossRef]
140. Ramteke, A.; Ting, H.; Agarwal, C.; Mateen, S.; Somasagara, R.; Hussain, A.; Graner, M.; Frederick, B.; Agarwal, R.; Deep, G. Exosomes secreted under hypoxia enhance invasiveness and stemness of prostate cancer cells by targeting adherens junction molecules: Hypoxic-exosomes role in pca aggressiveness. *Mol. Carcinog.* **2015**, *54*, 554–565. [CrossRef] [PubMed]
141. Franzen, C.A.; Blackwell, R.H.; Todorovic, V.; Greco, K.A.; Foreman, K.E.; Flanigan, R.C.; Kuo, P.C.; Gupta, G.N. Urothelial cells undergo epithelial-to-mesenchymal transition after exposure to muscle invasive bladder cancer exosomes. *Oncogenesis* **2015**, *4*, e163. [CrossRef] [PubMed]
142. Rahman, M.A.; Barger, J.F.; Lovat, F.; Gao, M.; Otterson, G.A.; Nana-Sinkam, P. Lung cancer exosomes as drivers of epithelial mesenchymal transition. *Oncotarget* **2016**, *7*, 54852–54866. [CrossRef] [PubMed]
143. Fabbri, M.; Paone, A.; Calore, F.; Galli, R.; Croce, C.M. A new role for microRNAs, as ligands of Toll-like receptors. *RNA Biol.* **2013**, *10*, 169–174. [CrossRef]
144. Kim, C.W.; Lee, H.M.; Lee, T.H.; Kang, C.; Kleinman, H.K.; Gho, Y.S. Extracellular membrane vesicles from tumor cells promote angiogenesis via sphingomyelin. *Cancer Res.* **2002**, *62*, 6312–6317. [PubMed]
145. Kholia, S.; Ranghino, A.; Garnieri, P.; Lopatina, T.; Deregibus, M.C.; Rispoli, P.; Brizzi, M.F.; Camussi, G. Extracellular vesicles as new players in angiogenesis. *Vasc. Pharmacol.* **2016**, *86*, 64–70. [CrossRef] [PubMed]
146. Lombardo, G.; Dentelli, P.; Togliatto, G.; Rosso, A.; Gili, M.; Gallo, S.; Deregibus, M.C.; Camussi, G.; Brizzi, M.F. Activated Stat5 trafficking Via Endothelial Cell-derived Extracellular Vesicles Controls IL-3 Pro-angiogenic Paracrine Action. *Sci. Rep.* **2016**, *6*, 25689. [CrossRef] [PubMed]
147. van Balkom, B.W.M.; de Jong, O.G.; Smits, M.; Brummelman, J.; den Ouden, K.; de Bree, P.M.; van Eijndhoven, M.A.J.; Pegtel, D.M.; Stoorvogel, W.; Würdinger, T.; et al. Endothelial cells require miR-214 to secrete exosomes that suppress senescence and induce angiogenesis in human and mouse endothelial cells. *Blood* **2013**, *121*, 3997–4006. [CrossRef] [PubMed]
148. Teng, X.; Chen, L.; Chen, W.; Yang, J.; Yang, Z.; Shen, Z. Mesenchymal Stem Cell-Derived Exosomes Improve the Microenvironment of Infarcted Myocardium Contributing to Angiogenesis and Anti-Inflammation. *Cell Physiol. Biochem.* **2015**, *37*, 2415–2424. [CrossRef]
149. Zeng, Z.; Li, Y.; Pan, Y.; Lan, X.; Song, F.; Sun, J.; Zhou, K.; Liu, X.; Ren, X.; Wang, F.; et al. Cancer-derived exosomal miR-25-3p promotes pre-metastatic niche formation by inducing vascular permeability and angiogenesis. *Nat. Commun.* **2018**, *9*, 5395. [CrossRef]
150. Fang, J.; Zhang, Z.; Shang, L.; Luo, Y.; Lin, Y.; Yuan, Y.; Zhuang, S. Hepatoma cell-secreted exosomal microRNA-103 increases vascular permeability and promotes metastasis by targeting junction proteins. *Hepatology* **2018**, *68*, 1459–1475. [CrossRef]

151. Lawson, J.; Dickman, C.; MacLellan, S.; Towle, R.; Jabalee, J.; Lam, S.; Garnis, C. Selective secretion of microRNAs from lung cancer cells via extracellular vesicles promotes CAMK1D-mediated tube formation in endothelial cells. *Oncotarget* **2017**, *8*, 83913–83924. [CrossRef] [PubMed]
152. Zhuang, G.; Wu, X.; Jiang, Z.; Kasman, I.; Yao, J.; Guan, Y.; Oeh, J.; Modrusan, Z.; Bais, C.; Sampath, D.; et al. Tumour-secreted miR-9 promotes endothelial cell migration and angiogenesis by activating the JAK-STAT pathway. *EMBO J.* **2012**, *31*, 3513–3523. [CrossRef]
153. Qu, J.; Lu, D.; Guo, H.; Miao, W.; Wu, G.; Zhou, M. MicroRNA-9 regulates osteoblast differentiation and angiogenesis via the AMPK signaling pathway. *Mol. Cell. Biochem.* **2016**, *411*, 23–33. [CrossRef] [PubMed]
154. Simon, T.; Pinioti, S.; Schellenberger, P.; Rajeeve, V.; Wendler, F.; Cutillas, P.R.; King, A.; Stebbing, J.; Giamas, G. Shedding of bevacizumab in tumour cells-derived extracellular vesicles as a new therapeutic escape mechanism in glioblastoma. *Mol. Cancer* **2018**, *17*, 132. [CrossRef] [PubMed]
155. Feng, Q.; Zhang, C.; Lum, D.; Druso, J.E.; Blank, B.; Wilson, K.F.; Welm, A.; Antonyak, M.A.; Cerione, R.A. A class of extracellular vesicles from breast cancer cells activates VEGF receptors and tumour angiogenesis. *Nat. Commun.* **2017**, *8*, 14450. [CrossRef]
156. Qu, Z.; Wu, J.; Wu, J.; Luo, D.; Jiang, C.; Ding, Y. Exosomes derived from HCC cells induce sorafenib resistance in hepatocellular carcinoma both in vivo and in vitro. *J. Exp. Clin. Cancer Res.* **2016**, *35*, 159. [CrossRef]
157. Takahashi, K.; Yan, I.K.; Kogure, T.; Haga, H.; Patel, T. Extracellular vesicle-mediated transfer of long non-coding RNA ROR modulates chemosensitivity in human hepatocellular cancer. *FEBS Open Bio* **2014**, *4*, 458–467. [CrossRef]
158. Stone, L. Exosome transmission of sunitinib resistance. *Nat. Rev. Urol.* **2016**, *13*, 297. [CrossRef]
159. Wieckowski, E.U.; Visus, C.; Szajnik, M.; Szczepanski, M.J.; Storkus, W.J.; Whiteside, T.L. Tumor-Derived Microvesicles Promote Regulatory T Cell Expansion and Induce Apoptosis in Tumor-Reactive Activated CD8 $^{+}$ T Lymphocytes. *J. Immunol.* **2009**, *183*, 3720–3730. [CrossRef]
160. Vu, L.T.; Peng, B.; Zhang, D.X.; Ma, V.; Mathey-Andrews, C.A.; Lam, C.K.; Kiomourtzis, T.; Jin, J.; McReynolds, L.; Huang, L.; et al. Tumor-secreted extracellular vesicles promote the activation of cancer-associated fibroblasts via the transfer of microRNA-125b. *J. Extracell. Vesicles* **2019**, *8*, 1599680. [CrossRef]
161. Poggio, M.; Hu, T.; Pai, C.-C.; Chu, B.; Belair, C.D.; Chang, A.; Montabana, E.; Lang, U.E.; Fu, Q.; Fong, L.; et al. Suppression of Exosomal PD-L1 Induces Systemic Anti-tumor Immunity and Memory. *Cell* **2019**, *177*, 414–427.e13. [CrossRef] [PubMed]
162. Del Re, M.; Marconcini, R.; Pasquini, G.; Rofi, E.; Vivaldi, C.; Bloise, F.; Restante, G.; Arrigoni, E.; Caparello, C.; Bianco, M.G.; et al. PD-L1 mRNA expression in plasma-derived exosomes is associated with response to anti-PD-1 antibodies in melanoma and NSCLC. *Br. J. Cancer* **2018**, *118*, 820–824. [CrossRef] [PubMed]
163. Theodoraki, M.-N.; Yerneni, S.; Gooding, W.E.; Ohr, J.; Clump, D.A.; Bauman, J.E.; Ferris, R.L.; Whiteside, T.L. Circulating exosomes measure responses to therapy in head and neck cancer patients treated with cetuximab, ipilimumab, and IMRT. *OncoImmunology* **2019**, *8*, e1593805. [CrossRef] [PubMed]
164. Yen, E.-Y.; Miaw, S.-C.; Yu, J.-S.; Lai, I.-R. Exosomal TGF-β1 is correlated with lymphatic metastasis of gastric cancers. *Am. J. Cancer Res.* **2017**, *7*, 2199–2208.
165. Que, R.; Lin, C.; Ding, G.; Wu, Z.; Cao, L. Increasing the immune activity of exosomes: The effect of miRNA-depleted exosome proteins on activating dendritic cell/cytokine-induced killer cells against pancreatic cancer. *J. Zhejiang Univ. Sci. B* **2016**, *17*, 352–360. [CrossRef]
166. Feitelson, M.A.; Arzumanyan, A.; Kulathinal, R.J.; Blain, S.W.; Holcombe, R.F.; Mahajna, J.; Marino, M.; Martinez-Chantar, M.L.; Nawroth, R.; Sanchez-Garcia, I.; et al. Sustained proliferation in cancer: Mechanisms and novel therapeutic targets. *Semin. Cancer Biol.* **2015**, *35*, S25–S54. [CrossRef]
167. Yin, Y.; Cai, X.; Chen, X.; Liang, H.; Zhang, Y.; Li, J.; Wang, Z.; Chen, X.; Zhang, W.; Yokoyama, S.; et al. Tumor-secreted miR-214 induces regulatory T cells: A major link between immune evasion and tumor growth. *Cell Res.* **2014**, *24*, 1164–1180. [CrossRef]
168. Cesi, G.; Philippidou, D.; Kozar, I.; Kim, Y.J.; Bernardin, F.; Van Niel, G.; Wienecke-Baldacchino, A.; Felten, P.; Letellier, E.; Dengler, S.; et al. A new ALK isoform transported by extracellular vesicles confers drug resistance to melanoma cells. *Mol. Cancer* **2018**, *17*, 145. [CrossRef]
169. Kalra, H.; Gangoda, L.; Fonseka, P.; Chitti, S.V.; Liem, M.; Keerthikumar, S.; Samuel, M.; Boukouris, S.; Al Saffar, H.; Collins, C.; et al. Extracellular vesicles containing oncogenic mutant β-catenin activate Wnt signalling pathway in the recipient cells. *J. Extracell. Vesicles* **2019**, *8*, 1690217. [CrossRef]

170. Yadav, A.K.; Desai, N.S. Cancer Stem Cells: Acquisition, Characteristics, Therapeutic Implications, Targeting Strategies and Future Prospects. *Stem Cell Rev. Rep.* **2019**, *15*, 331–355. [CrossRef]
171. Wang, L.; Yang, G.; Zhao, D.; Wang, J.; Bai, Y.; Peng, Q.; Wang, H.; Fang, R.; Chen, G.; Wang, Z.; et al. CD103-positive CSC exosome promotes EMT of clear cell renal cell carcinoma: Role of remote MiR-19b-3p. *Mol. Cancer* **2019**, *18*, 86. [CrossRef] [PubMed]
172. Sun, Z.; Wang, L.; Dong, L.; Wang, X. Emerging role of exosome signalling in maintaining cancer stem cell dynamic equilibrium. *J. Cell Mol. Med.* **2018**, *22*, 3719–3728. [CrossRef] [PubMed]
173. Gabrusiewicz, K.; Li, X.; Wei, J.; Hashimoto, Y.; Marisetty, A.L.; Ott, M.; Wang, F.; Hawke, D.; Yu, J.; Healy, L.M.; et al. Glioblastoma stem cell-derived exosomes induce M2 macrophages and PD-L1 expression on human monocytes. *OncoImmunology* **2018**, *7*, e1412909. [CrossRef]
174. Hardin, H.; Helein, H.; Meyer, K.; Robertson, S.; Zhang, R.; Zhong, W.; Lloyd, R.V. Thyroid cancer stem-like cell exosomes: Regulation of EMT via transfer of lncRNAs. *Lab. Invest.* **2018**, *98*, 1133–1142. [CrossRef] [PubMed]
175. Hwang, W.-L.; Lan, H.-Y.; Cheng, W.-C.; Huang, S.-C.; Yang, M.-H. Tumor stem-like cell-derived exosomal RNAs prime neutrophils for facilitating tumorigenesis of colon cancer. *J. Hematol. Oncol.* **2019**, *12*, 10. [CrossRef]
176. Bouvy, C.; Wannez, A.; Laloy, J.; Chatelain, C.; Dogné, J.-M. Transfer of multidrug resistance among acute myeloid leukemia cells via extracellular vesicles and their microRNA cargo. *Leuk. Res.* **2017**, *62*, 70–76. [CrossRef] [PubMed]
177. Aung, T.; Chapuy, B.; Vogel, D.; Wenzel, D.; Oppermann, M.; Lahmann, M.; Weinhage, T.; Menck, K.; Hupfeld, T.; Koch, R.; et al. Exosomal evasion of humoral immunotherapy in aggressive B-cell lymphoma modulated by ATP-binding cassette transporter A3. *Proc. Natl. Acad. Sci. USA* **2011**, *108*, 15336–15341. [CrossRef]
178. Bebawy, M.; Combes, V.; Lee, E.; Jaiswal, R.; Gong, J.; Bonhoure, A.; Grau, G.E.R. Membrane microparticles mediate transfer of P-glycoprotein to drug sensitive cancer cells. *Leukemia* **2009**, *23*, 1643–1649. [CrossRef]
179. Lu, J.F.; Luk, F.; Gong, J.; Jaiswal, R.; Grau, G.E.R.; Bebawy, M. Microparticles mediate MRP1 intercellular transfer and the re-templating of intrinsic resistance pathways. *Pharmacol. Res.* **2013**, *76*, 77–83. [CrossRef] [PubMed]
180. Lu, J.F.; Pokharel, D.; Bebawy, M. A novel mechanism governing the transcriptional regulation of ABC transporters in MDR cancer cells. *Drug Deliv. Transl. Res.* **2017**, *7*, 276–285. [CrossRef]
181. Dong, Y.; Pan, Q.; Jiang, L.; Chen, Z.; Zhang, F.; Liu, Y.; Xing, H.; Shi, M.; Li, J.; Li, X.; et al. Tumor endothelial expression of P-glycoprotein upon microvesicular transfer of TrpC5 derived from adriamycin-resistant breast cancer cells. *Biochem. Biophys. Res. Commun.* **2014**, *446*, 85–90. [CrossRef]
182. Zhang, Q.; Liu, R.-X.; Chan, K.-W.; Hu, J.; Zhang, J.; Wei, L.; Tan, H.; Yang, X.; Liu, H. Exosomal transfer of p-STAT3 promotes acquired 5-FU resistance in colorectal cancer cells. *J. Exp. Clin. Cancer Res.* **2019**, *38*, 320. [CrossRef] [PubMed]
183. Zhao, K.; Wang, Z.; Li, X.; Liu, J.; Tian, L.; Chen, J. Exosome-mediated transfer of CLIC1 contributes to the vincristine-resistance in gastric cancer. *Mol. Cell Biochem.* **2019**, *462*, 97–105. [CrossRef]
184. Gu, Y.Y.; Yu, J.; Zhang, J.F.; Wang, C. Suppressing the secretion of exosomal miR-19b by gw4869 could regulate oxaliplatin sensitivity in colorectal cancer. *Neoplasma* **2019**, *66*, 39–45. [CrossRef]
185. Ma, Y.; Yuwen, D.; Chen, J.; Zheng, B.; Gao, J.; Fan, M.; Xue, W.; Wang, Y.; Li, W.; Shu, Y.; et al. Exosomal Transfer Of Cisplatin-Induced miR-425-3p Confers Cisplatin Resistance In NSCLC Through Activating Autophagy. *Int. J. Nanomed.* **2019**, *14*, 8121–8132. [CrossRef]
186. Wu, H.; Zhou, J.; Mei, S.; Wu, D.; Mu, Z.; Chen, B.; Xie, Y.; Ye, Y.; Liu, J. Circulating exosomal microRNA-96 promotes cell proliferation, migration and drug resistance by targeting LMO7. *J. Cell. Mol. Med.* **2017**, *21*, 1228–1236. [CrossRef] [PubMed]
187. Wang, J.; Lv, B.; Su, Y.; Wang, X.; Bu, J.; Yao, L. Exosome-Mediated Transfer of lncRNA HOTTIP Promotes Cisplatin Resistance in Gastric Cancer Cells by Regulating HMGA1/miR-218 Axis. *OncoTargets Therapy* **2019**, *12*, 11325–11338. [CrossRef]
188. Cao, Y.-L.; Zhuang, T.; Xing, B.-H.; Li, N.; Li, Q. Exosomal DNMT1 mediates cisplatin resistance in ovarian cancer. *Cell Biochem. Funct.* **2017**, *35*, 296–303. [CrossRef]

189. Akao, Y.; Khoo, F.; Kumazaki, M.; Shinohara, H.; Miki, K.; Yamada, N. Extracellular Disposal of Tumor-Suppressor miRs-145 and -34a via Microvesicles and 5-FU Resistance of Human Colon Cancer Cells. *Int. J. Mol. Sci.* **2014**, *15*, 1392–1401. [CrossRef]
190. Yu, D.; Wu, Y.; Zhang, X.; Lv, M.; Chen, W.; Chen, X.; Yang, S.; Shen, H.; Zhong, S.; Tang, J.; et al. Exosomes from adriamycin-resistant breast cancer cells transmit drug resistance partly by delivering miR-222. *Tumor Biol.* **2016**, *37*, 3227–3235. [CrossRef]
191. Wei, Y.; Lai, X.; Yu, S.; Chen, S.; Ma, Y.; Zhang, Y.; Li, H.; Zhu, X.; Yao, L.; Zhang, J. Exosomal miR-221/222 enhances tamoxifen resistance in recipient ER-positive breast cancer cells. *Breast Cancer Res. Treat.* **2014**, *147*, 423–431. [CrossRef] [PubMed]
192. Mikamori, M.; Yamada, D.; Eguchi, H.; Hasegawa, S.; Kishimoto, T.; Tomimaru, Y.; Asaoka, T.; Noda, T.; Wada, H.; Kawamoto, K.; et al. MicroRNA-155 Controls Exosome Synthesis and Promotes Gemcitabine Resistance in Pancreatic Ductal Adenocarcinoma. *Sci. Rep.* **2017**, *7*, 42339. [CrossRef]
193. de Souza, P.S.; Cruz, A.L.S.; Viola, J.P.B.; Maia, R.C. Microparticles induce multifactorial resistance through oncogenic pathways independently of cancer cell type. *Cancer Sci.* **2015**, *106*, 60–68. [CrossRef] [PubMed]
194. Yin, J.; Zeng, A.; Zhang, Z.; Shi, Z.; Yan, W.; You, Y. Exosomal transfer of miR-1238 contributes to temozolomide-resistance in glioblastoma. *EBioMedicine* **2019**, *42*, 238–251. [CrossRef] [PubMed]
195. Powers, M.P. The ever-changing world of gene fusions in cancer: A secondary gene fusion and progression. *Oncogene* **2019**, *38*, 7197–7199. [CrossRef]
196. Li, Q.; Zhong, Z.; Zeng, C.; Meng, L.; Li, C.; Luo, Y.; Wang, H.; Li, W.; Wang, J.; Cheng, F.; et al. A clinical observation of Chinese chronic myelogenous leukemia patients after discontinuation of tyrosine kinase inhibitors. *Oncotarget* **2016**, *7*. [CrossRef]
197. Kumar-Sinha, C.; Tomlins, S.A.; Chinnaiyan, A.M. Recurrent gene fusions in prostate cancer. *Nat. Rev. Cancer* **2008**, *8*, 497–511. [CrossRef]
198. Fujita, K.; Nonomura, N. Urinary biomarkers of prostate cancer. *Int. J. Urol.* **2018**, *25*, 770–779. [CrossRef]
199. Wong, C.-H.; Chen, Y.-C. Clinical significance of exosomes as potential biomarkers in cancer. *World J. Clin. Cases* **2019**, *7*, 171–190. [CrossRef]
200. Schey, K.L.; Luther, J.M.; Rose, K.L. Proteomics characterization of exosome cargo. *Methods* **2015**, *87*, 75–82. [CrossRef]
201. Mohammadi, S.; Yousefi, F.; Shabaninejad, Z.; Movahedpour, A.; Mahjoubin Tehran, M.; Shafiee, A.; Moradizarmehri, S.; Hajighadimi, S.; Savardashtaki, A.; Mirzaei, H. Exosomes and cancer: From oncogenic roles to therapeutic applications. *IUBMB Life* **2020**, *72*, 724–748. [CrossRef] [PubMed]
202. Luan, X.; Sansanaphongpricha, K.; Myers, I.; Chen, H.; Yuan, H.; Sun, D. Engineering exosomes as refined biological nanoplatforms for drug delivery. *Acta Pharm. Sin.* **2017**, *38*, 754–763. [CrossRef] [PubMed]
203. Liang, G.; Kan, S.; Zhu, Y.; Feng, S.; Feng, W.; Gao, S. Engineered exosome-mediated delivery of functionally active miR-26a and its enhanced suppression effect in HepG2 cells. *Int. J. Nanomed.* **2018**, *13*, 585–599. [CrossRef]
204. Vader, P.; Mol, E.A.; Pasterkamp, G.; Schiffelers, R.M. Extracellular vesicles for drug delivery. *Adv. Drug Deliv. Rev.* **2016**, *106*, 148–156. [CrossRef]
205. Li, Y.; Zheng, Q.; Bao, C.; Li, S.; Guo, W.; Zhao, J.; Chen, D.; Gu, J.; He, X.; Huang, S. Circular RNA is enriched and stable in exosomes: A promising biomarker for cancer diagnosis. *Cell Res.* **2015**, *25*, 981–984. [CrossRef]
206. Wiklander, O.P.B.; Nordin, J.Z.; O'Loughlin, A.; Gustafsson, Y.; Corso, G.; Mäger, I.; Vader, P.; Lee, Y.; Sork, H.; Seow, Y.; et al. Extracellular vesicle in vivo biodistribution is determined by cell source, route of administration and targeting. *J. Extracell. Vesicles* **2015**, *4*, 26316. [CrossRef] [PubMed]
207. Johnsen, K.B.; Gudbergsson, J.M.; Skov, M.N.; Pilgaard, L.; Moos, T.; Duroux, M. A comprehensive overview of exosomes as drug delivery vehicles—Endogenous nanocarriers for targeted cancer therapy. *Biochim. Biophys. Acta (BBA) Rev. Cancer* **2014**, *1846*, 75–87. [CrossRef]
208. Ma, J.; Zhang, Y.; Tang, K.; Zhang, H.; Yin, X.; Li, Y.; Xu, P.; Sun, Y.; Ma, R.; Ji, T.; et al. Reversing drug resistance of soft tumor-repopulating cells by tumor cell-derived chemotherapeutic microparticles. *Cell Res.* **2016**, *26*, 713–727. [CrossRef]
209. Parolini, I.; Federici, C.; Raggi, C.; Lugini, L.; Palleschi, S.; De Milito, A.; Coscia, C.; Iessi, E.; Logozzi, M.; Molinari, A.; et al. Microenvironmental pH Is a Key Factor for Exosome Traffic in Tumor Cells. *J. Biol. Chem.* **2009**, *284*, 34211–34222. [CrossRef]

210. Kim, M.S.; Haney, M.J.; Zhao, Y.; Mahajan, V.; Deygen, I.; Klyachko, N.L.; Inskoe, E.; Piroyan, A.; Sokolsky, M.; Okolie, O.; et al. Development of exosome-encapsulated paclitaxel to overcome MDR in cancer cells. *Nanomed. Nanotechnol. Biol. Med.* **2016**, *12*, 655–664. [CrossRef]
211. Qiao, L.; Hu, S.; Huang, K.; Su, T.; Li, Z.; Vandergriff, A.; Cores, J.; Dinh, P.-U.; Allen, T.; Shen, D.; et al. Tumor cell-derived exosomes home to their cells of origin and can be used as Trojan horses to deliver cancer drugs. *Theranostics* **2020**, *10*, 3474–3487. [CrossRef] [PubMed]
212. Hao, D.; Li, Y.; Zhao, G.; Zhang, M. Soluble fms-like tyrosine kinase-1-enriched exosomes suppress the growth of small cell lung cancer by inhibiting endothelial cell migration. *Thorac. Cancer* **2019**, *10*, 1962–1972. [CrossRef] [PubMed]
213. Liu, T.; Zhang, X.; Du, L.; Wang, Y.; Liu, X.; Tian, H.; Wang, L.; Li, P.; Zhao, Y.; Duan, W.; et al. Exosome-transmitted miR-128-3p increase chemosensitivity of oxaliplatin-resistant colorectal cancer. *Mol. Cancer* **2019**, *18*, 43. [CrossRef] [PubMed]
214. Wan, F.-Z.; Chen, K.-H.; Sun, Y.-C.; Chen, X.-C.; Liang, R.-B.; Chen, L.; Zhu, X.-D. Exosomes overexpressing miR-34c inhibit malignant behavior and reverse the radioresistance of nasopharyngeal carcinoma. *J. Transl. Med.* **2020**, *18*, 12. [CrossRef]
215. Kobayashi, M.; Sawada, K.; Miyamoto, M.; Shimizu, A.; Yamamoto, M.; Kinose, Y.; Nakamura, K.; Kawano, M.; Kodama, M.; Hashimoto, K.; et al. Exploring the potential of engineered exosomes as delivery systems for tumor-suppressor microRNA replacement therapy in ovarian cancer. *Biochem. Biophys. Res. Commun.* **2020**, *527*, 153–161. [CrossRef]
216. Tian, H.; Li, W. Dendritic cell-derived exosomes for cancer immunotherapy: Hope and challenges. *Ann. Transl. Med.* **2017**, *5*, 221. [CrossRef]
217. Liu, C.; Guo, J.; Tian, F.; Yang, N.; Yan, F.; Ding, Y.; Wei, J.; Hu, G.; Nie, G.; Sun, J. Field-Free Isolation of Exosomes from Extracellular Vesicles by Microfluidic Viscoelastic Flows. *ACS Nano* **2017**, *11*, 6968–6976. [CrossRef]
218. Chen, T.; Arslan, F.; Yin, Y.; Tan, S.; Lai, R.; Choo, A.; Padmanabhan, J.; Lee, C.; de Kleijn, D.P.; Lim, S. Enabling a robust scalable manufacturing process for therapeutic exosomes through oncogenic immortalization of human ESC-derived MSCs. *J. Transl. Med.* **2011**, *9*, 47. [CrossRef]
219. Lener, T.; Gimona, M.; Aigner, L.; Börger, V.; Buzas, E.; Camussi, G.; Chaput, N.; Chatterjee, D.; Court, F.A.; del Portillo, H.A.; et al. Applying extracellular vesicles based therapeutics in clinical trials—An ISEV position paper. *J. Extracell. Vesicles* **2015**, *4*, 30087. [CrossRef]
220. Nie, W.; Wu, G.; Zhang, J.; Huang, L.; Ding, J.; Jiang, A.; Zhang, Y.; Liu, Y.; Li, J.; Pu, K.; et al. Responsive Exosome Nano-bioconjugates for Synergistic Cancer Therapy. *Angew. Chem. Int. Ed.* **2020**, *59*, 2018–2022. [CrossRef]
221. Yong, T.; Zhang, X.; Bie, N.; Zhang, H.; Zhang, X.; Li, F.; Hakeem, A.; Hu, J.; Gan, L.; Santos, H.A.; et al. Tumor exosome-based nanoparticles are efficient drug carriers for chemotherapy. *Nat. Commun.* **2019**, *10*, 3838. [CrossRef] [PubMed]
222. Bai, L.; Liu, Y.; Guo, K.; Zhang, K.; Liu, Q.; Wang, P.; Wang, X. Ultrasound Facilitates Naturally Equipped Exosomes Derived from Macrophages and Blood Serum for Orthotopic Glioma Treatment. *ACS Appl. Mater. Interfaces* **2019**, *11*, 14576–14587. [CrossRef]
223. Zhang, K.-L.; Wang, Y.-J.; Sun, J.; Zhou, J.; Xing, C.; Huang, G.; Li, J.; Yang, H. Artificial chimeric exosomes for anti-phagocytosis and targeted cancer therapy. *Chem. Sci.* **2019**, *10*, 1555–1561. [CrossRef]
224. Pomatto, M.A.C.; Bussolati, B.; D'Antico, S.; Ghiotto, S.; Tetta, C.; Brizzi, M.F.; Camussi, G. Improved Loading of Plasma-Derived Extracellular Vesicles to Encapsulate Antitumor miRNAs. *Mol. Ther. Methods Clin. Dev.* **2019**, *13*, 133–144. [CrossRef] [PubMed]
225. Gomari, H.; Forouzandeh Moghadam, M.; Soleimani, M. Targeted cancer therapy using engineered exosome as a natural drug delivery vehicle. *OncoTargets Therapy* **2018**, *11*, 5753–5762. [CrossRef] [PubMed]
226. Limoni, S.K.; Moghadam, M.F.; Moazzeni, S.M.; Gomari, H.; Salimi, F. Engineered Exosomes for Targeted Transfer of siRNA to HER2 Positive Breast Cancer Cells. *Appl. Biochem. Biotechnol.* **2019**, *187*, 352–364. [CrossRef]
227. O'Brien, K.P.; Khan, S.; Gilligan, K.E.; Zafar, H.; Lalor, P.; Glynn, C.; O'Flatharta, C.; Ingoldsby, H.; Dockery, P.; De Bhulbh, A.; et al. Employing mesenchymal stem cells to support tumor-targeted delivery of extracellular vesicle (EV)-encapsulated microRNA-379. *Oncogene* **2018**, *37*, 2137–2149. [CrossRef]

228. Kim, M.S.; Haney, M.J.; Zhao, Y.; Yuan, D.; Deygen, I.; Klyachko, N.L.; Kabanov, A.V.; Batrakova, E.V. Engineering macrophage-derived exosomes for targeted paclitaxel delivery to pulmonary metastases: In vitro and in vivo evaluations. *Nanomed. Nanotechnol. Biol. Med.* **2018**, *14*, 195–204. [CrossRef]
229. Kamerkar, S.; LeBleu, V.S.; Sugimoto, H.; Yang, S.; Ruivo, C.F.; Melo, S.A.; Lee, J.J.; Kalluri, R. Exosomes facilitate therapeutic targeting of oncogenic KRAS in pancreatic cancer. *Nature* **2017**, *546*, 498–503. [CrossRef]
230. Bellavia, D.; Raimondo, S.; Calabrese, G.; Forte, S.; Cristaldi, M.; Patinella, A.; Memeo, L.; Manno, M.; Raccosta, S.; Diana, P.; et al. Interleukin 3-receptor targeted exosomes inhibit in vitro and in vivo Chronic Myelogenous Leukemia cell growth. *Theranostics* **2017**, *7*, 1333–1345. [CrossRef]
231. Morishita, M.; Takahashi, Y.; Matsumoto, A.; Nishikawa, M.; Takakura, Y. Exosome-based tumor antigens–adjuvant co-delivery utilizing genetically engineered tumor cell-derived exosomes with immunostimulatory CpG DNA. *Biomaterials* **2016**, *111*, 55–65. [CrossRef] [PubMed]
232. Tian, Y.; Li, S.; Song, J.; Ji, T.; Zhu, M.; Anderson, G.J.; Wei, J.; Nie, G. A doxorubicin delivery platform using engineered natural membrane vesicle exosomes for targeted tumor therapy. *Biomaterials* **2014**, *35*, 2383–2390. [CrossRef] [PubMed]
233. Ohno, S.; Takanashi, M.; Sudo, K.; Ueda, S.; Ishikawa, A.; Matsuyama, N.; Fujita, K.; Mizutani, T.; Ohgi, T.; Ochiya, T.; et al. Systemically Injected Exosomes Targeted to EGFR Deliver Antitumor MicroRNA to Breast Cancer Cells. *Mol. Ther.* **2013**, *21*, 185–191. [CrossRef] [PubMed]

8

Microvesicles in Cancer: Small Size and Large Potential

Kerstin Menck [1,†], Suganja Sivaloganathan [1,†], Annalen Bleckmann [1,2] and Claudia Binder [2,*]

[1] Department of Medicine A, Hematology, Oncology, and Pneumology, University Hospital Münster, 48149 Münster, Germany; Kerstin.Menck@ukmuenster.de (K.M.); Suganja.Sivaloganathan@ukmuenster.de (S.S.); annalen.bleckmann@ukmuenster.de (A.B.)

[2] Department of Hematology/Medical Oncology, University Medical Center Göttingen, 37075 Göttingen, Germany

* Correspondence: claudia.binder@med.uni-goettingen.de;

† These authors contributed equally.

Abstract: Extracellular vesicles (EV) are secreted by all cell types in a tumor and its microenvironment (TME), playing an essential role in intercellular communication and the establishment of a TME favorable for tumor invasion and metastasis. They encompass a variety of vesicle populations, among them the well-known endosomal-derived small exosomes (Exo), but also larger vesicles (diameter > 100 nm) that are shed directly from the plasma membrane, the so-called microvesicles (MV). Increasing evidence suggests that MV, although biologically different, share the tumor-promoting features of Exo in the TME. Due to their larger size, they can be readily harvested from patients' blood and characterized by routine methods such as conventional flow cytometry, exploiting the plethora of molecules expressed on their surface. In this review, we summarize the current knowledge about the biology and the composition of MV, as well as their role within the TME. We highlight not only the challenges and potential of MV as novel biomarkers for cancer, but also discuss their possible use for therapeutic intervention.

Keywords: microvesicles; biomarker; cancer; tumor microenvironment; therapy

1. Introduction

Tumor development is a multistep process that is accompanied by various cellular reprogramming events in which cells acquire the hallmarks of cancer cells to gain and sustain abnormal growth and invasive capacity [1]. The complex process of tumor formation and spreading not only depends on the tumor cells themselves but also requires rewiring of the surrounding stromal cells. This can be induced by cell intrinsic events (genetic or epigenetic aberrations) or by external factors originating from direct cell–cell contact or from indirect cell communication. While soluble factors, such as chemokines, cytokines, and growth factors, have long been known for their role during tumor development, extracellular vesicles (EV) have recently attracted increasing attention as mediators of intercellular communication. Conserved from prokaryotes to eukaryotes, the secretion of EV is observed in all cell types [2]. The term EV encompasses all types of secreted membrane-enclosed vesicles, which are highly heterogeneous. On the basis of various characteristics, ranging from size, biogenesis, cell of origin, morphology, and content, EV are categorized into four main classes: endosomal-derived small exosomes (Exo) (50–150 nm), plasma membrane-derived middle-sized microvesicles (MV) (100–1000 nm), and large oncosomes (LO) (1000–10,000 nm), as well as apoptotic bodies (500–4000 nm) that are released from dying cells [3,4]. Although commonly used to categorize EV, this classification has been challenged by recent evidence demonstrating that, for instance, the Exo contain further subtypes with different biological and biochemical properties [5]. In order to avoid terminological ambiguities, the term "small EV" was coined for EV harvested at > 100,000 g. However, since most authors still use the term Exo, we will adhere to this nomination throughout the review.

In cancer, EV have been shown to be essential for various steps during tumor initiation and progression. By horizontal transfer of bioactive molecules between cancer cells and the neighboring stroma, EV interfere with signaling and regulation of gene expression in the recipient cell. Thus, the malignant cells can convert the phenotype of surrounding benign cells into a tumor-supporting one and create a favorable environment for cancer progression and metastatic spread. While much attention has been paid to the role of Exo in cancer, the function of the larger MV is still poorly defined. This seems surprising, since MV, in contrast to Exo, are easily accessible in patients' blood and are characterizable by routine methods that should make them ideal candidates for "liquid biopsies". Additionally, MV have long been known for their involvement in metastasis formation. This was initially attributed to their procoagulant activity, favoring the formation of microthrombi and facilitating the extravasation of the thus captured circulating tumor cells (reviewed in [6]). However, more recently, accumulating evidence points to a plethora of different ways in which MV are involved in the various steps of tumor progression.

In this review, we discuss the current knowledge about the role of MV during cancer development and shed light on how these insights can be exploited clinically for cancer diagnostics. Moreover, we highlight the divergent features of MV compared to Exo as cancer therapeutics and illustrate the various ways in which MV can either promote or counteract cancer therapy. For this purpose, we specifically included studies in which EV were harvested at 10,000–20,000 g and collectively refer to them as MV.

2. Preparation of MV

In order to decipher the role of MV in cancer development and progression, effective methods are required that allow for the stringent isolation of pure MV populations from different cell types and biological fluids. A major caveat in MV research is that the currently available isolation methods potentially co-isolate LO or Exo, yielding a mixed population of EV. This may explain many of the apparently conflicting results in the field of EV research. To address this major challenge, new technologies are under development, but they are not yet suitable for laboratory use. In an endeavor to standardize the experimental procedures and limit experimental variability in the field, scientists of the International Society of Extracellular Vesicles (ISEV) published a position paper indicating the appropriate methods for isolation of EV from cells or biological fluids and highlighting the current knowledge and major caveats of these procedures [7]. A variety of methods are available on the basis of different principles for enriching the various EV subpopulations, including density gradient centrifugation, size-exclusion chromatography, precipitation via volume-excluding polymers, immunoaffinity capture methods, high-pressure liquid chromatography, field flow fractionation, and flow cytometry [7,8].

To date, differential centrifugation is still the method of choice for isolating MV since it yields a reasonably good separation of MV from Exo with regard to protein and RNA content as well as function [9–11]. In principle, samples are spun down at 2000 g to initially pellet large EV such as LO, followed by centrifugation at 10,000–20,000 g to sediment the middle-sized MV, while ultracentrifugation at ≥100,000 g leads to a harvest of small Exo [8]. Although ultracentrifugation has been criticized for inducing vesicle aggregation and thus affecting downstream applications [12–14], these studies were conducted on the 100,000 g Exo pellet. It is unclear to what extent these problems also apply to the 10,000 g MV pellet as well. Another popular method to separate EV from contaminations with non-vesicular proteins or RNA aggregates is ultracentrifugation of EV samples on density gradients. However, the typically used sucrose gradients lack sufficient resolution to separate EV that have slightly different densities and are released by different mechanisms [15]. Since the density of MV and Exo is comparable, they cannot be separated on sucrose gradients easily [11]. Likewise, precipitation methods such as protein organic solvent precipitation (PROSPR) have been demonstrated to co-isolate MV, but the preparations were mainly enriched in smaller Exo, with MV being only a minor side population [16]. Immunoaffinity capture methods rely on antibodies for specific EV surface markers that are coupled to beads and used to isolate specific vesicle subpopulations. Due to the current lack of specific markers for MV, this method has been scarcely used for the preparation of MV. Currently, novel micro-/nano-based devices for MV isolation from clinical samples are being developed (reviewed in [17]). However, one major problem is that the thus isolated vesicle preparations have often been poorly characterized and it is unclear whether they yield MV with a purity comparable to sequential centrifugation protocols. The same applies to methods originally established for Exo isolation such as precipitation or size-exclusion chromatography. While none of the currently available isolation procedures for MV and Exo yield pure vesicle populations, but rather MV- or Exo-enriched fractions, more effective techniques that separate pure MV from Exo might help to shed light on their distinct cell biological features such as biogenesis, uptake, and cargo trafficking routes. Then again, from a clinical viewpoint, the isolation of pure MV populations might not be absolutely necessary for their use as biomarkers, since recent studies have demonstrated the potential of MV for cancer diagnostics or therapy monitoring, even when the preparations contain some smaller Exo.

No matter which method is chosen for EV isolation, the researchers of the ISEV community highly recommend validating the obtained EV by different techniques [7]. This includes the analysis of marker protein expression comprising (i) transmembrane, (GPI)-anchored, as well as cytosolic EV marker proteins (e.g., CD63, CD9, Alix, Syntenin, Rgap1, along with negative controls consisting of (ii) non-EV co-isolated proteins (e.g., ApoB, albumin), (iii) proteins typically present in non-EV subcellular structures such as the Golgi or endoplasmatic reticulum (e.g., GM130, Calreticulin, histones), and (iv) secreted proteins (e.g., collagen, epidermal growth factor (EGF), interleukins). Moreover, EV should be characterized by at least two distinct techniques including their visualization by, for instance, electron or atomic force microscopy, as well as the analysis of their biophysical properties via, for example, nanoparticle tracking analysis (NTA) or Raman spectroscopy. Nanoparticle tracking analysis (NTA) is the most suitable method for analyzing the isolated MV in terms of quantification and sizing. NTA tracks the Brownian movement of laser-illuminated particles and calculates the diameter based on the Stokes–Einstein equation [18]. While NTA remains the most commonly used method for quantitative MV analysis, other methods such as tunable resistive pulse sensing or dynamic light scattering are also available [19]. However, a major limitation of these methods is that they cannot efficiently analyze larger vesicles and that they do not yield any information on the molecular composition of the MV. They are, therefore, combined with other methods such as tunable resistive pulse sensing and Raman spectroscopy or NTA with fluorescence labelling to obtain more information about the isolated MV [19].

A major caveat in MV research is that the currently available isolation methods potentially co-isolate LO or Exo, yielding a mixed population of EV. This may explain many of the apparently conflicting results in the field of EV research. To address this major challenge, new technologies are under development, but are not yet suitable for laboratory use. In an endeavor to standardize the experimental procedures and limit experimental variability in the field, scientists of the International Society of Extracellular Vesicles (ISEV) published a position paper that indicated the appropriate methods for isolation of EV from cells or biological fluids and highlighted the current knowledge and major caveats of these procedures [7]. Furthermore, the researchers of the ISEV community highly recommend validating different techniques for various cell types and biological fluids. A crowdsourcing knowledge base was established to create further transparency with regards to experimental and methodological parameters of EV isolation (http://evtrack.org) [20]. This platform encourages researchers to upload published and unpublished experiments, thereby creating an informed dialogue among researchers about relevant experimental parameters. This represents a major step in facilitating standardization in EV, as well as MV, research.

3. Biogenesis of MV

MV directly bud off from the outer cell membrane. The shedding process comprises molecular rearrangements of the plasma membrane regarding lipid and protein composition as well as Ca^{2+} levels. Ca^{2+}-dependent aminophospholipid translocases, flippases, floppases, scramblases, and calpain drive the translocation of phosphatidylserine from the inner to the outer membrane leaflet, which is considered a typical characteristic of MV (comprehensively reviewed in [21]). Apoptotic bodies, which are larger in size, also externalize phosphatidylserine on their surface [22,23]. Therefore, the isolation of MV should be conducted solely from healthy and viable cells to avoid contamination with apoptotic bodies, which otherwise can be difficult to discriminate from MV. Ca^{2+} levels regulate membrane rigidity and curvature and maintain physical bending of the membrane, which leads to restructuring and contraction of the underlying actin cytoskeleton enabling MV formation and pinching (reviewed in [24]). MV formation and release are also affected by the lipids ceramide and cholesterol [25]. Neutral sphingomyelinase activity, which hydrolyses lipid sphingomyelin into phosphorylcholine and ceramide, was shown to be involved in Exo and MV release. Inhibition of the enzyme led to a reduction in Exo release, while simultaneously increasing MV budding [26], which suggests that the release of both EV subpopulations is interconnected, albeit on the basis of distinct biogenetic mechanisms.

In addition to lipids, enzyme machineries involved in cytoskeletal regulation play a key role in MV formation and budding. One example is the small GTPase protein ADP-ribosylation factor 6 (ARF6), which stimulates phospholipase D (PLD) that subsequently associates with extracellular signal-regulated kinase (ERK) at the plasma membrane. ERK activates a signaling cascade downstream of the myosin light chain kinase (MLCK) that results in contraction of actinomyosin and enables MV release [27]. Similar to MV, LO are thought to derive from the plasma membrane. However, in contrast to MV, the shedding of LO has been exclusively attributed to aggressive cancer cells that have acquired an amoeboid phenotype to facilitate motility and invasiveness ([4]). Other Rho family small GTPases such as RhoA and RHO-associated protein kinases are equally important regulators of actin dynamics relevant for MV formation [28]. In addition, the endosomal sorting complex required for transport (ESCRT), which is mainly known for its role in the biogenesis of Exo [29], is also involved in MV formation and the last phase of their release. Interaction of arrestin domain-containing protein 1 (ARRDC1) with the late endosomal protein tumor susceptibility gene 101 (TSG101) results in relocation of TSG101 from the endosomal to the plasma membrane, which then induces the release of MV [30].

4. Membrane Composition of MV

EV-mediated cell–cell communication requires targeting and uptake into the recipient cell to deliver the bioactive cargo, which then induces functional and phenotypical changes. These events depend on the composition of the EV membrane, as surface molecules on EV are responsible for binding and docking to recipient cells [31,32]. The molecular composition of the MV membrane closely resembles that of the parental cell [33]. It is enriched in phospholipid lysophosphatidylcholine, sphingolipid sphingomyelins, acylcarnitine, and fatty acyl esters of L-carnitine [34]. ARF6, a key regulator of MV biogenesis, was shown to mediate MV surface molecule selection by recruiting proteins such as ß1 integrin receptor, major histocompatibility complex (MHC) class I and II molecules, membrane type 1-matrix metalloproteinase (MT1-MMP), vesicular SNARE (v-SNARE), and vesicle-associated membrane protein 3 (VAMP3) to tumor MV [27]. Moreover, the bioactive cargo of MV depends on the conditions the parental cells are subjected to, such as inflammation or other stressors. An additional example is hypoxia, which induces recruitment of the RAS-related protein Rab22a to the site of MV budding in breast cancer cells, thus influencing MV formation and loading [35].

Since both LO and MV are derived from the plasma membrane, it is not surprising that some transmembrane proteins are present on either of these EV. Analysis of protein marker expression on LO and MV revealed a common signature, underlining the fact that the definition of MV-specific markers remains challenging. Of note, some of the markers initially thought to be specific for Exo, including tetraspanins (CD9, CD63, CD81, HSP60, HSP70, HSP90), membrane transporters and fusion proteins (annexin, flotillin), and multivesicular body (MVB) synthesis proteins (Alix, TSG101) were also found in varying amounts on MV and LO [9]. The fact that, despite their different routes of origin, EV share some common surface molecules and MV-specific markers are still lacking represents major challenges in EV research. It further emphasizes that to correctly characterize EV populations it is indispensable to combine a variety of parameters in addition to marker expression, such as size, sedimentation coefficient, and others.

5. Role of Cancer-Associated MV and Their Protein Cargo in Tumor Progression

EV-mediated cell–cell communication within the tumor microenvironment (TME) is highly complex and has recently been comprehensively reviewed in [36,37]. The role of EV in promoting tumor progression has mostly been elucidated in studies on mixed populations of EV without focusing on a specific subpopulation. Therefore, the specific contribution of MV in this complex process remained enigmatic for a long time. However, with increasing awareness of the presence of large MV, accumulating evidence has begun to unravel the tumor-promoting role of MV in TME communication, as described below by selected examples. The function of MV largely depends on their bioactive cargo, in particular the shuttling of tumor-specific proteins to the surrounding cells. While researchers initially concentrated on the role of nucleic acids transported via EV (e.g., DNA, mRNA, or miRNA), the focus has more recently shifted towards the analysis of the EV proteome. The protein content of MV has been found to be strikingly different from that of the Exo proteome and is enriched in proteins involved in microtubule/cortical actin and cytoskeleton networks, ARF6, its effector phospholipase D2, and parts of the ESCRT family (ESCRT-I) [27,38]. Carrying these bioactive cargo molecules, MV are able to influence either the adjacent tumor cells in an autologous way or the neighboring stromal cells in a heterologous kind of cell–cell communication.

MV-mediated autologous communication transfers oncogenic traits between tumor cells, resulting in enhanced tumor growth and progression. One example is multiple myeloma in which MV shed by the cancer cells were shown to enhance tumor cell proliferation and thereby stimulate tumor growth [39]. Interestingly, this effect was related to the enrichment of the extracellular matrix metalloproteinase inducer (EMMPRIN/CD147) on the tumor MV. This protein is known to be frequently overexpressed in solid tumors as well as in some lymphomas and leukemias [40]. In line with this finding, another study in breast cancer cells showed that the highly glycosylated isoform of EMMPRIN in particular is present in high levels on breast cancer cell-derived MV and stimulates tumor cell invasion via activation of p38/MAPK signaling [11]. Strikingly, a similar, high-EMMPRIN expression was also found on blood MV from patients with metastatic breast cancer where it was co-expressed with the tumor marker Mucin-1 (MUC1/CA 15-3) [11]. Another tumor-specific factor implicated in pro-tumorigenic tumor–tumor crosstalk via MV is the truncated oncogenic form of the epidermal growth factor receptor (EGFR), EGFRvIII, which is commonly expressed in aggressive brain tumor cells. It was shown that MV secreted by U373 glioma cells contained EGFRvIII, enabling them to transfer malignant characteristics from highly aggressive tumor cells to EGFRvIII-negative, more benign tumor cells, thereby promoting their oncogenic transformation [41]. Hence, MV are convenient communicators within the TME that can either mediate the horizontal transfer of oncogenic material or activate oncogenic signaling pathways in neighboring cancer cells, stimulating their proliferative, survival, mitogenic, and angiogenic potential and shifting them to a highly invasive phenotype.

In addition to tumor–tumor communication, MV were also found to mediate reciprocal crosstalk between the tumor and the surrounding stroma cells. Such heterologous interactions were observed, for instance, between tumor cells and surrounding immune cells, being seemingly essential for cancer immune evasion. As shown for breast cancer cells, the secretion of both tumor MV as well as Exo induced the expression of Wnt5a in tumor-associated macrophages. Macrophage Wnt5a was then, in turn, delivered to breast cancer cells via macrophage-derived MV and Exo, where it activated ß-catenin-independent Wnt signaling, leading to increased tumor invasion [42]. This example indicates that EV-based cell–cell communication can occur bidirectionally in a reciprocal loop and reprogram tumor-associated immune cells towards a tumor-supporting phenotype. The finding that MV-enriched preparations isolated at 50,000 g induced the differentiation of monocytes producing the anti-inflammatory cytokine IL-10 further supports this notion [43,44]. In line with this, early stimulation with tumor MV triggered macrophage polarization towards an anti-inflammatory phenotype with decreased anti-tumor cytotoxic potential [45].

Apart from macrophages, other immune cells can be affected by MV, such as the T cells. As T cells represent the first line of the immune defense, tumor cells seem to have established strategies to suppress T cell activity and dampen antitumoral immune response by exploiting MV-mediated cell–cell communication. For instance, leukemia-derived MV deliver miRNAs to T cells, which then interact with their targets, resulting in a T cell exhaustion phenotype [46]. Moreover, MV released by irradiated breast cancer cells were shown to carry an increased amount of immune-modulating proteins, such as programmed cell death ligand 1 (PD-L1). Via transfer of this immunosuppressive protein, tumor MV inhibited cytotoxic T cell activity and enabled tumor growth [47].

Tumor cell-derived MV were also shown to modulate fibroblasts in the tumor stroma. Stimulation of fibroblasts with prostate cancer MV converted them to an activated phenotype and triggered the release of tumor-promoting fibroblast MV [48]. Similar observations were made in oral squamous carcinoma when normal human fibroblasts were treated with tumor-derived MV [49]. In this study, the switch to cancer-associated fibroblasts (CAF) was mainly mediated via metabolic reprogramming of the fibroblasts to aerobic glycolysis, with an increase in glucose uptake and lactate secretion. Co-culturing the generated CAF with oral squamous cell carcinoma cells again led to enhanced cancer cell invasion and migration. Interestingly, MV-induced fibroblast activation and spreading seems to occur in particular in the stiff matrix environment that is typically found in the tumor periphery [50]. Tumor MV are also able to modulate angiogenesis within the primary tumor. In a study on a mixed population of MV and Exo, normal endothelial cells were found to endocytose tumor EV, which activated PI3K/Akt signaling and promoted the motility as well as the tube formation activity of the endothelial cells [51]. As the pro-angiogenic factor vascular endothelial growth factor (VEGF) has been found on MV and Exo, this might be one important factor contributing to the stimulatory effect of tumor MV on endothelial cells [52]. Of note, MV transport a unique 90 kDa form of VEGF that has only a weak affinity for bevacizumab, the clinically used monoclonal anti-VEGF antibody, which might thus be ineffective in blocking MV-mediated activation of the VEGF receptor (VEGFR) [53]. Similarly, MV derived from multiple myeloma cells were shown to transfer CD138, a myeloma cell marker, to endothelial cells. This stimulated the endothelial cells to proliferate, invade, and secrete the angiogenic factors IL-6 and VEGF to promote tube formation [54]. Taken together, these observations emphasize the important role of MV in tumor–stroma crosstalk during tumor progression.

Using the same mechanisms as discussed above, MV were found to modulate the surrounding environment beyond the primary tumor by shaping the formation of pre-metastatic niches over long distances. In pancreatic cancer, tumor MV were able to enter the liver microcirculation and extravasate through the vessel wall in a CD36-dependent manner where they were taken up by perivascular macrophages and primed the liver metastatic niche [55]. Furthermore, the highly metastatic melanoma cell line B16F10 releases large amounts of tumor MV into its surroundings. These MV were found to be able to induce metastasis formation in BALB/c mice, which are normally resistant to the B1610 tumor cell line [56]. Metastatic niche formation might be essentially influenced by tumor MV uptake into organ-specific macrophage populations, and this uptake seems to be further influenced by systemic changes such as inflammation [57]. However, further in vivo studies are required to characterize the uptake kinetics and to clarify which cells are specifically targeted by tumor MV during systemic spreading. In summary, an increasing number of studies confirms that MV play a critical role in cell–cell communication within the TME and support local tumor growth and progression as well as tumor spreading to distant sites through the priming of metastatic niches (Figure 1).

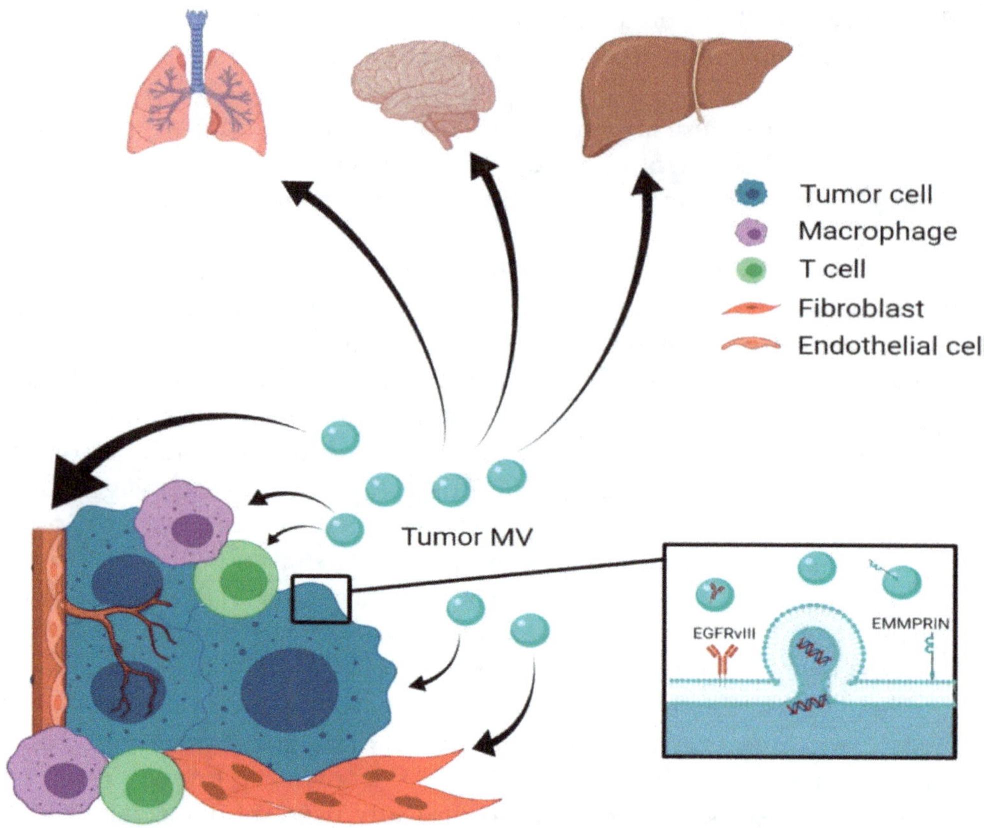

Figure 1. Microvesicles (MV) shaping the local tumor microenvironment (TME) and distant metastatic niches to promote tumor growth. On the one hand, MV shed from the plasma membrane of tumor cells act in an autologous way on other tumor cells by transferring oncogenic cargo and stimulating pro-tumorigenic signaling. On the other hand, MV are able to mediate heterologous cell–cell communication by acting on cells present within the TME, such as macrophages, T cells, fibroblasts, or endothelial cells. As a result, these cells become activated and change their phenotype to enhance tumor growth. Once the tumor has gained access to the circulation, tumor MV can be distributed throughout the body and mediate cargo transfer and cell–cell communication at distant sites. This induces the formation of pre-metastatic niches that offer a favorable environment for subsequent seeding of tumor cells in secondary organs.

6. Analysis of MV in Peripheral Blood

MV have been successfully detected in almost all body fluids, including saliva [58], urine [59], cerebrospinal fluid [60], breast milk [61], ejaculate [62], synovial fluid [63], bronchoalveolar lavages [64], and blood, wherein they were described as early as 1967 by Peter Wolf as "platelet dust" [65]. The advantage of body fluids apart from blood is that they contain MV specifically enriched for the cancers drained by them. Nevertheless, some of these liquids are either difficult to obtain or provide additional methodological problems. For instance, MV in urine are of special interest as biomarkers for urothelial cancers. However, their isolation poses two additional challenges: (1) the most abundant protein in urine, the Tamm–Horsfall protein, associates and co-precipitates with EV, and (2) standard protocols pre-clear urine samples at 17,000 g to deplete the urinary sediment, and thus also deplete the MV contained within. Therefore, the number of studies on MV in urine meeting the criteria mentioned in the introduction is very limited. Since blood is the most thoroughly studied body fluid, as well as the most likely application route for using MV with regard to diagnostic and therapeutic interventions, we chose to focus this review on MV in blood.

When isolating MV from blood, several studies have shown that the choice of anticoagulant used for blood collection has a critical influence on the obtained MV, although the results seem to be contradictory. While the use of protease inhibitors, either sodium heparin or a mix of hirudin supplemented with a factor Xa inhibitor, was initially recommended because of the high number of preserved MV [66], other reports have argued for the use of sodium citrate since the lower numbers of MV might point to less artificial MV shedding [67]. In contrast, Jamaly et al. recently used various anticoagulants to isolate MV and measure their total plasma concentrations by NTA without finding any significant differences in numbers [68]. Apart from total MV numbers, the isolation of MV from serum drawn into lithium heparin tubes seems to result in the aggregation of platelet-derived MV, with sticky vesicle pellets and a reduction in platelet MV numbers, thereby specifically affecting certain MV subpopulations [69,70]. In addition to the anticoagulant, other preanalytical factors such as centrifugation, temperature, freeze–thaw cycles, or agitation can influence MV in blood samples, which should be taken into account when setting up isolation protocols for clinical studies [66,67,71]. However, long-term storage of plasma samples without repeated freeze–thaw cycles seems feasible without significant MV loss [67,69,72]. Another important factor for the analysis of large MV in blood is the initial removal of platelets and blood cells to obtain platelet-poor plasma (PPP). Since PPP is often prepared by centrifugation at 2500–3000 $\times g$, this step might result in a significant loss of larger MV, which already pellet at this force [9]. It has been proposed that two centrifugation steps at 1500$\times g$ might be preferable for MV isolation [70]; however, more elaborate sorting methods might be required to specifically isolate larger EV from blood.

A current study by Brennan et al. compared different EV isolation methods, including ultracentrifugation, density gradient centrifugation, and size exclusion chromatography, as well as different combinations thereof, for their potential to isolate EV from plasma [73]. While none were found to be clearly superior, the authors showed that the isolation method significantly affected the yield, size, and degree of contamination with serum (lipo)proteins [73]. This must therefore be considered carefully on the basis of the planned downstream application. Unfortunately, the study was conducted on whole EV preparations, and thus it is again unclear whether the results are also applicable for larger MV. A novel approach to specifically reduce the contamination of plasma EV preparations with lipoproteins is to use anti-apolipoprotein B (ApoB) antibody-coated magnetic beads prior to EV isolation [74]. However, the method might lead to a high loss of vesicles.

The preferred method for analyzing MV from clinical samples is flow cytometry because of its wide availability in most labs and clinical centers, its capability for rapid preparation and acquisition of a large number of samples, as well as its potential for standardization. Since the size threshold of standard flow cytometers is ~300 nm (see [75] for a current instrument comparison), this allows the quick measurement and quantification of MV protein expression in contrast to Exo, which have to be coupled prior to analysis to larger latex beads following a time-consuming protocol. Protocols exist for staining of single or multiple markers on plasma-derived MV for flow cytometry [76,77]. Even though flow cytometry might not allow the detection of very small MV <300 nm and thus not reproduce the whole spectrum of MV in blood, several studies have successfully demonstrated the potential of flow cytometry for measuring MV-associated cancer biomarkers, as discussed in detail below. However, analysis of MV samples by flow cytometry bears several pitfalls, and the settings should be chosen with care to avoid misinterpretation of the results [78]. When measuring MV by flow cytometry, their presence should be confirmed by additional methods such as NTA or electron microscopy as the correct sizing of vesicles by flow cytometry can be challenging [79]. When discriminating MV from background noise using annexin V staining, the use of saline was recommended since the routinely used PBS seemed to create artificial nano-sized vesicles and could thus lead to false positive results [80]. Another pitfall is that the insoluble fraction of immune complexes and MV have overlapping size profiles, which might interfere with their flow cytometric analysis. To confirm that MV and not immune complexes are measured, samples can be treated with low concentrations of detergents (e.g., 0.05% Triton X-100), which lyse MV, thus resulting in a loss of signal in contrast to immune complexes that keep their fluorescent signal [63]. Recently, another label-free approach called flow cytometry scatter ratio (Flow-SR) has been introduced to discriminate MV <500 nm from lipoprotein particles, which are highly abundant in plasma samples [67].

Apart from flow cytometry, there are protocols for surface profiling of plasma MV by antibody microarrays. However, this is a semiquantitative method and thus more useful for initial screening of patient samples for the expression of potential novel biomarkers [81]. In summary, the current studies have shown that isolation and analysis of MV from human biofluids, such as plasma, is feasible using standard laboratory equipment. However, it is clear that transparent reporting and standardization of protocols are highly important for implementing and translating MV as novel biomarkers and therapeutics into the clinic [82,83].

7. MV Levels in Cancer Patients

Most studies have focused on measuring total MV levels in the plasma of cancer patients. In this context, it must be kept in mind that tumors can secrete large numbers of apoptotic bodies under therapy due to the massive induction of apoptosis in the tumor tissue. Apoptotic bodies can have a similar size as MV and are equally characterized by an externalization of phosphatidylserine, a marker that is often used to identify MV by flow cytometry. Hence, care should be taken to include only patients prior to treatment into these studies. While the observed results are very contradictory, the majority of reports have documented increased MV levels in cancer patients compared with healthy controls. Augmented MV levels have been detected by flow cytometric quantification, mostly using TruCount Beads, in the plasma of patients suffering from gastric cancer [84], lung cancer [85], breast and pancreatic cancer [86,87], and colorectal cancer [88], as well as in most hematologic malignancies including chronic lymphocytic leukemia (CLL) and multiple myeloma [89–91]. The same trend has been confirmed for hepatocellular carcinoma by measuring MV protein levels by bicinchoninic acid assay (BCA) [92]. In contrast, two studies investigating heterogeneous cohorts of patients with different types of advanced cancers did not detect significant differences in total plasma MV numbers by flow cytometric quantification or analysis of MV protein content [69,93]. Similar observations were made by a novel capture/imaging approach in glioblastoma patients [94]. Another report quantified the total number of annexin V+ MV by flow cytometry and found a significant decrease of MV levels in patients with colorectal cancer and benign colorectal diseases compared with healthy controls [95]. The discrepancy between the observations might be due to the methods used for MV quantification, the composition of the patient and control cohorts, or diverging effects in different cancer subtypes. Patient selection criteria and the clinical setting of sample acquisition (before/during/after treatment) are not always conclusively reported in the studies, which could be another reason for the heterogeneous results.

While the increased levels of MV in hematologic malignancies could be explained by additional vesicle shedding from the large numbers of cancer cells in blood, the source of the elevated MV counts in solid tumor patients remains undefined. In these patients, plasma levels of tumor MV might depend on the access of the tumor to the circulation or might originate from other blood cell populations as a reaction to the growing tumor. Other limitations are that all the studies are either single center studies; include a very small number of patients; or have compared MV levels between cancer patients and younger, healthy controls. It was recently found that total MV levels were higher in healthy, untreated elderly than in younger individuals, which might have biased the results [96].

An argument for a specific increase in MV levels in cancer patients is that MV levels seem to increase with advanced tumor stages and have been found to be elevated in late stage metastatic breast cancer [86,87], colorectal cancer [88], hepatocellular carcinoma [92], or CLL [90] compared with early stage patients. Of note, a significant increase was already observed in stage I colorectal cancer patients compared with healthy controls [88]. In line with this, total MV levels decreased 1 month after surgery in hepatocellular carcinoma patients and have thus been suggested as a potential diagnostic biomarker [92]. Moreover, higher MV levels in blood might additionally serve as a prognostic biomarker since they were found to correlate with a poor clinical outcome in glioblastoma or CLL [90,97]. In contrast, in lung cancer, an increase in MV in blood was associated with better progression-free survival [85]. An interesting finding was the detection of very large vesicles (1–14 μm), measurable with the established CellSearch system used for the analysis of circulating tumor cells (CTCs) [98]. These epithelial cell adhesion molecule (EpCAM)+/cytokeratin CK+/CD45- large vesicles without nuclei were present at frequencies one order of magnitude higher than CTCs and were elevated in the plasma of prostate, breast, and colorectal, but not lung cancer patients, with high numbers correlating with shorter overall survival [99]. While these vesicles have not yet been validated to be "true EV" using the MISEV guidelines, their discovery might represent one additional opportunity of translating EV as biomarkers into the clinic using an instrument already applied in diagnostics.

8. MV as Cancer Biomarkers

The growing number of in vitro studies demonstrating that cancer cells shed large numbers of EV with tumor-specific material has fueled interest in using EV as biomarkers in the clinic. Most investigations thus far have focused on the detection of tumor-derived vesicles within the complex mixture of plasma EV. This seems to be a promising strategy for gaining information about the genetic makeup of the tumor and identifying targetable characteristics, in particular for solid tumors, which are not directly accessible without invasive procedures (e.g., pancreatic cancer, glioblastoma). In contrast to the classical soluble "tumor markers" in the serum, MV also allow for detection of non-secreted, membrane-bound, or even intracellular molecules. Thus, the MV-based liquid biopsy enlarges the spectrum of detectable tumor-associated factors that could help not only with regard to diagnosis but also with accurate patient risk stratification and therapy monitoring. Since tumors seem to gain access to the circulation quite early [100], tumor MV in blood are believed to be of potential interest as early diagnostic biomarkers. In contrast to the extremely sparse CTCs, they are present in significantly greater numbers, which might improve detection sensitivity. Recently, tumor EV, consisting of a mixture of small and large vesicles, were shown to be able to pass the intact blood–brain barrier in an orthotopic glioblastoma xenotransplant model of human cancer stem cells as well as in glioma patients [101]. In line with this finding, EV carrying tumor-specific alterations were detected in the blood of patients with low-grade glioma, most of whom still had an intact blood–brain barrier [101], thus further supporting the concept of using EV as early cancer biomarkers. Another important point is that tumors are known to be highly heterogeneous [102], a circumstance that currently gains more and more attention with the novel single cell analysis technologies. While a biopsy taken from a limited area within the tumor only partly mirrors the molecular composition of the whole tumor, EV are believed to give a more comprehensive picture. This might apply also to patients with advanced, metastatic cancer, considering that metastases often differ from the primary tumor, as well as among each other, and might thus express different markers or potential therapeutic targets [103,104]. Interestingly, an early study in glioblastoma described the detection of oncogenic EGFRvIII transcripts in plasma-derived EV in two patients who had been diagnosed as EGFRvIII-negative on the basis of a tumor tissue biopsy [105]. Thus, it may be speculated that EV might mirror tumor heterogeneity more accurately than limited biopsies.

When aiming at the specific detection of tumor-derived MV in blood, the first step is to identify cancer-specific factors that are not expressed on MV shed by other cell populations. Therefore, a thorough characterization of tumor MV is necessary, which usually has been performed on cancer cell line-derived vesicles for later translation to patient samples. A recent report has suggested that MV reflect the cellular transcriptomic landscape better than Exo and can thus give valuable information about mutation status or gene amplifications that can be used for cancer diagnostics [106]. In line with this, hallmark oncogenic fusion transcripts (e.g., BCR-ABL1, TEL-AML1, MLL-AF6) as well as other specific tumorigenic transcripts (e.g., HOXA9, MEIS1) were found in MV of leukemia cell lines [107,108]. Similarly, in a glioblastoma model, genomic DNA (gDNA) sequences of tumor-associated genes were identified with a distinct distribution in either apoptotic bodies, MV or Exo isolated from human cancer stem cells, or from the blood of xenotransplanted mice, including specific detection of PIK3CA, EGFR, AKT1, and MDM2 sequences in MV [101]. However, the question remains as to whether analysis of vesicle-associated transcripts is superior to mutation detection based on cell-free DNA. Encouraging results have been obtained in this context by the group of Krug et al., who demonstrated increased sensitivity in detecting EGFR mutations in the plasma of lung cancer patients with a combined analysis of circulating tumor and EV DNA compared to circulating tumor DNA alone [109].

MV and Exo have been shown to differ in their RNA composition [10]. While several population-specific microRNAs have been identified for both MV and Exo, there seems to be a trend that miRNAs are relatively more enriched in Exo and mRNAs in MV [106,110,111]. In addition, an enrichment compared with the whole cell has been observed as several miRNAs that were almost undetectable in colorectal cancer cell lysates (<5 transcript per million (TPM)) were found at high levels (>1000 TPM) in the respective MV and Exo from the cells [111]. However, it is currently unclear whether sufficient amounts of RNA molecules are shuttled to EV in order to functionally influence recipient cells, as several critical reports have revealed that most RNA species are present at frequencies of much less than 1 copy per EV [106,110,112]. Therefore, the focus of the rest of this review is on MV-associated proteins that have the additional benefit that they could be used not only as biomarkers, but also for specific drug delivery and targeting. The protein content of EV has been evaluated largely by Western blotting, flow cytometry, as well as different proteomic approaches. While shot-gun proteomics was initially used to thoroughly characterize the whole EV proteome, novel targeted proteomic acquisition strategies such as selected (or multiple) reaction monitoring (SRM/MRM) allow the detection of a predetermined selection of specific target peptides with high sensitivity and quantitative accuracy [113]. These methods thus facilitate the discovery and large-scale validation of biomarker pipelines in complex biological samples (e.g., blood) [114]. In a recent study, SRM/MRM was successfully employed to identify novel potential biomarkers for prostate cancer on Exo in urine [115], demonstrating its potential for future EV-related biomarker studies.

The MV proteome has been proposed to be more similar to the cellular proteome than that of Exo [35]. However, not all proteins from the cellular plasma membrane are found on MV, and some are highly enriched, which suggests specific sorting mechanisms [116]. An overview of MV-associated tumor antigens that have been successfully detected in the blood of cancer patients and used as biomarkers is presented in Table 1. In hematologic malignancies, MV carrying the respective malignancy-associated antigen (e.g., CD38 for multiple myeloma, CD30 for Hodgkin lymphoma) were detected in cancer patients but were almost undetectable in healthy controls [91]. Similar observations have been made for solid tumors. In one of the earliest studies, the oncogenic receptor human epidermal growth factor receptor 2 (Her2/Neu) was identified and significantly elevated on blood-derived MV from gastric cancer patients [84]. Meanwhile this finding has been confirmed for colorectal cancer [117]. In the largest study to date of a cohort of 330 cancer patients with advanced solid tumors and 103 healthy, non-cancer controls, the number of MV carrying the tumor antigen EMMPRIN were found to be increased with advanced tumor stage [69]. Expanding the analyses on MV co-expressing EMMPRIN together with other tumor-associated molecules, such as epithelial cell adhesion molecule (EpCAM/CD326), MUC1/CA 15-3, or EGFR1, the numbers were not only significantly elevated in cancer patients versus controls but also associated with poor overall survival, thus suggesting that MV might indeed represent valuable prognostic cancer biomarkers. EpCAM is among the most studied antigens as it is known to be overexpressed on tumors of epithelial origin and is thus a bona fide tumor marker. The combination of EpCAM with other tissue-specific markers such as, for instance, the asialoglycoprotein receptor 1 (ASGPR1), exclusively expressed in liver cells, allowed the separation of patients with liver disorders into patients with or without liver tumors (including hepatocellular or cholangiocellular carcinoma) [118]. In line with this, the EpCAM+/EMMPRIN+ MV levels correlated with tumor size in colorectal cancer patients and decreased after surgical removal of the tumor [118,119]. In yet another study performed on plasma samples from metastatic prostate cancer patients, caveolin-1 (CAV1) was found on 5–10% of large EV (1–10 μm) isolated by filtration, but with a size profile similar to the 10,000 g pellet, whereas it was barely detectable in healthy controls [120].

Table 1. Common tumor antigens detected on patient-derived MV.

Marker	Cancer Subtype	Method	Reference
CD13	Leukemia (AML/CML) Myelodysplastic syndrome	Flow cytometry	[91]
CD138	Multiple myeloma	Flow cytometry	[89]
CD19	Non-Hodgkin lymphoma Leukemia (CLL) Morbus Waldenström	Flow cytometry	[91]
CD30	Hodgkin lymphoma	Flow cytometry	[91]
CD38	Multiple myeloma	Flow cytometry	[91]
c-Met	Gastric cancer	Western blot	[84]
Caveolin-1	Prostate cancer	Flow cytometry	[120]
EGFR	Colorectal	Flow cytometry Western blot	[76] [117]
	GBM	Immunofluorescence	[94]
EGFRvIII	GBM	Immunofluorescence	[94]
EMMPRIN	Various	Flow cytometry Western blot	[11] [76]
	Colorectal cancer Lung cancer Pancreas carcinoma	Flow cytometry	[119]
	Gastric cancer	Western blot	[84]
EpCAM	Breast cancer Lung cancer Head and neck cancer Colorectal cancer Pancreatic cancer	Flow cytometry	[76] [119]
	GBM	Immunofluorescence	[94]
EpCAM/EMMPRIN	Colorectal cancer Lung cancer Pancreas carcinoma	Flow cytometry	[119]
EpCAM/ASGPR1	Hepatocellular carcinoma Cholangiocarcinoma	Flow cytometry	[118]
FAK	Breast cancer	Western blot	[87]
HepPar1	Hepatocellular carcinoma	Flow cytometry	[121]
Her2/Neu	Gastric cancer	Flow cytometry	[84]
	Colorectal cancer	Flow cytometry Western blot	[117]
MUC1	Breast cancer Pancreatic cancer	Flow cytometry	[86]
	Breast cancer	Flow cytometry Western blot	[11]
	Breast cancer Lung cancer Head and neck cancer	Flow cytometry	[76]

From the thus far published MV biomarker studies, one can deduce that the population of tumor MV carrying the respective tumor antigens constitutes up to 5–10% of the total MV in blood of advanced cancer patients [69,94,120]. A recent study has profiled MV at the single vesicle level and proposed that there is a huge heterogeneity among MV, and that not all MV, even from the same cell line, carry the respective tumor markers [94], which further hampers the detection of tumor MV in blood.

Moreover, especially low-abundance tumor MV might be concealed in the heterogeneous mixture of blood MV, which might necessitate the development of novel protocols for specific enrichment of tumor MV populations. One suggestion is to remove the platelet-derived MV prior to analysis [70] since they represent the largest MV population in blood. A protocol was recently published for isolating different subpopulations of small and large EV from solid tumor tissue, in this example from melanoma [122,123]. This is an important step that will now allow for isolation of MV from primary tumor tissues and metastases to define not only tumor-specific but even patient-specific biomarkers. The comparison of MV profiles from the tumor tissues (primary tumor/metastases) with MV present in body fluids might then provide information on clonal evolution and serve as the basis for therapy selection and monitoring.

9. MV Biomarker Signatures

The studies conducted thus far have shown that while single MV-associated biomarkers are often elevated in the blood of cancer patients, each of the antigens alone failed to reliably distinguish cancer patients from healthy controls, mainly due to low sensitivity. In contrast, combinations of several markers have proved to be of superior diagnostic value, such as the combination of EMMPRIN/EpCAM/MUC1/EGFR, which reliably separated cancer patients from healthy controls with a high AUC value of 0.85 [69]. It seems likely that more comprehensive signatures comprising cancer subtype-specific markers can even further increase sensitivity and specificity. Importantly, it has to be considered that the overexpression of certain oncogenes might impact MV protein profiles as was shown for breast cancer for the oncogene Src for Exo [124] or Her2/Neu, which was shuttled along with its associated proteins onto MV and Exo [125]. The resulting specific signatures have already been successfully used for the correct determination of breast cancer subtypes from serum EV [126]. Compared to soluble cancer biomarkers, which often suffer from low specificity and a high rate of false positive results, the strength of MV is their potential as biomarker platforms that simultaneously carry multiple markers, allowing for the recognition and detection of specific patterns.

10. The Microvesicle Reactome in Blood

Several conditions, such as hypoxia, inflammation, exercise, or nutrition, influence levels of MV subpopulations in blood (reviewed in [127]). It might thus be worth considering the fact that not only the analysis of tumor MV in blood but also changes in blood MV composition or the markers on distinct blood MV subpopulations might give information about cellular transformation (Figure 2). Although, for example, endothelial cell-derived MV were detected at elevated levels in some hematologic malignancies [91], a slight increase was observed in platelet- as well as leukocyte-derived MV in solid tumor patients [69,88,128]. An analysis of the phospho-proteome of MV and Exo isolated from plasma samples of breast cancer patients by ultracentrifugation described 156 unique phosphosites, which differed significantly between the vesiculome of healthy individuals and patients [129]. More specifically, in the blood of gastric cancer patients, the levels of MV carrying the chemokine receptor CCR6 were elevated. The cell type of origin, however, remained unclear [84].

Among the EV subpopulations, MV are considered the main source of tissue factor (TF) activity [130]. Higher amounts of MV with TF activity were detected in the plasma of metastatic cancer patients, suggesting their use as novel markers for venous thromboembolism, a common complication in advanced cancer patients [131]. In addition to TF, phosphatidylserine, whose exposure on the outer membrane is a typical feature of all MV, can further induce thrombin generation [132]. This potential side effect should therefore be carefully taken into account when considering the application of MV as therapeutics, although first studies in mice did not show an effect of the injected MV on coagulation [133].

Taken together, analysis of the whole mixture of plasma vesicles does not only detect the presence or absence of tumor EV but yields more comprehensive information since it additionally mirrors the reaction of immune and other cells against the tumor EV.

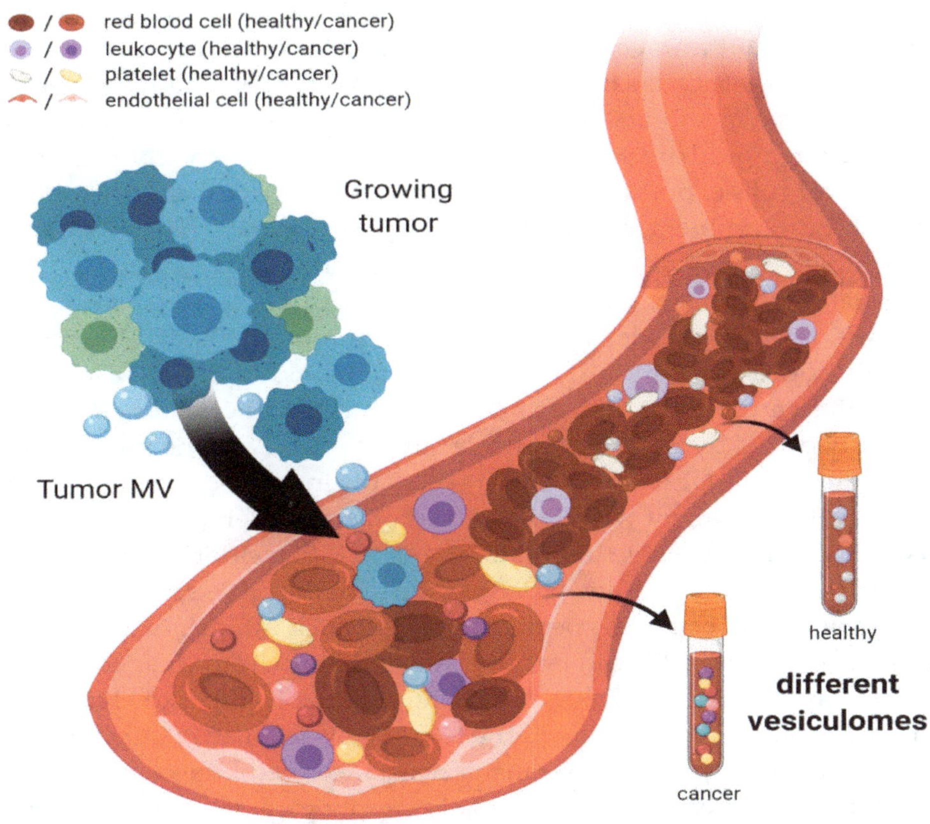

Figure 2. The microvesiculome in blood. The growth of a tumor is accompanied by secretion of circulating tumor cells (CTCs) as well as of tumor-derived soluble factors and MV. When encountering any of these components, benign blood cells can become activated and react by shedding their own, altered MV. Taken together, this results in changes in the composition as well as expression pattern of MV in the blood of cancer patients. Liquid biopsy-based MV sampling should therefore not only be focused on the detection of tumor MV, but also on alterations in non-tumor blood MV.

11. MV for Therapy Monitoring

First studies have revealed the potential of MV for monitoring therapeutic responses in cancer patients. Krishnan et al. reported that plasma levels of CD138+ MV corresponded with therapeutic response in individual multiple myeloma patients [89]. In general, total levels of MV were found to decrease after surgical removal of the tumor, which was generally associated with a better outcome [92,97]. In line with this, the number of EpCAM+ MV decreased 10 days after surgery in colorectal cancer patients, although the effect was not seen for EpCAM+/EMMPRIN+ MV [119]. This might be explained by the additional presence of EMMPRIN on platelet- and immune cell-derived MV [11], which are directly influenced by the surgery. Prolonged high levels of total EV or endothelial cell-derived CD144+ MV after chemotherapy seem to predict a poor clinical response and were found to be associated with shorter progression-free and overall survival [134,135]. The same trend was observed in glioblastoma patients showing tumor progression after radiochemotherapy who generally exhibited higher MV levels compared with patients with pseudoprogression or stable disease [136]. Moreover, an increased concentration of large and small EV prior to therapy was associated with failure of chemotherapy in breast cancer patients [134]. Taken together, these results demonstrate that MV levels not only mirror the presence of tumor, but that MV might also have a protective role against chemotherapeutic agents.

12. MV and Cancer Therapy

As discussed above, MV collected by liquid biopsies can give valuable information about targetable characteristics in the growing tumor and thus be used as an innovative read-out for targeted therapy decisions and patient stratification for personalized medicine. However, it has also become clear that MV can interfere with cancer treatment and limit therapeutic success. On the one hand, MV have been identified as transporters for P-glycoprotein (MDR1) and multidrug resistance-associated protein 1 (MRP1), two plasma membrane multidrug efflux transporters, and have thus been recognized as vehicles for spreading multidrug resistance from resistant to sensitive tumor clones [137,138]. On the other hand, cancer cells export cytotoxic agents via their MV during chemotherapy as has been shown, for instance, for doxorubicin, cisplatin, gemcitabine, or docetaxel [133,139,140]. Cancer cells seem to shed higher numbers of MV during treatment, and it has been suggested that the potential of a cell to shed MV correlates with its resistance to treatment [140]. Therefore, it seemed to be a promising strategy to inhibit MV release in order to increase treatment efficiency. Indeed, the combination of chloramine and bisindolylmaleimide-I, two inhibitors blocking the release of both MV and Exo, highly increased apoptosis in cancer cell lines treated with chemotherapeutic agents [141]. In another study, reducing MV shedding by inhibition of calpain via siRNA or the specific inhibitor calpeptin significantly elevated intracellular levels of docetaxel. This increased the susceptibility of the cell to chemotherapy, allowing for treatment with lower doses of chemotherapeutic agent at higher efficiency in vitro as well as in a xenograft mouse model in vivo [139]. Considering the toxic effects of general inhibition of EV secretion, it still remains elusive as to whether this strategy is feasible for human patients or whether it will be possible to specifically target such inhibitors to the tumor tissue.

Another mechanism as to how EV can impede cancer therapy is that cancer cells have been shown to export therapeutic targets such as CD20, Her2/Neu, or PD-L1 onto Exo [142–144] that capture the administered therapeutic antibodies and shield the tumor cells from attack, resulting in therapy failure. Moreover, Exo have been recently shown to transfer resistance-mediating cargo from stromal to cancer cells [145]. Whether the same mechanisms indeed also apply for large MV is currently unknown.

Considering the evidence for a tumor-supporting role of MV, they should also be regarded as novel targets for therapeutic approaches. This notion is supported by a study by Keklikoglou et al. showing in a mixed population of MV and Exo that chemotherapy induces the release of Annexin A6 (ANXA6+) EV from murine breast cancer cells. When these chemotherapy-induced ANXA6+ EV were injected into the tail vein of mice, they activated endothelial cells to produce the chemokine Ccl2, which led to the expansion of Ly6C+CCR2+ lung monocytes and thus pre-conditioned the lungs for the seeding of breast cancer metastases [146]. Similar observations of EV-mediated pre-metastatic niche formation have been made for melanoma [147]. In line with these findings, MV isolated from cancer patients induced a tumor-supporting phenotype in human macrophages ex vivo and highly increased the invasive potential of benign breast cancer cells, while MV isolated from healthy controls had no such effect [69].

13. MV as Cancer Vaccines

Although the studies presented thus far argue for an unfavorable role of MV in hampering cancer therapy, the final two chapters discuss how MV could also be exploited positively with regard to therapeutic interventions. In a first study, Zhang et al. reported that while the immunization of mice with PBS or tumor cell lysates failed to prevent the growth of subsequently injected tumor cells, mice treated with tumor Exo or MV developed tumors in only 87.5% or 50% of the animals, respectively [148]. These initial results suggested that tumor MV, which are enriched in a multitude of tumor-specific proteins (see Table 1), could serve as a novel, cell-free source of tumor antigens that are more immunogenic than other parts of the tumor cells, including tumor Exo. A possible explanation for their superior efficacy compared with Exo might be the presence of DNA as well as mitochondrial fragments inside MV that discriminate them from Exo and might provide an additional immunostimulatory signal for innate immune cells. However, more studies are required to confirm the differential immunogenic effect of MV and Exo and explain the potential difference.

In general, the anti-tumor immune response is believed to occur largely via cytotoxic CD8+ T cells activated by tumor antigen-presenting dendritic cells. Since immunization with tumor MV prior to tumor growth is hardly possible in human patients, strategies will have to be developed to treat already existing tumors. Indeed, in the above-mentioned study, the injection of tumor MV into mice with already growing tumors failed to induce a sufficient activation of T cells [148]. Instead, the injection of dendritic cells primed with tumor MV overcame the problem of inadequate T cell activation, enhancing T cell infiltration in the tumor tissue as well as tumor cell killing [148]. Mechanistically, incubation of dendritic cells with tumor MV induced dendritic cell maturation with concordant upregulation of the co-stimulatory molecules CD80 and CD86, increasing their homing to the tumor-draining lymph nodes [148,149]. Further evidence that the use of MV as tumor vaccines might indeed lead to successful anti-tumor immune responses has been provided by a recent study by Pineda et al. who harvested MV from C6 rat glioma cells and subsequently administered them to rats, which had been inoculated with a subcutaneous C6 glioma. MV vaccination increased the number of infiltrating T cells in the tumor tissue, induced tumor cell apoptosis, and led to a reduction of tumor growth [150]. Interestingly, it might even be possible to deliver future MV vaccines orally, as it was discovered that tumor MV injected intragastrically into mice are taken up by intestinal epithelial cells in the ileum and transferred by transcytosis to dendritic cells at the basolateral side that can subsequently induce T cell activation and anti-tumor immune responses [151]. However, this treatment was only successful with concordant anti-acidic agents, as MV are sensitive to degradation by gastric acid.

Thus far, clinical studies testing the efficiency of tumor cell- or Exo-based vaccines have shown only limited efficiency and it remains to be seen whether this trend is different for MV. The efficiency of tumor MV vaccines will critically depend on whether it will be possible, on the one hand, to find ways of boosting their efficacy in eliciting an anti-tumor immune response, while, on the other hand, limiting their immunosuppressive functions.

14. MV as Vehicles for Drug Delivery

The finding that EV transport bioactive molecules (e.g., proteins, metabolites, or nucleic acids) raised interest early on in terms of their use as a novel drug delivery system. Compared with cell-based therapeutics, EV have the advantages that they do not possess the potential for transformation or unlimited growth, they are known to cross biological barriers such as the blood–brain barrier, and as non-living material they can be stored or transported more easily and be modified with more aggressive manipulation techniques. Another positive aspect is their natural origin, which makes them more biocompatible than artificial liposomal drug carriers, with enhanced stability, less immunogenicity, and less liver toxicity [152]. In general, EV as drug carriers offer protection for their content, such as from enzymatic degradation, which enhances cargo stability in biofluids. EV can be modified with several approaches to increase their bioactive potential, including either manipulation strategies for the secreting cell, such as genetic manipulation, cell stress, or hypoxia, or post-isolation loading methods, such as electroporation, sonication, heat-shock, or EV transfection. A current overview of the techniques, their advantages, and their limitations is presented in [153,154]. Although most methods have been solely tested for Exo, the majority should be equally applicable to MV, considering the similarities in their biophysical makeup.

The first report on MV as a drug delivery system was published by Tang et al., who successfully used chemotherapy-loaded MV to inhibit tumor growth in a murine hepatocarcinoma ascites model, as well as in severe combined immunodeficient (SCID) mice injected with ovarian cancer cells [133]. Delivery was even successful to solid tumors when MV were injected intravenously [133]. In a first in vivo approach, the same group tested the therapeutic potential of intrathoracic injections of cisplatin-loaded tumor MV in six end-stage, cisplatin-resistant lung cancer patients. The injections greatly reduced the number of tumor cells (>95%) in the metastatic malignant pleural effusions in three of the six patients, suggesting an efficacy of the MV injections and also their ability to reverse drug resistance [155]. However, the treatment did not show any benefit in the other three patients, raising the question about the underlying reasons for the treatment failure. Two clinical trials are currently registered to further explore the potential of MV as therapeutic tool in cancer: a phase I/II study investigating the use of red blood cell-derived MV loaded with methotrexate in the treatment of malignant ascites (clinicaltrials.gov identifier: NCT03230708), and a phase II study aimed at generating methotrexate-loaded autologous tumor MV from malignant effusion and testing their effect on tumor growth and immune regulation (clinicaltrials.gov identifier: NCT02657460).

The majority of the conducted studies thus far has used EV from mesenchymal stem cells as native vehicles for therapeutic applications, and none of them has observed significant side effects [156]. Alternatively, red blood cell-derived EV have been suggested especially for the delivery of RNA drugs since they are easily available, do not contain DNA, and can be easily modified by electroporation [157]. However, in mouse models, their systemic administration as drug carrier was found to only be feasible for the therapy of leukemias, while successful delivery to solid tumors required intratumoral injections due to the otherwise low target organ specificity [157]. In general, EV biodistribution seems to depend on the kind of cell used for EV production, the EV delivery route, the dosage, and the uptake efficiency [158]. After systemic injection, a major part of the injected EV is trapped in the liver, spleen, or lung [158,159], which on the one hand might limit EV distribution to other organs, but on the other hand might be used for specific cargo delivery to these organs. Several engineering strategies are currently being tested to improve EV circulation kinetics and increase targeting to specific cell populations [160]. A hallmark of MV is that phosphatidylserine is exposed on their outer membrane surface, a feature which distinguishes them from Exo. Since externalized phosphatidylserine is a recognition signal for macrophages and triggers phagocytosis [161], MV might be superior for drug delivery to this specific cell population.

Interestingly, a recent report suggested that the choice of MV or Exo as drug carrier influences the intracellular delivery route of the transported cargo. While MV-loaded paclitaxel was mostly delivered by membrane fusion and endocytosis, Exo-loaded paclitaxel was predominantly taken up by endocytosis into prostate cancer cells [162]. Moreover, MV seem to promote drug entry into the nucleus, even into highly resistant tumor-repopulating cells [155]. Since the mechanism of internalization and intracellular trafficking significantly affects the functional efficiency of the delivered drug, further research aiming at elucidating the different fates of MV- and Exo-delivered cargo will help to choose the most accurate EV delivery system for a given drug.

While MV are larger and therefore allow for packaging larger amounts of drug molecules, it could be more difficult for them to penetrate into the tissue due to their size. One solution that has been presented suggests isolating MV from tumor cells grown in and adapted to soft fibrin gels, which resulted in the shedding of less stiff MV [163]. These softer MV with a higher capacity for deformation showed enhanced extravasation into the tumor with deep tissue penetration and, as a result, enhanced treatment efficacy when loaded with the chemotherapeutic agent doxorubicin [163]. Moreover, a weaker extravasation was observed in non-tumor tissue compared with MV isolated from cells cultured in 2D [163], suggesting less toxicity. A problem for both MV and Exo seems to be their limited lifespan in circulation. For fluorescently labeled pancreatic tumor MV injected into healthy mice, the signal was cleared after 60 min, while the majority of MV were already lost within 15 min after injection [164].

Taken together, while both Exo and large MV have been shown to possess the potential for drug delivery, more comparative studies are required to evaluate common as well as diverging features of both EV populations as therapeutic vehicles. Both still suffer from the lack of scalable, specific, cost-effective production and isolation methods as well as poor drug loading efficiency, problems that need to be solved for their standardized use in the clinic. Moreover, heterogeneity and batch-to-batch variability are known to strongly influence their therapeutic efficacy [165].

15. Concluding Remarks and Future Perspectives

Over the past decade, an increasing number of reports has indicated that plasma membrane-derived MV are a distinct vesicle population with some common but also several diverging features compared with the smaller Exo. While most researchers have focused on the analysis of Exo, MV offer several advantages as diagnostic and therapeutic tools. In particular, their isolation is easier and less time-consuming and does not necessarily require ultracentrifugation with its associated problems of vesicle aggregation, damage and loss, and instrument availability. Additionally, protocols are available for their rapid and thorough characterization by standard flow cytometry, a technique that is already well-established in routine clinical diagnostics. Interesting findings also include the possibility for delivery of larger amounts of drugs or distinct classes of biomolecules (e.g., mRNA) by MV compared to Exo, as well as more successful targeting of MV cargo to the nuclear or cytosolic compartment, aspects that surely require further exploration.

It has become clear that MV have a multifaceted influence on cancer therapy, as summarized in Figure 3. They can not only serve as innovative tools for targeted drug delivery but should also be considered as important therapeutic targets when developing novel treatment strategies. Therefore, it will be highly advantageous to further elucidate the molecular mechanisms underlying MV formation and secretion in order to learn how MV biogenesis, especially in cancer cells, can be specifically inhibited or modulated therapeutically. However, at present, the lack of standardization, multicenter studies, and technical challenges, such as storage, production, quality control, and targeted delivery, still hamper the use of MV as diagnostic and therapeutic tools in cancer and must be overcome for them to be exploitable in the clinic.

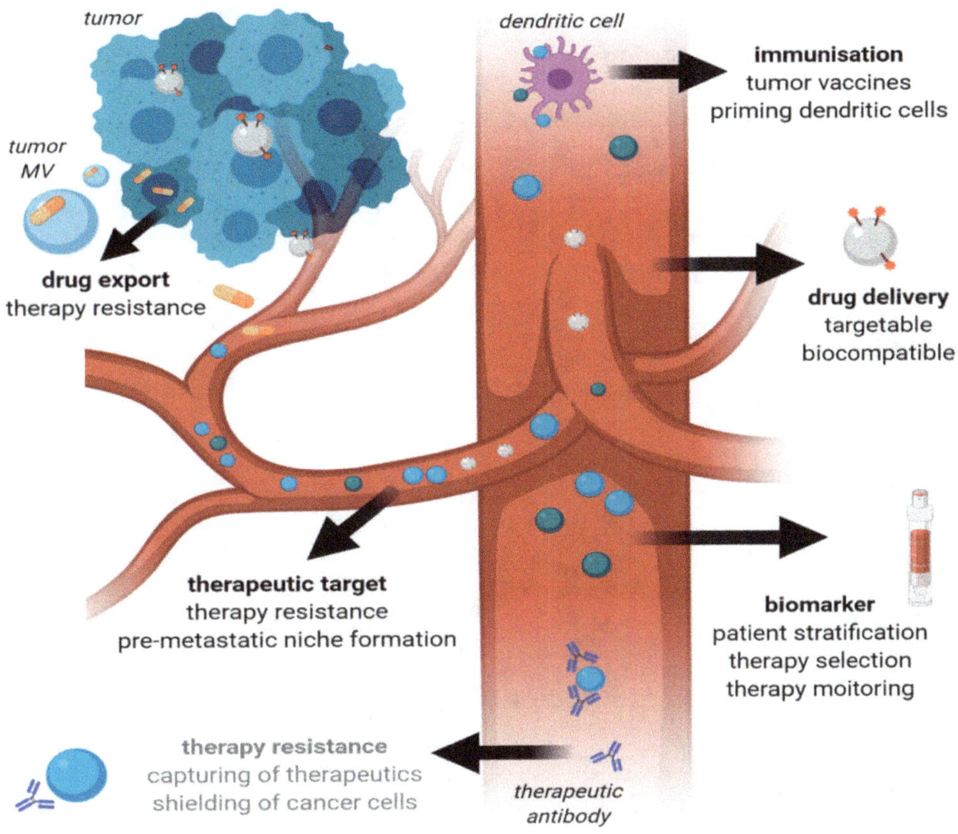

Figure 3. MV in cancer therapy. The figure summarizes the roles of MV in cancer therapy. Once a tumor has gained access to the circulation, circulating tumor MV in blood can be used as biomarkers in liquid biopsies. Moreover, they can mediate therapy resistance by capturing or exporting the administered anti-cancer drugs. Due to their role in therapy resistance and tumor microenvironment crosstalk resulting in pre-metastatic niche formation, they should be regarded as a novel therapeutic target. However, MV can also have a beneficial role in cancer therapy as they represent promising novel drug delivery systems with the benefits of high biocompatibility as well as opportunities for specific targeting and crossing of biological barriers. Vesicular functions, which have only been reported for endosomal-derived small exosomes (Exo) and lack experimental proof for MV, are highlighted in grey.

Author Contributions: Conceptualization, C.B., K.M., and A.B.; writing—original draft preparation K.M. and S.S. writing—review and editing, C.B., A.B., K.M., and S.S.; supervision, C.B., A.B., and K.M. All authors have read and agreed to the published version of the manuscript.

Acknowledgments: Suganja Sivaloganathan is a member of CiM-IMPRS, the joint graduate school of the Cells-in-Motion Interfaculty Centre, University of Münster, Germany, and the International Max Planck Research School-Molecular Biomedicine, Münster, Germany.

Abbreviations

Extracellular vesicle	EV
Tumor microenvironment	TME
Exosomes	Exo
Microvesicles	MV
Large oncosomes	LO
Nanoparticle tracking analysis	NTA
International society of extracellular vesicles	ISEV
ADP-ribosylation factor 6	ARF6
Phospholipase D	PLD
Myosin light chain kinase	MLCK
Endosomal sorting complex required for transport	ESCRT
Arrestin domain-containing protein 1	ARRDC1
Tumor susceptibility gene 101	TSG101
Membrane type 1-matrix metalloproteinase	MT1-MMP
Vesicular SNARE	v-SNARE
Vesicle-associated membrane protein 3	VAMP3
Extracellular matrix metalloproteinase inducer	EMMPRIN
Epidermal growth factor receptor	EGFR
Programmed cell death ligand 1	PD-L1
Cancer-associated fibroblasts	CAF
Platelet-poor plasma	PPP
Circulating tumor cells	CTCs
Chronic lymphocytic leukemia	CLL
Transcripts per million	TPM
Epithelial cell adhesion molecule	EpCAM
Asialoglycoprotein receptor 1	ASGPR1
Caveolin-1	CAV1
Tissue factor	TF
P-glycoprotein	MDR1
Multidrug resistance-associated protein 1	MRP1
Severe combined immunodeficient	SCID
Major histocompatibility complex	MHC
Vascular endothelial growth factor	VEGF
Vascular endothelial growth factor receptor	VEGFR
Transcripts per million	TPM

References

1. Hanahan, D.; Weinberg, R.A. Hallmarks of Cancer: The Next Generation. *Cell* **2011**, *144*, 646–674. [CrossRef] [PubMed]
2. Colombo, M.; Raposo, G.; Théry, C. Biogenesis, Secretion, and Intercellular Interactions of Exosomes and Other Extracellular Vesicles. *Annu. Rev. Cell Dev. Biol.* **2014**, *30*, 255–289. [CrossRef] [PubMed]

3. Yáñez-Mó, M.; Siljander, P.R.-M.; Andreu, Z.; Bedina Zavec, A.; Borràs, F.E.; Buzas, E.I.; Buzas, K.; Casal, E.; Cappello, F.; Carvalho, J.; et al. Biological properties of extracellular vesicles and their physiological functions. *J. Extracell. Vesicles* **2015**, *4*, 27066. [CrossRef] [PubMed]
4. Minciacchi, V.R.; Freeman, M.R.; Di Vizio, D. Extracellular Vesicles in Cancer: Exosomes, Microvesicles and the Emerging Role of Large Oncosomes. *Semin. Cell Dev. Biol.* **2015**, *40*, 41–51. [CrossRef] [PubMed]
5. Lee, S.-S.; Won, J.-H.; Lim, G.J.; Han, J.; Lee, J.Y.; Cho, K.-O.; Bae, Y.-K. A novel population of extracellular vesicles smaller than exosomes promotes cell proliferation. *Cell Commun. Signal.* **2019**, *17*, 95. [CrossRef]
6. Mezouar, S.; Mege, D.; Darbousset, R.; Farge, D.; Debourdeau, P.; Dignat-George, F.; Panicot-Dubois, L.; Dubois, C. Involvement of platelet-derived microparticles in tumor progression and thrombosis. *Semin. Oncol.* **2014**, *41*, 346–358. [CrossRef]
7. Théry, C.; Witwer, K.W.; Aikawa, E.; Alcaraz, M.J.; Anderson, J.D.; Andriantsitohaina, R.; Antoniou, A.; Arab, T.; Archer, F.; Atkin-Smith, G.K.; et al. Minimal information for studies of extracellular vesicles 2018 (MISEV2018): A position statement of the International Society for Extracellular Vesicles and update of the MISEV2014 guidelines. *J. Extracell. Vesicles* **2018**, *7*, 1535750. [CrossRef]
8. Mateescu, B.; Kowal, E.J.K.; van Balkom, B.W.M.; Bartel, S.; Bhattacharyya, S.N.; Buzás, E.I.; Buck, A.H.; de Candia, P.; Chow, F.W.N.; Das, S.; et al. Obstacles and opportunities in the functional analysis of extracellular vesicle RNA—An ISEV position paper. *J. Extracell. Vesicles* **2017**, *6*, 1286095. [CrossRef] [PubMed]
9. Kowal, J.; Arras, G.; Colombo, M.; Jouve, M.; Morath, J.P.; Primdal-Bengtson, B.; Dingli, F.; Loew, D.; Tkach, M.; Théry, C. Proteomic comparison defines novel markers to characterize heterogeneous populations of extracellular vesicle subtypes. *Proc. Natl. Acad. Sci. USA* **2016**, *113*, E968–E977. [CrossRef]
10. Crescitelli, R.; Lässer, C.; Szabó, T.G.; Kittel, A.; Eldh, M.; Dianzani, I.; Buzás, E.I.; Lötvall, J. Distinct RNA profiles in subpopulations of extracellular vesicles: Apoptotic bodies, microvesicles and exosomes. *J. Extracell. Vesicles* **2013**, *2*, 20677. [CrossRef] [PubMed]
11. Menck, K.; Scharf, C.; Bleckmann, A.; Dyck, L.; Rost, U.; Wenzel, D.; Dhople, V.M.; Siam, L.; Pukrop, T.; Binder, C.; et al. Tumor-derived microvesicles mediate human breast cancer invasion through differentially glycosylated EMMPRIN. *J. Mol. Cell Biol.* **2015**, *7*, 143–153. [CrossRef] [PubMed]
12. Linares, R.; Tan, S.; Gounou, C.; Arraud, N.; Brisson, A.R. High-speed centrifugation induces aggregation of extracellular vesicles. *J. Extracell. Vesicles* **2015**, *4*. [CrossRef] [PubMed]
13. Nordin, J.Z.; Lee, Y.; Vader, P.; Mäger, I.; Johansson, H.J.; Heusermann, W.; Wiklander, O.P.B.; Hällbrink, M.; Seow, Y.; Bultema, J.J.; et al. Ultrafiltration with size-exclusion liquid chromatography for high yield isolation of extracellular vesicles preserving intact biophysical and functional properties. *Nanomed. Nanotechnol. Biol. Med.* **2015**, *11*, 879–883. [CrossRef] [PubMed]
14. Mol, E.A.; Goumans, M.-J.; Doevendans, P.A.; Sluijter, J.P.G.; Vader, P. Higher functionality of extracellular vesicles isolated using size-exclusion chromatography compared to ultracentrifugation. *Nanomed. Nanotechnol. Biol. Med.* **2017**, *13*, 2061–2065. [CrossRef] [PubMed]
15. Bobrie, A.; Colombo, M.; Krumeich, S.; Raposo, G.; Théry, C. Diverse subpopulations of vesicles secreted by different intracellular mechanisms are present in exosome preparations obtained by differential ultracentrifugation. *J. Extracell. Vesicles* **2012**, *1*. [CrossRef]
16. Gallart-Palau, X.; Serra, A.; Wong, A.S.W.; Sandin, S.; Lai, M.K.P.; Chen, C.P.; Kon, O.L.; Sze, S.K. Extracellular vesicles are rapidly purified from human plasma by PRotein Organic Solvent PRecipitation (PROSPR). *Sci. Rep.* **2015**, *5*. [CrossRef]
17. Ko, J.; Carpenter, E.; Issadore, D. Detection and isolation of circulating exosomes and microvesicles for cancer monitoring and diagnostics using micro-/nano-based devices. *Analyst* **2016**, *141*, 450–460. [CrossRef]
18. Vestad, B.; Llorente, A.; Neurauter, A.; Phuyal, S.; Kierulf, B.; Kierulf, P.; Skotland, T.; Sandvig, K.; Haug, K.B.F.; Øvstebø, R. Size and concentration analyses of extracellular vesicles by nanoparticle tracking analysis: A variation study. *J. Extracell. Vesicles* **2017**, *6*, 1344087. [CrossRef]
19. Gardiner, C.; Vizio, D.D.; Sahoo, S.; Théry, C.; Witwer, K.W.; Wauben, M.; Hill, A.F. Techniques used for the isolation and characterization of extracellular vesicles: Results of a worldwide survey. *J. Extracell. Vesicles* **2016**, *5*, 32945. [CrossRef]
20. EV-TRACK Consortium; Van Deun, J.; Mestdagh, P.; Agostinis, P.; Akay, Ö.; Anand, S.; Anckaert, J.; Martinez, Z.A.; Baetens, T.; Beghein, E.; et al. EV-TRACK: Transparent reporting and centralizing knowledge in extracellular vesicle research. *Nat. Methods* **2017**, *14*, 228–232. [CrossRef]

21. Taylor, J.; Bebawy, M. Proteins Regulating Microvesicle Biogenesis and Multidrug Resistance in Cancer. *Proteomics* **2019**, *19*, 1800165. [CrossRef] [PubMed]
22. Battistelli, M.; Falcieri, E. Apoptotic Bodies: Particular Extracellular Vesicles Involved in Intercellular Communication. *Biology* **2020**, *9*, 21. [CrossRef] [PubMed]
23. Birge, R.B.; Boeltz, S.; Kumar, S.; Carlson, J.; Wanderley, J.; Calianese, D.; Barcinski, M.; Brekken, R.A.; Huang, X.; Hutchins, J.T.; et al. Phosphatidylserine is a global immunosuppressive signal in efferocytosis, infectious disease, and cancer. *Cell Death Differ.* **2016**, *23*, 962–978. [CrossRef] [PubMed]
24. van Niel, G.; D'Angelo, G.; Raposo, G. Shedding light on the cell biology of extracellular vesicles. *Nat. Rev. Mol. Cell Biol.* **2018**, *19*, 213–228. [CrossRef]
25. Sedgwick, A.E.; D'Souza-Schorey, C. The biology of extracellular microvesicles. *Traffic* **2018**, *19*, 319–327. [CrossRef]
26. Menck, K.; Sönmezer, C.; Worst, T.S.; Schulz, M.; Dihazi, G.H.; Streit, F.; Erdmann, G.; Kling, S.; Boutros, M.; Binder, C.; et al. Neutral sphingomyelinases control extracellular vesicles budding from the plasma membrane. *J. Extracell. Vesicles* **2017**, *6*, 1378056. [CrossRef]
27. Muralidharan-Chari, V.; Clancy, J.; Plou, C.; Romao, M.; Chavrier, P.; Raposo, G.; D'Souza-Schorey, C. ARF6-Regulated Shedding of Tumor Cell-Derived Plasma Membrane Microvesicles. *Curr. Biol.* **2009**, *19*, 1875–1885. [CrossRef]
28. Li, B.; Antonyak, M.A.; Zhang, J.; Cerione, R.A. RhoA triggers a specific signaling pathway that generates transforming microvesicles in cancer cells. *Oncogene* **2012**, *31*, 4740–4749. [CrossRef]
29. Colombo, M.; Moita, C.; van Niel, G.; Kowal, J.; Vigneron, J.; Benaroch, P.; Manel, N.; Moita, L.F.; Théry, C.; Raposo, G. Analysis of ESCRT functions in exosome biogenesis, composition and secretion highlights the heterogeneity of extracellular vesicles. *J. Cell Sci.* **2013**, *126*, 5553–5565. [CrossRef]
30. Nabhan, J.F.; Hu, R.; Oh, R.S.; Cohen, S.N.; Lu, Q. Formation and release of arrestin domain-containing protein 1-mediated microvesicles (ARMMs) at plasma membrane by recruitment of TSG101 protein. *Proc. Natl. Acad. Sci. USA* **2012**, *109*, 4146–4151. [CrossRef]
31. Hoshino, A.; Costa-Silva, B.; Shen, T.-L.; Rodrigues, G.; Hashimoto, A.; Tesic Mark, M.; Molina, H.; Kohsaka, S.; Di Giannatale, A.; Ceder, S.; et al. Tumour exosome integrins determine organotropic metastasis. *Nature* **2015**, *527*, 329–335. [CrossRef] [PubMed]
32. Nolte-'t Hoen, E.N.M.; Buschow, S.I.; Anderton, S.M.; Stoorvogel, W.; Wauben, M.H.M. Activated T cells recruit exosomes secreted by dendritic cells via LFA-1. *Blood* **2009**, *113*, 1977–1981. [CrossRef] [PubMed]
33. Cocucci, E.; Racchetti, G.; Meldolesi, J. Shedding microvesicles: Artefacts no more. *Trends Cell Biol.* **2009**, *19*, 43–51. [CrossRef] [PubMed]
34. Haraszti, R.A.; Didiot, M.-C.; Sapp, E.; Leszyk, J.; Shaffer, S.A.; Rockwell, H.E.; Gao, F.; Narain, N.R.; DiFiglia, M.; Kiebish, M.A.; et al. High-resolution proteomic and lipidomic analysis of exosomes and microvesicles from different cell sources. *J. Extracell. Vesicles* **2016**, *5*, 32570. [CrossRef]
35. Wang, T.; Gilkes, D.M.; Takano, N.; Xiang, L.; Luo, W.; Bishop, C.J.; Chaturvedi, P.; Green, J.J.; Semenza, G.L. Hypoxia-inducible factors and RAB22A mediate formation of microvesicles that stimulate breast cancer invasion and metastasis. *Proc. Natl. Acad. Sci. USA* **2014**, *111*, E3234–E3242. [CrossRef]
36. Maacha, S.; Bhat, A.A.; Jimenez, L.; Raza, A.; Haris, M.; Uddin, S.; Grivel, J.-C. Extracellular vesicles-mediated intercellular communication: Roles in the tumor microenvironment and anti-cancer drug resistance. *Mol. Cancer* **2019**, *18*, 55. [CrossRef]
37. Han, L.; Lam, E.W.-F.; Sun, Y. Extracellular vesicles in the tumor microenvironment: Old stories, but new tales. *Mol. Cancer* **2019**, *18*, 59. [CrossRef]
38. Xu, R.; Greening, D.W.; Zhu, H.-J.; Takahashi, N.; Simpson, R.J. Extracellular vesicle isolation and characterization: Toward clinical application. *J. Clin. Investig.* **2016**, *126*, 1152–1162. [CrossRef]
39. Arendt, B.K.; Walters, D.K.; Wu, X.; Tschumper, R.C.; Jelinek, D.F. Multiple myeloma cell-derived microvesicles are enriched in CD147 expression and enhance tumor cell proliferation. *Oncotarget* **2014**, *5*, 5686–5699. [CrossRef]
40. Riethdorf, S.; Reimers, N.; Assmann, V.; Kornfeld, J.-W.; Terracciano, L.; Sauter, G.; Pantel, K. High incidence of EMMPRIN expression in human tumors. *Int. J. Cancer* **2006**, *119*, 1800–1810. [CrossRef]
41. Al-Nedawi, K.; Meehan, B.; Micallef, J.; Lhotak, V.; May, L.; Guha, A.; Rak, J. Intercellular transfer of the oncogenic receptor EGFRvIII by microvesicles derived from tumour cells. *Nat. Cell Biol.* **2008**, *10*, 619–624. [CrossRef]

42. Menck, K.; Klemm, F.; Gross, J.C.; Pukrop, T.; Wenzel, D.; Binder, C. Induction and transport of Wnt 5a during macrophage-induced malignant invasion is mediated by two types of extracellular vesicles. *Oncotarget* **2013**, *4*, 2057–2066. [CrossRef] [PubMed]
43. Baj-Krzyworzeka, M.; Szatanek, R.; Węglarczyk, K.; Baran, J.; Zembala, M. Tumour-derived microvesicles modulate biological activity of human monocytes. *Immunol. Lett.* **2007**, *113*, 76–82. [CrossRef]
44. Lenart, M.; Rutkowska-Zapała, M.; Szatanek, R.; Węglarczyk, K.; Stec, M.; Bukowska-Strakova, K.; Gruca, A.; Czyż, J.; Siedlar, M. Alterations of TRIM21-mRNA expression during monocyte maturation. *Immunobiology* **2017**, *222*, 494–498. [CrossRef] [PubMed]
45. Baj-Krzyworzeka, M.; Mytar, B.; Szatanek, R.; Surmiak, M.; Węglarczyk, K.; Baran, J.; Siedlar, M. Colorectal cancer-derived microvesicles modulate differentiation of human monocytes to macrophages. *J. Transl. Med.* **2016**, *14*, 36. [CrossRef]
46. Timaner, M.; Kotsofruk, R.; Raviv, Z.; Magidey, K.; Shechter, D.; Kan, T.; Nevelsky, A.; Daniel, S.; de Vries, E.G.E.; Zhang, T.; et al. Microparticles from tumors exposed to radiation promote immune evasion in part by PD-L1. *Oncogene* **2020**, *39*, 187–203. [CrossRef] [PubMed]
47. Cui, J.; Li, Q.; Luo, M.; Zhong, Z.; Zhou, S.; Jiang, L.; Shen, N.; Geng, Z.; Cheng, H.; Meng, L.; et al. Leukemia cell-derived microvesicles induce T cell exhaustion via miRNA delivery. *OncoImmunology* **2018**, *7*, e1448330. [CrossRef]
48. Castellana, D.; Zobairi, F.; Martinez, M.C.; Panaro, M.A.; Mitolo, V.; Freyssinet, J.-M.; Kunzelmann, C. Membrane Microvesicles as Actors in the Establishment of a Favorable Prostatic Tumoral Niche: A Role for Activated Fibroblasts and CX3CL1-CX3CR1 Axis. *Cancer Res.* **2009**, *69*, 785–793. [CrossRef] [PubMed]
49. Jiang, E.; Xu, Z.; Wang, M.; Yan, T.; Huang, C.; Zhou, X.; Liu, Q.; Wang, L.; Chen, Y.; Wang, H.; et al. Tumoral microvesicle–activated glycometabolic reprogramming in fibroblasts promotes the progression of oral squamous cell carcinoma. *FASEB J.* **2019**, *33*, 5690–5703. [CrossRef] [PubMed]
50. Matrix Stiffness Regulates Microvesicle-Induced Fibroblast Activation|American Journal of Physiology-Cell Physiology. Available online: https://journals.physiology.org/doi/abs/10.1152/ajpcell.00418.2018 (accessed on 26 June 2020).
51. Kawamoto, T.; Ohga, N.; Akiyama, K.; Hirata, N.; Kitahara, S.; Maishi, N.; Osawa, T.; Yamamoto, K.; Kondoh, M.; Shindoh, M.; et al. Tumor-Derived Microvesicles Induce Proangiogenic Phenotype in Endothelial Cells via Endocytosis. *PLoS ONE* **2012**, *7*, e34045. [CrossRef]
52. Giusti, I.; Delle Monache, S.; Di Francesco, M.; Sanità, P.; D'Ascenzo, S.; Gravina, G.L.; Festuccia, C.; Dolo, V. From glioblastoma to endothelial cells through extracellular vesicles: Messages for angiogenesis. *Tumour Biol. J. Int. Soc. Oncodev. Biol. Med.* **2016**, *37*, 12743–12753. [CrossRef]
53. Feng, Q.; Zhang, C.; Lum, D.; Druso, J.E.; Blank, B.; Wilson, K.F.; Welm, A.; Antonyak, M.A.; Cerione, R.A. A class of extracellular vesicles from breast cancer cells activates VEGF receptors and tumour angiogenesis. *Nat. Commun.* **2017**, *8*, 14450. [CrossRef] [PubMed]
54. Liu, Y.; Zhu, X.; Zeng, C.; Wu, P.; Wang, H.; Chen, Z.; Li, Q. Microvesicles secreted from human multiple myeloma cells promote angiogenesis. *Acta Pharmacol. Sin.* **2014**, *35*, 230–238. [CrossRef] [PubMed]
55. Pfeiler, S.; Thakur, M.; Grünauer, P.; Megens, R.T.A.; Joshi, U.; Coletti, R.; Samara, V.; Müller-Stoy, G.; Ishikawa-Ankerhold, H.; Stark, K.; et al. CD36-triggered cell invasion and persistent tissue colonization by tumor microvesicles during metastasis. *FASEB J.* **2019**, *33*, 1860–1872. [CrossRef] [PubMed]
56. Lima, L.G.; Chammas, R.; Monteiro, R.Q.; Moreira, M.E.C.; Barcinski, M.A. Tumor-derived microvesicles modulate the establishment of metastatic melanoma in a phosphatidylserine-dependent manner. *Cancer Lett.* **2009**, *283*, 168–175. [CrossRef]
57. O'Dea, K.P.; Tan, Y.Y.; Shah, S.; Patel, B.V.; Tatham, K.C.; Wilson, M.R.; Soni, S.; Takata, M. Monocytes mediate homing of circulating microvesicles to the pulmonary vasculature during low-grade systemic inflammation. *J. Extracell. Vesicles* **2020**, *9*, 1706708. [CrossRef]
58. Xiao, H.; Wong, D.T.W. Proteomic analysis of microvesicles in human saliva by gel electrophoresis with liquid chromatography-mass spectrometry. *Anal. Chim. Acta* **2012**, *723*, 61–67. [CrossRef]
59. Viñuela-Berni, V.; Doníz-Padilla, L.; Figueroa-Vega, N.; Portillo-Salazar, H.; Abud-Mendoza, C.; Baranda, L.; González-Amaro, R. Proportions of several types of plasma and urine microparticles are increased in patients with rheumatoid arthritis with active disease. *Clin. Exp. Immunol.* **2015**, *180*, 442–451. [CrossRef]

60. Verderio, C.; Muzio, L.; Turola, E.; Bergami, A.; Novellino, L.; Ruffini, F.; Riganti, L.; Corradini, I.; Francolini, M.; Garzetti, L.; et al. Myeloid microvesicles are a marker and therapeutic target for neuroinflammation. *Ann. Neurol.* **2012**, *72*, 610–624. [CrossRef]
61. Zonneveld, M.I.; Brisson, A.R.; van Herwijnen, M.J.C.; Tan, S.; van de Lest, C.H.A.; Redegeld, F.A.; Garssen, J.; Wauben, M.H.M.; Hoen, E.N.M.N.-'t. Recovery of extracellular vesicles from human breast milk is influenced by sample collection and vesicle isolation procedures. *J. Extracell. Vesicles* **2014**, *3*, 24215. [CrossRef]
62. Höög, J.L.; Lötvall, J. Diversity of extracellular vesicles in human ejaculates revealed by cryo-electron microscopy. *J. Extracell. Vesicles* **2015**, *4*, 28680. [CrossRef] [PubMed]
63. György, B.; Módos, K.; Pállinger, É.; Pálóczi, K.; Pásztói, M.; Misják, P.; Deli, M.A.; Sipos, Á.; Szalai, A.; Voszka, I.; et al. Detection and isolation of cell-derived microparticles are compromised by protein complexes resulting from shared biophysical parameters. *Blood* **2011**, *117*, e39–e48. [CrossRef] [PubMed]
64. Lyberg, T.; Nakstad, B.; Hetland, O.; Boye, N.P. Procoagulant (thromboplastin) activity in human bronchoalveolar lavage fluids is derived from alveolar macrophages. *Eur. Respir. J.* **1990**, *3*, 61–67. [PubMed]
65. Wolf, P. The nature and significance of platelet products in human plasma. *Br. J. Haematol.* **1967**, *13*, 269–288. [CrossRef]
66. Jayachandran, M.; Miller, V.M.; Heit, J.A.; Owen, W.G. Methodology for Isolation, Identification and Characterization of Microvesicles in Peripheral Blood. *J. Immunol. Methods* **2012**, *375*, 207–214. [CrossRef] [PubMed]
67. Lacroix, R.; Judicone, C.; Poncelet, P.; Robert, S.; Arnaud, L.; Sampol, J.; Dignat-George, F. Impact of pre-analytical parameters on the measurement of circulating microparticles: Towards standardization of protocol. *J. Thromb. Haemost.* **2012**, *10*, 437–446. [CrossRef]
68. Jamaly, S.; Ramberg, C.; Olsen, R.; Latysheva, N.; Webster, P.; Sovershaev, T.; Brækkan, S.K.; Hansen, J.-B. Impact of preanalytical conditions on plasma concentration and size distribution of extracellular vesicles using Nanoparticle Tracking Analysis. *Sci. Rep.* **2018**, *8*, 17216. [CrossRef]
69. Menck, K.; Bleckmann, A.; Wachter, A.; Hennies, B.; Ries, L.; Schulz, M.; Balkenhol, M.; Pukrop, T.; Schatlo, B.; Rost, U.; et al. Characterisation of tumour-derived microvesicles in cancer patients' blood and correlation with clinical outcome. *J. Extracell. Vesicles* **2017**, *6*. [CrossRef]
70. Belov, L.; Hallal, S.; Matic, K.; Zhou, J.; Wissmueller, S.; Ahmed, N.; Tanjil, S.; Mulligan, S.P.; Best, O.G.; Simpson, R.J.; et al. Surface Profiling of Extracellular Vesicles from Plasma or Ascites Fluid Using DotScan Antibody Microarrays. *Methods Mol. Biol. Clifton* **2017**, *1619*, 263–301. [CrossRef]
71. Wisgrill, L.; Lamm, C.; Hartmann, J.; Preißing, F.; Dragosits, K.; Bee, A.; Hell, L.; Thaler, J.; Ay, C.; Pabinger, I.; et al. Peripheral blood microvesicles secretion is influenced by storage time, temperature, and anticoagulants. *Cytometry A* **2016**, *89*, 663–672. [CrossRef]
72. Yuana, Y.; Böing, A.N.; Grootemaat, A.E.; van der Pol, E.; Hau, C.M.; Cizmar, P.; Buhr, E.; Sturk, A.; Nieuwland, R. Handling and storage of human body fluids for analysis of extracellular vesicles. *J. Extracell. Vesicles* **2015**, *4*, 29260. [CrossRef]
73. Brennan, K.; Martin, K.; FitzGerald, S.P.; O'Sullivan, J.; Wu, Y.; Blanco, A.; Richardson, C.; Mc Gee, M.M. A comparison of methods for the isolation and separation of extracellular vesicles from protein and lipid particles in human serum. *Sci. Rep.* **2020**, *10*, 1039. [CrossRef] [PubMed]
74. Mørk, M.; Handberg, A.; Pedersen, S.; Jørgensen, M.M.; Bæk, R.; Nielsen, M.K.; Kristensen, S.R. Prospects and limitations of antibody-mediated clearing of lipoproteins from blood plasma prior to nanoparticle tracking analysis of extracellular vesicles. *J. Extracell. Vesicles* **2017**, *6*, 1308779. [CrossRef] [PubMed]
75. van der Pol, E.; Sturk, A.; van Leeuwen, T.; Nieuwland, R.; Coumans, F. Standardization of extracellular vesicle measurements by flow cytometry through vesicle diameter approximation. *J. Thromb. Haemost.* **2018**, *16*, 1236–1245. [CrossRef] [PubMed]
76. Menck, K.; Bleckmann, A.; Schulz, M.; Ries, L.; Binder, C. Isolation and Characterization of Microvesicles from Peripheral Blood. *J. Vis. Exp. JoVE* **2017**. [CrossRef]
77. Lukacs-Kornek, V.; Julich-Haertel, H.; Urban, S.K.; Kornek, M. Multi-Surface Antigen Staining of Larger Extracellular Vesicles. *Methods Mol. Biol. Clifton* **2017**, *1660*, 201–208. [CrossRef]
78. Chandler, W.L. Measurement of microvesicle levels in human blood using flow cytometry. *Cytometry B Clin. Cytom.* **2016**, *90*, 326–336. [CrossRef]
79. Erdbrügger, U.; Lannigan, J. Analytical challenges of extracellular vesicle detection: A comparison of different techniques. *Cytometry A* **2016**, *89*, 123–134. [CrossRef]

80. Xin, X.; Zhang, P.; Fu, X.; Mao, X.; Meng, F.; Tian, M.; Zhu, X.; Sun, H.; Meng, L.; Zhou, J. Saline is a more appropriate solution for microvesicles for flow cytometric analyses. *Oncotarget* **2017**, *8*, 34576–34585. [CrossRef]
81. Belov, L.; Matic, K.J.; Hallal, S.; Best, O.G.; Mulligan, S.P.; Christopherson, R.I. Extensive surface protein profiles of extracellular vesicles from cancer cells may provide diagnostic signatures from blood samples. *J. Extracell. Vesicles* **2016**, *5*. [CrossRef]
82. Witwer, K.W.; Buzás, E.I.; Bemis, L.T.; Bora, A.; Lässer, C.; Lötvall, J.; Hoen, E.N.N.-'t.; Piper, M.G.; Sivaraman, S.; Skog, J.; et al. Standardization of sample collection, isolation and analysis methods in extracellular vesicle research. *J. Extracell. Vesicles* **2013**, *2*, 20360. [CrossRef] [PubMed]
83. Clayton, A.; Boilard, E.; Buzas, E.I.; Cheng, L.; Falcón-Perez, J.M.; Gardiner, C.; Gustafson, D.; Gualerzi, A.; Hendrix, A.; Hoffman, A.; et al. Considerations towards a roadmap for collection, handling and storage of blood extracellular vesicles. *J. Extracell. Vesicles* **2019**, *8*, 1647027. [CrossRef] [PubMed]
84. Baran, J.; Baj-Krzyworzeka, M.; Weglarczyk, K.; Szatanek, R.; Zembala, M.; Barbasz, J.; Czupryna, A.; Szczepanik, A.; Zembala, M. Circulating tumour-derived microvesicles in plasma of gastric cancer patients. *Cancer Immunol. Immunother. CII* **2010**, *59*, 841–850. [CrossRef] [PubMed]
85. Fleitas, T.; Martínez-Sales, V.; Vila, V.; Reganon, E.; Mesado, D.; Martín, M.; Gómez-Codina, J.; Montalar, J.; Reynés, G. Circulating Endothelial Cells and Microparticles as Prognostic Markers in Advanced Non-Small Cell Lung Cancer. *PLoS ONE* **2012**, *7*. [CrossRef] [PubMed]
86. Tesselaar, M.E.T.; Romijn, F.P.H.T.M.; Linden, I.K.V.D.; Prins, F.A.; Bertina, R.M.; Osanto, S. Microparticle-associated tissue factor activity: A link between cancer and thrombosis? *J. Thromb. Haemost.* **2007**, *5*, 520–527. [CrossRef]
87. Galindo-Hernandez, O.; Villegas-Comonfort, S.; Candanedo, F.; González-Vázquez, M.-C.; Chavez-Ocaña, S.; Jimenez-Villanueva, X.; Sierra-Martinez, M.; Salazar, E.P. Elevated concentration of microvesicles isolated from peripheral blood in breast cancer patients. *Arch. Med. Res.* **2013**, *44*, 208–214. [CrossRef]
88. Zhao, L.; Bi, Y.; Kou, J.; Shi, J.; Piao, D. Phosphatidylserine exposing-platelets and microparticles promote procoagulant activity in colon cancer patients. *J. Exp. Clin. Cancer Res.* **2016**, *35*. [CrossRef]
89. Krishnan, S.R.; Luk, F.; Brown, R.D.; Suen, H.; Kwan, Y.; Bebawy, M. Isolation of Human CD138+ Microparticles from the Plasma of Patients with Multiple Myeloma. *Neoplasia* **2016**, *18*, 25–32. [CrossRef]
90. De Luca, L.; D'Arena, G.; Simeon, V.; Trino, S.; Laurenzana, I.; Caivano, A.; La Rocca, F.; Villani, O.; Mansueto, G.; Deaglio, S.; et al. Characterization and prognostic relevance of circulating microvesicles in chronic lymphocytic leukemia. *Leuk. Lymphoma* **2017**, *58*, 1424–1432. [CrossRef]
91. Caivano, A.; Laurenzana, I.; De Luca, L.; La Rocca, F.; Simeon, V.; Trino, S.; D'Auria, F.; Traficante, A.; Maietti, M.; Izzo, T.; et al. High serum levels of extracellular vesicles expressing malignancy-related markers are released in patients with various types of hematological neoplastic disorders. *Tumour Biol. J. Int. Soc. Oncodev. Biol. Med.* **2015**, *36*, 9739–9752. [CrossRef]
92. Wang, W.; Li, H.; Zhou, Y.; Jie, S. Peripheral blood microvesicles are potential biomarkers for hepatocellular carcinoma. *Cancer Biomark. Sect. Dis. Markers* **2013**, *13*, 351–357. [CrossRef]
93. Ender, F.; Freund, A.; Quecke, T.; Steidel, C.; Zamzow, P.; von Bubnoff, N.; Gieseler, F. Tissue factor activity on microvesicles from cancer patients. *J. Cancer Res. Clin. Oncol.* **2020**, *146*, 467–475. [CrossRef] [PubMed]
94. Fraser, K.; Jo, A.; Giedt, J.; Vinegoni, C.; Yang, K.S.; Peruzzi, P.; Chiocca, E.A.; Breakefield, X.O.; Lee, H.; Weissleder, R. Characterization of single microvesicles in plasma from glioblastoma patients. *Neuro Oncol.* **2019**, *21*, 606–615. [CrossRef] [PubMed]
95. Mege, D.; Panicot-Dubois, L.; Ouaissi, M.; Robert, S.; Sielezneff, I.; Sastre, B.; Dignat-George, F.; Dubois, C. The origin and concentration of circulating microparticles differ according to cancer type and evolution: A prospective single-center study. *Int. J. Cancer* **2016**, *138*, 939–948. [CrossRef] [PubMed]
96. Alique, M.; Ruíz-Torres, M.P.; Bodega, G.; Noci, M.V.; Troyano, N.; Bohórquez, L.; Luna, C.; Luque, R.; Carmona, A.; Carracedo, J.; et al. Microvesicles from the plasma of elderly subjects and from senescent endothelial cells promote vascular calcification. *Aging* **2017**, *9*, 778–789. [CrossRef]
97. Evans, S.M.; Putt, M.; Yang, X.-Y.; Lustig, R.A.; Martinez-Lage, M.; Williams, D.; Desai, A.; Wolf, R.; Brem, S.; Koch, C.J. Initial evidence that blood-borne microvesicles are biomarkers for recurrence and survival in newly diagnosed glioblastoma patients. *J. Neurooncol.* **2016**, *127*, 391–400. [CrossRef]

98. Coumans, F.a.W.; Doggen, C.J.M.; Attard, G.; de Bono, J.S.; Terstappen, L.W.M.M. All circulating EpCAM+CK+CD45- objects predict overall survival in castration-resistant prostate cancer. *Ann. Oncol.* **2010**, *21*, 1851–1857. [CrossRef]
99. Nanou, A.; Coumans, F.A.W.; van Dalum, G.; Zeune, L.L.; Dolling, D.; Onstenk, W.; Crespo, M.; Fontes, M.S.; Rescigno, P.; Fowler, G.; et al. Circulating tumor cells, tumor-derived extracellular vesicles and plasma cytokeratins in castration-resistant prostate cancer patients. *Oncotarget* **2018**, *9*, 19283–19293. [CrossRef]
100. Klein, C.A. Selection and adaptation during metastatic cancer progression. *Nature* **2013**, *501*, 365–372. [CrossRef]
101. García-Romero, N.; Carrión-Navarro, J.; Esteban-Rubio, S.; Lázaro-Ibáñez, E.; Peris-Celda, M.; Alonso, M.M.; Guzmán-De-Villoria, J.; Fernández-Carballal, C.; de Mendivil, A.O.; García-Duque, S.; et al. DNA sequences within glioma-derived extracellular vesicles can cross the intact blood-brain barrier and be detected in peripheral blood of patients. *Oncotarget* **2016**, *8*, 1416–1428. [CrossRef]
102. Ramón, Y.; Cajal, S.; Sesé, M.; Capdevila, C.; Aasen, T.; De Mattos-Arruda, L.; Diaz-Cano, S.J.; Hernández-Losa, J.; Castellví, J. Clinical implications of intratumor heterogeneity: Challenges and opportunities. *J. Mol. Med.* **2020**, *98*, 161–177. [CrossRef] [PubMed]
103. Ellsworth, R.E.; Blackburn, H.L.; Shriver, C.D.; Soon-Shiong, P.; Ellsworth, D.L. Molecular heterogeneity in breast cancer: State of the science and implications for patient care. *Semin. Cell Dev. Biol.* **2017**, *64*, 65–72. [CrossRef] [PubMed]
104. Ng, C.K.Y.; Bidard, F.-C.; Piscuoglio, S.; Geyer, F.C.; Lim, R.S.; de Bruijn, I.; Shen, R.; Pareja, F.; Berman, S.H.; Wang, L.; et al. Genetic Heterogeneity in Therapy-Naïve Synchronous Primary Breast Cancers and Their Metastases. *Clin. Cancer Res. Off. J. Am. Assoc. Cancer Res.* **2017**, *23*, 4402–4415. [CrossRef] [PubMed]
105. Skog, J.; Wurdinger, T.; van Rijn, S.; Meijer, D.; Gainche, L.; Sena-Esteves, M.; Curry, W.T.; Carter, R.S.; Krichevsky, A.M.; Breakefield, X.O. Glioblastoma microvesicles transport RNA and protein that promote tumor growth and provide diagnostic biomarkers. *Nat. Cell Biol.* **2008**, *10*, 1470–1476. [CrossRef]
106. Wei, Z.; Batagov, A.O.; Schinelli, S.; Wang, J.; Wang, Y.; El Fatimy, R.; Rabinovsky, R.; Balaj, L.; Chen, C.C.; Hochberg, F.; et al. Coding and noncoding landscape of extracellular RNA released by human glioma stem cells. *Nat. Commun.* **2017**, *8*. [CrossRef] [PubMed]
107. Zhu, X.; You, Y.; Li, Q.; Zeng, C.; Fu, F.; Guo, A.; Zhang, H.; Zou, P.; Zhong, Z.; Wang, H.; et al. BCR-ABL1-positive microvesicles transform normal hematopoietic transplants through genomic instability: Implications for donor cell leukemia. *Leukemia* **2014**, *28*, 1666–1675. [CrossRef]
108. Milani, G.; Lana, T.; Bresolin, S.; Aveic, S.; Pastò, A.; Frasson, C.; Kronnie, G. te Expression Profiling of Circulating Microvesicles Reveals Intercellular Transmission of Oncogenic Pathways. *Mol. Cancer Res.* **2017**, *15*, 683–695. [CrossRef]
109. Krug, A.K.; Enderle, D.; Karlovich, C.; Priewasser, T.; Bentink, S.; Spiel, A.; Brinkmann, K.; Emenegger, J.; Grimm, D.G.; Castellanos-Rizaldos, E.; et al. Improved EGFR mutation detection using combined exosomal RNA and circulating tumor DNA in NSCLC patient plasma. *Ann. Oncol.* **2018**, *29*, 700–706. [CrossRef]
110. Akers, J.C.; Ramakrishnan, V.; Kim, R.; Phillips, S.; Kaimal, V.; Mao, Y.; Hua, W.; Yang, I.; Fu, C.-C.; Nolan, J.; et al. miRNA contents of cerebrospinal fluid extracellular vesicles in glioblastoma patients. *J. Neurooncol.* **2015**, *123*, 205–216. [CrossRef]
111. Chen, M.; Xu, R.; Rai, A.; Suwakulsiri, W.; Izumikawa, K.; Ishikawa, H.; Greening, D.W.; Takahashi, N.; Simpson, R.J. Distinct shed microvesicle and exosome microRNA signatures reveal diagnostic markers for colorectal cancer. *PLoS ONE* **2019**, *14*. [CrossRef]
112. Chevillet, J.R.; Kang, Q.; Ruf, I.K.; Briggs, H.A.; Vojtech, L.N.; Hughes, S.M.; Cheng, H.H.; Arroyo, J.D.; Meredith, E.K.; Gallichotte, E.N.; et al. Quantitative and stoichiometric analysis of the microRNA content of exosomes. *Proc. Natl. Acad. Sci. USA* **2014**, *111*, 14888–14893. [CrossRef] [PubMed]
113. Rosa-Fernandes, L.; Rocha, V.B.; Carregari, V.C.; Urbani, A.; Palmisano, G. A Perspective on Extracellular Vesicles Proteomics. *Front. Chem.* **2017**, *5*. [CrossRef]
114. Whiteaker, J.R.; Lin, C.; Kennedy, J.; Hou, L.; Trute, M.; Sokal, I.; Yan, P.; Schoenherr, R.M.; Zhao, L.; Voytovich, U.J.; et al. A targeted proteomics-based pipeline for verification of biomarkers in plasma. *Nat. Biotechnol.* **2011**, *29*, 625–634. [CrossRef]
115. Fujita, K.; Kume, H.; Matsuzaki, K.; Kawashima, A.; Ujike, T.; Nagahara, A.; Uemura, M.; Miyagawa, Y.; Tomonaga, T.; Nonomura, N. Proteomic analysis of urinary extracellular vesicles from high Gleason score prostate cancer. *Sci. Rep.* **2017**, *7*, 42961. [CrossRef]

116. Crompot, E.; Van Damme, M.; Duvillier, H.; Pieters, K.; Vermeesch, M.; Perez-Morga, D.; Meuleman, N.; Mineur, P.; Bron, D.; Lagneaux, L.; et al. Avoiding False Positive Antigen Detection by Flow Cytometry on Blood Cell Derived Microparticles: The Importance of an Appropriate Negative Control. *PLoS ONE* **2015**, *10*. [CrossRef]
117. Stec, M.; Baj-Krzyworzeka, M.; Baran, J.; Węglarczyk, K.; Zembala, M.; Barbasz, J.; Szczepanik, A.; Zembala, M. Isolation and characterization of circulating micro(nano)vesicles in the plasma of colorectal cancer patients and their interactions with tumor cells. *Oncol. Rep.* **2015**, *34*, 2768–2775. [CrossRef] [PubMed]
118. Julich-Haertel, H.; Urban, S.K.; Krawczyk, M.; Willms, A.; Jankowski, K.; Patkowski, W.; Kruk, B.; Krasnodębski, M.; Ligocka, J.; Schwab, R.; et al. Cancer-associated circulating large extracellular vesicles in cholangiocarcinoma and hepatocellular carcinoma. *J. Hepatol.* **2017**, *67*, 282–292. [CrossRef] [PubMed]
119. Willms, A.; Müller, C.; Julich, H.; Klein, N.; Schwab, R.; Güsgen, C.; Richardsen, I.; Schaaf, S.; Krawczyk, M.; Krawczyk, M.; et al. Tumour-associated circulating microparticles: A novel liquid biopsy tool for screening and therapy monitoring of colorectal carcinoma and other epithelial neoplasia. *Oncotarget* **2016**, *7*, 30867–30875. [CrossRef] [PubMed]
120. Morello, M.; Minciacchi, V.R.; de Candia, P.; Yang, J.; Posadas, E.; Kim, H.; Griffiths, D.; Bhowmick, N.; Chung, L.W.; Gandellini, P.; et al. Large oncosomes mediate intercellular transfer of functional microRNA. *Cell Cycle* **2013**, *12*, 3526–3536. [CrossRef]
121. Abbate, V.; Marcantoni, M.; Giuliante, F.; Vecchio, F.; Gatto, I.; Mele, C.; Saviano, A.; Arciuolo, D.; Gaetani, E.; Ferrari, M.; et al. HepPar1-Positive Circulating Microparticles Are Increased in Subjects with Hepatocellular Carcinoma and Predict Early Recurrence after Liver Resection. *Int. J. Mol. Sci.* **2017**, *18*, 1043. [CrossRef]
122. Jang, S.C.; Crescitelli, R.; Cvjetkovic, A.; Belgrano, V.; Olofsson Bagge, R.; Sundfeldt, K.; Ochiya, T.; Kalluri, R.; Lötvall, J. Mitochondrial protein enriched extracellular vesicles discovered in human melanoma tissues can be detected in patient plasma. *J. Extracell. Vesicles* **2019**, *8*. [CrossRef] [PubMed]
123. Crescitelli, R.; Lässer, C.; Jang, S.C.; Cvjetkovic, A.; Malmhäll, C.; Karimi, N.; Höög, J.L.; Johansson, I.; Fuchs, J.; Thorsell, A.; et al. Subpopulations of extracellular vesicles from human metastatic melanoma tissue identified by quantitative proteomics after optimized isolation. *J. Extracell. Vesicles* **2020**, *9*. [CrossRef]
124. Imjeti, N.S.; Menck, K.; Egea-Jimenez, A.L.; Lecointre, C.; Lembo, F.; Bouguenina, H.; Badache, A.; Ghossoub, R.; David, G.; Roche, S.; et al. Syntenin mediates SRC function in exosomal cell-to-cell communication. *Proc. Natl. Acad. Sci. USA* **2017**, *114*, 12495–12500. [CrossRef] [PubMed]
125. Amorim, M.; Fernandes, G.; Oliveira, P.; Martins-de-Souza, D.; Dias-Neto, E.; Nunes, D. The overexpression of a single oncogene (ERBB2/HER2) alters the proteomic landscape of extracellular vesicles. *Proteomics* **2014**, *14*, 1472–1479. [CrossRef] [PubMed]
126. Rontogianni, S.; Synadaki, E.; Li, B.; Liefaard, M.C.; Lips, E.H.; Wesseling, J.; Wu, W.; Altelaar, M. Proteomic profiling of extracellular vesicles allows for human breast cancer subtyping. *Commun. Biol.* **2019**, *2*. [CrossRef] [PubMed]
127. Ayers, L.; Nieuwland, R.; Kohler, M.; Kraenkel, N.; Ferry, B.; Leeson, P. Dynamic microvesicle release and clearance within the cardiovascular system: Triggers and mechanisms. *Clin. Sci. Lond. Engl. 1979* **2015**, *129*, 915–931. [CrossRef]
128. Kim, H.K.; Song, K.S.; Park, Y.S.; Kang, Y.H.; Lee, Y.J.; Lee, K.R.; Kim, H.K.; Ryu, K.W.; Bae, J.M.; Kim, S. Elevated levels of circulating platelet microparticles, VEGF, IL-6 and RANTES in patients with gastric cancer: Possible role of a metastasis predictor. *Eur. J. Cancer* **2003**, *39*, 184–191. [CrossRef]
129. Chen, I.-H.; Xue, L.; Hsu, C.-C.; Paez, J.S.P.; Pan, L.; Andaluz, H.; Wendt, M.K.; Iliuk, A.B.; Zhu, J.-K.; Tao, W.A. Phosphoproteins in extracellular vesicles as candidate markers for breast cancer. *Proc. Natl. Acad. Sci. USA* **2017**, *114*, 3175–3180. [CrossRef]
130. Gamperl, H.; Plattfaut, C.; Freund, A.; Quecke, T.; Theophil, F.; Gieseler, F. Extracellular vesicles from malignant effusions induce tumor cell migration: Inhibitory effect of LMWH tinzaparin. *Cell Biol. Int.* **2016**, *40*, 1050–1061. [CrossRef]
131. Zwicker, J.I.; Liebman, H.A.; Neuberg, D.; Lacroix, R.; Bauer, K.A.; Furie, B.C.; Furie, B. Tumor-Derived Tissue Factor-Bearing Microparticles are Associated with Venous Thromboembolic Events in Malignancy. *Clin. Cancer Res. Off. J. Am. Assoc. Cancer Res.* **2009**, *15*, 6830–6840. [CrossRef]
132. Tripisciano, C.; Weiss, R.; Eichhorn, T.; Spittler, A.; Heuser, T.; Fischer, M.B.; Weber, V. Different Potential of Extracellular Vesicles to Support Thrombin Generation: Contributions of Phosphatidylserine, Tissue Factor, and Cellular Origin. *Sci. Rep.* **2017**, *7*. [CrossRef] [PubMed]

133. Tang, K.; Zhang, Y.; Zhang, H.; Xu, P.; Liu, J.; Ma, J.; Lv, M.; Li, D.; Katirai, F.; Shen, G.-X.; et al. Delivery of chemotherapeutic drugs in tumour cell-derived microparticles. *Nat. Commun.* **2012**, *3*, 1282. [CrossRef] [PubMed]
134. König, L.; Kasimir-Bauer, S.; Bittner, A.-K.; Hoffmann, O.; Wagner, B.; Santos Manvailer, L.F.; Kimmig, R.; Horn, P.A.; Rebmann, V. Elevated levels of extracellular vesicles are associated with therapy failure and disease progression in breast cancer patients undergoing neoadjuvant chemotherapy. *Oncoimmunology* **2017**, *7*. [CrossRef] [PubMed]
135. García Garre, E.; Luengo Gil, G.; Montoro García, S.; Gonzalez Billalabeitia, E.; Zafra Poves, M.; García Martinez, E.; Roldán Schilling, V.; Navarro Manzano, E.; Ivars Rubio, A.; Lip, G.Y.H.; et al. Circulating small-sized endothelial microparticles as predictors of clinical outcome after chemotherapy for breast cancer: An exploratory analysis. *Breast Cancer Res. Treat.* **2018**, *169*, 83–92. [CrossRef]
136. Koch, C.J.; Lustig, R.A.; Yang, X.-Y.; Jenkins, W.T.; Wolf, R.L.; Martinez-Lage, M.; Desai, A.; Williams, D.; Evans, S.M. Microvesicles as a Biomarker for Tumor Progression versus Treatment Effect in Radiation/Temozolomide-Treated Glioblastoma Patients. *Transl. Oncol.* **2014**, *7*, 752–758. [CrossRef]
137. Bebawy, M.; Combes, V.; Lee, E.; Jaiswal, R.; Gong, J.; Bonhoure, A.; Grau, G.E.R. Membrane microparticles mediate transfer of P-glycoprotein to drug sensitive cancer cells. *Leukemia* **2009**, *23*, 1643–1649. [CrossRef]
138. Lu, J.F.; Luk, F.; Gong, J.; Jaiswal, R.; Grau, G.E.R.; Bebawy, M. Microparticles mediate MRP1 intercellular transfer and the re-templating of intrinsic resistance pathways. *Pharmacol. Res.* **2013**, *76*, 77–83. [CrossRef]
139. Jorfi, S.; Ansa-Addo, E.A.; Kholia, S.; Stratton, D.; Valley, S.; Lange, S.; Inal, J. Inhibition of microvesiculation sensitizes prostate cancer cells to chemotherapy and reduces docetaxel dose required to limit tumor growth in vivo. *Sci. Rep.* **2015**, *5*. [CrossRef]
140. Muralidharan-Chari, V.; Kohan, H.G.; Asimakopoulos, A.G.; Sudha, T.; Sell, S.; Kannan, K.; Boroujerdi, M.; Davis, P.J.; Mousa, S.A. Microvesicle removal of anticancer drugs contributes to drug resistance in human pancreatic cancer cells. *Oncotarget* **2016**, *7*, 50365–50379. [CrossRef]
141. Kosgodage, U.S.; Trindade, R.P.; Thompson, P.R.; Inal, J.M.; Lange, S. Chloramidine/Bisindolylmaleimide-I-Mediated Inhibition of Exosome and Microvesicle Release and Enhanced Efficacy of Cancer Chemotherapy. *Int. J. Mol. Sci.* **2017**, *18*, 1007. [CrossRef]
142. Aung, T.; Chapuy, B.; Vogel, D.; Wenzel, D.; Oppermann, M.; Lahmann, M.; Weinhage, T.; Menck, K.; Hupfeld, T.; Koch, R.; et al. Exosomal evasion of humoral immunotherapy in aggressive B-cell lymphoma modulated by ATP-binding cassette transporter A3. *Proc. Natl. Acad. Sci. USA* **2011**, *108*, 15336–15341. [CrossRef] [PubMed]
143. Ciravolo, V.; Huber, V.; Ghedini, G.C.; Venturelli, E.; Bianchi, F.; Campiglio, M.; Morelli, D.; Villa, A.; Della Mina, P.; Menard, S.; et al. Potential role of HER2-overexpressing exosomes in countering trastuzumab-based therapy. *J. Cell. Physiol.* **2012**, *227*, 658–667. [CrossRef] [PubMed]
144. Chen, G.; Huang, A.C.; Zhang, W.; Zhang, G.; Wu, M.; Xu, W.; Yu, Z.; Yang, J.; Wang, B.; Sun, H.; et al. Exosomal PD-L1 Contributes to Immunosuppression and is Associated with anti-PD-1 Response. *Nature* **2018**, *560*, 382–386. [CrossRef] [PubMed]
145. Sansone, P.; Berishaj, M.; Rajasekhar, V.K.; Ceccarelli, C.; Chang, Q.; Strillacci, A.; Savini, C.; Shapiro, L.; Bowman, R.L.; Mastroleo, C.; et al. Evolution of cancer stem-like cells in endocrine-resistant metastatic breast cancers is mediated by stromal microvesicles. *Cancer Res.* **2017**, *77*, 1927–1941. [CrossRef]
146. Keklikoglou, I.; Cianciaruso, C.; Güç, E.; Squadrito, M.L.; Spring, L.M.; Tazzyman, S.; Lambein, L.; Poissonnier, A.; Ferraro, G.B.; Baer, C.; et al. Chemotherapy elicits pro-metastatic extracellular vesicles in breast cancer models. *Nat. Cell Biol.* **2019**, *21*, 190–202. [CrossRef]
147. Peinado, H.; Alečković, M.; Lavotshkin, S.; Matei, I.; Costa-Silva, B.; Moreno-Bueno, G.; Hergueta-Redondo, M.; Williams, C.; García-Santos, G.; Ghajar, C.; et al. Melanoma exosomes educate bone marrow progenitor cells toward a pro-metastatic phenotype through MET. *Nat. Med.* **2012**, *18*, 883–891. [CrossRef]
148. Zhang, H.; Tang, K.; Zhang, Y.; Ma, R.; Ma, J.; Li, Y.; Luo, S.; Liang, X.; Ji, T.; Gu, Z.; et al. Cell-free Tumor Microparticle Vaccines Stimulate Dendritic Cells via cGAS/STING Signaling. *Cancer Immunol. Res.* **2015**, *3*, 196–205. [CrossRef]
149. Ma, J.; Wei, K.; Zhang, H.; Tang, K.; Li, F.; Zhang, T.; Liu, J.; Xu, P.; Yu, Y.; Sun, W.; et al. Mechanisms by Which Dendritic Cells Present Tumor Microparticle Antigens to CD8+ T Cells. *Cancer Immunol. Res.* **2018**, *6*, 1057–1068. [CrossRef]

150. Pineda, B.; García, F.J.S.; Olascoaga, N.K.; de la Cruz, V.P.; Salazar, A.; Moreno-Jiménez, S.; Pedro, N.H.; Márquez-Navarro, A.; Plata, A.O.; Sotelo, J. Malignant Glioma Therapy by Vaccination with Irradiated C6 Cell-Derived Microvesicles Promotes an Antitumoral Immune Response. *Mol. Ther.* **2019**, *27*, 1612–1620. [CrossRef]
151. Dong, W.; Zhang, H.; Yin, X.; Liu, Y.; Chen, D.; Liang, X.; Jin, X.; Lv, J.; Ma, J.; Tang, K.; et al. Oral delivery of tumor microparticle vaccines activates NOD2 signaling pathway in ileac epithelium rendering potent antitumor T cell immunity. *Oncoimmunology* **2017**, *6*. [CrossRef]
152. Wu, T.; Tang, M. Review of the effects of manufactured nanoparticles on mammalian target organs. *J. Appl. Toxicol. JAT* **2018**, *38*, 25–40. [CrossRef] [PubMed]
153. Armstrong, J.P.; Holme, M.N.; Stevens, M.M. Re-Engineering Extracellular Vesicles as Smart Nanoscale Therapeutics. *ACS Nano* **2017**, *11*, 69–83. [CrossRef] [PubMed]
154. Sil, S.; Dagur, R.S.; Liao, K.; Peeples, E.S.; Hu, G.; Periyasamy, P.; Buch, S. Strategies for the use of Extracellular Vesicles for the Delivery of Therapeutics. *J. Neuroimmune Pharmacol.* **2019**. [CrossRef] [PubMed]
155. Ma, J.; Zhang, Y.; Tang, K.; Zhang, H.; Yin, X.; Li, Y.; Xu, P.; Sun, Y.; Ma, R.; Ji, T.; et al. Reversing drug resistance of soft tumor-repopulating cells by tumor cell-derived chemotherapeutic microparticles. *Cell Res.* **2016**, *26*, 713–727. [CrossRef] [PubMed]
156. Lener, T.; Gimona, M.; Aigner, L.; Börger, V.; Buzas, E.; Camussi, G.; Chaput, N.; Chatterjee, D.; Court, F.A.; del Portillo, H.A.; et al. Applying extracellular vesicles based therapeutics in clinical trials—An ISEV position paper. *J. Extracell. Vesicles* **2015**, *4*. [CrossRef]
157. Usman, W.M.; Pham, T.C.; Kwok, Y.Y.; Vu, L.T.; Ma, V.; Peng, B.; Chan, Y.S.; Wei, L.; Chin, S.M.; Azad, A.; et al. Efficient RNA drug delivery using red blood cell extracellular vesicles. *Nat. Commun.* **2018**, *9*, 2359. [CrossRef]
158. Wiklander, O.P.B.; Nordin, J.Z.; O'Loughlin, A.; Gustafsson, Y.; Corso, G.; Mäger, I.; Vader, P.; Lee, Y.; Sork, H.; Seow, Y.; et al. Extracellular vesicle in vivo biodistribution is determined by cell source, route of administration and targeting. *J. Extracell. Vesicles* **2015**, *4*. [CrossRef]
159. Gangadaran, P.; Li, X.J.; Lee, H.W.; Oh, J.M.; Kalimuthu, S.; Rajendran, R.L.; Son, S.H.; Baek, S.H.; Singh, T.D.; Zhu, L.; et al. A new bioluminescent reporter system to study the biodistribution of systematically injected tumor-derived bioluminescent extracellular vesicles in mice. *Oncotarget* **2017**, *8*, 109894–109914. [CrossRef]
160. Murphy, D.E.; de Jong, O.G.; Brouwer, M.; Wood, M.J.; Lavieu, G.; Schiffelers, R.M.; Vader, P. Extracellular vesicle-based therapeutics: Natural versus engineered targeting and trafficking. *Exp. Mol. Med.* **2019**, *51*, 1–12. [CrossRef]
161. Tanaka, Y.; Schroit, A.J. Insertion of fluorescent phosphatidylserine into the plasma membrane of red blood cells. Recognition by autologous macrophages. *J. Biol. Chem.* **1983**, *258*, 11335–11343.
162. Saari, H.; Lisitsyna, E.; Rautaniemi, K.; Rojalin, T.; Niemi, L.; Nivaro, O.; Laaksonen, T.; Yliperttula, M.; Vuorimaa-Laukkanen, E. FLIM reveals alternative EV-mediated cellular up-take pathways of paclitaxel. *J. Control. Release* **2018**, *284*, 133–143. [CrossRef] [PubMed]
163. Liang, Q.; Bie, N.; Yong, T.; Tang, K.; Shi, X.; Wei, Z.; Jia, H.; Zhang, X.; Zhao, H.; Huang, W.; et al. The softness of tumour-cell-derived microparticles regulates their drug-delivery efficiency. *Nat. Biomed. Eng.* **2019**, *3*, 729–740. [CrossRef] [PubMed]
164. Thomas, G.M.; Panicot-Dubois, L.; Lacroix, R.; Dignat-George, F.; Lombardo, D.; Dubois, C. Cancer cell-derived microparticles bearing P-selectin glycoprotein ligand 1 accelerate thrombus formation in vivo. *J. Exp. Med.* **2009**, *206*, 1913–1927. [CrossRef] [PubMed]
165. Kordelas, L.; Rebmann, V.; Ludwig, A.-K.; Radtke, S.; Ruesing, J.; Doeppner, T.R.; Epple, M.; Horn, P.A.; Beelen, D.W.; Giebel, B. MSC-derived exosomes: A novel tool to treat therapy-refractory graft-versus-host disease. *Leukemia* **2014**, *28*, 970–973. [CrossRef] [PubMed]

9

Suppression of Ovarian Cancer Cell Growth by AT-MSC Microvesicles

Agnieszka Szyposzynska, Aleksandra Bielawska-Pohl, Agnieszka Krawczenko, Olga Doszyn, Maria Paprocka and Aleksandra Klimczak *

Laboratory of Biology of Stem and Neoplastic Cells, Hirszfeld Institute of Immunology and Experimental Therapy, Polish Academy of Sciences, R. Weigla 12, 53-114 Wroclaw, Poland; agnieszka.szyposzynska@hirszfeld.pl (A.S.); aleksandra.bielawska-pohl@hirszfeld.pl (A.B.-P.); agnieszka.krawczenko@hirszfeld.pl (A.K.); olga.doszyn@hirszfeld.pl (O.D.); maria.paprocka@hirszfeld.pl (M.P.)

* Correspondence: aleksandra.klimczak@hirszfeld.pl;

Abstract: Transport of bioactive cargo of microvesicles (MVs) into target cells can affect their fate and behavior and change their microenvironment. We assessed the effect of MVs derived from human immortalized mesenchymal stem cells of adipose tissue-origin (HATMSC2-MVs) on the biological activity of the ovarian cancer cell lines ES-2 (clear cell carcinoma) and OAW-42 (cystadenocarcinoma). The HATMSC2-MVs were characterized using dynamic light scattering (DLS), transmission electron microscopy, and flow cytometry. The anti-tumor properties of HATMSC2-MVs were assessed using MTT for metabolic activity and flow cytometry for cell survival, cell cycle progression, and phenotype. The secretion profile of ovarian cancer cells was evaluated with a protein antibody array. Both cell lines internalized HATMSC2-MVs, which was associated with a decreased metabolic activity of cancer cells. HATMSC2-MVs exerted a pro-apoptotic and/or necrotic effect on ES-2 and OAW-42 cells and increased the expression of anti-tumor factors in both cell lines compared to control. In conclusion, we confirmed an effective transfer of HATMSC2-MVs into ovarian cancer cells that resulted in the inhibition of cell proliferation via different pathways, apoptosis and/or necrosis, which, with high likelihood, is related to the presence of different anti-tumor factors secreted by the ES-2 and OAW-42 cells.

Keywords: ovarian cancer cells; ES-2; OAW-42; microvesicles; adipose tissue origin mesenchymal stem cells

---

1. Introduction

Today, ovarian cancer is one of the most dangerous types of cancer in women. This is associated with a lack of screening tests and late diagnosis. Moreover, the disease has no symptoms in the early stages. Currently, various ovarian cancer therapies are used depending on the histological type of ovarian cancer, its stage, and the patient's predisposition. Standard treatment is a surgery combined with platinum-based chemotherapy [1]. Clinical trials focus primarily on an anti-angiogenic strategy [Vascular endothelial growth factor (VEGF) inhibition] or on modulating the immune system [2]. An extremely important field in oncology is research focused on cancer stem cells (CSCs). Cancer stem cells constitute a small population of tumor cells and play an important role in metastasis. Moreover, these cells are resistant to widely used drugs, which often leads to tumor recurrence [3]. Thus, a search for effective factors is needed that inhibit the biological activity of CSCs.

Mesenchymal stem/stromal cells (MSCs) are multipotent cells that reside in the majority of human tissues and organs, and in steady-state conditions, are responsible for the maintenance of tissue homeostasis [4,5]. Cells with MSC characteristics can be isolated from various source tissues, such as bone marrow, adipose tissue, dental pulp, skin, skeletal muscle, or perinatal tissues, including the umbilical cord, cord blood, Warton's jelly, and amniotic fluid. The tissue source of MSCs affects their cellular phenotype and biological properties [6]. MSCs and their derivates are a promising tool in clinical applications thanks to their high proliferative potential, longevity, and immunomodulatory properties [7,8].

Extracellular vesicles (EVs), such as exosomes and microvesicles (MVs), play an important role as mediators of cell-to-cell communication [9]. EVs are released by all normal, apoptotic, and neoplastic cells [10]. The transport of bioactive cargo, such as proteins, lipids, or nucleic acids, into the recipient cells may affect their phenotype and biological activity [11].

The tumor microenvironment consist not only of tumor cells, but also fibroblasts, smooth muscle cells, immune cells, endothelial cells, and mesenchymal stem cells [12]. Cell-to-cell communication in tumor niches takes place through direct contact between the surrounding cells and gap junctions or through the paracrine activity of the cells (e.g., the release of soluble factors or EVs).

The effect of EVs derived from MSCs of different tissue origin on cancer cells is not well understood. Different studies have confirmed the pro-tumorigenic [13] or anti-tumorigenic activity [14] of EVs derived from MSCs on ovarian cancer cells. This effect depends on the origin of the MSCs, methods of EV isolation, and tumor type [15].

The aim of this study was to examine the effect of MVs derived from immortalized human MSCs of adipose tissue origin (HATMSC2-MVs) on the biological activity of two ovarian cancer cell lines: ES-2, representing poorly differentiated ovarian clear cell carcinoma, and OAW-42, representing ovarian cystadenocarcinoma, with different genetic backgrounds and therapeutic responses. These two cell lines were characterized according to their phenotype and the secretion profile of cytokines and trophic factors released in response to MV treatment. Moreover, we investigated the proliferation and cell death processes/pathways (apoptosis and necrosis) of ovarian cancer cells in the presence of different ratios of HATMSC2-MVs and target cells.

2. Results

2.1. *Characterization of HATMSC2-Derived MVs*

Size distribution of MVs was analyzed using dynamic light scattering (DLS). In the histogram, a single distinct peak characteristic for MVs was observed, confirming the presence of a homogenous population of MVs. The average size of MVs, assessed using DLS, was 456 nm (Figure 1a). The size of individual MVs was confirmed using transmission electron microscopy (TEM) imaging (Figure 1b).

Isolation efficiency, using the flow cytometry method, revealed that the average number of MVs was 172×10^6 MVs/mL (Figure 1c). A Bradford assay was performed to estimate the protein concentration within the MVs, and the average protein concentration was assessed at 169.8 μg/mL.

2.2. *Surface Marker Analysis of HATMSC2 Cells and HATMSC2-MVs*

The HATMSC2 cells and HATMSC2-MVs were tested for the presence of MSC markers CD73, CD90, CD105, the HLA ABC and HLA DR antigens, and the leukocyte marker CD45. The analysis confirmed the presence of CD73, CD90, and CD105 on the surface of HATMSC2 cells. The cells were also positive for the HLA ABC antigen and negative for the HLA DR antigen and for the pan-leukocyte antigen CD45 [16]. Importantly, HATMSC2-MVs expressed surface markers typical for MSCs, including CD73 ($50.50 \pm 2.18\%$ of the population), CD90 ($90.67 \pm 5.36\%$ of the population), CD105 ($45.32 \pm 3.24\%$ of the population), and HLA ABC ($88.20 \pm 4.61\%$ of the population). HATMSC2-MVs did not express HLA DR or CD45 (Figure 2).

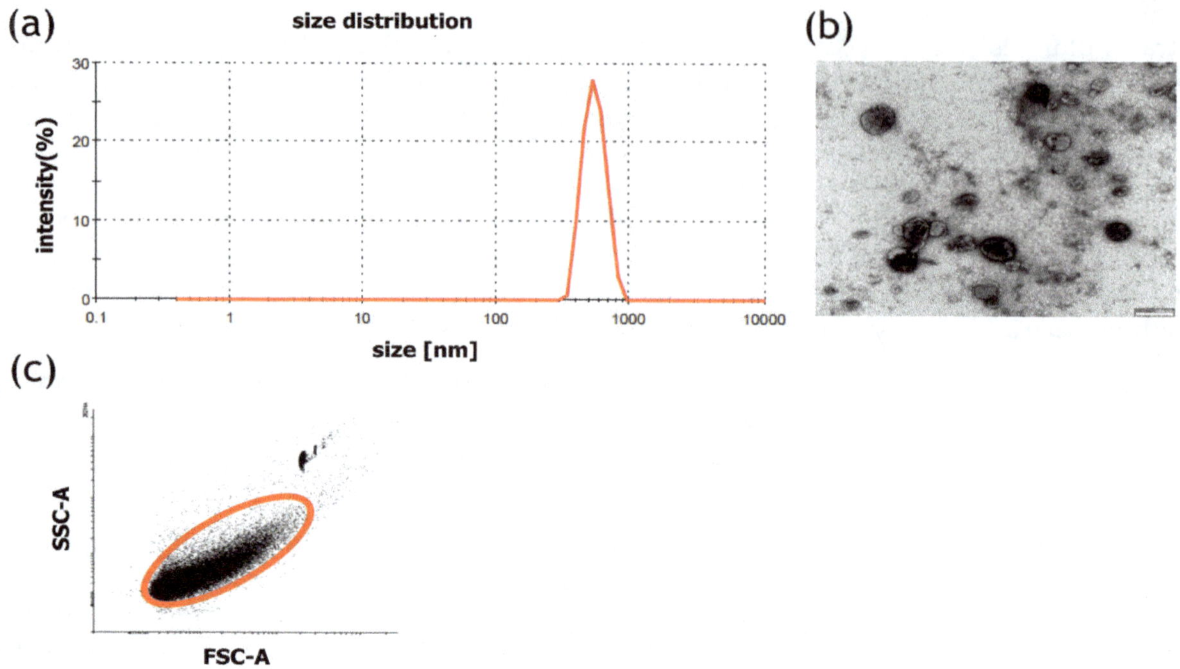

Figure 1. Characteristics of HATMSC2-MVs (**a**) Representative HATMSC2-MV size distribution histogram obtained using dynamic light scattering analysis. (**b**) Representative transmission electron microscopy image of HATMSC2-MVs, bars represent 500 nm. (**c**) Representative dot plot showing forward scatter (FSC) vs. side scatter (SSC). Gate R-1 shows the population of HATMSC2-MVs, and gate R-2 represents the counting beads. HATMSC2-MVs—microvesicles derived from immortalized human mesenchymal stem cells of adipose tissue origin.

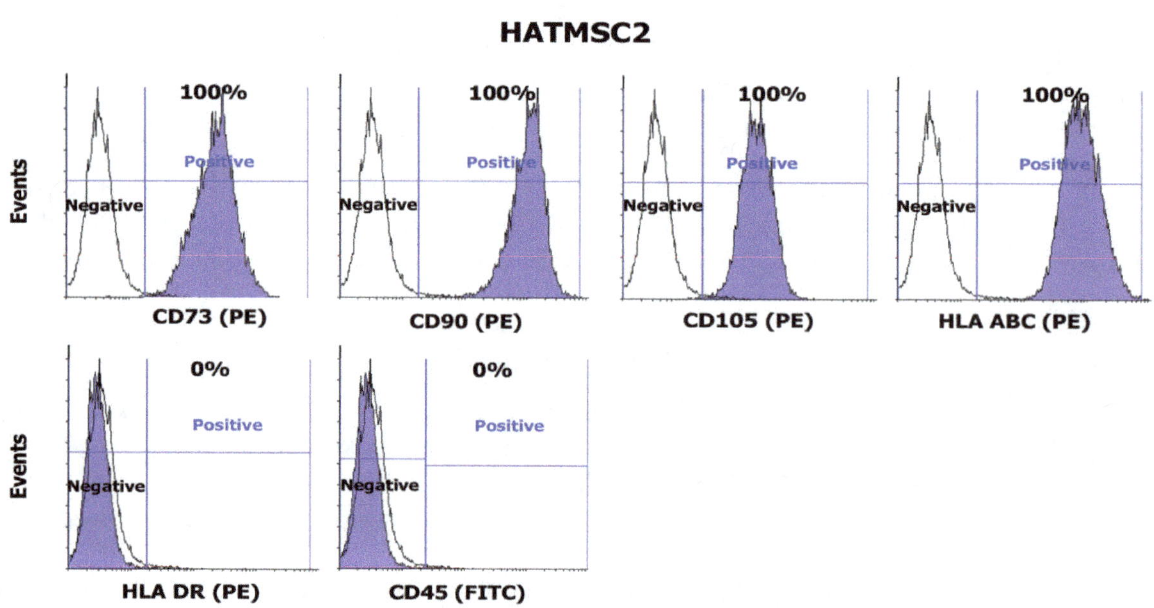

Figure 2. *Cont.*

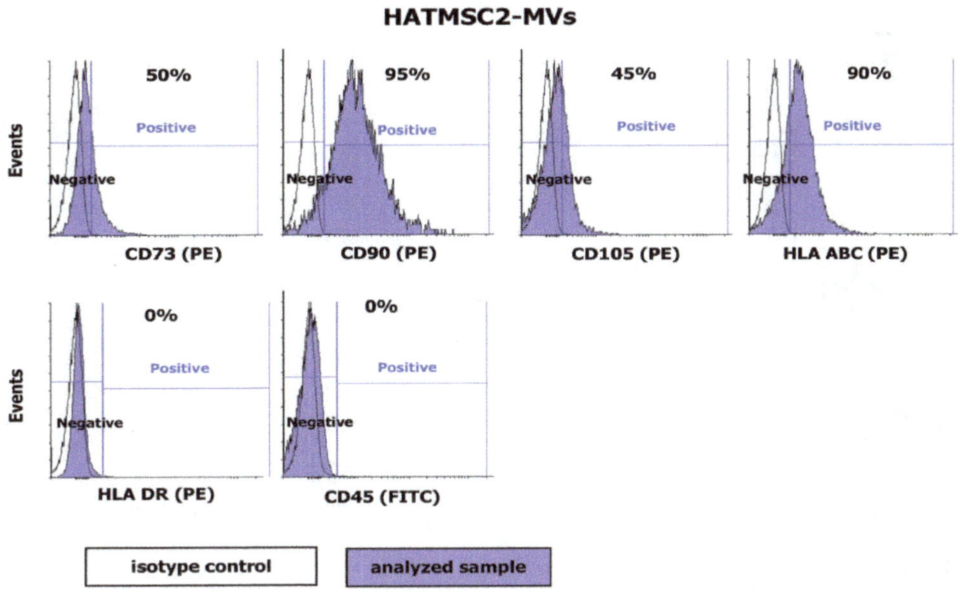

Figure 2. Representative flow cytometry histograms of flow cytometry analysis of mesenchymal stem cells markers (CD73, CD90, CD105, HLA ABC, HLA DR, and CD45) on HATMSC2 cells and HATMSC2-MVs from three independent experiments. Cells and MVs were stained with selected antibodies conjugated with fluorochromes. Blue filled histograms correspond to HATMSC2 cells and HATMSC2-MVs labeled with defined antibodies, and empty histograms represent the isotype controls. HATMSC2-MVs—microvesicles derived from immortalized human mesenchymal stem cells of adipose tissue origin.

2.3. Internalization of HATMSC2-MVs into Ovarian Cancer Cells

The internalization of fluorescently-labeled HATMSC2-MVs into ovarian cancer cell lines was analyzed using three-dimensional microscopic imaging. HATMSC2-MV co-culture with target cells (ES-2 and OAW-42 cell lines) for 24 h resulted in the incorporation of the MVs into ES-2 and OAW-42 cells, as shown by green fluorescence (DiO) expression in the cytoplasm of the target cells across the Z-stack slices (Figure 3a).

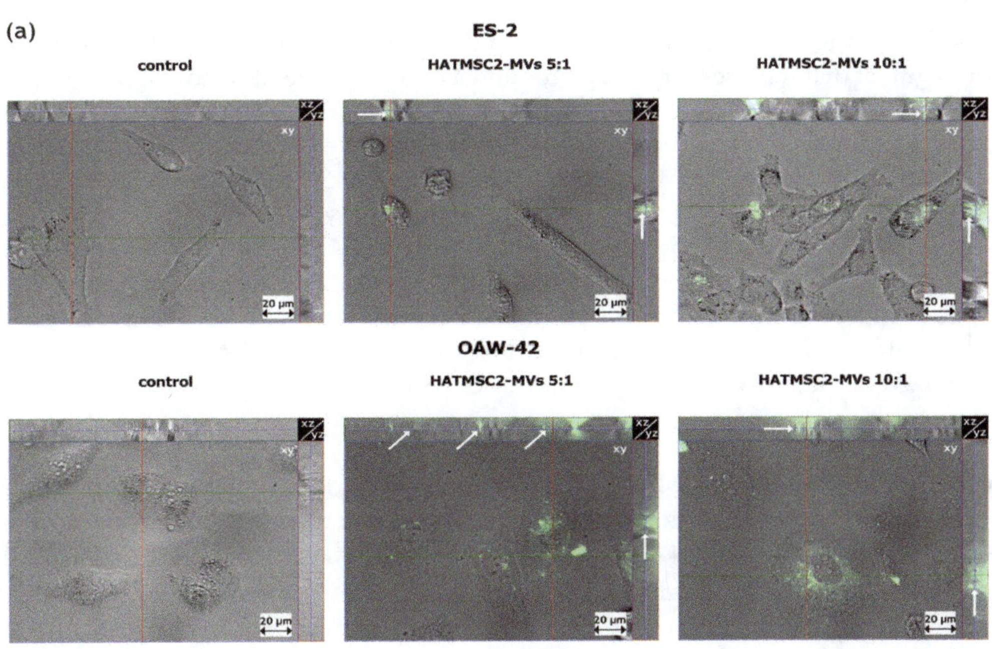

Figure 3. *Cont.*

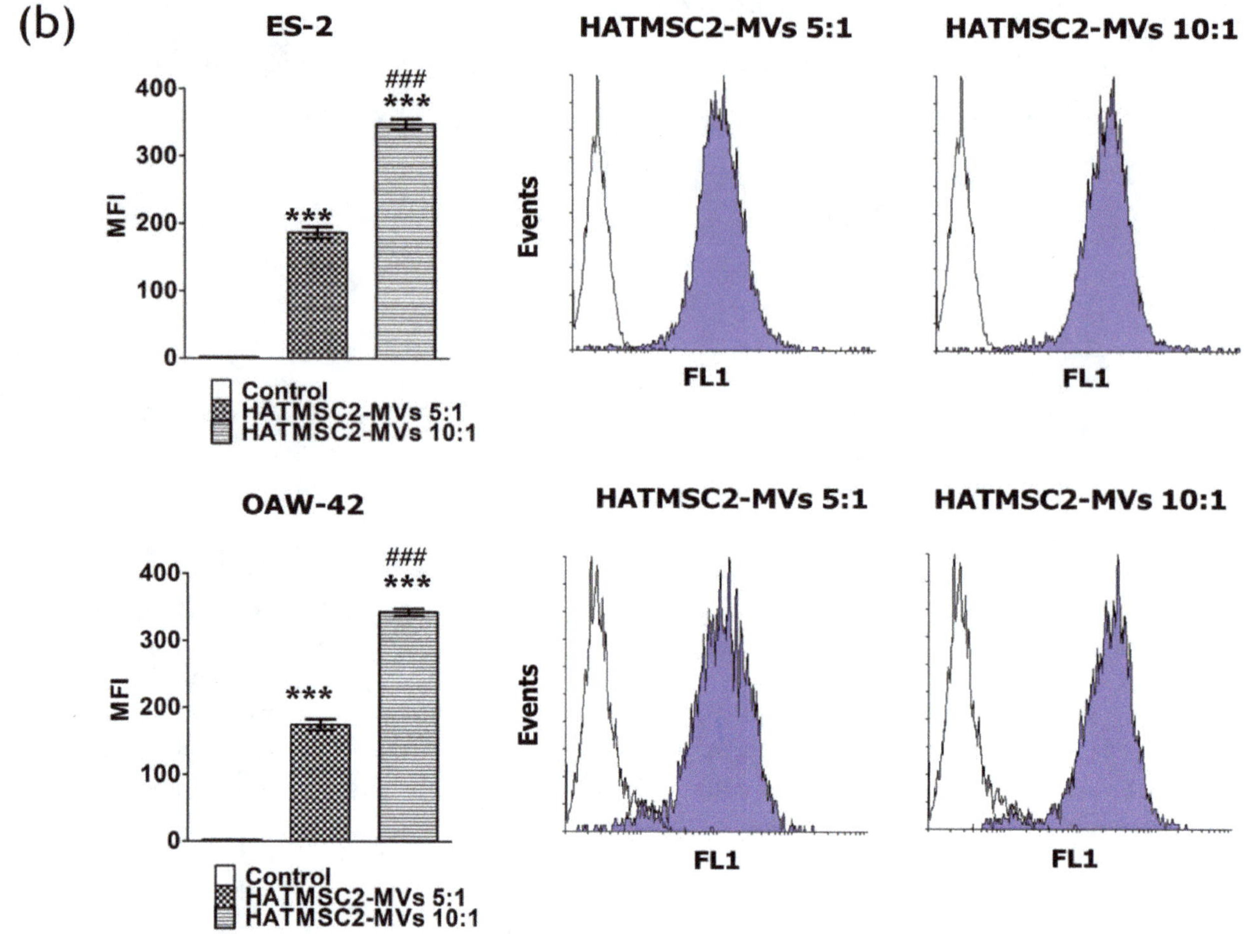

Figure 3. Internalization of HATMSC2-MVs into ovarian cancer cells. (**a**) Images from three-dimensional microscopic analysis of HATMSC2-MV internalization into ES-2 and OAW-42 cells at different ratios. The images were taken using an inverted microscope after 24 h of incubation with fluorescently-labeled HATMSC2-MVs (scale bar: 20 μm). A set of representative orthogonal slices is shown. Each image in a group consists of a large middle segment that represents the midpoint of the Z-stack in the *xy* plane; a narrow top segment that represents the *xz* plane; and a narrow segment on the right that represents the *yz* plane. The arrows point to MVs that have been taken up into the cell. (**b**) Bottom left panel, the bar graph represents the mean fluorescence intensity (MFI) of ES-2 cells treated with fluorescently labeled HATMSC2-MVs at different ratios. Untreated cells without MVs served as a control. Right panel, flow cytometry analysis of HATMSC2-MV internalization. Empty histograms represent the control for untreated cells, and blue filled histograms show the green fluorescence of ovarian cancer cells ES-2 and OAW-42 after HATMSC2-MV internalization at different ratios. The data represent mean ± SEM values from three independent experiments performed in duplicate. *** $p < 0.001$ calculated vs. control, ### $p < 0.001$ calculated vs. The HATMSC2-MVs 5:1 treatment. HATMSC2-MVs—microvesicles derived from immortalized human mesenchymal stem cells of adipose tissue origin.

Furthermore, the uptake of HATMSC2-MVs by ovarian cancer cells was confirmed by an increase in mean fluorescence intensity (MFI), as analyzed using flow cytometry. The results showed a significant increase in MFI in the ES-2 and OAW-42 cell lines treated with HATMSC2-MVs for both the ratios of 5:1 and 10:1 (the number of MVs to one target cell) compared to the control groups ($p < 0.001$). Moreover, this effect was dose-dependent, and significant differences between the ratios 5:1 and 10:1 ($p < 0.001$) were observed (Figure 3b).

2.4. *Anti-Proliferative Activity of HATMSC2-MVs*

The anti-proliferative activity of HATMSC2-MVs was analyzed using the MTT assay. ES-2 and OAW-42 cells were treated with MVs at four different ratios: 1:1, 5:1, 10:1, and 100:1. The HATMSC-MV treatment caused a significant decrease in OAW-42 cell proliferation on day 3 ($p < 0.01$) at a ratio of 100:1 (Figure 4a). The anti-proliferative activity of the MVs used at a ratio of 100:1 in OAW-42 cells on day 3 was also shown on a microscopic images of calcein-stained ovarian cancer cells (Figure 4b).

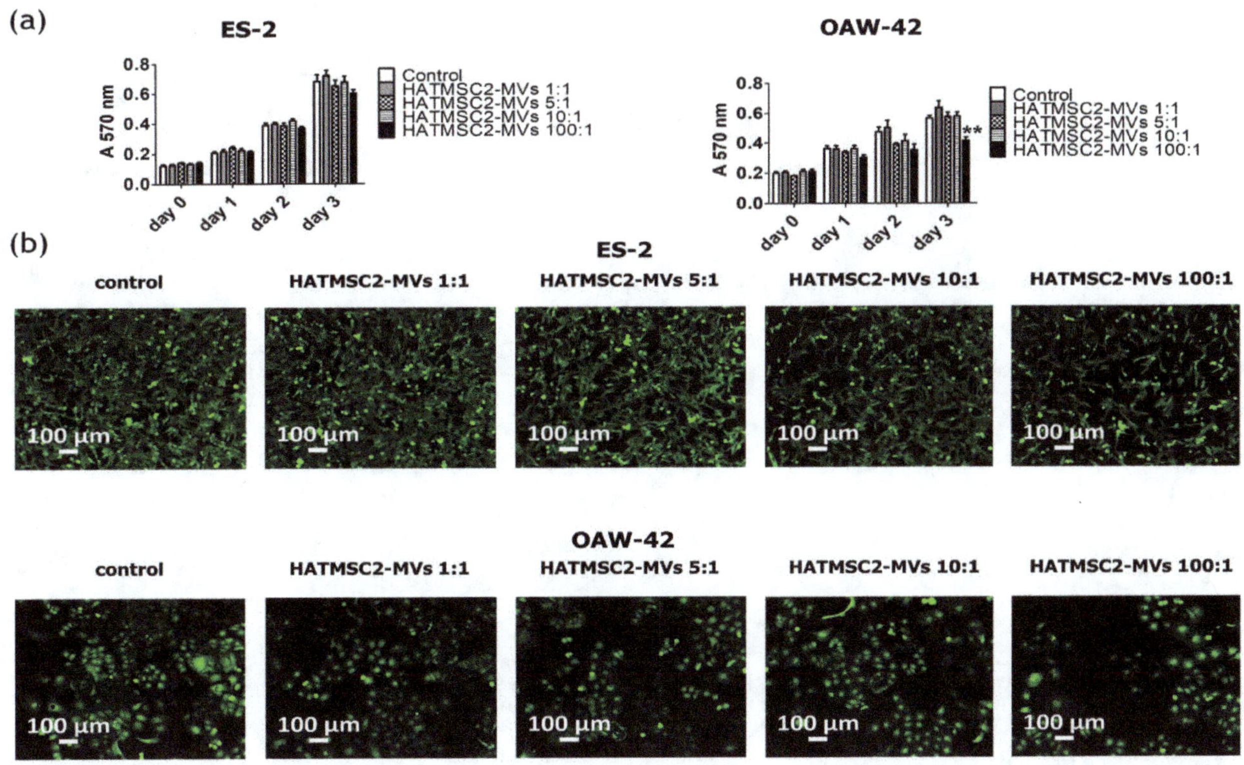

Figure 4. Effect of HATMSC2-MVs on the proliferation activity of ovarian cancer cells. (**a**) Proliferation activity of ES-2 and OAW-42 cells cultured in standard conditions was measured using an MTT assay on day 0, 1, 2, and 3 following treatment with HATMSC2-MVs at different ratios. Untreated cells without MVs served as a control. The data represent mean ± SEM values from four independent experiments performed in triplicate. ** $p < 0.01$ calculated vs. control on a given day. (**b**) Representative images from microscopic analysis of the morphology of ovarian cancer cells treated with HATMSC2-MVs at different ratios. ES-2 and OAW-42 cells were co-incubated with HATMSC2-MVs for 72 h. Afterwards, the cells were stained with Calcein AM and images were taken using an inverted microscope (scale bar: 100 µm). HATMSC2-MVs—microvesicles derived from immortalized human mesenchymal stem cells of adipose tissue origin.

2.5. *Effect of HATMSC2-MVs on Cell Cycle Progression*

The effect of HATMSC2-MVs on cell cycle progression was tested using flow cytometry analysis of ES-2 and OAW-42 cells stained with propidium iodide. We observed an increase in the percentage of cells in the sub-G1 phase (dead cells) in the samples treated with the MV ratio of 100:1 in ES-2 cells, compared to the control group (mean 2.57 ± 0.54% vs. 0.79 ± 0.05%; $p < 0.01$). Similarly, in OAW-42 cells treated with the MVs ratio of 100:1, the percentage of cells in the sub-G1 phase increased to 15.66 ± 2.86% compared to 2.74 ± 0.48% in control group ($p < 0.001$). Moreover, in OAW-42 cells treated with an MVs ratio of 100:1, the percentage of cells in the G0/G1 phase decreased from 63.06 ± 1.49% in the control group to 55.87 ± 1.37% in the test group ($p < 0.01$), (Figure 5).

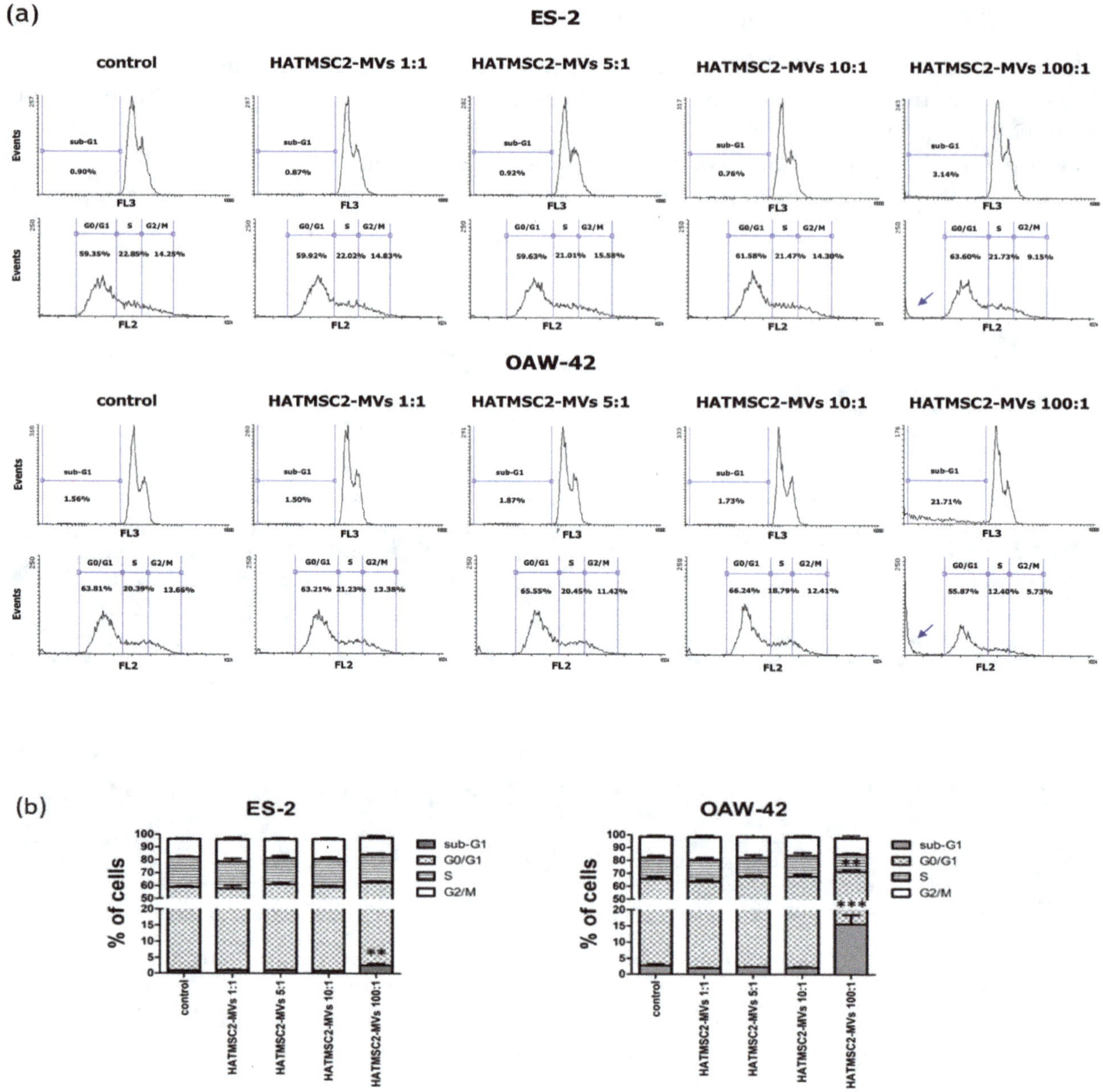

Figure 5. Effect of HATMSC2-MVs on the cell cycle progression of ovarian cancer cells. (**a**) Representative flow cytometry histograms showing cell cycle progression in ES-2 and OAW-42 cells treated with HATMSC2-MVs at different ratios. Untreated cells without MVs served as a control. Arrows represent the increased peaks in the sub-G1 phase of the cell cycle. (**b**) Percentages of cells in the sub-G1, G0/G1, S, and G2/M phases were determined using Flowing Software 2. The data represent mean ± SEM values from three independent experiments performed in duplicate. *** $p < 0.001$, ** $p < 0.01$ calculated vs. control. HATMSC2-MVs—microvesicles derived from immortalized human mesenchymal stem cells of adipose tissue origin

2.6. Proapoptotic Activity of HATMSC2-MVs

We examined the impact of HATMSC2-MVs on ovarian cancer cell survival. Cell death processes, such as apoptosis and necrosis, were assessed using flow cytometry after 72 h of co-culture of HATMSC2-MVs with ES-2 and OAW-42 cells at the ratios of 1:1, 5:1, 10:1, and 100:1. Untreated cells served as a control. The obtained results revealed that the HATMSC2-MVs treatment affected cell viability depending on the ratio of HATMSC2-MVs and cancer cells. The ratio of MVs 100:1 had the greatest impact on cell viability, both in the ES-2 and OAW-42 cells (Figure 6).

In ES-2 cells, the percentage of live cells treated with HATMSC2-MVs decreased to 61.23 ± 7.71% for the ratio of 100:1 vs. control (89.65 ± 0.99%; $p < 0.001$). The average percentage of early apoptotic cells increased to 8.22 ± 1.32% vs. control 1.95 ± 0.41%, ($p < 0.001$), whereas the average percentage of late apoptotic cells increased to 11.75 ± 3.22% vs. control 2.79 ± 0.32%, ($p < 0.01$). The average percentage of ES-2 necrotic cells increased to 18.81 ± 4.79% vs. control 5.72 ± 0.81%, ($p < 0.01$).

In OAW-42 cells, the percentage of live cells treated with HATMSC2-MVs decreased to 47.78 ± 10.11% for the ratio of 100:1 vs. control (86.17 ± 2.12%, $p < 0.001$). The average percentage of late apoptotic cells increased to 18.53 ± 5.17% vs. control 1.57 ± 0.34%, ($p < 0.001$). The average percentage of OAW-42 necrotic cells increased to 27.51 ± 5.04% vs. control 10.70 ± 2.02%, ($p < 0.001$).

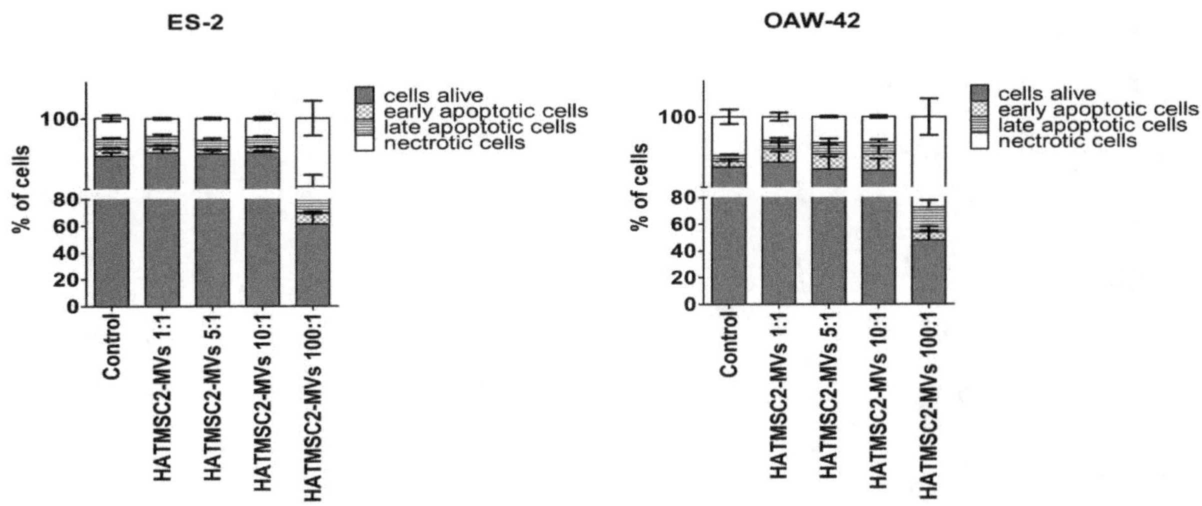

ES-2	HATMSC2-MVs 100:1
cells alive	***
early apoptotic cells	***
late apoptotic cells	**
necrotic cells	**

OAW-42	HATMSC2-MVs 100:1
cells alive	***
early apoptotic cells	ns
late apoptotic cells	***
necrotic cells	***

Figure 6. Quantification of cell viability after treatment with HATMSC2-MVs for 72 h at different ratios, determined using flow cytometric analysis of the apoptotic and necrotic cells via the double-staining of ES-2 and OAW-42 cells with propidium iodide and Annexin V. The percentages of alive, early apoptotic, late apoptotic, and necrotic cells were determined using Flowing Software 2. The data represent mean ± SEM values from five independent experiments performed in duplicate. *** $p < 0.001$, ** $p < 0.01$ calculated vs. control, ns: non-significant results. HATMSC2-MVs—microvesicles derived from immortalized human mesenchymal stem cells of adipose tissue origin.

2.7. Effect of HATMSC2-MVs on the Phenotype of Ovarian Cancer Cell Lines

To determine the effect of HATMSC2 -MVs on the phenotype of ovarian cancer cell lines treated at the ratios of 10:1 and 100:1, we tested the presence of the CD34, CD44, CD133, SSEA4, CD73, CD90, and CD105 markers using flow cytometry. The results showed that both cell lines, ES-2 and OAW-42, were positive for the adhesion molecule CD44 (Figure 7). However, the expression of the CD44 marker was higher (97.80% ± 2.20% of the population) for the ES-2 cells compared to OAW-42 cells (78.36% ± 3.20% of the population). Both ES-2 and OAW-42 cell lines were negative for CD34 and CD133 and for the pluripotency-related marker SSEA4. An analysis of the expression of MSC markers showed that both ES-2 and OAW-42 were positive for CD73 and CD90, whereas the CD105 marker was detected in ES-2 cells (98.20% ± 1.79% of the population), but not in OAW-42 cells (Figure 7a). The HATMSC2-MVs treatment did not affect the phenotype of ES-2 and OAW-42 cells (Figure 7b).

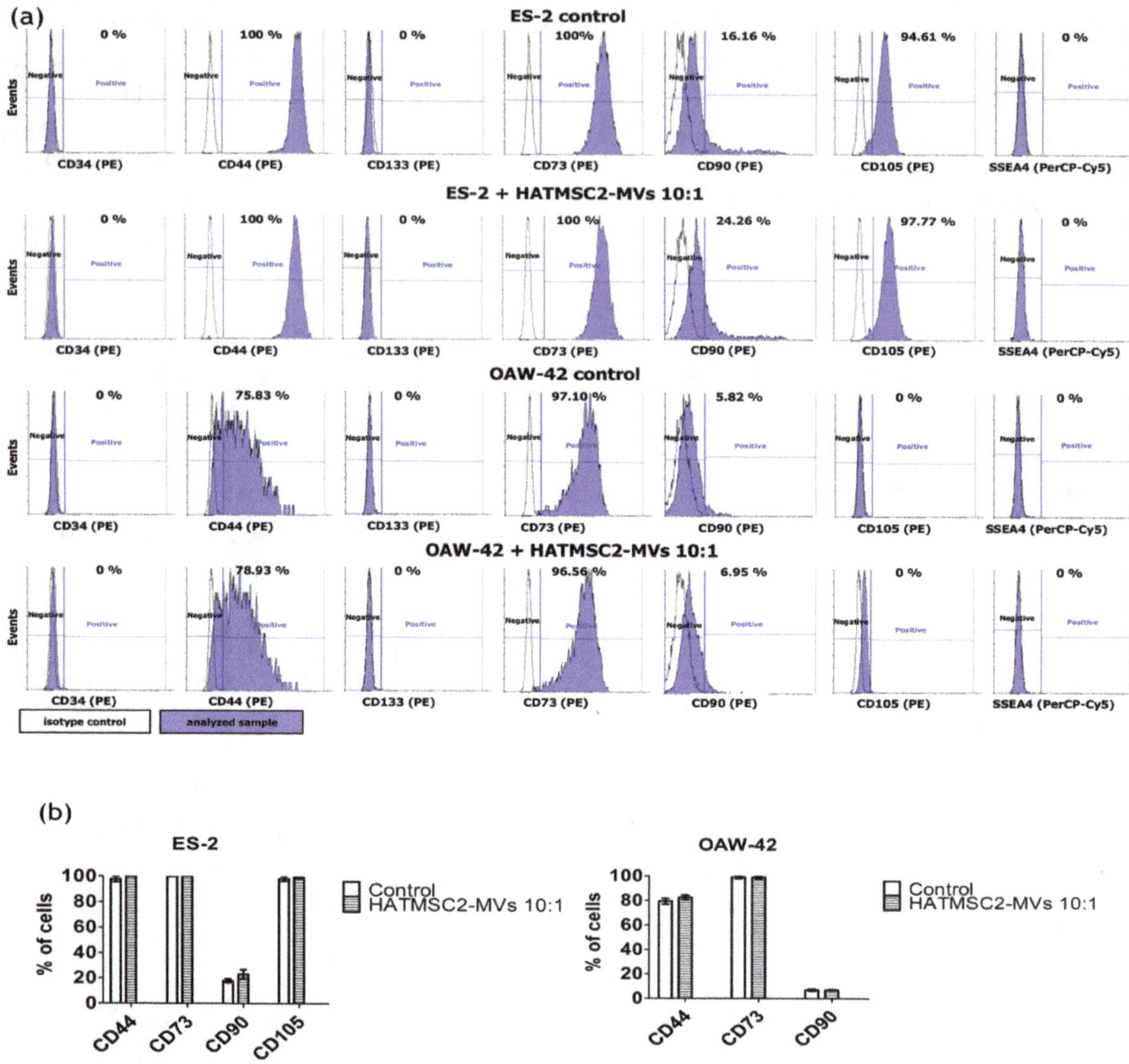

Figure 7. Characteristics of human ovarian cancer cell lines before and after HATMSC2-MV treatment at different ratios. (**a**) Representative histograms of flow cytometry analysis for surface markers (CD34, CD44, CD133, SSEA4, CD73, CD90, and CD105) on the ES-2 and OAW-42 cell lines. The cells were stained with selected antibodies conjugated with fluorochromes. Blue filled histograms correspond to ES-2 and OAW-42 cells, labeled with defined antibodies, and empty histograms represent the isotype controls. (**b**) The percentages of cells positive for selected markers were determined using Flowing Software 2. The data represent mean ± SEM values from three independent experiments. HATMSC2-MVs—microvesicles derived from immortalized human mesenchymal stem cells of adipose tissue origin.

2.8. *Effect of HATMSC2-MVs on the Secretion Profile of Ovarian Cancer Cell Lines*

The effect of HATMSC2-MVs on the biological properties of ovarian cancer cells was determined in all experiments at different ratios of MVs to cancer cells (1:1, 5:1, 10:1 and 100:1). However, the best effect was seen when a ratio of 100:1 was used. Therefore, to determine the effect of HATMSC2-MVs on the secretion profile of the ES-2 and OAW-42 cell lines, only the ratio of 100:1 was used. The secretion profile was evaluated using a human cytokine antibody array (Figure 8a). Most of the 120 cytokines and trophic factors identified in this analysis affect cancer cells either by promoting cancer growth or through their anti-tumor properties. For the ES-2 cells treated with HATMSC2-MVs, among cancer-promoting cytokines and trophic factors, a decrease was observed for angiogenesis-related cytokines, such as growth related oncogene-alpha (GRO-alpha) (by 53%), angiopoietin 2 (by 10%), VEGF (by 44%), and VEGFD (by 8%), and for the pro-angiogenic and pro-inflammatory cytokines IL-6 (by 15%), IL-8 (by 36%), MIP-1α (by 7%), and MIP-1β (by 8%). The levels of apoptosis-related TRAIL-R3 and TRAIL-R4 also decreased (by 10% and 14%, respectively). However, an increase was observed in the levels of the pro-inflammatory cytokines IL-1α (by 18%), IL-1β (by 21%), MCP-1 (by 43%), MIG (by 28%), TNFα (by 13%), IL-13 (by 6%), eotaxin (by 16%), and eotaxin-2 (by 21%), and growth factors bFGF (by 9%), EGF (by 33%), and HGF (by 21%). On the other hand, we observed an increase in the levels of several anti-cancer cytokines, such as IL-1 receptor antagonist (IL-1ra) (by 42%), IL-2 by (21%), IL-2 receptor alpha chain (IL-2Ra) (by 13%), IL-12 (by 6% and 10% for the p40 and p70 subunits, respectively), IL-15 (by 34%), and IFN-γ (by 43%). Nevertheless, the overall number of expressed cytokines in the ES-2 cells was 92; however, the expression of a majority of these cells differed before and after the HATMSC2-MV treatment (Figure 8b).

For the OAW-42 cells, the overall number of expressed cytokines increased from 58 in the control group to 87 after the MV treatment. However, most of these cytokines did not show a major difference in expression between the control and test groups. Among the tumor-promoting cytokines, we observed a decrease in the levels of the pro-angiogenic and pro-inflammatory cytokines GRO-alpha (by 192%), VEGF (by 34%), IL-6 (by 176%), IL-8 (by 204%), and RANTES (by 16%), and an increase in other cytokines, such as eotaxin-2 (by 8%), eotaxin-3 (by 6%), IL-1β (by 24%), MIG (by 18%), TNFα (by 36%), TGF-beta1 (by 21%), and -beta3 (by 34%), bFGF (by 15%), TRAIL-R3 (by 11%), and TRAIL-R4 (by 15%). An increase in the levels of anti-cancer cytokines was similar as observed for the ES-2 cells. All data were presented as a heat map (Figure 8b); selected cytokines which exhibited the largest differences between the treated cells and the control group were additionally shown on a column graph (Figure 8c).

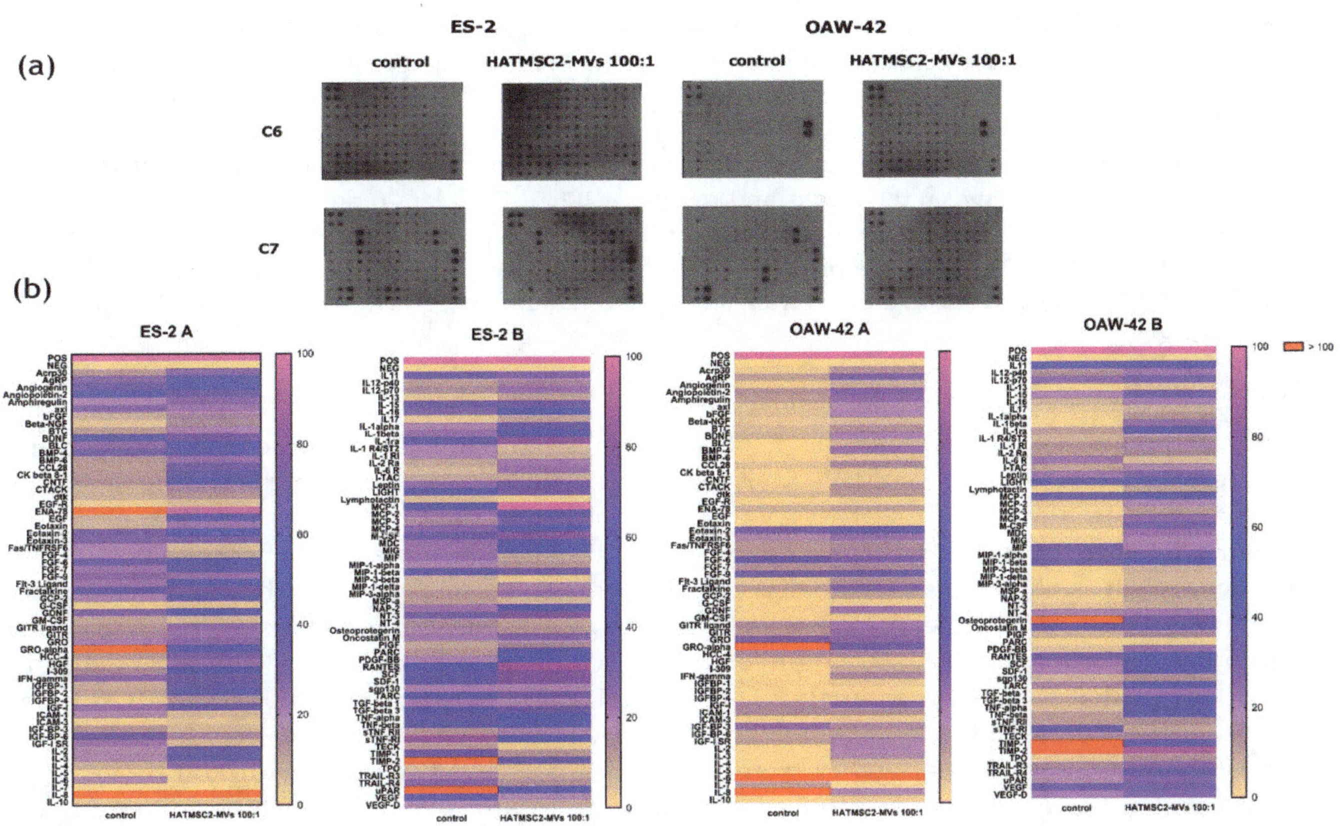

Figure 8. *Cont.*

(c)

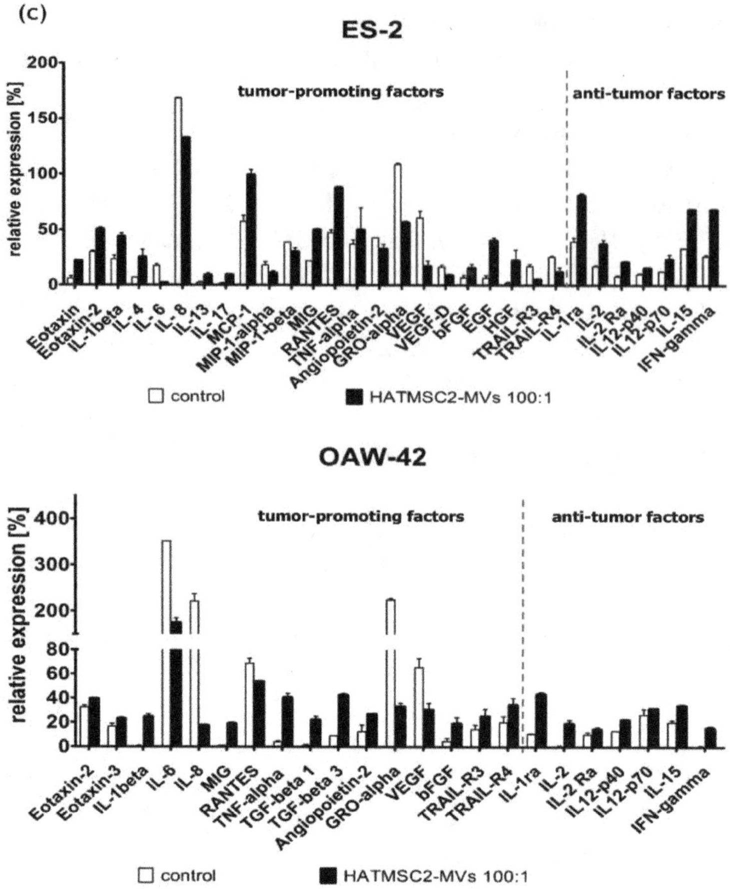

Figure 8. Effect of HATMSC2-MVs on the secretion profiles of ovarian cancer cell lines. (**a**) Scans of representative antibody arrays for ES-2 and OAW-42 supernatants. Untreated cells served as a control. The signal intensity for each antibody spot is proportional to the relative concentration of the antigen in the sample. (**b**) Heat map of cytokine levels for the ES-2 and OAW-42 supernatants; magenta and yellow indicate higher and lower expression limits, respectively. Outstanding values (above 100% of positive control) are depicted in red. Data are normalized to the internal positive control spots, which are consistent between the arrays and represent 100%. The data represent the mean from a duplicate assessment. (**c**) Column graph representing selected proteins (equal to or above 10% of the positive control). The data are presented as mean ± SEM values from a duplicate assessment. HATMSC2-MVs—microvesicles derived from immortalized human mesenchymal stem cells of adipose tissue origin.

3. Discussion

The paracrine activity of cells via EVs is an important link in cell-to-cell communication. Recent research has shown that EVs derived from MSCs play an important role in tumor microenvironment. Tumor cells secrete EVs to reprogram the mesenchymal stem cells present in the tumor microenvironment. The reprogrammed MSCs release exosomes that affect other cells in the tumor niche, such as fibroblasts, endothelial cells, and immune cells, inducing their pro-tumorigenic activity [17]. However, the effect of MVs derived from outside the tumor microenvironment, e.g., from the MSCs of adipose tissue origin, on cancer cells is not well understood and still debatable. The purpose of this study was to analyze the biological behavior of two histologically different ovarian cancer cell lines, ES-2 and OAW-42, in response to HATMSC2-MV treatment. In this study, we investigated whether MVs derived from human immortalized MSCs of adipose tissue origin may represent a new form of supportive therapy in ovarian cancer treatment. Flow cytometry and microscopic analysis confirmed the internalization of HATMSC2-MVs into target cells. Moreover, in all functional experiments, we used untouched MVs, but not the MVs lysate tested by different research groups [14]. We showed that treatment with HATMSC2-MVs gradually decreased the proliferation of ES-2 and OAW-42 cells, depending on the dose; however, a significant effect was observed on day 3, only for the OAW-42 cell line, when the highest ratio of HATMSC2-MVs of 100:1 (100 MVs per cell) was used. Similar results were obtained by Reza et al. [14], who reported an anti-proliferative and pro-apoptotic effect of ATMSC exosomes on ovarian cancer cells. The main mechanism involved in the action of the exosomes was the transfer of different miRNAs into the recipient cells. However, Reza et al. used protease or RNase-digested exosomes. The anti-proliferative activity of MVs derived from BM-MSCs was also confirmed in vitro on the SKOV3 cell line and in vivo through an intra-tumor injection of MVs into an established tumor generated by a subcutaneous injection of these cells into SCID mice [18]. On the other hand, Dong et al. [19] showed that EVs derived from MSCs isolated from the umbilical cord increased the proliferation of lung adenocarcinoma cells by transporting miR-410 to target cells.

Further experiments involving an analysis of cell cycle progression and cell death processes, such as apoptosis and necrosis, confirmed that HATMSC2-MV treatment affected cancer cell viability. The cell cycle analysis showed that treatment with HATMSC2-MVs, at the ratio of 100:1, significantly increased the percentage of cells of both cell lines in the sub-G1 phase compared to control; however, the increase of OAW-42 cells in the sub-G1 phase was significantly higher compared to the ES-2 cell line. The increased number of cells in the sub-G1 phase suggests that both ovarian cancer cell lines underwent cell death via apoptosis. A similar pro-apoptotic effect of the bioactive factors derived from MSCs isolated from human Wharton's jelly and applied in the form of a conditioned medium or Wharton's jelly-MSCs lysate was observed in a study on the OVCAR3 and SCOV3 ovarian cancer cell lines, confirming the anti-cancer properties of the MSC secretome [20]. The pro-apoptotic effect of HATMSC2-MVs on the examined ovarian cancer cell lines was also confirmed through flow cytometry analysis of cell death processes, distinguishing between apoptosis and necrosis. When HATMSC2-MVs were cultured with ovarian cancer cell lines at the ratio of 100:1, we observed a significant increase in the percentage of early and late apoptotic cells for the ES-2 cells, whereas in the OAW-42 cells, a substantial increase was observed for late apoptotic cells. Moreover, in both cells lines, HATMSC2-MVs significantly increased the percentage of necrotic cells. These results suggest that HATMSC2-MVs at a ratio of 100:1 induce mechanisms governing ovarian cancer cell death via both apoptosis and necrosis. Studies on the anti-cancer properties of the MSC secretome report that co-culture of MSCs of different tissue origin with ovarian cancer cell lines increases apoptosis with varying effects [21]. Interestingly, the percentage of apoptotic cells was higher when the supernatant derived from AT-MSCs was applied compared to the supernatants derived from BM-MSCs and UC-MSCs [21].

Additionally, we assessed the secretion profile of ovarian cancer cell lines and the effect of HATMSC2-MVs on the presence of the produced cytokines and trophic factors with different functions; one set of bioactive factors is known to promote cancer cells growth and metastasis, and the second set of cytokines is associated with anti-tumor properties. The differences in the secretion profiles of the examined ovarian cancer cell lines correlated with the histological type of the tumor. ES-2 cells were derived from clear cell carcinoma, with a good prognosis for the patient when diagnosed at an early stage of the disease and poor survival when diagnosed at an advanced stage, because this type of ovarian cancer is often more resistant to chemotherapy than serous cystadenocarcinoma, represented by the OAW-42 cell line [22]. The presence of cancer-promoting cytokines and chemokines, such as IL-6, IL-8, GRO-alpha, MIP-1α, MIP-1β, angiopoetin-2, and VEGF, which are associated with tumor growth, metastatic properties, and a poor prognosis, was detected in the supernatants collected from both ES-2 cells and OAW-42 cells. The application of HATMSC2-MVs resulted in a substantial decrease in IL-6, IL-8, GRO-alpha, and VEGF secretion in both cell lines. IL-8 is a pleiotropic chemokine with a dual function, which acts as a chemoattractant for neutrophils, inducing innate immune responses, whereas in the ovarian cancer environment, it contributes to the pro-survival activity of tumor cells and resistance to chemotherapy. A high production of IL-8 correlates with faster proliferation and increases the potential of angiogenesis, adhesion, and invasion of platinum sensitive (PEA1 and PEO14) and platinum resistant (PEA2 and PEO23) cell lines [23]. Browne et al. demonstrated a significant increase in the expression of IL-8 in specimens of the serous type of ovarian cancer compared to clear cancer ovarian carcinoma tissue [24]. The HATMSC2-MVs markedly inhibit IL-8 production in both examined cell lines in vitro. This effect can be used as a potential supportive therapy for ovarian cancer treatment. IL-6 was present in the OAW-42 culture supernatant at a very high level, in contrast to the ES-2 cells. It is well known that IL-6 plays a crucial role in the stimulation of inflammatory cytokine production, tumor angiogenesis, cell proliferation, and tumor macrophage infiltration [25]. A high production of IL-6 by ovarian cancer cells contributes to tumor progression and correlates with a poor prognosis [26]. The HATMSC2-MVs inhibit the activity of IL-6, and in conjunction with a decreased level of IL-8, may exert suppressive effects on the ovarian cancer cell line. GRO-alpha and VEGF were produced by ES-2 and OAW-42 cells; however, the GRO-alpha level markedly exceeded the VEGF level. Both growth factors, GRO-alpha and VEGF, are important for tumor growth and metastasis, especially in terms of supporting cancer angiogenesis. The diverse production of cytokines and growth factors by ovarian cancer cells is associated with the biological activity of cancer cells and may affect tumor progression, as reported in a study performed simultaneously on a set of 120 cytokines in ovarian cancer ascites [25]. Our study determined that HATMSC2-MVs substantially reduced the secretion of GRO-alpha and VEGF in both cell lines. Numerous growth factors and cytokines, including those assessed in our study, such as IL-6, IL-8, MCP-1, RANTES, GRO-alpha, and VEGF, are involved in promoting tumor growth and ovarian cancer cell aggressiveness. Therefore, characterizing cytokine secretion may provide information on the functional profile of cancer cells. This may help to create targeted therapy for ovarian cancer, in which angiogenesis is inhibited by a blockage of NF-κB, suppressing VEGF and IL-8 activity [27], or by targeting CXCR2, the key receptor for the GRO-alpha and IL-8 chemokine activity [28]. RANTES (CCL5) level decreased only in OAW-42 cells after the HATMSC2-MVs treatment. RANTES is involved in trafficking immune cells into the inflammation site and acts as a co-activator of T cells promoting the polarization of the immune response towards the Th1 profile. In the ovarian cancer microenvironment, RANTES acts through paracrine or autocrine signaling to promote tumor cell migration, invasion, and metastasis [29]. In contrast, MCP-1 was detected in supernatants collected from ES-2 cells, and its level increased after the HATMSC2-MV treatment. The main function of MCP-1 in the tumor microenvironment is to attract tumor-associated monocytes (TAMs) [30]. Research performed by Furukawa et al. [30] demonstrated that the MCP-1 chemokine promoted the invasion and adhesion of the ovarian cancer cell line SKOV3, contributing to the progression of tumors. Tumor necrosis factor-related apoptosis-inducing ligand (TRAIL)-R3 and –R4 are known as the negative regulators of TRAIL-mediated apoptosis in cancer cells [31–33]. The internalization of HATMSC2-MVs by the examined cell lines exerts a different effect on ovarian cancer cells, and decrease the level of TRAIL-R3 and TRAIL-R4 in ES-2 cells and increase their level in OAW-42 cell lines. The downregulation of TRAIL-R3 and TRAIL-R4 is associated with an increased level of early apoptotic cells in the ES-2 cell line treated with the 100:1 ratio. A very recent study, performed on a murine xenograft model, documented that the EVs isolated from the TRAIL expressing cell line 293T in combination with cyclin-dependent kinase inhibitor (dinaciclib) successfully inhibited the growth of human lung cancer cell lines NCI-H727 and A549 and the human breast adenocarcinoma cell line MDAMB231 by inducing apoptosis [34].

The HATMSC2-MVs used in this study affect both histologically different cell lines, the ES-2 cells and the OAW-42 cells, by increasing the production of tumor-suppressive cytokines, such as IL-1ra, IL-2, IL-2Ra, IL-12-p40, IL12-p70, IL-15, and IFN-γ. Studies that used bioactive factors released to the culture medium from Wharton's jelly MSCs led to a similar inhibition of the proliferation of the ovarian cancer cell line OVCAR3 through a decreased expression of oncogenic cytokines and growth factors and an increased expression of anti-tumor related cytokines [35]. The anti-inflammatory properties of IL-1ra, a naturally occurring inhibitor to IL-1, contribute to tumor growth inhibition by competitive binding to IL-1 receptors blocking cancer-promoting activity of IL-1 [36]. The anti-proliferative effect of ovarian cancer cell lines can be also supported by an increased production of IFN-γ following the exposure of cells to HATMSC2-MVs, as documented in both examined cell lines. It has been reported that IFN-α and IFN-γ, applied in combination with IL-4 fused to Pseudomonas exotoxin, inhibit tumor growth in an experimental mouse model of human ovarian cancer. The anti-tumor effect was accomplished by the activation of the IFN signaling pathways and the subsequent activation of molecules inducing apoptotic cell death [37]. Characterization of a wide range of tumor-promoting factors and anti-tumor cytokines after ovarian cancer cell expose to HATMSC2-MVs provides information on how they affect the production of functional cytokines and shed light on the mechanism altering the behavior of ovarian cancer cells in response to MV treatment.

Consequently, we characterized the ovarian cancer cell lines ES-2 and OAW-42 for the presence of CSC and MSC markers. The results showed that the ES-2 and OAW-42 cells were positive for CD44 and negative for CD133, and that the application of HATMSC2-MVs had no effect on the expression of these markers. Similar results were reported by Tudrej et al., who showed that CD44 expression was higher for ES-2 cells compared to OAW-42 cell lines [38]. They found a small subpopulation of ES-2 cells positive for the CD133 marker expression (around 0.2%). However, we did not observe any CD133 positive cells in our study. To our best knowledge, the expression of specific MSC markers on ovarian cancer cell lines has been studied in the form of a single MSC marker as a potential therapeutic target [39–42], whereas a complete analysis of MSC markers has been performed in a limited number of studies concerning the biological activity of ovarian cancer cell lines [21]. Our results revealed that both ES-2 and OAW-42 cells were strongly positive for CD73, and that HATMSC2-MVs had no impact on CD73 expression. CD73, also known as cell surface nucleotidase, is an immunosuppressive enzyme involved in tumor progression and metastasis, and its expression is associated with a poor prognosis for high-grade serous ovarian cancer [42]. The functional inhibition of CD73 via either a chemical compound or a neutralizing antibody reduced the tumorigenesis of primary high-grade serous epithelial ovarian cancer cells [41]. In contrast, CD90 was present on a limited population of both examined cell lines, and co-culture of ES-2 and OAW-42 cells with HATMSC2-MVs at a ratio of 1:10 did not increase the expression of this marker. It was reported previously that the overexpression of CD90 inhibited the sphere-forming ability of SKOV3 cell lines and increased cell apoptosis. These studies also suggest that CD90 may decrease cell growth through a downregulation of the expression of other CSC markers, including CD133 and CD24 [40]. Interestingly, the CD105 molecule was detected only on poorly-differentiated ES-2 cells, but not on the better-differentiated OAW-42 cells. This finding confirmed the mesenchymal phenotype of ES-2 cells, which is associated with increased aggressiveness and metastatic potential. The HATMSC2-MVs have no marked impact on CD105 expression in either of the cell lines. Studies on the biological role of CD105 in ovarian cancer revealed that high CD105, CD44, or CD106 expression was associated with drug resistance, an advanced stage of the disease, poor differentiation, and high rate of cancer relapse [43]. The downregulation of CD105 expression with a clinically relevant CD105-neutralizing mAb (TRC105) inhibited high-grade serous cancer metastasis, reduced ascites, and hampered the growth of abdominal tumor nodules in animal models of ovarian cancer [39].

A systematic review, introducing the impact of experimental anti-tumor cellular therapies involving MSCs of different human tissue origin, also highlights the possibility to use MSC secretome, in the form of a conditioned medium or EVs, as a cell-free therapy to inhibit cancer growth [44]. Thus, MVs may serve as a carrier for the delivery of therapeutic agents to target cells. A modification of primary MSCs for the secretion of inhibitory growth factors and pro-apoptotic factors may be employed to prepare the MVs carrying the pro-apoptotic signal and transport them to target ovarian cancer cells. Thus, MVs may be applied as a supportive therapy to enhance the therapeutic effect of chemotherapy, especially for multidrug resistant cancers.

In conclusion, we confirmed an effective transfer of HATMSC2-MVs into target ovarian cancer cells, which affected the biological behavior of these cells. Our results revealed that HATMSC2-MVs inhibit tumor cell proliferation in the two histologically distinct ovarian cancer cell lines via different pathways, apoptosis and/or necrosis. This phenomenon, with high likelihood, is related to the secretion of the different anti-tumor factors by the ES-2 (representing poorly differentiated ovarian clear cell carcinoma), and OAW-42 (representing ovarian cystadenocarcinoma) cell lines treated with HATMSC2-MVs. However, further studies are needed to determine the possible mechanisms involved in HATMSC2-MV-mediated effect on target cells, as well as to validate their anti-tumorigenic potential with respect to cancer cells isolated from human tissues. Therefore, understanding the mechanisms involved in the bilateral interaction between the MVs and ovarian cancer cells may be help to design new treatment modalities for an effective anti-tumor cell-free therapy.

4. Materials and Methods

4.1. Cell Culture

The ES-2 cell line was purchased from ATCC (American Type Culture Collection, Manassas, VA, USA) (catalog number: CRL-1978™). The cells were cultured in the DMEM and OptiMEM GlutaMax media, mixed in equal proportions. The DMEM medium was supplemented with 10% FBS (Gibco, Thermo Scientific, Carlsbad, CA, USA), a 1% penicillin/streptomycin solution (Gibco, Thermo Scientific, Carlsbad, CA, USA) and L-glutamine (Gibco, Thermo Scientific, Carlsbad, CA, USA). The OptiMEM GlutaMax medium was supplemented with 3% FBS (Gibco, Thermo Scientific, Carlsbad, CA, USA) and a 1% penicillin/streptomycin solution (Gibco, Thermo Scientific, Carlsbad, CA, USA).

The OAW-42 cell line was purchased from ECACC (European Collection of Authenticated Cell Cultures, Salisbury, United Kingdom) (catalog number: 85073102). The cells were cultured in the same media conditions, mixed in equal proportions, and additionally supplemented with a 10 µg/mL insulin solution (Sigma-Aldrich, St. Louis, MO, USA).

All cells were cultured in standard conditions (21% O_2, 5% CO_2, 95% humidity, 37 °C temperature). Upon reaching 70–80% confluence, the cells were harvested with a 0.05% trypsin/0.02% EDTA (*w/v*) solution (IIET, Wroclaw, Poland) and seeded onto new culture flasks.

The human mesenchymal stem cell line HATMSC2 was established in our laboratory using the hTERT and pSV402 plasmids, as described in a previous study [16].

4.2. MV Isolation Using Sequential Centrifugation

MVs were isolated according to the well-established protocol developed in our laboratory [45] based on the procedure introduced in the previous study [46]. HATMSC2 cells were cultured in multi-layer cell culture flasks (Nunc TripleFlasks, Thermo Scientific, Carlsbad, CA, USA) using DMEM + 10% FBS until they reached 75% confluence. Next, the cells were cultured in serum-free media in hypoxic conditions (1% O_2) for 48 h to enhance the release of MVs. The conditioned media collected from the HATMSC2 cultures were mixed to obtain a homogenous starting material before the isolation of MVs. In the next step, the conditioned media were centrifuged at 300× *g* for 10 min at 4 °C, and at 2000× *g* for 10 min at 4 °C, in order to remove cellular debris and apoptotic bodies. Subsequently, the supernatants were subjected to double centrifugation at 12,000× *g* for 30 min at 4 °C using a Sorvall LYNX 6000 ultracentrifuge (Thermo Scientific, Carlsbad, CA, USA) with an intermediate washing step in PBS (IIET, Wroclaw, Poland). The obtained MV pellets were resuspended in 150 µL of PBS and stored at −80 °C.

4.3. Analysis of MVs

The size distribution of MVs was measured with DLS (Malvern Zetasizer, Malvern, UK). The measurement was performed for 2 min at 25 °C. PBS was used to disperse the samples. Moreover, to confirm the size of the MVs, the samples were analyzed using transmission electron microscopy (TEM). The PBS-suspended MVs were placed on a carbon-coated copper grid (400 mesh) and incubated for 1 min, and the excess liquid was removed with filter paper. Next, the samples were stained with 2% uranyl acetate, dried, and examined with a transmission electron microscope (JEOL, Peabody, MA, USA) at 80 kV. The number of MVs was calculated using a BD Fortessa Flow Cytometer (BD Biosciences,

San Jose, CA, USA) and fluorescent counting beads (CountBright™ Absolute Counting Beads for flow cytometry, Thermo Scientific, Carlsbad, CA, USA). Prior to analysis, 10 μL of the MV sample was diluted in PBS to a final volume of 300 μL, after which 50 μL of counting beads were added. The threshold for the forward scatter (FSC) was set at 200. To determine the number of MVs in the samples, 5000 counting beads were collected using a BD Fortessa Flow Cytometer (BD Biosciences, San Jose, CA, USA). The data were analyzed using the BD FACSDiva Software (BD Biosciences, San Jose, CA, USA). The number of MVs was calculated according to the CountBright™ Absolute Counting Beads manufacturer's instructions, using the ratio of MV events and the number of counting bead events. The protein concentration of MVs was determined with a Bradford assay (Thermo Scientific, Carlsbad, CA, USA) according to the vendor's instructions. The MV samples or the BSA standard were briefly incubated with the Bradford reagent for 5 min on a 96-well plate. The absorbance was measured with a Synergy H4 plate reader (Biotek, Winooski, VT, USA) at 595 nm.

4.4. Flow Cytometry Analysis of HATMSC2 Cells and HATMSC2-MVs

HATMSC2 cells were detached using the trypsin/EDTA solution and incubated with PE-conjugated antibodies specific for the human CD73 (clone AD2), CD90 (clone 5E10), CD105 (clone 266), HLA ABC (clone G46-2.6), HLA DR (clone G46-6) (BD Biosciences, San Jose, CA, USA), and FITC-conjugated CD45 antibody (clone 2D1) (BD Biosciences, San Jose, CA, USA) and with the appropriate isotype controls (BD Biosciences, San Jose, CA, USA) for 30 min at 4 °C. Afterwards, the labeled cells were washed with PBS (IIET, Wroclaw, Poland) and analyzed using a FACSCalibur flow cytometer (BD Biosciences, San Jose, CA, USA). The obtained data were processed using the CellQuest software (BD Biosciences, San Jose, CA, USA). The histograms were created using Flowing Software 2. The surface markers of the MVs were analyzed using a BD Fortessa Flow Cytometer (BD Biosciences, San Jose, CA, USA) after staining with specific fluorophore-conjugated antibodies. MVs suspended in PBS were incubated with PE-conjugated antibodies specific for human CD73, CD90, and CD105, HLA ABC, HLA DR, and FITC-conjugated antibody for CD45 and with the appropriate isotype controls for 30 min at 4 °C. The percentage of positive MVs was calculated using the BD FACSDiva Software (BD Biosciences, San Jose, CA, USA).

4.5. Internalization of MVs

ES-2 and OAW-42 cells were seeded into a Lab-Tek II Chambered # 1.5 Coverglass system (Nalge Nunc International, Naperville, IL, USA) at a density of 15×10^3 cells per chamber. Fluorescence staining of the MVs was performed, as established in our recent study [45]. After washing with PBS, the MVs were resuspended in the DMEM + 10% FBS and OptiMEM GlutaMax + 3% FBS media (mixed in equal proportions), and added to the cells at a ratio of 5:1 (5 MVs per cell) and 10:1 (10 MVs per cell). The cells were incubated with MVs for 24 h and washed twice with PBS prior to imaging. The internalization of the MVs into target cells was immediately analyzed at 37 °C using an Axio Observer inverted microscope equipped with a dry 63x objective (Zeiss, Gottingen, Germany). The labeled MVs were detected using an EGFP Filter set. Thirty Z-sections with a 0.6-μm interval were recorded simultaneously in the brightfield and fluorescence channel. Optical orthogonal sectioning was performed in order to visualize the internalization of the MVs. Images were obtained and processed using the Zen Blue Software (Zeiss, Gottingen, Germany). A similar analysis of EV internalization using fluorescence microscopy was previously described by Adamiak et. al. [47]. After 24 h of incubation with MVs, the cells were washed once with PBS, detached using the trypsin/EDTA solution, washed once more with PBS, and analyzed using flow cytometry with FACSCalibur (BD Biosciences, San Jose, CA, USA). The cells were detected using the FL1 channel (480 nm). The histograms were created using Flowing Software 2 (Perttu Terho, Turku Centre for Biotechnology, Finland).

4.6. Proliferation Activity

The proliferation activity of ES-2 and OAW-42 treated with HATMSC2-MVs was measured using the MTT assay. The cells were seeded on a 96-well plate at a concentration of 2×10^3 cells/well in the DMEM + 10% FBS and OptiMEM GlutaMax + 3% FBS media (mixed in equal proportions); MVs at a ratio of 1:1 (1 MV per cell), 5:1 (5 MVs per cell), 10:1 (10 MVs per cell), and 100:1 (100 MVs per cell) were added to the cells. ES-2 and OAW-42 cells without MVs were used as a control. After 4 h, 24 h, 48 h, and 72 h, the absorbance of the formazan dye produced by living cells was measured using a Wallac 1420 Victor2 Microplate Reader (Perkin Elmer, Waltham, MA, USA) at 570 nm. After 72 h of co-incubation with HATMSC2-MVs, the ovarian cancer cells were stained using Calcein AM (Thermo Fisher, Carlsbad, USA). 100 μL of Calcein AM (1 μM solution) were added to each well. The cells were incubated for 15 min at room temperature. Images were obtained using an Axio Observer inverted microscope equipped with a dry 10x objective (Zeiss, Gottingen, Germany). The labeled cells were detected using an Alexa Fluor 488 Filter set. The images were processed with the Zen Blue software (Zeiss, Gottingen, Germany).

4.7. Cell Cycle Analysis

The cell cycle analysis was performed based on previously published method [20,48]. The cells were seeded on 24-well plates at a concentration of 12×10^3 in the DMEM + 10% FBS and OptiMEM GlutaMax + 3% FBS media (mixed in equal proportions). MVs at a ratio of 1:1 (1 MV per cell), 5:1 (5 MVs per cell), 10:1 (10 MVs per cell), and 100:1 (100 MVs per cell) were added to the cells. ES-2 and OAW-42 cells without MVs were used as a control. After 72 h, the cells were detached using the trypsin/EDTA solution; the conditioned media were also collected and mixed with the respective cell suspension samples. The samples were centrifuged at 1400 rpm for 4 min at 4 °C. After the supernatant was removed, the cells were resuspended in ice-cold 70% ethanol and incubated for 30 min on ice at 4 °C. Afterwards, PBS Ca^{2+} Mg^{2+} + 2.5% FBS was added to the cells, and the samples were centrifuged at 1400 rpm for 5 min. This step was repeated twice. The cells were then resuspended in a solution of propidium iodide in PBS (50 μg/mL) and RNase (20 μg/mL) and incubated overnight at 4 °C. The cells were analyzed using a FACSCalibur flow cytometer (BD Biosciences, San Jose, CA, USA). The obtained data were analyzed using Flowing Software 2 (Perttu Terho, Turku Centre for Biotechnology, Finland).

4.8. Cell Viability and Apoptosis Analysis Using Flow Cytometry

In order to determine the effect of HATMSC2-MVs on the viability of the cells, an Annexin V Apoptosis Detection Kit (Thermo Scientific, Carlsbad, CA, USA) was used. ES-2 and OAW-42 cells were seeded in a 24-well plate at a density of 25×10^3 in the DMEM + 10% FBS and OptiMEM GlutaMax + 3% FBS media (mixed in equal proportions). Before the analysis, the cells were treated with MVs at a ratio of 1:1 (1 MV per cell), 5:1 (5 MVs per cell), 10:1 (10 MVs per cell), and 100:1 (100 MVs per cell) for 72 h. ES-2 and OAW-42 cells without MVs were used as a negative control. After incubation with MVs, the cells were stained with Annexin V and propidium iodide according to the manufacturer's recommendations. The cells were analyzed for live (Annexin V negative and propidium iodide negative), early apoptotic (Annexin V positive and propidium iodide negative), late apoptotic (Annexin V positive and propidium iodide positive), and necrotic cells (Annexin V negative and propidium iodide positive) using a FACSCalibur flow cytometer (BD Biosciences, San Jose, CA, USA). The analysis was performed using Flowing Software 2 (Perttu Terho, Turku Centre for Biotechnology, Finland).

4.9. Flow Cytometry Analysis of Ovarian Cancer Cell Lines

In order to determine the effect of HATMSC2-MVs on the phenotype of ovarian cancer cell lines, flow cytometry analysis was performed. ES-2 and OAW-42 cells were treated for 72 h with MVs at a ratio of 10:1. ES-2 and OAW-42 cells were seeded in a 6-well plate at a density of 6×10^4 per well in the DMEM + 10% FBS and OptiMEM GlutaMax + 3% FBS media (mixed in equal proportions). MVs were added to the cells at a ratio of 10:1. ES-2 and OAW-42 cells without MVs were used as a control. After 72 h, the cells were washed with PBS, and the culture medium was replaced with DMEM without FBS. Following a subsequent 24 h of culture in DMEM without FBS, the cells were detached using the trypsin/EDTA solution and incubated with PE-conjugated antibodies specific for the human CD34 (clone 8G12), CD44 (clone 515), CD133 (clone W6B3C1), CD73 (clone AD2), CD90 (clone 5E10), and CD105 (clone 266) molecules and the PerCP-Cy5.5—SSEA4 antibody (clone MC813-70) (all from BD Biosciences, San Jose, CA, USA) and with the appropriate isotype controls (BD Biosciences, San Jose, CA, USA) for 30 min at 4 °C. Afterwards, the labeled cells were washed with PBS (IIET, Wroclaw, Poland) and analyzed using a FACSCalibur flow cytometer (BD Biosciences, San Jose, CA, USA). The obtained data were processed using the CellQuest software (BD Biosciences, San Jose, CA, USA). The histograms were created using Flowing Software 2 (Perttu Terho, Turku Centre for Biotechnology, Finland).

4.10. Secretion Profiles of Ovarian Cancer Cell Lines

In order to determine the effect of HATMSC2-MVs on the secretion profiles of ovarian cancer cell lines, a C-Series Human Cytokine Antibody Array C1000 (RayBio®, Norcross, GA, USA) was used. ES-2 and OAW-42 cells were seeded in a 6-well plate at a density of 6×10^4 per well in the DMEM + 10% FBS and OptiMEM GlutaMax + 3% FBS media (mixed in equal proportions). MVs were added to the cells at a ratio of 100:1. The ES-2 and OAW-42 cells without MVs were used as a control. After 72 h, the cells were washed with PBS, and the culture medium was replaced with DMEM without FBS. Following the subsequent 24 h of culture in DMEM without FBS, the conditioned medium was collected and centrifuged for 10 min at 300× g to remove cellular debris, and the cells were used for flow cytometry analysis (see 4.9. Briefly, 2 mL of blocking buffer were applied onto the membrane and incubated for 30 min at room temperature. Next, 2 mL of the supernatant collected from the control and treated cells were incubated with the membrane overnight at 4 °C. Following a series of washes, a biotinylated antibody cocktail was applied onto the membrane and incubated for 2 h at room temperature. Unbound antibodies were removed with a series of washes, and the membrane was placed in HRP-streptavidin and incubated for 2 h at room temperature. Following a third series of washes, chemiluminescence was detected, and the bound proteins were visualized using an X-ray film. Signal intensities were compared using the ImageJ software (MosaicJ, Philippe Thevenaz): relative differences in the expression levels of each analyzed sample were measured and normalized to the intensities of the positive control using the Protein Array Analyzer plugin. The obtained data were analyzed automatically using the Microsoft® Excel-based Analysis Software Tool for Human Cytokine Antibody Array C1000. The results were calculated as a percentage of expression, with positive control set to 100% and negative control set to 0% (relative expression). The threshold was set to 10%. All results equal to or above 10% were considered as real expression.

4.11. Statistical Analysis

All statistical analyses were performed using GraphPad Prism version 7 (GraphPad Software Inc., San Diego, CA, USA). The data were compared using the one-way ANOVA test with Dunnett's multiple comparison. All results are presented as mean ± SEM values.

Author Contributions: Conceptualization, A.K. (Aleksandra Klimczak) and A.S.; methodology, A.S., A.B.-P., A.K. (Agnieszka Krawczenko), M.P., O.D. and A.K. (Aleksandra Klimczak); formal analysis, A.S., A.B.-P. and A.K. (Aleksandra Klimczak); investigation, A.S., A.B.-P., A.K. (Agnieszka Krawczenko) and O.D.; data curation, A.S., A.B.-P. and A.K. (Aleksandra Klimczak); visualization, A.S. and O.D.; supervision, A.K. (Aleksandra Klimczak); funding acquisition, A.K. (Aleksandra Klimczak); writing—original draft preparation, A.S. and A.B.-P.; writing—review and editing, A.K. (Aleksandra Klimczak). All authors have read and agreed to the published version of the manuscript.

Acknowledgments: The authors are grateful to Sylwia Nowak and Krzysztof Pawlik from the Hirszfeld Institute of Immunology and Experimental Therapy, Polish Academy of Sciences, for their support in analyzing the MVs using transmission electron microscopy.

Abbreviations

ATMSC	Adipose tissue-derived mesenchymal stem cell
BM-MSC	Bone marrow -derived mesenchymal stem cell
CSC	Cancer stem cells
DMEM	Dulbecco's modified Eagle's medium
EGF	Epidermal growth factor
FBS	Fetal bovine serum
FGF	Fibroblast growth factor
FITC	Fluorescein isothiocyanate
GRO	Growth- regulated oncogene, CXCL1 chemokine
HATMSC	Human adipose tissue mesenchymal stem cell line
IL	Interleukin
MCP-1	Macrophage chemoattractant protein-1, CCL2 chemokine
MVs	Microvesicles
MTT	(3-(4,5-Dimetylthiazol-2-yl)-2,5-Diphenyltetrazolium Bromide
PE	Phycoerythrin
RANTES	Regulated on Activation, Normal T Cell Expressed and Secreted, CCL5 chemokine
SSEA4	Stage Specific Embryonic Antigen 4
UC-MSC	Umbilical cord -derived mesenchymal stem cell
VEGF	Vascular endothelial growth factor

References

1. Al-Alem, L.F.; Pandya, U.M.; Baker, A.T.; Bellio, C.; Zarrella, B.D.; Clark, J.; DiGloria, C.M.; Rueda, B.R. Ovarian cancer stem cells: What progress have we made? *Int. J. Biochem. Cell Biol.* **2019**, *107*, 92–103. [CrossRef] [PubMed]
2. Cortez, A.J.; Tudrej, P.; Kujawa, K.A.; Lisowska, K.M. Advances in ovarian cancer therapy. *Cancer Chemother. Pharmacol.* **2018**, *81*, 17–38. [CrossRef] [PubMed]
3. Bregenzer, M.E.; Horst, E.N.; Mehta, P.; Novak, C.M.; Repetto, T.; Mehta, G. The Role of Cancer Stem Cells and Mechanical Forces in Ovarian Cancer Metastasis. *Cancers* **2019**, *11*, 1008. [CrossRef] [PubMed]
4. Keating, A. Mesenchymal stromal cells. *Curr. Opin. Hematol.* **2006**, *13*, 419–425. [CrossRef] [PubMed]
5. Klimczak, A.; Kozlowska, U. Mesenchymal Stromal Cells and Tissue-Specific Progenitor Cells: Their Role in Tissue Homeostasis. *Stem Cells Int.* **2016**, *2016*, 4285215. [CrossRef] [PubMed]
6. Kozlowska, U.; Krawczenko, A.; Futoma, K.; Jurek, T.; Rorat, M.; Patrzalek, D.; Klimczak, A. Similarities and differences between mesenchymal stem/progenitor cells derived from various human tissues. *World J. Stem Cells* **2019**, *11*, 347–374. [CrossRef]
7. Hass, R.; Kasper, C.; Bohm, S.; Jacobs, R. Different populations and sources of human mesenchymal stem cells (MSC): A comparison of adult and neonatal tissue-derived MSC. *Cell Commun. Signal.* **2011**, *9*, 12. [CrossRef]
8. Varderidou-Minasian, S.; Lorenowicz, M.J. Mesenchymal stromal/stem cell-derived extracellular vesicles in tissue repair: Challenges and opportunities. *Theranostics* **2020**, *10*, 5979–5997. [CrossRef]
9. Camussi, G.; Deregibus, M.C.; Bruno, S.; Cantaluppi, V.; Biancone, L. Exosomes/microvesicles as a mechanism of cell-to-cell communication. *Kidney Int.* **2010**, *78*, 838–848. [CrossRef]
10. Nawaz, M.; Fatima, F.; Vallabhaneni, K.C.; Penfornis, P.; Valadi, H.; Ekstrom, K.; Kholia, S.; Whitt, J.D.; Fernandes, J.D.; Pochampally, R.; et al. Extracellular Vesicles: Evolving Factors in Stem Cell Biology. *Stem Cells Int.* **2016**, *2016*, 1073140. [CrossRef]
11. Bobis-Wozowicz, S.; Kmiotek, K.; Sekula, M.; Kedracka-Krok, S.; Kamycka, E.; Adamiak, M.; Jankowska, U.; Madetko-Talowska, A.; Sarna, M.; Bik-Multanowski, M.; et al. Human Induced Pluripotent Stem Cell-Derived Microvesicles Transmit RNAs and Proteins to Recipient Mature Heart Cells Modulating Cell Fate and Behavior. *Stem Cells* **2015**, *33*, 2748–2761. [CrossRef] [PubMed]

12. Albini, A.; Sporn, M.B. The tumour microenvironment as a target for chemoprevention. *Nat. Rev. Cancer* **2007**, *7*, 139–147. [CrossRef] [PubMed]
13. Du, T.; Ju, G.; Wu, S.; Cheng, Z.; Cheng, J.; Zou, X.; Zhang, G.; Miao, S.; Liu, G.; Zhu, Y. Microvesicles derived from human Wharton's jelly mesenchymal stem cells promote human renal cancer cell growth and aggressiveness through induction of hepatocyte growth factor. *PLoS ONE* **2014**, *9*, e96836. [CrossRef] [PubMed]
14. Reza, A.; Choi, Y.J.; Yasuda, H.; Kim, J.H. Human adipose mesenchymal stem cell-derived exosomal-miRNAs are critical factors for inducing anti-proliferation signalling to A2780 and SKOV-3 ovarian cancer cells. *Sci. Rep.* **2016**, *6*, 38498. [CrossRef]
15. Lindoso, R.S.; Collino, F.; Vieyra, A. Extracellular vesicles as regulators of tumor fate: Crosstalk among cancer stem cells, tumor cells and mesenchymal stem cells. *Stem Cell Investig.* **2017**, *4*, 75. [CrossRef]
16. Kraskiewicz, H.; Paprocka, M.; Bielawska-Pohl, A.; Krawczenko, A.; Panek, K.; Kaczynska, J.; Szyposzynska, A.; Psurski, M.; Kuropka, P.; Klimczak, A. Can supernatant from immortalized adipose tissue MSC replace cell therapy? An in vitro study in chronic wounds model. *Stem Cell Res. Ther.* **2020**, *11*, 29. [CrossRef]
17. Whiteside, T.L. Exosome and mesenchymal stem cell cross-talk in the tumor microenvironment. *Semin. Immunol.* **2018**, *35*, 69–79. [CrossRef]
18. Bruno, S.; Collino, F.; Deregibus, M.C.; Grange, C.; Tetta, C.; Camussi, G. Microvesicles derived from human bone marrow mesenchymal stem cells inhibit tumor growth. *Stem Cells Dev.* **2013**, *22*, 758–771. [CrossRef]
19. Dong, L.; Pu, Y.; Zhang, L.; Qi, Q.; Xu, L.; Li, W.; Wei, C.; Wang, X.; Zhou, S.; Zhu, J.; et al. Human umbilical cord mesenchymal stem cell-derived extracellular vesicles promote lung adenocarcinoma growth by transferring miR-410. *Cell Death Dis.* **2018**, *9*, 218. [CrossRef]
20. Kalamegam, G.; Sait, K.H.W.; Ahmed, F.; Kadam, R.; Pushparaj, P.N.; Anfinan, N.; Rasool, M.; Jamal, M.S.; Abu-Elmagd, M.; Al-Qahtani, M. Human Wharton's Jelly Stem Cell (hWJSC) Extracts Inhibit Ovarian Cancer Cell Lines OVCAR3 and SKOV3 in vitro by Inducing Cell Cycle Arrest and Apoptosis. *Front. Oncol.* **2018**, *8*, 592. [CrossRef]
21. Khalil, C.; Moussa, M.; Azar, A.; Tawk, J.; Habbouche, J.; Salameh, R.; Ibrahim, A.; Alaaeddine, N. Anti-proliferative effects of mesenchymal stem cells (MSCs) derived from multiple sources on ovarian cancer cell lines: An in-vitro experimental study. *J. Ovarian Res.* **2019**, *12*, 70. [CrossRef] [PubMed]
22. Lisio, M.A.; Fu, L.; Goyeneche, A.; Gao, Z.H.; Telleria, C. High-Grade Serous Ovarian Cancer: Basic Sciences, Clinical and Therapeutic Standpoints. *Int. J. Mol. Sci.* **2019**, *20*, 952. [CrossRef] [PubMed]
23. Stronach, E.A.; Cunnea, P.; Turner, C.; Guney, T.; Aiyappa, R.; Jeyapalan, S.; de Sousa, C.H.; Browne, A.; Magdy, N.; Studd, J.B.; et al. The role of interleukin-8 (IL-8) and IL-8 receptors in platinum response in high grade serous ovarian carcinoma. *Oncotarget* **2015**, *6*, 31593–31603. [CrossRef]
24. Browne, A.; Sriraksa, R.; Guney, T.; Rama, N.; Van Noorden, S.; Curry, E.; Gabra, H.; Stronach, E.; El-Bahrawy, M. Differential expression of IL-8 and IL-8 receptors in benign, borderline and malignant ovarian epithelial tumours. *Cytokine* **2013**, *64*, 413–421. [CrossRef] [PubMed]
25. Matte, I.; Lane, D.; Laplante, C.; Rancourt, C.; Piche, A. Profiling of cytokines in human epithelial ovarian cancer ascites. *Am. J. Cancer Res.* **2012**, *2*, 566–580. [PubMed]
26. Coward, J.; Kulbe, H.; Chakravarty, P.; Leader, D.; Vassileva, V.; Leinster, D.A.; Thompson, R.; Schioppa, T.; Nemeth, J.; Vermeulen, J.; et al. Interleukin-6 as a therapeutic target in human ovarian cancer. *Clin. Cancer Res.* **2011**, *17*, 6083–6096. [CrossRef]
27. Huang, S.; Robinson, J.B.; Deguzman, A.; Bucana, C.D.; Fidler, I.J. Blockade of nuclear factor-kappaB signaling inhibits angiogenesis and tumorigenicity of human ovarian cancer cells by suppressing expression of vascular endothelial growth factor and interleukin 8. *Cancer Res.* **2000**, *60*, 5334–5339.
28. Yung, M.M.; Tang, H.W.; Cai, P.C.; Leung, T.H.; Ngu, S.F.; Chan, K.K.; Xu, D.; Yang, H.; Ngan, H.Y.; Chan, D.W. GRO-alpha and IL-8 enhance ovarian cancer metastatic potential via the CXCR2-mediated TAK1/NFkappaB signaling cascade. *Theranostics* **2018**, *8*, 1270–1285. [CrossRef]
29. Long, H.; Xie, R.; Xiang, T.; Zhao, Z.; Lin, S.; Liang, Z.; Chen, Z.; Zhu, B. Autocrine CCL5 signaling promotes invasion and migration of CD133+ ovarian cancer stem-like cells via NF-kappaB-mediated MMP-9 upregulation. *Stem Cells* **2012**, *30*, 2309–2319. [CrossRef]

30. Furukawa, S.; Soeda, S.; Kiko, Y.; Suzuki, O.; Hashimoto, Y.; Watanabe, T.; Nishiyama, H.; Tasaki, K.; Hojo, H.; Abe, M.; et al. MCP-1 promotes invasion and adhesion of human ovarian cancer cells. *Anticancer Res.* **2013**, *33*, 4785–4790.
31. Heilmann, T.; Vondung, F.; Borzikowsky, C.; Szymczak, S.; Kruger, S.; Alkatout, I.; Wenners, A.; Bauer, M.; Klapper, W.; Rocken, C.; et al. Heterogeneous intracellular TRAIL-receptor distribution predicts poor outcome in breast cancer patients. *J. Mol. Med.* **2019**, *97*, 1155–1167. [CrossRef] [PubMed]
32. Abdollahi, T. Potential for TRAIL as a therapeutic agent in ovarian cancer. *Vitam. Horm.* **2004**, *67*, 347–364. [PubMed]
33. Braga Lda, C.; Alvares da Silva Ramos, A.P.; Traiman, P.; Silva, L.M.; Lopes da Silva-Filho, A. TRAIL-R3-related apoptosis: Epigenetic and expression analyses in women with ovarian neoplasia. *Gynecol. Oncol.* **2012**, *126*, 268–273. [CrossRef] [PubMed]
34. Ke, C.; Hou, H.; Li, J.; Su, K.; Huang, C.; Lin, Y.; Lu, Z.; Du, Z.; Tan, W.; Yuan, Z. Extracellular Vesicle Delivery of TRAIL Eradicates Resistant Tumor Growth in Combination with CDK Inhibition by Dinaciclib. *Cancers* **2020**, *12*, 1157. [CrossRef]
35. Kalamegam, G.; Sait, K.H.W.; Anfinan, N.; Kadam, R.; Ahmed, F.; Rasool, M.; Naseer, M.I.; Pushparaj, P.N.; Al-Qahtani, M. Cytokines secreted by human Wharton's jelly stem cells inhibit the proliferation of ovarian cancer (OVCAR3) cells in vitro. *Oncol. Lett.* **2019**, *17*, 4521–4531. [CrossRef]
36. Lewis, A.M.; Varghese, S.; Xu, H.; Alexander, H.R. Interleukin-1 and cancer progression: The emerging role of interleukin-1 receptor antagonist as a novel therapeutic agent in cancer treatment. *J. Transl. Med.* **2006**, *4*, 48. [CrossRef]
37. Green, D.S.; Husain, S.R.; Johnson, C.L.; Sato, Y.; Han, J.; Joshi, B.; Hewitt, S.M.; Puri, R.K.; Zoon, K.C. Combination immunotherapy with IL-4 Pseudomonas exotoxin and IFN-alpha and IFN-gamma mediate antitumor effects in vitro and in a mouse model of human ovarian cancer. *Immunotherapy* **2019**, *11*, 483–496. [CrossRef]
38. Tudrej, P.; Olbryt, M.; Zembala-Nozynska, E.; Kujawa, K.A.; Cortez, A.J.; Fiszer-Kierzkowska, A.; Piglowski, W.; Nikiel, B.; Glowala-Kosinska, M.; Bartkowska-Chrobok, A.; et al. Establishment and Characterization of the Novel High-Grade Serous Ovarian Cancer Cell Line OVPA8. *Int. J. Mol. Sci.* **2018**, *19*, 2080. [CrossRef]
39. Bai, S.; Zhu, W.; Coffman, L.; Vlad, A.; Schwartz, L.E.; Elishaev, E.; Drapkin, R.; Buckanovich, R.J. CD105 Is Expressed in Ovarian Cancer Precursor Lesions and Is Required for Metastasis to the Ovary. *Cancers* **2019**, *11*, 1710. [CrossRef]
40. Chen, W.C.; Hsu, H.P.; Li, C.Y.; Yang, Y.J.; Hung, Y.H.; Cho, C.Y.; Wang, C.Y.; Weng, T.Y.; Lai, M.D. Cancer stem cell marker CD90 inhibits ovarian cancer formation via beta3 integrin. *Int. J. Oncol.* **2016**, *49*, 1881–1889. [CrossRef]
41. Lupia, M.; Angiolini, F.; Bertalot, G.; Freddi, S.; Sachsenmeier, K.F.; Chisci, E.; Kutryb-Zajac, B.; Confalonieri, S.; Smolenski, R.T.; Giovannoni, R.; et al. CD73 Regulates Stemness and Epithelial-Mesenchymal Transition in Ovarian Cancer-Initiating Cells. *Stem Cell Rep.* **2018**, *10*, 1412–1425. [CrossRef] [PubMed]
42. Turcotte, M.; Spring, K.; Pommey, S.; Chouinard, G.; Cousineau, I.; George, J.; Chen, G.M.; Gendoo, D.M.; Haibe-Kains, B.; Karn, T.; et al. CD73 is associated with poor prognosis in high-grade serous ovarian cancer. *Cancer Res.* **2015**, *75*, 4494–4503. [CrossRef] [PubMed]
43. Zhang, J.; Yuan, B.; Zhang, H.; Li, H. Human epithelial ovarian cancer cells expressing CD105, CD44 and CD106 surface markers exhibit increased invasive capacity and drug resistance. *Oncol. Lett.* **2019**, *17*, 5351–5360. [CrossRef] [PubMed]
44. Christodoulou, I.; Goulielmaki, M.; Devetzi, M.; Panagiotidis, M.; Koliakos, G.; Zoumpourlis, V. Mesenchymal stem cells in preclinical cancer cytotherapy: A systematic review. *Stem Cell Res. Ther.* **2018**, *9*, 336. [CrossRef] [PubMed]
45. Krawczenko, A.; Bielawska-Pohl, A.; Paprocka, M.; Kraskiewicz, H.; Szyposzynska, A.; Wojdat, E.; Klimczak, A. Microvesicles from Human Immortalized Cell Lines of Endothelial Progenitor Cells and Mesenchymal Stem/Stromal Cells of Adipose Tissue Origin as Carriers of Bioactive Factors Facilitating Angiogenesis. *Stem Cells Int.* **2020**, *2020*, 1289380. [CrossRef] [PubMed]
46. Witwer, K.W.; Buzas, E.I.; Bemis, L.T.; Bora, A.; Lasser, C.; Lotvall, J.; Nolte-'t Hoen, E.N.; Piper, M.G.; Sivaraman, S.; Skog, J.; et al. Standardization of sample collection, isolation and analysis methods in extracellular vesicle research. *J. Extracell. Vesicles* **2013**, *2*. [CrossRef]

47. Adamiak, M.; Cheng, G.; Bobis-Wozowicz, S.; Zhao, L.; Kedracka-Krok, S.; Samanta, A.; Karnas, E.; Xuan, Y.T.; Skupien-Rabian, B.; Chen, X.; et al. Induced Pluripotent Stem Cell (iPSC)-Derived Extracellular Vesicles Are Safer and More Effective for Cardiac Repair Than iPSCs. *Circ. Res.* **2018**, *122*, 296–309. [CrossRef]
48. Matuszyk, J.; Cebrat, M.; Kalas, W.; Strzadala, L. HA1004, an inhibitor of serine/threonine protein kinases, restores the sensitivity of thymic lymphomas to Ca^{2+}-mediated apoptosis through a protein kinase A-independent mechanism. *Int. Immunopharmacol.* **2002**, *2*, 435–442. [CrossRef]

10

Diagnostic and Therapeutic Applications of Exosomes in Cancer with a Special Focus on Head and Neck Squamous Cell Carcinoma (HNSCC)

Eliane Ebnoether [1,2] and Laurent Muller [1,2,*]

1 Department of Biomedicine, University of Basel, 4031 Basel, Switzerland; Eliane.Ebnoether@usb.ch
2 Department of Otorhinolaryngology, Head and Neck Surgery, University Hospital of Basel, 4051 Basel, Switzerland
* Correspondence: laurent.muller@usb.ch

Abstract: Exosomes are nanovesicles part of a recently described intercellular communication system. Their properties seem promising as a biomarker in cancer research, where more sensitive monitoring and therapeutic applications are desperately needed. In the case of head and neck squamous cell carcinoma (HNSCC), overall survival often remains poor, although huge technological advancements in the treatment of this disease have been made. In the following review, diagnostic and therapeutic properties are highlighted and summarised. Impressive first results have been obtained but more research is needed to implement these innovative techniques into daily clinical routines.

Keywords: head and neck squamous cell carcinoma (HNSCC); exosomes; cancer; biomarker; diagnostic; therapy; liquid biopsy

1. Introduction

After the discovery of exosomes decades ago, they have moved into spotlight for various applications. Their potential as promising biomarkers, especially in cancer, has led to new research approaches making personalised medicine more reachable. In the field of head and neck cancer, biomarkers that simplify diagnosis and treatment are scarce.

Head and neck cancer is the sixth leading cancer by incidence worldwide with a poor five-year overall survival rate of 50% [1]. More than 90% are histologically head and neck squamous cell carcinoma (HNSCC). Although progress has been made and therapies have been enhanced, survival rate has not improved significantly. Eventually a lot of patients die due to metastasis and recurrences.

As diagnosis is tedious, needing a lot of clinical experience as well as resources, and the available therapies are toxic and have a huge impact on quality of life, validated biomarkers would support treatment stratification with improved outcome and reduced toxicity.

In this review, interaction within the immune system, potential diagnostic and therapeutic applications of exosomes with focus on HNSCC are highlighted. Changes of exosomal levels with cancer, under radiation and chemotherapy are discussed and therapeutic options like vaccination, immunomodulation and elimination are studied.

2. Intercellular Communication by Exosomes

The first descriptions of exosomes date back more than 30 years. In 1983, a study group discovered the "blebbing" of membranes from transferrin into red blood cells due to maturation of reticulocytes [2]. It was only later that the term exosome was born to describe this "blebbing".

Exosomes were largely dismissed as means of cellular waste and had fallen into oblivion for many decades. Things changed when their ability for intercellular communication, interaction with the immune system and possible use as a vector of drugs were identified [3]. To date, the exact definition of exosomes has led to a lot of discussions. Different types of vesicles, such as exosomes, extracellular vesicles (EVs) and microvesicles (MVs), have been used to describe similar properties intermixing. Up to now, differences between them and exact functions have not been universally defined [4]. The current opinion describes exosomes as double layered nanovesicles (30–50 nm) originating from the endosomal pathway, thus carrying, for example, tsg101 as a typical marker. They are released by all cell types of the body and are present in extracellular spaces and liquids (e.g., blood, saliva, urine) [5]. They can be directly visualised by electron microscopy (see Figure 1). As seen in Figure 1, differential centrifugation and size-exclusion chromatography, usually yields exosomes of different sizes. Typical surface markers are tetraspanins (CD9, CD63, CD81) and Rab proteins (Ras-related in brain) [6].

Exosomes could be compared to hemerodromes in ancient Greece. Pheidippides is said to be the first marathon runner and covered around 240 km in two days to ask for military support in the fight of Marathon. Exosomes are, like Pheidippides, a messenger system allowing intercellular communication. After this huge run, the hemerodrome Pheidippides is said to have collapsed and died [7]. This ancient story may compare to exosomes that are mostly used up after transmission of the message.

Cancer cells especially are thought to use this communication system avidly. As such, they are known to massively produce exosomes, called tumour-derived exosomes (TEX). During their formation, they acquire lipids, nucleic acids and proteins from the parental cell [8]. Thakur discovered in 2014 the representation of the whole cancer genomic DNA in exosomes [9]. TEX carry double-stranded DNA representing the entire genome and reflecting the mutational status of parental tumour cells. From our own results and that of others, we know that exosomes are able to interact with immune cells and assist in tumour immune escape [10,11]. TEX take part in the development, progression and response to treatment in cancer by alternating intercellular communication.

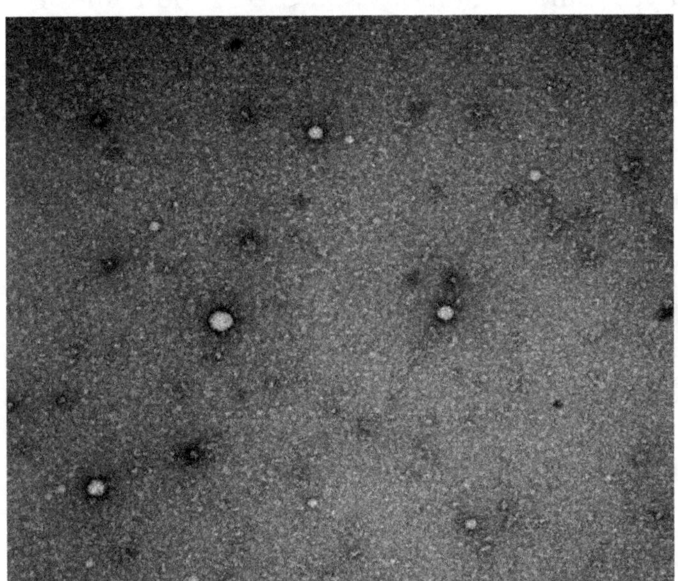

Figure 1. Representative electron microscopy picture by negative staining showing their typical appearance and size range. The exosomes were isolated as described previously from the plasma of a head and neck squamous cell carcinoma (HNSCC) patient [11].

Because of the described properties above, exosomes are ideal candidates to be used as biomarkers and for implementation in new cancer therapies.

3. Exosomes in Cancer

Exosomes have gained attention in various cancer types as possible biomarkers. In pancreatic cancer RNA, proteins and DNA in cancer-derived exosomes have been studied. Researchers found 3000 proteins secreted in exosomes derived from pancreatic cancer [12]. Epidermal growth factor receptor (EGFR) seemed to play an important role as binding leads to an increased activity of carcinogenesis signal transduction pathway and binding of associated ligands (EGF, TGF-alpha) is enhanced in most of pancreatic cancer types [13]. Through binding of EGF to EGFR tumour aggressiveness can increase and enhance cell proliferation, migration and probably metastasis [14]. In 2015, Melo et al. described glypican-1 (GPC1) as a specific surface marker for pancreatic cancer. It additionally was able to distinguish normal control subjects from subjects with benign pancreatic lesions [15].

Kahlert et al. assessed possible participation of genomic DNA in cancer-derived exosomes of patients with pancreatic ductal adenocarcinoma (PDAC). Fragments of genomic DNA could be harvested in exosomes and KRAS as well as p53 mutations were detected, giving an opportunity for profiling cancer. These findings underline their use to forecast prognosis and suited management [16].

In breast cancer, exosomal levels of CEA and CA 153 are linked with cancer progression [17,18]. The amount of miRNA in exosomes correlates with malignancy and prognosis [19,20]. A variety of miRNA have been studied and showed an association with breast tumour subtype as well as stage [21].

Concerning the most frequent cancer worldwide, the protein NY-ESO-1 was found to have a significant correlation with survival in lung cancer [22]. In non-small-cell lung cancer miRNA, miR-21 and miR-4257 in exosomes were upregulated in case of recurrence [23]. Decline in miR-51 and miR-373 in lung cancer patients was a factor of poor prognosis [24]. TEX also demonstrated a higher affinity to EGFR resulting in activation of Akt/protein kinase B pathway and overexpression of VEGF and finally augmented tumour vascularity [25]. Ueda et al. described CD91 as specific exosomal marker for lung adenocarcinoma. A panel of exosomal surface marker CD91, CD317 and EGFR was able to differentiate 75% of patients [26,27].

Studies of exosomes in patients with brain tumor showed that exosomes, depending on their cargo, can be used to analyse tumor mutations and predict outcomes of therapy [28]. Analysis of mRNA and total protein expression of patients with glioma after a vaccination trial could show a relation to immunopathological and clinical parameters [29].

Peinado et al. found in 2012 that the size distribution and number of exosomes did not alternate with clinical stage in patients with melanoma, though protein concentration showed a tumour-stage-dependent rise, being highest at the WHO IV stage compared to normal controls and other stages. Protein-poor exosomes in stage IV patients (< 50 μg/mL) showed a survival advantage [30]. Not only could changes in protein levels be detected, but also protein accumulation patterns. Exosomes marked with radioactive iodine-131 in breast-cancer-bearing mice concentrated in the field of the primary tumor as well as in the lung as a premetastastic niche. Moreover, after receiving cancer treatment radioactivity in lung cells diminished and cancer promoting factors declined (VEGFR2, ICAM-1, bFGF, KC, TNF-α, IL-6, IL-12, IL-10 and IL-13) showing a diminution of metastatic competency of the exosomes [31].

These findings implicate similar results. Although a good approach is made, a well-founded, stable biomarker in mentioned carcinomas has not been established up to now.

4. Diagnostic Application of Exosomes

4.1. Exosomal Blood Levels Correlate With Clinical Disease .

To date, for diagnosis of HNSCC and many other cancers, we have to rely on clinical, histopathological and radiological findings. Often a biopsy in the operating room with possible side-effects is inevitable; however, a minimal tumour growth is needed for detection. Exosomes have moved into scope as possible biomarkers, showing early changes in cell properties in real time and being noninvasive.

Skog and his research group showed in 2008 a similar mutational pattern in blood exosomes and glioblastoma cells [32]. In breast cancer, a correlation between rising levels in CEA and CA 153 in exosomes and cancer progression could be shown [9,18,33]. In 2012, Peinado was able to demonstrate an increase in metastatic behaviour in advanced melanoma and exosomal levels in blood correlated well with disease stage [30].

Coming to HNSCC results are very sparse. Studies could show a disease-activity-dependent level of exosomes in HNSCC with advanced-stage tumours exhibiting an elevated level [34]. Other authors showed similar results [35]. Patients had higher exosomal levels in plasma compared to healthy controls. This is the first study describing action of exosomes in this kind of cancer [34]. Finally, the amount of exosomes almost doubled with progression of cancer stage [36].

Aggravated stress situations like inflammation as well as cancer both lead to an increase of exosomes as part of an enhanced intercellular communication, complicating the differentiation between both [37].

Exosomes produced by normal hematopoietic cells mediate antitumour responses and maintain homeostasis, whereas TEX profile changes during the course of disease and treatment. TEX are able to reprogram the bone marrow, cell differentiation migration and organise metastatic tumour spread. These results raised thoughts of a need to subcategorize exosomes as "immune-system-induced" and "tumour-induced", as functions differ [38]. The Whiteside group separated exosomes in $CD3^{positive}$-fractions (descending from T-lymphocytes) and $CD3^{negative}$-fractions. Hereby, the fraction of immune-derived and tumour-derived exosomes could be studied. In the course of disease, $CD3^{positive}$-exosomes alter their cargo and the ratio of $CD3^{negative}$-exosomes increases, being a marker for cancer progression [38,39]. Clinical response shows an association with found alternations in exosomal cargo.

It has therefore to be considered that TEX show immunosuppressive functions and interference with immune cells, and are used by cancer for tumour escape. A correlation with disease stage and activity could be shown [10,36].

4.2. Isolation of Exosomes

Exosomes can be obtained from many body fluids, such as urine, liquor, ascites, saliva or blood of patients [34,40,41]. However, the isolation process is still time-consuming [42–44]. Different research groups may use various isolation techniques, which makes their comparison difficult [45]. The international society for extracellular vesicles designed an online course to support with methods [46]. In 2013, a consensus paper for standardisation and isolation was released [47]. Our group uses initial differential centrifugation and size-exclusion chromatography by means of columns. If exosomes will be implicated in daily clinical routine, a standardised, time-saving procedure needs to be developed, which is less prone to errors. Our group is working at the moment on rapid bead-based capturing of exosomes for an easier isolation of exosomes, leading to a drastic time reduction and less vulnerability.

In the future, reproducible and reliable techniques are needed as monitoring tools in the clinical setting. Some starting points have been set but still more results are needed. A huge potential will be, as shown by Hong et al., the increased sensitivity of exosomes compared to standard diagnosis techniques [48,49].

4.3. Advantages and Drawbacks of Exosomal Applications

As mentioned above, diagnosis of HNSCC is still strenuous, invasive and linked with great medical expertise. Exosomes, as a potential new diagnostic tool, open up the era of liquid biopsy. As blood collection is a standard procedure in clinical settings, it would be easily applicable to the clinical routine. Further methods like flow cytometry, Western blot or electron microscopy are already established and can be used for continuative examinations. Together with circulating tumor DNA (ct DNA) and circulating tumor cells (CTC), a very sensitive panel for liquid biopsy can be built up, leading to presumably higher sensitivity and earlier diagnosis compared to biopsy or radiography. Initial promising results have been shown in colorectal cancer as well as breast cancer [50,51]. Scholer et al. showed appearance of ct DNA in patients postoperatively after colorectal cancer and detected relapse 9.4 months earlier than CT scan [52]. Compared to ct DNA and CTCs, exosomes seem to be of higher abundance and more stable, which makes them easier to apply into the busy clinical routine.

A persisting main drawback is the lack of standardisation of isolation techniques, which would be important for comparison of results. Besides, platelets in blood seem to secrete exosomes avidly. Already venepuncture and shear forces can lead to release of platelet-derived exosomes and falsification of data [47]. Another not negligible influencing factor is alteration of exosomal levels by infections and other diseases. Owing to the small size of exosomes ranging between 30–150 nm, direct analysis is difficult.

Summing up, the potential of exosomes as diagnostic markers in terms of liquid biopsy is great but prior to clinical application some questions have to be clarified.

4.4. Exosome Tropism and Building of a Premetastatic Niche

Biodistribution of TEX shows organ-specific colonisation due to exosomal integrins. Based on this organotropism TEX seem to prepare a premetastatic niche. Clinical results showed that profiles of integrin expression in plasma TEX could be prognostic factors to predict locations for future metastasis. Exosomal integrins are proposed to promote adhesion, trigger signalling pathways and inflammatory responses in target cells leading to educating the organ for expansion of metastatic cells [53]. Garofalo et al. could not support the theory of exosomal integrins responsible for building of the premetastatic niche in their work. They assume a special "receptor/ligand" combination on the exosomes responsible for exosome tropism [54].

Further studies are needed to focus on the mechanism underlying cancer tropism in exosomes as better understanding is important for use of exosome in delivery as theranostic agents and exosomal integrins as a probable diagnostic factor.

4.5. Exosomal Alterations with Active Therapy

Changes of plasmatic level of exosomes could not just be shown in cancer but also under active therapy. Patients who underwent surgery had a drastic reduction of plasmatic levels of $CD63^{+}$-exosomes, leading to the assumption, that the tumor mass is responsible for a high number of circulating exosomes. In contrast, exosomes containing Caveolin-1 (CAV-1) increased after surgery with the possible cause directed by inflammation [55]. According to Zorrilla et al. a lower level of exosomes prior and after surgery is linked with increased life expectancy in oral squamous cell carcinoma [55].

Interesting results were found by the group of Lisa Mutschelknaus. As radiation is a common therapy in HNSCC, they discovered a radiation-induced change in the exosomal cargo. It promoted AKT-dependent migration and motility in recipient cancer cells. The AKT pathway is one of the most frequently mutated oncogenic pathways [56–58]. The effect was dose-dependent and is an explanation for tumour immune escape and potential driving force for metastatic progression under radiation [59].

Our recent results showed a dramatic decrease of exosomal level seven weeks after primary, combined radio-chemotherapy of HNSCC patients (see Figure 2, results unpublished). This preliminary data demonstrates a positive correlation between clinical response and exosomal blood levels. Although more patients have to be analysed these encouraging results points toward their possible successful use to monitor the therapy.

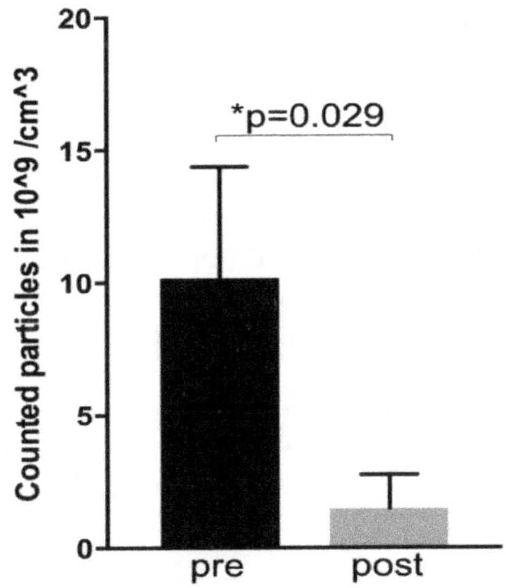

Figure 2. *Preliminary results:* Mean exosomal levels in blood of three patients with active HNSCC before (pre—week 0) and after (post—week 7) primary radio-chemotherapy. The exosomes were isolated as described before [11]. Measurements were performed by Zetaviewer (Particle Metrix, Duesseldorf, Germany). Statistical significance was calculated by a Student *t*-test.

5. Therapeutic Approaches Based On Exosomes

When it comes to therapeutic options, research results are even scarcer. At this point we want to express a warning. Since we are still very early in this research field and the biology behind exosomes is incompletely understood, the application of these new and innovative techniques, although of high research interest, may cause unforeseen harm to patients, and so experiments should be well planned.

5.1. Modulation Of Immune System

An imbalance of the immune system might lead to overaction of the inflammatory pathway. Inflammation gets induced through tissue damage or recognition of pathogens. Chronic inflammation has been associated with an increased risk of malignant cell transformation [60,61]. If progression of cancer takes place, different modulations within the immune system are required: apoptosis, proliferation and angiogenesis. The culmination of inflammatory mediators (Stat3, IL-6, TNF-alpha) guides to an immuno-suppression and following progression of cancer. Exosomes are able to participate in this regulation and alteration of immune players with either inhibiting or promoting action.

Exosomes isolated from malignant ascites of ovarian cancer modulated the function of monocytic cells [62]. Production and secretion of proinflammatory cytokines was induced, like interleukin, tumour necrosis factor alpha via TLR2 and 4, leading to an activated nuclear factor as well as signal transducer and activator of transcription (Stat3). The appearance of activated nuclear factor is often observed in cancer cells and results in a more aggressive phenotype with tissue invasion, metastasis and resistance to growth inhibition. [63] Communication between Stat3 and nuclear factor is important for regulation among inflammatory and cancer cells. Apoptosis, angiogenesis and tumour invasion is regulated making resistance towards immune surveillance possible [64].

Apoptosis, as one of the first steps for cancer progression, can be seen through activated T-cells via expression of death ligands (FasL [65] and TRAIL [66]), leading to an evasion of immune surveillance. Concluding modulation with release of FasL-positive exosomes is able to guide immune escape. A regulation of T-cell apoptosis was exemplary shown in human colorectal cancer [67]. Adding up impaired dendritic cell differentiation and suppression of natural killer cells has a similar effect [68,69]. In metastatic ovarian cancer, Yokoi et al. proved that cancer-derived exosomes induced apoptosis after uptake by mesothelial cells perforated the peritoneum resulting in peritoneal dissemination [70].

Concerning proliferation as an important modulation, exosomes derived from thrombin-activated platelets showed an exciting potential for survival, proliferation and chemotaxis of hematopoietic cells [71]. Besides they were capable of activating monocytes and B-cells [72].

Angiogenesis plays an important role during tumour growth, supplying oxygen and nutrition to the cancer and showing effect on an exosomal level [73].

The induction of angiogenesis has been described by various research groups. In 2008, a promotion of angiogenesis could be seen in exosomes secreted by glioblastoma cells [32]. In the same disease, Svensson et al. noted activation of PAR-2 also in vascular endothelial cells. They have arisen through induced angiogenesis after hypoxic treatment [74]. Later, mRNA of exosomes demonstrated proliferation of vascular endothelial cells [75]. An impact while studying renal cancer cell line was seen as CD105-positive exosomes led after affection of vascular endothelial cells to a promotion of growth. CD105-negative cells maintained no effect [76]. Subsequently, promotion of angiogenesis was shown in breast cancer, as miR-210 contained by exosomes seemed to have an impact on it and reversion by suppression of exosomes secretion stated the contrary [77]. Similar findings were reported by Li et al. in liver cancer cells [78].

The migration and invasion of cancer cells are subsequent important factors in metastasis. In colon carcinoma, a transfer of TEX induced a rise of cell proliferation and chemoresistance [79]. In melanoma cells of patients with advanced stages of carcinoma, bone marrow-derived cells were stimulated by exosomes forming a metastatic niche and leading to a prometastatic phenotype [30]. Also, HSP90alpha, as an exosomal surface marker, is able to activate plasminogen to lead to enhanced invasive capacity of tumor cells [80].

Exosomes were studied in relation to T- and NK-cells, as well as macrophages. Exosomes possess the ability of downregulating CD69 expression on $CD4^{+}$ T-cells and interfere with T-cell activation. Exosomes of cancer patients showed even higher immune suppression by decreased levels of CD69 expression. [34] These findings confirm results of other groups [11,81–83]. It is suspected that mRNA expression and the translation into inhibitory proteins gets enhanced by TEX.

When looking at $CD8^{+}$ T-cells, coincubation with TEX lets them undergo apoptosis [11].

TEX also exert effects on Treg-cells. The adenosine pathway is used by Treg-cells to operate suppression. A downregulation of mRNA coding for this pathway genes was shown in Treg-cells after coincubation with TEX. Also, TEX appear to activate resting Treg-cells through the increase of CD39 as well as CD73 production. Furthermore, calcium signalling is important for T-cell activation and T-cell-dependent immune responses. Incubation with TEX showed a calcium influx in T-cells, which was stronger than the response induced by dendritic cells [11].

The contribution of B-cells was, to date, not well respected. Results of our experiments could show a suppression of activated B-cells by TEX, which seems to be another way TEX may exert protumour effects (unpublished results).

The exact mechanism of immunomodulation through TEX is still unclear. An important part seems to be the uptake of small noncoding RNA (microRNA, miRNA), which leads to alteration of RNA expression and reprogramming of recipient cells [38]. This was, in particular, shown in immune cells, which take up exosomes as B-cells, monocytes and dendritic cells [82]. miRNA leads to destabilisation of RMA after binding and consecutive reduction of expression of the encoded protein [84]. Some of exosomes-mediated post-transcriptional alternations are associated with horizontal transfer of miRNA and oncogenic proteins from tumour cells to endothelial cells.

Exosomes having miRNA as their cargo are able to shape immune responses and seem responsible for protumorigenic functions in the recipient cells.

TEX signalling happens through receptor binding either on the vesicle surface or an internalisation by the recipient cell takes place. Our work could show that a direct surface interaction of exosomes with T-cells is sufficient to transfer signals. There is no internalisation needed as seen in phagocytic cells [11]. Internalisation and binding of TEX by B-cells and monocytes appeared to be time-dependent, dose-dependent and recipient cell type dependent.

Taken together, TEX take effect in antitumor immune functions and promotion of tumour progression leading to a gain or loss of function in recipient cells and being an option used by tumours as immune escape [30,38,85]. TEX play a dual part, having proinflammatory as well as anti-inflammatory properties. Exosomes of healthy donors did not induce immunosuppression [86].

5.2. Loaded Exosomes as Targeted Drug Delivery

Yang et al. showed that unmodified exosomes are able to cross the blood–brain barrier to some extent [87]. Brain exposure can, however, be increased by using certain brain-targeting ligands [88,89].

Exosomes can be modified using various techniques after isolation (exogenous loading): freeze–thaw cycles, incubation, sonication, electroporation and extrusion.

On the contrary, modification can also be made during the formation of exosomes (endogenous loading).

As a breakthrough, Batrakova managed to load exosomes exogenous with the chemotherapeutic agent paclitaxel as the drug-delivery technique. Therapy was, with respect to lung metastasis within mice and in vitro, a lot more successful, having effect even on multiresistant cells and reducing required medication 50 times [90]. Loading drugs into exosomes seems a powerful way of delivering anticancer drugs, but it may lead to increased toxicity, which has to be evaluated carefully.

Another study group loaded glioblastoma-derived exosomes with curcumin (an anti-inflammatory drug) and administered it intranasal to animals with encephalitis. It led to a clinical amelioration with induction of immune tolerance and apoptosis of activated immune cells [91].

Several groups studied the delivery of small interfering RNA (siRNA) into cells using exosomes loaded with electroporation. Some report functional siRNA, others were unsuccessful in proceeding with electroporation [89,92–94].

The endogenous approach uses direct transfection of RNA in interest cells, from which exosomes are later derived. Loading with miRNA and siRNA was possible and functional delivery was shown by various groups [94–98]. At the moment, limiting factors of this method are unknown alterations made to the exosome donor cell due to transfection leading to the changed cargo of exosomes. Besides, when conducting transfection, the produced side-products are copurified during isolation, features of exosomes can be changed leading to false-positive effects.

These are promising results for using exosomes as drug delivery vehicles and could be the solution to therapy-resistant cells as well as tumour immune escape.

A huge advantage, the possibility of exosomes to cross the blood–brain barrier, has to be highlighted; this leads to a higher bioavailability regarding the therapeutic use of exosomes if drugs of biologics are administered to them. Additionally, the toxicological effect is selective and targeted, reducing systemic effects compared to conventional free drugs.

However, research is in the fledgling stages and loading of the vesicles can be challenging, with probable alteration of exosomal membrane and function. Another known limitation is well known in cancer research, that once the successful research moves on from animal models to humans, the observed effects may be fading.

5.3. Modified Exosomes as Vaccinations

Already two decades ago, dendritic-cell derived exosomes were pulsed with tumour-cell peptides and injected in immune-competent mice. An immunogenic rejection of the tumour was obtained [99]. Due to promising results two clinical trials against melanoma and non-small-cell lung cancer were launched. Regrettably just a small number of patients benefitted from the vaccines.

Some years later, a phase I clinical trial was conducted in China. Extracellular vesicles (EV) from ascites of colorectal cancer patients were taken and combined with granulocyte-macrophage colony-stimulating factor (GM-CSF) to enhance antitumour dendritic cell activity. Patients benefited from GM-CSF combined with EV but not from EV alone [100].

The results show the successful use of exosomes to induce an antitumour immune response. From previous vaccination trials performed in cancer, it is known that the antitumour response is not enough to get rid of the cancer and new studies are being performed to combine immune-modulatory drugs with vaccines to increase their efficiency.

5.4. Engineered/Designer Exosomes

The first approaches for delivery drug technologies in the context of loading already existing exosomes have already been described above. This technology bears some problems like probable missing tissue specificity and immunogenic potential.

Alvarez-Erviti et al. engineered exosomes with self-derived immature dendritic cells to reduce immunogenicity and conducted them to express Lamp2b to realise tissue-specificity. Purified exosomes were loaded with siRNA by electroporation fused to a peptide important in the treatment of Alzheimer's disease and injected intravenously. Results showed specific delivery to the brain with a knockdown of mRNA and peptid by 60% [89].

Rashid et al. highlights further engineering of exosomes and their accumulated future hopes and problems. The perfect cell used to engineer exosomes should be nonimmunogenic, have an unlimited cell passage capacity without mutating, have an ability to produce itself abundantly and be easily modified. They believe to have found these potentials in HEK293-Tcells. Mice with metastatic breast cancer showed slower tumor growth after injection with engineered exosomes combined with a better survival compared to control.

Based on HEK293-Tcells, Kojima et al. even took it a step further and designed exosomes completely out of single components. Therapeutic catalase mRNA as cargo in exosomes showed attenuated neuroinflammation in Parkinson's disease as a positive results [101].

Designer exosomes are still in their fledgling stages but deliver promising results for further research as drug-delivering technologies.

5.5. Exosomes in Immune-Modulatory Therapies

Therapy in HNSCC has been mainly limited to surgery and radio-chemotherapy in the past few years. With growing understanding of the immune system and discovery of CTLA-4, PD-1, its inhibiting functions and suitable antibodies through James Allison and Tasuku Honjo check-point inhibition was found. Antibodies of CTLA-4 showed auspicious results in metastasised melanoma. PD-1 was further studied in HNSCC.

PD-L1 is found on the surface of many cell types being a membrane-bound ligand and upregulated in inflammation or malign situations. After binding to the PD-1 receptor on immune T-cells an activation of T-cells gets suppressed important to keep inflammatory responses regulated, being a checkpoint inhibitor. Tumor cells can adapt the mechanism to evade immune destruction [102]. After the discovery of Honjo, an immune-modulatory treatment with anti-PD-1 treatment was developed, showing some response in metastasised HNSCC. To date, it is the only FDA-approved immune-modulatory therapy with successful results.

One of the great drawbacks remains the poor response rate of 10–30% with anti-PD-1 treatment [103]. In 2017, isolated exosomes of patients with HNSCC could be shown to carry PD-L1 and PD-1 [36]. PD-L1-carrying exosomes correlate with clinical stage, disease activity and prognosis. Higher numbers of PD-L1^{positive} exosomes leads to poorer prognosis. In contrast, PD-1-containing exosomes did not match with the clinicopathological profile.

Plasma-derived exosomes in HNSCC showed biologically active PD-L1, which managed T-cell dysfunction. Anti-PD-1 was able to reverse this suppression. Maybe exosomal PD-L1, but not PD-1, is clinically relevant, possibly explaining reduced response rate with anti-PD-1 treatment and stratifying clinical responders from nonresponders. [104,105]. Besides, the membrane-bound form PD-L1 was also shown to be freely soluble in blood and other liquids (sPD-L1). Comparing exosome-bound PD-L1 and sPD-L1 levels in HNSCC, the soluble form did not correlate with clinicopathological findings. Results of other studies showed alternating results [106,107].

Exosomal PD-L1 recently moved into the spotlight as a possible therapeutic target, as suppression induces antitumour immunity and memory. Membrane-bound PD-L1 leads to diminished activity of T-cells in lymph nodes. Blocking of exosomal PD-L1 leads to a reversion and suppression of growth of local tumour cells.

Summing up, exosomal PD-L1 is an important regulator of tumour progression with the ability to suppress T-cell activation. Inhibition can lead to antitumour immunity [108].

5.6. Blocking/Elimination of Exosomes

A group from Massachusetts observed an interesting finding: incubation of glioma-cell derived exosomes with heparin led to a decreased transfer of exosomes between the parental and recipient cell [109,110]. The blockage happened on the surface and not internally. Further studies concerning the blocking of exosomes or removing them have, to our knowledge, not been conducted.

5.7. Challenges to Overcome Application of Exosomes

Exosomes may undergo morphological and functional changes during carcinogenesis, making them elusive for therapeutic applications. In peripheral blood, tumour-associated antigens (TAA) can be found also. Circulating anti-TAA antibodies appear years before clinical diagnosis and are present in all tumor types [111–113]. Despite vast research, no specific TAA has been found for HNSCC. Maybe a panel of expression markers will increase sensitivity and specificity.

6. Ongoing Clinical Trials with Exosomes in Cancer

In recent years, exosomes have risen to be a trending topic, and a lot of clinical trials have emerged. At the moment, there are 85 active studies listed (www.clinicaltrials.gov, 05 June 2020). Owing to the amount, we omit to list them all. Four studies are listed in the field of HNSCC and exosomes. Further details are mentioned in Figure 3.

Status	Study Title	Conditions	Interventions	Locations
Active, not recruiting	**Edible Plant Exosome** Ability to Prevent Oral Mucositis Associated With Chemoradiation Treatment of Head and Neck Cancer	- Head and Neck Cancer - Oral Mucositis	Dietary Supplement: Grape extract Drug: Lortab, Fentanyl patch, mouthwash	James Graham Brown Cancer Center Louisville, Kentucky, United States
Active, not recruiting Has results	**Metformin** Hydrochloride in Affecting Cytokines and **Exosomes** in Patients With Head and Neck Cancer	- Larynx - Lip - Oral Cavity - Pharynx	Radiation: External Beam Radiation Therapy Drug: Metformin Hydrochloride Other: Placebo	Sidney Kimmel Cancer Center at Thomas Jefferson University Philadelphia, Pennsylvania, United States
Recruiting	**Nivolumab and BMS986205** in Treating Patients With Stage II-IV Squamous Cell Cancer of the Head and Neck	- Lip - Oral Cavity Squamous Cell Carcinoma - Pharynx	Biological: Nivolumab Biological: IDO1 Inhibitor (BMS-986205) Procedure: Therapeutic Conventional Surgery Other: Questionnaire Administration	- Ohio State University Columbus, Ohio, United States - Sidney Kimmel Cancer Center at Thomas Jefferson University Philadelphia, Pennsylvania, United States
Recruiting	**Exosome Testing** as a Screening Modality for **Human Papillomavirus-Positive** Oropharyngeal Squamous Cell Carcinoma	- Oropharyngeal Cancer		University of New Mexico Cancer Center Albuquerque, New Mexico, United States

Figure 3. Ongoing clinical trials in cancer involving exosomes in HNSCC.

The first study investigates grape exosomes as a possible treatment of oral mucositis and examines production of cytokines and immune responses to tumour exosomal antigens and metabolic and molecular markers. The second study examines effects of metformin treatment during radiotherapy on mucositis and change in exosomes and cytokines. The third study is led by Brystol-Myers Squibb. A new biological is tested (BMS986205) for advanced-stage HNSCC. Exosomes are studied in a sidearm and do not belong to the direct treatment. The last study conducted by University of New Mexico analyses a prophylactical swab similar to the cervix PAP-swab and compares it with exosomal levels.

7. Conclusions

Since the discovery of exosomes some decades ago, an extended research occurred in this field. Regarding diagnostic possibilities there have been potential approaches, especially within monitoring of HNSCC shown for a follow-up marker (Figure 4). However, for implication in the clinical setting an amendment, stabilisation and standardisation of isolation techniques is needed. Moreover, techniques need to be quick and adequate to present a high sensitivity and specificity in clinical daily routine. There are already ongoing studies.

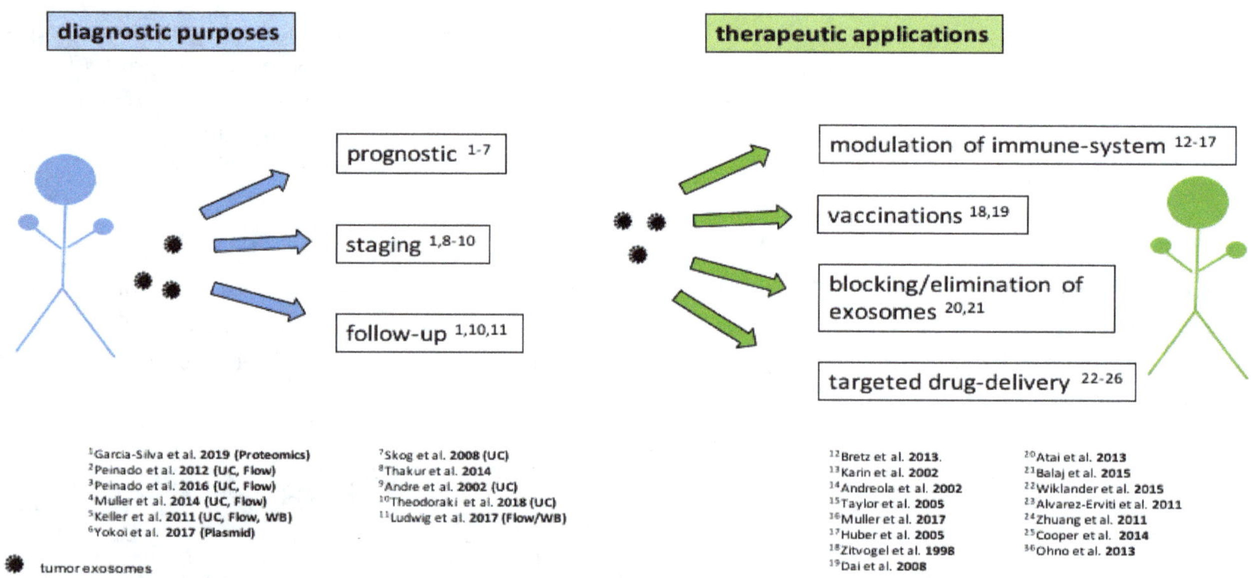

Figure 4. Applications of exosomes in cancer treatment. Isolation technique and method of analysis are written in brackets for the diagnostic purposes. Most research groups used size exclusion chromatography with ultracentrifugation (UC) or flow cytometry (Flow). Some used Western blotting (WB), plasmid transfection (Plasmid) or Fluorescence Activated Cell Sorting (FACS). List of papers consists of a selection.

As in the case of diagnosing recurrences, there is as a complicating factor the hazard of differentiation between recurrence and inflammation (e.g., as result to immunotherapy) as both show increase in exosomal levels and exosomes play a key role in both pathways. This factor aggravates the interpretation of immunotherapy either just being inflammation, tumor necrosis or active recurrence.

In respect of therapeutic options, a great need of action is required. Immunotherapy by PD-1/PD-L1 inhibitors showed already the right direction. A targeted-oriented therapy would be the aim.

Prevailing setbacks in work with exosomes is the not completely understood biology behind them. Therefore, application as targeted drug therapy is delicate as accurate changes and interactions are not known.

Summing up, exosomes seem promising agents to be used for diagnostic or therapeutic purposes in the future, but further studies, including of larger patient cohorts, are needed.

Author Contributions: E.E. wrote the original manuscript, revised and finalized the manuscript. L.M. provided supervision, reviewed and edited the original manuscript. All authors have read and agreed to the published version of the manuscript.

References

1. Bray, F.; Ferlay, J.; Soerjomataram, I.; Siegel, R.L.; Torre, L.A.; Jemal, A. Global cancer statistics 2018: GLOBOCAN estimates of incidence and mortality worldwide for 36 cancers in 185 countries. *CA Cancer J. Clin.* **2018**, *68*, 394–424. [CrossRef] [PubMed]
2. Harding, C.; Heuser, J.; Stahl, P. Receptor-mediated endocytosis of transferrin and recycling of the transferrin receptor in rat reticulocytes. *J. Cell Biol.* **1983**, *97*, 329–339. [CrossRef] [PubMed]
3. Raposo, G.; Nijman, H.W.; Stoorvogel, W.; Liejendekker, R.; Harding, C.V.; Melief, C.J.; Geuze, H.J. B lymphocytes secrete antigen-presenting vesicles. *J. Exp. Med.* **1996**, *183*, 1161–1172. [CrossRef] [PubMed]
4. Mathieu, M.; Martin-Jaular, L.; Lavieu, G.; Thery, C. Specificities of secretion and uptake of exosomes and other extracellular vesicles for cell-to-cell communication. *Nat. Cell Biol.* **2019**, *21*, 9–17. [CrossRef]
5. Keller, S.; Ridinger, J.; Rupp, A.K.; Janssen, J.W.; Altevogt, P. Body fluid derived exosomes as a novel template for clinical diagnostics. *J. Transl. Med.* **2011**, *9*, 86. [CrossRef]
6. Colombo, M.; Moita, C.; van Niel, G.; Kowal, J.; Vigneron, J.; Benaroch, P.; Manel, N.; Moita, L.F.; Thery, C.; Raposo, G. Analysis of ESCRT functions in exosome biogenesis, composition and secretion highlights the heterogeneity of extracellular vesicles. *J. Cell Sci.* **2013**, *126*, 5553–5565. [CrossRef]
7. TEOE, B. Battle of Marathon. In *Encylopaedia Britannica*; Encyclopaedia Britannica Inc.: Chicago, IL, USA, 1998.
8. Xin, T.; Greco, V.; Myung, P. Hardwiring stem cell communication through tissue structure. *Cell* **2016**, *164*, 1212–1225. [CrossRef]
9. Thakur, B.K.; Zhang, H.; Becker, A.; Matei, I.; Huang, Y.; Costa-Silva, B.; Zheng, Y.; Hoshino, A.; Brazier, H.; Xiang, J.; et al. Double-stranded DNA in exosomes: A novel biomarker in cancer detection. *Cell Res.* **2014**, *24*, 766–769. [CrossRef]
10. Muller, L.; Mitsuhashi, M.; Simms, P.; Gooding, W.E.; Whiteside, T.L. Tumor-derived exosomes regulate expression of immune function-related genes in human T cell subsets. *Sci. Rep.* **2016**, *6*, 20254. [CrossRef]
11. Muller, L.; Simms, P.; Hong, C.S.; Nishimura, M.I.; Jackson, E.K.; Watkins, S.C.; Whiteside, T.L. Human tumor-derived exosomes (TEX) regulate Treg functions via cell surface signaling rather than uptake mechanisms. *Oncoimmunology* **2017**, *6*, e1261243. [CrossRef]
12. Adamczyk, K.A.; Klein-Scory, S.; Tehrani, M.M.; Warnken, U.; Schmiegel, W.; Schnolzer, M.; Schwarte-Waldhoff, I. Characterization of soluble and exosomal forms of the EGFR released from pancreatic cancer cells. *Life Sci.* **2011**, *89*, 304–312. [CrossRef] [PubMed]
13. Bloomston, M.; Bhardwaj, A.; Ellison, E.C.; Frankel, W.L. Epidermal growth factor receptor expression in pancreatic carcinoma using tissue microarray technique. *Dig. Surg.* **2006**, *23*, 74–79. [CrossRef] [PubMed]
14. Nuzhat, Z.; Kinhal, V.; Sharma, S.; Rice, G.E.; Joshi, V.; Salomon, C. Tumour-derived exosomes as a signature of pancreatic cancer - liquid biopsies as indicators of tumour progression. *Oncotarget* **2017**, *8*, 17279–17291. [CrossRef] [PubMed]
15. Melo, S.A.; Luecke, L.B.; Kahlert, C.; Fernandez, A.F.; Gammon, S.T.; Kaye, J.; LeBleu, V.S.; Mittendorf, E.A.; Weitz, J.; Rahbari, N.; et al. Glypican-1 identifies cancer exosomes and detects early pancreatic cancer. *Nature* **2015**, *523*, 177–182. [CrossRef]
16. Kahlert, C.; Melo, S.A.; Protopopov, A.; Tang, J.; Seth, S.; Koch, M.; Zhang, J.; Weitz, J.; Chin, L.; Futreal, A.; et al. Identification of double-stranded genomic DNA spanning all chromosomes with mutated KRAS and p53 DNA in the serum exosomes of patients with pancreatic cancer. *J. Biol. Chem.* **2014**, *289*, 3869–3875. [CrossRef]
17. Dreyer, F.; Baur, A. Biogenesis and Functions of Exosomes and Extracellular Vesicles. *Methods Mol. Biol.* **2016**, *1448*, 201–216. [CrossRef]

18. Andre, F.; Schartz, N.E.; Movassagh, M.; Flament, C.; Pautier, P.; Morice, P.; Pomel, C.; Lhomme, C.; Escudier, B.; Le Chevalier, T.; et al. Malignant effusions and immunogenic tumour-derived exosomes. *Lancet* **2002**, *360*, 295–305. [CrossRef]
19. Tetta, C.; Ghigo, E.; Silengo, L.; Deregibus, M.C.; Camussi, G. Extracellular vesicles as an emerging mechanism of cell-to-cell communication. *Endocrine* **2013**, *44*, 11–19. [CrossRef]
20. Joyce, D.P.; Kerin, M.J.; Dwyer, R.M. Exosome-encapsulated microRNAs as circulating biomarkers for breast cancer. *Int. J. Cancer* **2016**, *139*, 1443–1448. [CrossRef]
21. Jia, Y.; Chen, Y.; Wang, Q.; Jayasinghe, U.; Luo, X.; Wei, Q.; Wang, J.; Xiong, H.; Chen, C.; Xu, B.; et al. Exosome: Emerging biomarker in breast cancer. *Oncotarget* **2017**, *8*, 41717–41733. [CrossRef]
22. Sandfeld-Paulsen, B.; Jakobsen, K.R.; Baek, R.; Folkersen, B.H.; Rasmussen, T.R.; Meldgaard, P.; Varming, K.; Jorgensen, M.M.; Sorensen, B.S. Exosomal proteins as diagnostic biomarkers in lung cancer. *J. Thorac. Oncol.* **2016**, *11*, 1701–1710. [CrossRef] [PubMed]
23. Dejima, H.; Iinuma, H.; Kanaoka, R.; Matsutani, N.; Kawamura, M. Exosomal microRNA in plasma as a non-invasive biomarker for the recurrence of non-small cell lung cancer. *Oncol. Lett.* **2017**, *13*, 1256–1263. [CrossRef]
24. Alipoor, S.D.; Adcock, I.M.; Garssen, J.; Mortaz, E.; Varahram, M.; Mirsaeidi, M.; Velayati, A. The roles of miRNAs as potential biomarkers in lung diseases. *Eur. J. Pharmacol.* **2016**, *791*, 395–404. [CrossRef] [PubMed]
25. Reclusa, P.; Sirera, R.; Araujo, A.; Giallombardo, M.; Valentino, A.; Sorber, L.; Bazo, I.G.; Pauwels, P.; Rolfo, C. Exosomes genetic cargo in lung cancer: A truly Pandora's box. *Transl. Lung Cancer Res.* **2016**, *5*, 483–491. [CrossRef] [PubMed]
26. Ueda, K.; Ishikawa, N.; Tatsuguchi, A.; Saichi, N.; Fujii, R.; Nakagawa, H. Antibody-coupled monolithic silica microtips for highthroughput molecular profiling of circulating exosomes. *Sci. Rep.* **2014**, *4*, 6232. [CrossRef] [PubMed]
27. Jakobsen, K.R.; Paulsen, B.S.; Baek, R.; Varming, K.; Sorensen, B.S.; Jorgensen, M.M. Exosomal proteins as potential diagnostic markers in advanced non-small cell lung carcinoma. *J. Extracell. Vesicles* **2015**, *4*, 26659. [CrossRef] [PubMed]
28. Redzic, J.S.; Ung, T.H.; Graner, M.W. Glioblastoma extracellular vesicles: Reservoirs of potential biomarkers. *Pharmgenomics Pers. Med.* **2014**, *7*, 65–77. [CrossRef]
29. Muller, L.; Muller-Haegele, S.; Mitsuhashi, M.; Gooding, W.; Okada, H.; Whiteside, T.L. Exosomes isolated from plasma of glioma patients enrolled in a vaccination trial reflect antitumor immune activity and might predict survival. *Oncoimmunology* **2015**, *4*, e1008347. [CrossRef]
30. Peinado, H.; Aleckovic, M.; Lavotshkin, S.; Matei, I.; Costa-Silva, B.; Moreno-Bueno, G.; Hergueta-Redondo, M.; Williams, C.; Garcia-Santos, G.; Ghajar, C.; et al. Melanoma exosomes educate bone marrow progenitor cells toward a pro-metastatic phenotype through MET. *Nat. Med.* **2012**, *18*, 883–891, corrected in *Nat. Med.* **2016**, *22*, 1502, doi:10.1038/nm1216-1502b. [CrossRef]
31. Rashid, M.H.; Borin, T.F.; Ara, R.; Angara, K.; Cai, J.; Achyut, B.R.; Liu, Y.; Arbab, A.S. Differential in vivo biodistribution of (131)I-labeled exosomes from diverse cellular origins and its implication for theranostic application. *Nanomedicine* **2019**, *21*, 102072. [CrossRef]
32. Skog, J.; Wurdinger, T.; van Rijn, S.; Meijer, D.H.; Gainche, L.; Sena-Esteves, M.; Curry, W.T., Jr.; Carter, B.S.; Krichevsky, A.M.; Breakefield, X.O. Glioblastoma microvesicles transport RNA and proteins that promote tumour growth and provide diagnostic biomarkers. *Nat. Cell Biol.* **2008**, *10*, 1470–1476. [CrossRef] [PubMed]
33. Garcia-Silva, S.; Benito-Martin, A.; Sanchez-Redondo, S.; Hernandez-Barranco, A.; Ximenez-Embun, P.; Nogues, L.; Mazariegos, M.S.; Brinkmann, K.; Amor Lopez, A.; Meyer, L.; et al. Use of extracellular vesicles from lymphatic drainage as surrogate markers of melanoma progression and BRAF (V600E) mutation. *J. Exp. Med.* **2019**, *216*, 1061–1070. [CrossRef] [PubMed]
34. Muller, L.; Hong, C.S.; Stolz, D.B.; Watkins, S.C.; Whiteside, T.L. Isolation of biologically-active exosomes from human plasma. *J. Immunol. Methods* **2014**, *411*, 55–65. [CrossRef] [PubMed]
35. Theodoraki, M.N.; Hoffmann, T.K.; Jackson, E.K.; Whiteside, T.L. Exosomes in HNSCC plasma as surrogate markers of tumour progression and immune competence. *Clin. Exp. Immunol.* **2018**, *194*, 67–78. [CrossRef]
36. Ludwig, S.; Floros, T.; Theodoraki, M.N.; Hong, C.S.; Jackson, E.K.; Lang, S.; Whiteside, T.L. Suppression of Lymphocyte Functions by Plasma Exosomes Correlates with Disease Activity in Patients with Head and Neck Cancer. *Clin. Cancer Res.* **2017**, *23*, 4843–4854. [CrossRef] [PubMed]

37. Isola, A.L.; Chen, S. Exosomes: The Messengers of Health and Disease. *Curr. Neuropharmacol.* **2017**, *15*, 157–165. [CrossRef]
38. Whiteside, T.L. Tumor-Derived Exosomes and Their Role in Cancer Progression. *Adv. Clin. Chem.* **2016**, *74*, 103–141. [CrossRef]
39. Theodoraki, M.N.; Hoffmann, T.K.; Whiteside, T.L. Separation of plasma-derived exosomes into CD3((+)) and CD3((-)) fractions allows for association of immune cell and tumour cell markers with disease activity in HNSCC patients. *Clin. Exp. Immunol.* **2018**, *192*, 271–283. [CrossRef]
40. Hong, C.S.; Funk, S.; Whiteside, T.L. Isolation of Biologically Active Exosomes from Plasma of Patients with Cancer. *Methods Mol. Biol.* **2017**, *1633*, 257–265. [CrossRef]
41. Nonaka, T.; Wong, D.T.W. Liquid Biopsy in Head and Neck Cancer: Promises and Challenges. *J. Dent. Res.* **2018**, *97*, 701–708. [CrossRef]
42. Whiteside, T.L. Extracellular vesicles isolation and their biomarker potential: Are we ready for testing? *Ann. Transl. Med.* **2017**, *5*, 54. [CrossRef] [PubMed]
43. Thery, C.; Amigorena, S.; Raposo, G.; Clayton, A. Isolation and characterization of exosomes from cell culture supernatants and biological fluids. *Curr. Protoc. Cell Biol.* **2006**, *3*, 3.22.1–3.22.29. [CrossRef] [PubMed]
44. Taylor, D.D.; Shah, S. Methods of isolating extracellular vesicles impact down-stream analyses of their cargoes. *Methods* **2015**, *87*, 3–10. [CrossRef]
45. Van Deun, J.; Mestdagh, P.; Sormunen, R.; Cocquyt, V.; Vermaelen, K.; Vandesompele, J.; Bracke, M.; De Wever, O.; Hendrix, A. The impact of disparate isolation methods for extracellular vesicles on downstream RNA profiling. *J. Extracell. Vesicles* **2014**, *3*. [CrossRef] [PubMed]
46. Lasser, C.; Thery, C.; Buzas, E.I.; Mathivanan, S.; Zhao, W.; Gho, Y.S.; Lotvall, J. The International Society for Extracellular Vesicles launches the first massive open online course on extracellular vesicles. *J. Extracell. Vesicles* **2016**, *5*, 34299. [CrossRef] [PubMed]
47. Witwer, K.W.; Buzas, E.I.; Bemis, L.T.; Bora, A.; Lasser, C.; Lotvall, J.; Nolte-'t Hoen, E.N.; Piper, M.G.; Sivaraman, S.; Skog, J.; et al. Standardization of sample collection, isolation and analysis methods in extracellular vesicle research. *J. Extracell. Vesicles* **2013**, *2*. [CrossRef] [PubMed]
48. Hong, C.S.; Danet-Desnoyers, G.; Shan, X.; Sharma, P.; Whiteside, T.L.; Boyiadzis, M. Human acute myeloid leukemia blast-derived exosomes in patient-derived xenograft mice mediate immune suppression. *Exp. Hematol.* **2019**, *76*, 60–66.e2. [CrossRef] [PubMed]
49. Hong, C.S.; Sharma, P.; Yerneni, S.S.; Simms, P.; Jackson, E.K.; Whiteside, T.L.; Boyiadzis, M. Circulating exosomes carrying an immunosuppressive cargo interfere with cellular immunotherapy in acute myeloid leukemia. *Sci. Rep.* **2017**, *7*, 14684. [CrossRef] [PubMed]
50. Jia, S.; Zhang, R.; Li, Z.; Li, J. Clinical and biological significance of circulating tumor cells, circulating tumor DNA, and exosomes as biomarkers in colorectal cancer. *Oncotarget* **2017**, *8*, 55632–55645. [CrossRef] [PubMed]
51. Alimirzaie, S.; Bagherzadeh, M.; Akbari, M.R. Liquid biopsy in breast cancer: A comprehensive review. *Clin. Genet.* **2019**, *95*, 643–660. [CrossRef]
52. Scholer, L.V.; Reinert, T.; Orntoft, M.W.; Kassentoft, C.G.; Arnadottir, S.S.; Vang, S.; Nordentoft, I.; Knudsen, M.; Lamy, P.; Andreasen, D.; et al. Clinical Implications of Monitoring Circulating Tumor DNA in Patients with Colorectal Cancer. *Clin. Cancer Res.* **2017**, *23*, 5437–5445. [CrossRef]
53. Hoshino, A.; Costa-Silva, B.; Shen, T.L.; Rodrigues, G.; Hashimoto, A.; Tesic Mark, M.; Molina, H.; Kohsaka, S.; Di Giannatale, A.; Ceder, S.; et al. Tumour exosome integrins determine organotropic metastasis. *Nature* **2015**, *527*, 329–335. [CrossRef] [PubMed]
54. Garofalo, M.; Villa, A.; Crescenti, D.; Marzagalli, M.; Kuryk, L.; Limonta, P.; Mazzaferro, V.; Ciana, P. Heterologous and cross-species tropism of cancer-derived extracellular vesicles. *Theranostics* **2019**, *9*, 5681–5693. [CrossRef] [PubMed]
55. Rodriguez Zorrilla, S.; Perez-Sayans, M.; Fais, S.; Logozzi, M.; Gallas Torreira, M.; Garcia Garcia, A. A Pilot Clinical Study on the Prognostic Relevance of Plasmatic Exosomes Levels in Oral Squamous Cell Carcinoma Patients. *Cancers* **2019**, *11*, 429. [CrossRef] [PubMed]
56. Knowles, J.A.; Golden, B.; Yan, L.; Carroll, W.R.; Helman, E.E.; Rosenthal, E.L. Disruption of the AKT pathway inhibits metastasis in an orthotopic model of head and neck squamous cell carcinoma. *Laryngoscope* **2011**, *121*, 2359–2365. [CrossRef] [PubMed]

57. Bussink, J.; van der Kogel, A.J.; Kaanders, J.H. Activation of the PI3-K/AKT pathway and implications for radioresistance mechanisms in head and neck cancer. *Lancet Oncol.* **2008**, *9*, 288–296. [CrossRef]
58. Liu, Q.; Turner, K.M.; Alfred Yung, W.K.; Chen, K.; Zhang, W. Role of AKT signaling in DNA repair and clinical response to cancer therapy. *Neuro Oncol.* **2014**, *16*, 1313–1323. [CrossRef]
59. Mutschelknaus, L.; Azimzadeh, O.; Heider, T.; Winkler, K.; Vetter, M.; Kell, R.; Tapio, S.; Merl-Pham, J.; Huber, S.M.; Edalat, L.; et al. Radiation alters the cargo of exosomes released from squamous head and neck cancer cells to promote migration of recipient cells. *Sci. Rep.* **2017**, *7*, 12423. [CrossRef]
60. Landskron, G.; De la Fuente, M.; Thuwajit, P.; Thuwajit, C.; Hermoso, M.A. Chronic inflammation and cytokines in the tumor microenvironment. *J. Immunol. Res.* **2014**, *2014*, 149185. [CrossRef]
61. Bermejo-Martin, J.F.; Martin-Loeches, I.; Bosinger, S. Inflammation and infection in critical care medicine. *Mediators Inflamm.* **2014**, *2014*, 456256. [CrossRef]
62. Bretz, N.P.; Ridinger, J.; Rupp, A.K.; Rimbach, K.; Keller, S.; Rupp, C.; Marme, F.; Umansky, L.; Umansky, V.; Eigenbrod, T.; et al. Body fluid exosomes promote secretion of inflammatory cytokines in monocytic cells via Toll-like receptor signaling. *J. Biol. Chem.* **2013**, *288*, 36691–36702. [CrossRef] [PubMed]
63. Karin, M.; Cao, Y.; Greten, F.R.; Li, Z.W. NF-kappaB in cancer: From innocent bystander to major culprit. *Nat. Rev. Cancer* **2002**, *2*, 301–310. [CrossRef] [PubMed]
64. Bromberg, J.; Darnell, J.E., Jr. The role of STATs in transcriptional control and their impact on cellular function. *Oncogene* **2000**, *19*, 2468–2473. [CrossRef] [PubMed]
65. Andreola, G.; Rivoltini, L.; Castelli, C.; Huber, V.; Perego, P.; Deho, P.; Squarcina, P.; Accornero, P.; Lozupone, F.; Lugini, L.; et al. Induction of lymphocyte apoptosis by tumor cell secretion of FasL-bearing microvesicles. *J. Exp. Med.* **2002**, *195*, 1303–1316. [CrossRef]
66. Taylor, D.D.; Gercel-Taylor, C. Tumour-derived exosomes and their role in cancer-associated T-cell signalling defects. *Br. J. Cancer* **2005**, *92*, 305–311. [CrossRef] [PubMed]
67. Huber, V.; Fais, S.; Iero, M.; Lugini, L.; Canese, P.; Squarcina, P.; Zaccheddu, A.; Colone, M.; Arancia, G.; Gentile, M.; et al. Human colorectal cancer cells induce T-cell death through release of proapoptotic microvesicles: Role in immune escape. *Gastroenterology* **2005**, *128*, 1796–1804. [CrossRef]
68. Clayton, A.; Mason, M.D. Exosomes in tumour immunity. *Curr. Oncol.* **2009**, *16*, 46–49. [CrossRef]
69. Clayton, A.; Mitchell, J.P.; Court, J.; Mason, M.D.; Tabi, Z. Human tumor-derived exosomes selectively impair lymphocyte responses to interleukin-2. *Cancer Res.* **2007**, *67*, 7458–7466. [CrossRef]
70. Yokoi, A.; Yoshioka, Y.; Yamamoto, Y.; Ishikawa, M.; Ikeda, S.I.; Kato, T.; Kiyono, T.; Takeshita, F.; Kajiyama, H.; Kikkawa, F.; et al. Malignant extracellular vesicles carrying MMP1 mRNA facilitate peritoneal dissemination in ovarian cancer. *Nat. Commun.* **2017**, *8*, 14470. [CrossRef]
71. Baj-Krzyworzeka, M.; Majka, M.; Pratico, D.; Ratajczak, J.; Vilaire, G.; Kijowski, J.; Reca, R.; Janowska-Wieczorek, A.; Ratajczak, M.Z. Platelet-derived microparticles stimulate proliferation, survival, adhesion, and chemotaxis of hematopoietic cells. *Exp. Hematol.* **2002**, *30*, 450–459. [CrossRef]
72. Sprague, D.L.; Elzey, B.D.; Crist, S.A.; Waldschmidt, T.J.; Jensen, R.J.; Ratliff, T.L. Platelet-mediated modulation of adaptive immunity: Unique delivery of CD154 signal by platelet-derived membrane vesicles. *Blood* **2008**, *111*, 5028–5036. [CrossRef] [PubMed]
73. Folkman, J. Anti-angiogenesis: New concept for therapy of solid tumors. *Ann. Surg.* **1972**, *175*, 409–416. [CrossRef] [PubMed]
74. Svensson, K.J.; Kucharzewska, P.; Christianson, H.C.; Skold, S.; Lofstedt, T.; Johansson, M.C.; Morgelin, M.; Bengzon, J.; Ruf, W.; Belting, M. Hypoxia triggers a proangiogenic pathway involving cancer cell microvesicles and PAR-2-mediated heparin-binding EGF signaling in endothelial cells. *Proc. Natl. Acad. Sci. USA* **2011**, *108*, 13147–13152. [CrossRef] [PubMed]
75. Hong, B.S.; Cho, J.H.; Kim, H.; Choi, E.J.; Rho, S.; Kim, J.; Kim, J.H.; Choi, D.S.; Kim, Y.K.; Hwang, D.; et al. Colorectal cancer cell-derived microvesicles are enriched in cell cycle-related mRNAs that promote proliferation of endothelial cells. *BMC Genom.* **2009**, *10*, 556. [CrossRef]
76. Grange, C.; Tapparo, M.; Collino, F.; Vitillo, L.; Damasco, C.; Deregibus, M.C.; Tetta, C.; Bussolati, B.; Camussi, G. Microvesicles released from human renal cancer stem cells stimulate angiogenesis and formation of lung premetastatic niche. *Cancer Res.* **2011**, *71*, 5346–5356. [CrossRef]
77. Kosaka, N.; Iguchi, H.; Hagiwara, K.; Yoshioka, Y.; Takeshita, F.; Ochiya, T. Neutral sphingomyelinase 2 (nSMase2)-dependent exosomal transfer of angiogenic microRNAs regulate cancer cell metastasis. *J. Biol. Chem.* **2013**, *288*, 10849–10859. [CrossRef]

78. Li, R.; Wang, Y.; Zhang, X.; Feng, M.; Ma, J.; Li, J.; Yang, X.; Fang, F.; Xia, Q.; Zhang, Z.; et al. Exosome-mediated secretion of LOXL4 promotes hepatocellular carcinoma cell invasion and metastasis. *Mol. Cancer* **2019**, *18*, 18. [CrossRef]
79. Soldevilla, B.; Rodriguez, M.; San Millan, C.; Garcia, V.; Fernandez-Perianez, R.; Gil-Calderon, B.; Martin, P.; Garcia-Grande, A.; Silva, J.; Bonilla, F.; et al. Tumor-derived exosomes are enriched in DeltaNp73, which promotes oncogenic potential in acceptor cells and correlates with patient survival. *Hum. Mol. Genet.* **2014**, *23*, 467–478. [CrossRef]
80. McCready, J.; Sims, J.D.; Chan, D.; Jay, D.G. Secretion of extracellular hsp90alpha via exosomes increases cancer cell motility: A role for plasminogen activation. *BMC Cancer* **2010**, *10*, 294. [CrossRef]
81. Kim, J.W.; Wieckowski, E.; Taylor, D.D.; Reichert, T.E.; Watkins, S.; Whiteside, T.L. Fas ligand-positive membranous vesicles isolated from sera of patients with oral cancer induce apoptosis of activated T lymphocytes. *Clin. Cancer Res.* **2005**, *11*, 1010–1020.
82. Czystowska, M.; Han, J.; Szczepanski, M.J.; Szajnik, M.; Quadrini, K.; Brandwein, H.; Hadden, J.W.; Signorelli, K.; Whiteside, T.L. IRX-2, a novel immunotherapeutic, protects human T cells from tumor-induced cell death. *Cell Death Differ.* **2009**, *16*, 708–718. [CrossRef] [PubMed]
83. Wieckowski, E.U.; Visus, C.; Szajnik, M.; Szczepanski, M.J.; Storkus, W.J.; Whiteside, T.L. Tumor-derived microvesicles promote regulatory T cell expansion and induce apoptosis in tumor-reactive activated CD8+ T lymphocytes. *J. Immunol.* **2009**, *183*, 3720–3730. [CrossRef] [PubMed]
84. Fanini, F.; Fabbri, M. Cancer-derived exosomic microRNAs shape the immune system within the tumor microenvironment: State of the art. *Semin. Cell Dev. Biol.* **2017**, *67*, 23–28. [CrossRef]
85. Zhang, H.G.; Grizzle, W.E. Exosomes: A novel pathway of local and distant intercellular communication that facilitates the growth and metastasis of neoplastic lesions. *Am. J. Pathol.* **2014**, *184*, 28–41. [CrossRef] [PubMed]
86. Whiteside, T.L. Immune modulation of T-cell and NK (natural killer) cell activities by TEXs (tumour-derived exosomes). *Biochem. Soc. Trans.* **2013**, *41*, 245–251. [CrossRef] [PubMed]
87. Yang, T.; Martin, P.; Fogarty, B.; Brown, A.; Schurman, K.; Phipps, R.; Yin, V.P.; Lockman, P.; Bai, S. Exosome delivered anticancer drugs across the blood-brain barrier for brain cancer therapy in Danio rerio. *Pharm. Res.* **2015**, *32*, 2003–2014. [CrossRef]
88. Wiklander, O.P.; Nordin, J.Z.; O'Loughlin, A.; Gustafsson, Y.; Corso, G.; Mager, I.; Vader, P.; Lee, Y.; Sork, H.; Seow, Y.; et al. Extracellular vesicle in vivo biodistribution is determined by cell source, route of administration and targeting. *J. Extracell. Vesicles* **2015**, *4*, 26316. [CrossRef]
89. Alvarez-Erviti, L.; Seow, Y.; Yin, H.; Betts, C.; Lakhal, S.; Wood, M.J. Delivery of siRNA to the mouse brain by systemic injection of targeted exosomes. *Nat. Biotechnol.* **2011**, *29*, 341–345. [CrossRef]
90. Kim, M.S.; Haney, M.J.; Zhao, Y.; Mahajan, V.; Deygen, I.; Klyachko, N.L.; Inskoe, E.; Piroyan, A.; Sokolsky, M.; Okolie, O.; et al. Development of exosome-encapsulated paclitaxel to overcome MDR in cancer cells. *Nanomedicine* **2016**, *12*, 655–664. [CrossRef]
91. Zhuang, X.; Xiang, X.; Grizzle, W.; Sun, D.; Zhang, S.; Axtell, R.C.; Ju, S.; Mu, J.; Zhang, L.; Steinman, L.; et al. Treatment of brain inflammatory diseases by delivering exosome encapsulated anti-inflammatory drugs from the nasal region to the brain. *Mol. Ther.* **2011**, *19*, 1769–1779. [CrossRef]
92. Cooper, J.M.; Wiklander, P.B.; Nordin, J.Z.; Al-Shawi, R.; Wood, M.J.; Vithlani, M.; Schapira, A.H.; Simons, J.P.; El-Andaloussi, S.; Alvarez-Erviti, L. Systemic exosomal siRNA delivery reduced alpha-synuclein aggregates in brains of transgenic mice. *Mov. Disord.* **2014**, *29*, 1476–1485. [CrossRef] [PubMed]
93. Wahlgren, J.; De, L.K.T.; Brisslert, M.; Vaziri Sani, F.; Telemo, E.; Sunnerhagen, P.; Valadi, H. Plasma exosomes can deliver exogenous short interfering RNA to monocytes and lymphocytes. *Nucleic Acids Res.* **2012**, *40*, e130. [CrossRef]
94. Ohno, S.; Takanashi, M.; Sudo, K.; Ueda, S.; Ishikawa, A.; Matsuyama, N.; Fujita, K.; Mizutani, T.; Ohgi, T.; Ochiya, T.; et al. Systemically injected exosomes targeted to EGFR deliver antitumor microRNA to breast cancer cells. *Mol. Ther.* **2013**, *21*, 185–191. [CrossRef] [PubMed]
95. Kosaka, N.; Iguchi, H.; Yoshioka, Y.; Takeshita, F.; Matsuki, Y.; Ochiya, T. Secretory mechanisms and intercellular transfer of microRNAs in living cells. *J. Biol. Chem.* **2010**, *285*, 17442–17452. [CrossRef] [PubMed]
96. Zhou, Y.; Xiong, M.; Fang, L.; Jiang, L.; Wen, P.; Dai, C.; Zhang, C.Y.; Yang, J. miR-21-containing microvesicles from injured tubular epithelial cells promote tubular phenotype transition by targeting PTEN protein. *Am. J. Pathol.* **2013**, *183*, 1183–1196. [CrossRef]

97. Zhang, Y.; Liu, D.; Chen, X.; Li, J.; Li, L.; Bian, Z.; Sun, F.; Lu, J.; Yin, Y.; Cai, X.; et al. Secreted monocytic miR-150 enhances targeted endothelial cell migration. *Mol. Cell* **2010**, *39*, 133–144. [CrossRef]
98. Zhang, Y.; Li, L.; Yu, J.; Zhu, D.; Zhang, Y.; Li, X.; Gu, H.; Zhang, C.Y.; Zen, K. Microvesicle-mediated delivery of transforming growth factor beta1 siRNA for the suppression of tumor growth in mice. *Biomaterials* **2014**, *35*, 4390–4400. [CrossRef]
99. Zitvogel, L.; Regnault, A.; Lozier, A.; Wolfers, J.; Flament, C.; Tenza, D.; Ricciardi-Castagnoli, P.; Raposo, G.; Amigorena, S. Eradication of established murine tumors using a novel cell-free vaccine: Dendritic cell-derived exosomes. *Nat. Med.* **1998**, *4*, 594–600. [CrossRef]
100. Dai, S.; Wei, D.; Wu, Z.; Zhou, X.; Wei, X.; Huang, H.; Li, G. Phase I clinical trial of autologous ascites-derived exosomes combined with GM-CSF for colorectal cancer. *Mol. Ther.* **2008**, *16*, 782–790. [CrossRef]
101. Kojima, R.; Bojar, D.; Rizzi, G.; Hamri, G.C.; El-Baba, M.D.; Saxena, P.; Auslander, S.; Tan, K.R.; Fussenegger, M. Designer exosomes produced by implanted cells intracerebrally deliver therapeutic cargo for Parkinson's disease treatment. *Nat. Commun.* **2018**, *9*, 1305. [CrossRef]
102. Topalian, S.L.; Drake, C.G.; Pardoll, D.M. Immune checkpoint blockade: A common denominator approach to cancer therapy. *Cancer Cell* **2015**, *27*, 450–461. [CrossRef] [PubMed]
103. Page, D.B.; Postow, M.A.; Callahan, M.K.; Allison, J.P.; Wolchok, J.D. Immune modulation in cancer with antibodies. *Annu. Rev. Med.* **2014**, *65*, 185–202. [CrossRef] [PubMed]
104. Theodoraki, M.N.; Yerneni, S.S.; Hoffmann, T.K.; Gooding, W.E.; Whiteside, T.L. Clinical Significance of PD-L1(+) Exosomes in Plasma of Head and Neck Cancer Patients. *Clin. Cancer Res.* **2018**, *24*, 896–905. [CrossRef] [PubMed]
105. Chen, G.; Huang, A.C.; Zhang, W.; Zhang, G.; Wu, M.; Xu, W.; Yu, Z.; Yang, J.; Wang, B.; Sun, H.; et al. Exosomal PD-L1 contributes to immunosuppression and is associated with anti-PD-1 response. *Nature* **2018**, *560*, 382–386. [CrossRef]
106. Zhang, J.; Gao, J.; Li, Y.; Nie, J.; Dai, L.; Hu, W.; Chen, X.; Han, J.; Ma, X.; Tian, G.; et al. Circulating PD-L1 in NSCLC patients and the correlation between the level of PD-L1 expression and the clinical characteristics. *Thorac. Cancer* **2015**, *6*, 534–538. [CrossRef]
107. Okuma, Y.; Hosomi, Y.; Nakahara, Y.; Watanabe, K.; Sagawa, Y.; Homma, S. High plasma levels of soluble programmed cell death ligand 1 are prognostic for reduced survival in advanced lung cancer. *Lung Cancer* **2017**, *104*, 1–6. [CrossRef]
108. Poggio, M.; Hu, T.; Pai, C.C.; Chu, B.; Belair, C.D.; Chang, A.; Montabana, E.; Lang, U.E.; Fu, Q.; Fong, L.; et al. Suppression of Exosomal PD-L1 Induces Systemic Anti-tumor Immunity and Memory. *Cell* **2019**, *177*, 414–427.e13. [CrossRef]
109. Atai, N.A.; Balaj, L.; van Veen, H.; Breakefield, X.O.; Jarzyna, P.A.; Van Noorden, C.J.; Skog, J.; Maguire, C.A. Heparin blocks transfer of extracellular vesicles between donor and recipient cells. *J. Neurooncol.* **2013**, *115*, 343–351. [CrossRef]
110. Balaj, L.; Atai, N.A.; Chen, W.; Mu, D.; Tannous, B.A.; Breakefield, X.O.; Skog, J.; Maguire, C.A. Heparin affinity purification of extracellular vesicles. *Sci. Rep.* **2015**, *5*, 10266. [CrossRef]
111. Chapman, C.J.; Thorpe, A.J.; Murray, A.; Parsy-Kowalska, C.B.; Allen, J.; Stafford, K.M.; Chauhan, A.S.; Kite, T.A.; Maddison, P.; Robertson, J.F. Immunobiomarkers in small cell lung cancer: Potential early cancer signals. *Clin. Cancer Res.* **2011**, *17*, 1474–1480. [CrossRef]
112. Zhong, L.; Coe, S.P.; Stromberg, A.J.; Khattar, N.H.; Jett, J.R.; Hirschowitz, E.A. Profiling tumor-associated antibodies for early detection of non-small cell lung cancer. *J. Thorac. Oncol.* **2006**, *1*, 513–519. [CrossRef] [PubMed]
113. Li, G.; Miles, A.; Line, A.; Rees, R.C. Identification of tumour antigens by serological analysis of cDNA expression cloning. *Cancer Immunol. Immunother.* **2004**, *53*, 139–143. [CrossRef] [PubMed]

11

Large Extracellular Vesicles can be Characterised by Multiplex Labelling using Imaging Flow Cytometry

Suzanne M Johnson [1,*], Antonia Banyard [2], Christopher Smith [1], Aleksandr Mironov [3] and Martin G. McCabe [1,4]

1 Children's Cancer Group, Division of Cancer Sciences, School of Medical Sciences, Faculty of Biology Medicine and Health, University of Manchester, Oglesby Cancer Research Building, Manchester Academic Health Science Centre, Manchester Cancer Research Centre, Manchester M20 4GJ, UK; Christopher.Smith-5@manchester.ac.uk (C.S.); Martin.McCabe@manchester.ac.uk (M.G.M.)

2 Flow Cytometry Core Facility, Cancer Research UK Manchester Institute, University of Manchester, Alderley Park, Macclesfield, UK; Antonia.Banyard@manchester.ac.uk

3 Electron Microscopy Core Facility, School of Biological Sciences, Faculty of Biology Medicine and Health, University of Manchester, Manchester M13 9PT, UK; Aleksandr.Mironov@manchester.ac.uk

4 Manchester Children's Brain Tumour Research Network Royal Manchester Children's Hospital, Manchester M13 9WL, UK

* Correspondence: suzanne.johnson@manchester.ac.uk

Abstract: Extracellular vesicles (EVs) are heterogeneous in size (30 nm–10 μm), content (lipid, RNA, DNA, protein), and potential function(s). Many isolation techniques routinely discard the large EVs at the early stages of small EV or exosome isolation protocols. We describe here a standardised method to isolate large EVs from medulloblastoma cells and examine EV marker expression and diameter using imaging flow cytometry. Our approach permits the characterisation of each large EVs as an individual event, decorated with multiple fluorescently conjugated markers with the added advantage of visualising each event to ensure robust gating strategies are applied. Methods: We describe step-wise isolation and characterisation of a subset of large EVs from the medulloblastoma cell line UW228-2 assessed by fluorescent light microscopy, transmission electron microscopy (TEM) and tunable resistance pulse sensing (TRPS). Viability of parent cells was assessed by Annexin V exposure by flow cytometry. Imaging flow cytometry (Imagestream Mark II) identified EVs by direct fluorescent membrane labelling with Cell Mask Orange (CMO) in conjunction with EV markers. A stringent gating algorithm based on side scatter and fluorescence intensity was applied and expression of EV markers CD63, CD9 and LAMP 1 assessed. Results: UW228-2 cells prolifically release EVs of up to 6 μm. We show that the Imagestream Mark II imaging flow cytometer allows robust and reproducible analysis of large EVs, including assessment of diameter. We also demonstrate a correlation between increasing EV size and co-expression of markers screened. Conclusions: We have developed a labelling and stringent gating strategy which is able to explore EV marker expression (CD63, CD9, and LAMP1) on individual EVs within a widely heterogeneous population. Taken together, data presented here strongly support the value of exploring large EVs in clinical samples for potential biomarkers, useful in diagnostic screening and disease monitoring.

Keywords: extracellular vesicles; imaging flow cytometry; biomarker reservoirs; cancer diagnostics; disease monitoring; large EVs

---

1. Introduction

The term extracellular vesicle (EV) refers to particles released from cells which are delimited by a lipid bilayer, but do not contain a nucleus [1]. EVs are heterogeneous in biogenesis [2], size, and content.

They range in size from 30 nm exosomes [3] to oncosomes up to 10 μM [4], and contain cargo of all biomolecule categories [5]. Attempts to categorise EVs primarily by surface marker expression have been confounded by the recognition that many of the markers previously considered to be subset- or derivation-specific, are actually present on multiple or even all classes of EVs [6]. Delineation and characterisation of specific EV subsets is an essential goal to achieve a better understanding of EV biology [7], yet there are no techniques that accurately quantify EVs across the full EV size range, or combine quantification with the ability to screen for EV marker expression or EV content.

Small EVs are characteristically isolated by high-speed centrifugation at 100,000× *g* [8], and include both EVs derived intracellularly from late endosomes and released by exocytosis (exosomes), and other small EVs not derived from endosomes (3). Multiple commercial solutions exist for the isolation of exosomes from a variety of biological fluids including tissue culture supernatant, plasma, and urine. In contrast, there are no commercially available solutions for the isolation of large EVs. As a result, isolation methods vary, and knowledge of large EV content and function in biological samples is relatively lacking. Large EVs, defined as >200 nm by recent guidelines set out by the International Society of Extracellular Vesicles [1], include cancer cell-derived oncosomes, dead cell-derived apoptotic bodies and platelets, and are visible by light microscopy [9]. In published literature, EVs larger than 1 μm have historically been assumed to be apoptotic bodies [10]. However, we and others [4] have demonstrated that viable cell cultures produce large EVs which do not have the ultrastructural features reminiscent of fragments of apoptotic cells.

EVs are released by all cells providing an efficient mechanism of cell to cell communication [11]. Increasing evidence points to key roles for EVs in cancer diagnosis, prognostication, and surveillance of cancer [12]. Large EVs from prostate cancer cells were shown to contain tumour-specific biomarkers [4,13] and mediate intercellular transfer of bioactive molecules including miRNA [14]. We previously reported a population of large EVs released by leukaemic cells which were actin-rich and contained intact organelles [15]. These large EVs could be internalised by normal stromal cells and induced a switch in the preferred metabolic pathway of the recipient cells [9]. Additionally, we found that leukaemia-derived EVs expressed a surface marker indicative of their parent cell (CD19) and could be detected in the peripheral blood of murine models and patient bone marrow plasma [9]. Taken together, our previous work and existing large EV literature suggest that large EVs, often discarded in techniques to isolate smaller EVs and exosomes, could be considered as extensive reservoirs of biomolecules useful to study EV biogenesis and function, and to identify clinically relevant biomarkers for disease detection and treatment monitoring [16,17].

The principle advantage of characterising large EVs as single events by imaging flow cytometry is the potential for simultaneously identifying parent cell, EV, and tumour markers. We report for the first time a characterisation of size distribution and EV marker expression in this heterogeneous EV population, undertaken in accordance with the most recent international consensus guidelines for EV research from the International Society of Extracellular Vesicles [1].

In this proof of concept study, we set out to: (1) highlight the abundance of large EVs produced by cells derived from the malignant brain tumour medulloblastoma in vitro; (2) describe variations in the expression of established EV markers in the large EV population; (3) describe how the Imagestream (ISX) can address sample heterogeneity by facilitating high throughput, single event EV analyses. We describe a protocol to isolate intact large EVs without cell contamination, from cells growing in serum-free medium, using gravity flow filtration combined with low-speed centrifugation. Our data show the breadth of heterogeneity in the size and marker expression of large EVs isolated from a single cell line and serves to highlight the importance of sample purity, isolation techniques and experimental controls as we seek to identify tumour-specific EV markers for use in the clinic.

2. Results

Large EVs from the medulloblastoma cell line UW228-2 are visible with light microscopy, can be isolated from viable, serum-free cell culture supernatant with intact membranes, and contain a polymerised actin cytoskeleton. A proportion of these large EVs express reported EV markers.

In this study the large EV isolation SOP was applied to the SHH-driven medulloblastoma cell line UW228-2. In accordance with the recently published international consensus MISEV2018 guidelines [1], we demonstrated the existence and membrane integrity of large EVs using both light and electron microscopy (Figure 1A,B). The pan-EV marker CD63 was expressed by parent medulloblastoma cells and a proportion of large EVs in viable cultures (Figure 1A). Large EVs also exhibited an active cytoskeleton indicated by the polymerised actin marker Phalloidin, but did not stain for DAPI, indicating that they did not contain nuclear double-stranded DNA. EV membrane integrity, size, and intra-vesicle content was examined by transmission electron microscopy (TEM) (Figure 1B). Isolated EVs were spiked with medulloblastoma parent cells for comparison and adhered to ACLAR film coated with CellTak prior to TEM. Serial sections demonstrated that the EVs had a limiting membrane, internal organelles but no nucleus, and were independent from cells. By contrast, cells had internal organelles including a nucleus and cytoplasmic protrusions indicative of filopodia.

Cells were grown in serum-free medium for 24 h prior to EV isolation, staining, and analysis. Experimental conditions were optimised to eliminate false positive membrane labelling or bovine EV contamination [18]. To assess whether growth in serum-free medium resulted in increased cell apoptosis, phosphatidyl serine exposure on the surface of cells was examined. Figure 1C indicates the quadrant gating strategy applied using positive controls and the comparison of cells grown in standard culture versus serum-free conditions. Imaging flow cytometry was also performed to visualise Annexin V/PI staining on parent cells (Figure 1D). In triplicate experiments, there was no significant difference between the viability of cells cultured in complete or serum-free media (Figure 1E). Figure 1F provides a workflow for the isolation and characterisation of large EVs used in this study. At the time of each EV isolation, total parent cell count and viability was assessed using trypan blue exclusion and found to be 7.9×10^6 cells (98% viable), 8.3×10^6 cells (99% viable), and 6.8×10^6 cells (98% viable).

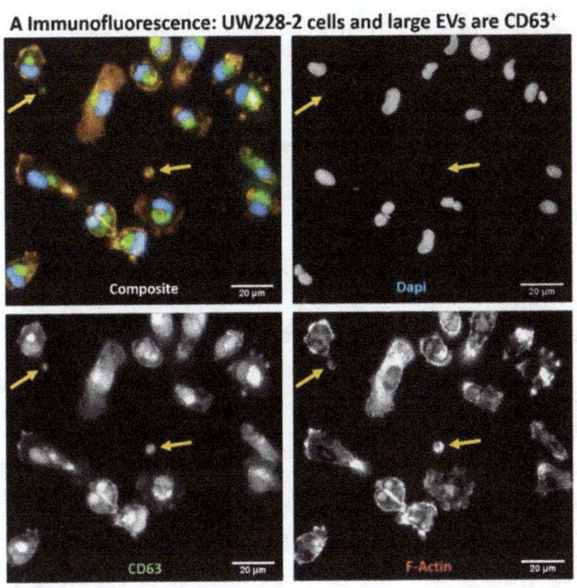

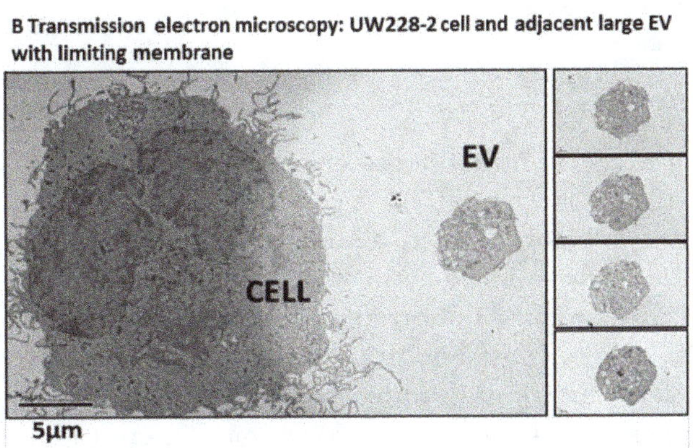

Figure 1. *Cont.*

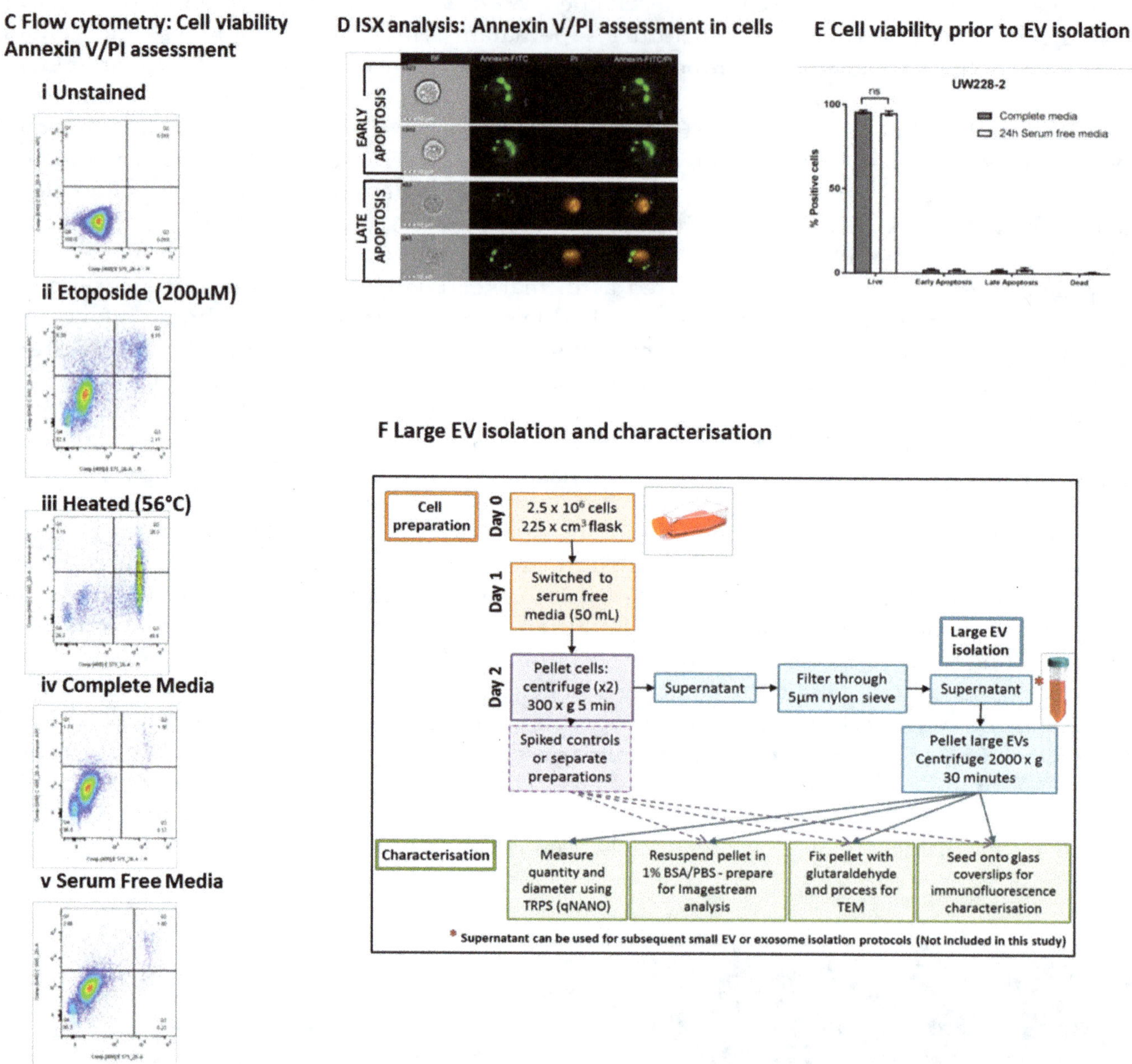

Figure 1. Immunofluorescence: UW228-2 cells and large extracellular vesicles (EVs) are CD63$^+$. Medulloblastoma cell lines produce large extracellular vesicles (EVs) which can be harvested from serum-free conditioned media after 24 h. (**A**) Medulloblastoma cells produce EVs in vitro which contain polymerised actin and CD63, but no nucleus. The medulloblastoma cell line UW228-2 was cultured on glass bottom plates for 24 h; fixed and probed for the tetraspanin and EV marker CD63 (FITC—green). Polymerised actin was labelled using Alexa Fluor 555 phalloidin (f-ACTIN—red). Nuclei were counterstained with DAPI (DAPI—blue). Images of large EVs (yellow arrows) and cells were captured using the Perkin Elmer Operetta system at ×40 magnification. The individual images were captured in black and white using the appropriate emission filters and a coloured composite image created using Columbus software. Scale bar represents 20 μM. (**B**) Transmission electron microscopy: UW228-2 cell and adjacent large EV with limiting membrane. Large EVs have a limiting membrane, contain organelles but no nucleus and are < 6 μm in size. Cells and EVs were applied to Poly-D-lysine coated ACLAR film and fixed with glutaraldehyde before processing for Transmission Electron

Microscopy. Serial sections were taken to assess sample purity, EV membrane integrity, size, and content comparative to the parent cell. Images were captured using a Biotwin Philips TECNAI G2 microscope at ×1900 magnification. Scale bar represents 5 μM. Left panel shows a cell and a large EV. Right panel: serial sections taken through the entire EV (4 images shown). The wide-field image shows an EV of 5.5 μM diameter at its central plane of focus, alongside a parent medulloblastoma cell of 18 μM. (**C**) Flow cytometry: Cell viability Annexin V/PI assessment. Cell apoptosis was assessed by flow cytometry. Unstained cells (i) and cells treated with 200 μM Etoposide (ii; Annexin V positive gate) or heated (iii; PI positive gate) provided positive controls which were used to assign accurate quadrant gates to bivariate scatter plots showing PI versus Annexin V APC. Live cells appear in Q4, Annexin only in Q2 (Early apoptosis), dual labelled in Q3 (Late apoptosis) and PI only (Dead cells). UW228-2 cells were cultured for 24 h in complete media (iv; DMEM with 10% FBS) or serum-free media (v; DMEM only). Representative plots of triplicate experiments shown. (**D**) Imagestream (ISX) analysis: Annexin V/PI assessment in cells. Imaging flow cytometry allows visual distinction between early and late stages of apoptosis. Cells in standard culture conditions were labelled using an Annexin FITC/PI kit and examined using the Imagestream (ISX) Mark II. Images were captured at x60 magnification and representative gallery images demonstrate Annexin V FITC only membrane labelling, indicating early apoptosis (upper panels) and dual Annexin V FITC/PI labelling, indicating late apoptosis (lower panels). (**E**) Cell viability prior to EV isolation. UW228-2 cells can be cultured in serum-free conditions for 24 h with no loss in viability. No difference in viability could be seen between culture conditions: Percentage unlabelled cells (live), Annexin V positive only (Early apoptosis), Annexin V and PI positive (Late Apoptosis) or PI positive only (DEAD) after 24 h in complete (grey bars) or serum-free medium (open bars). (**F**) Large EV isolation and characterisation. EV isolation protocol from cultured cells was optimised to harvest large EVs with minimal processing. Cells were seeded into large flasks (225 cm^3) and allowed to adhere. Media was changed after 24 h for 50 mL serum-free DMEM and cells cultured at 37 °C for a further 24 h. Trypsinised cells were collected by 2 successive centrifugation steps at 300× *g* 5 min keeping the supernatant each time. EV containing supernatant was filtered using a double layered 5 μm pore nylon membrane by gravity. Filtered, cell-free supernatant was centrifuged at 2000× *g* for 30 min using a bench top centrifuge with swing out buckets. The resulting cell and EV pellets were processed appropriately for the downstream technique: For flow cytometry cells were processed and analysed separately, whilst cells were spiked into wells for direct comparison by microscopy.

2.1. EVs are Highly Heterogeneous and Differentially Express EV Markers

A reliable fluorescence marker was essential to demarcate EVs from background scatter events. Cell mask orange (CMO) is a fluorescent plasma membrane label composed of amphipathic molecules comprising a lipophilic moiety for membrane loading and a negatively charged hydrophilic dye for anchoring of the probe in the plasma membrane. We performed a titration using the parent cells with serial dilutions of the dye in serum-free media ranging from 1 in 1000 (5 μg/mL) to 1 in 100,000 (50 ng/mL) (Figure 2A) and found that 2.5 μg/mL (1 in 2000) was an optimal concentration providing high median fluorescence intensity without saturation. Using the Imagestream Mark II (ISX) we confirmed that the final concentration of CMO labelled EVs in a typical preparation did not result in coincidence events or swarm which could lead to false positive results when looking at multi-colour labelling. Serial dilutions of CMO labelled EVs showed a linear decrease in objects per mL (Figure 2B(i)) with an increasing dilution factor, whilst the fluorescence intensity remained stable across the dilutions (Figure 2B(ii)). To facilitate reproducible reporting across experiments and to offer a means to standardise EV measurements, we determined the minimum MESF value for CMO+ events which would distinguish between unstained and CMO labelled EVs using our staining protocol, as described elsewhere [19]. Figure 2C shows a representative bivariate plot (i) of the low, medium-low, medium-high, and high fluorescence bead populations. The histogram (ii) was used to determine the MFI of each peak and log fluorescence intensities were converted to MESF values using information supplied by the manufacturer (iii). Linear regression of log MFI versus log MESF

(iv) was used to calculate the MESF corresponding to the minimum fluorescence intensity of events within the designated CMO+ gate as follows. The maximum MESF values of the unstained EVs in each experiment were: Replicate 1: 271.42 = MESF 48.23; Replicate 2: 271.0 = MESF 49.27; Replicate 3: 272.88 = MESF 42.48. Therefore, to set a standardised lower threshold of detection which could be used across replicate experiments we assigned a lower MESF threshold for CMO^+ events of 50 (Example shown in Figure 2C(v)). We therefore report here CMO^+ events as number of events >MESF 50.

Previous reports have identified that lipid dye aggregates can mimic EVs when using fluorescence lipid markers [20]. We used Triton-X 100 treatment [21] of fully labeled EVs to disrupt EV signals demonstrated by a loss of CMO + events (Figure 2D(i)). Figure 2D(ii) shows an overlay of unstained EVs, fully stained EVs and fully stained EVs treated with Triton-X 100. The MFI of the post-Triton-X 100 treated sample is reduced to the level of the US-EVs.

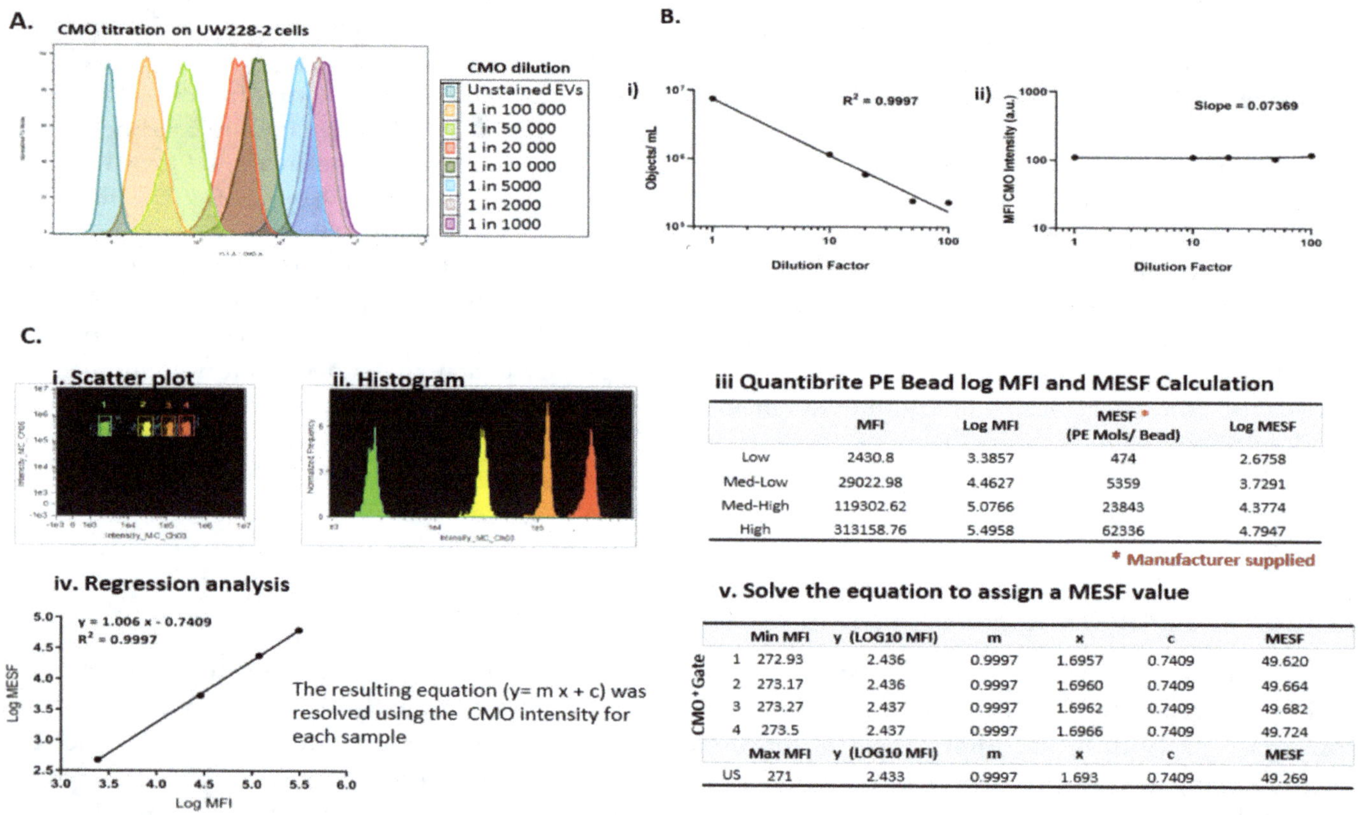

	MFI	Log MFI	MESF * (PE Mols/ Bead)	Log MESF
Low	2430.8	3.3857	474	2.6758
Med-Low	29022.98	4.4627	5359	3.7291
Med-High	119302.62	5.0766	23843	4.3774
High	313158.76	5.4958	62336	4.7947

	Min MFI	y (LOG10 MFI)	m	x	c	MESF
1	272.93	2.436	0.9997	1.6957	0.7409	49.620
2	273.17	2.436	0.9997	1.6960	0.7409	49.664
3	273.27	2.437	0.9997	1.6962	0.7409	49.682
4	273.5	2.437	0.9997	1.6966	0.7409	49.724
	Max MFI	**y (LOG10 MFI)**	**m**	**x**	**c**	**MESF**
US	271	2.433	0.9997	1.693	0.7409	49.269

Figure 2. *Cont.*

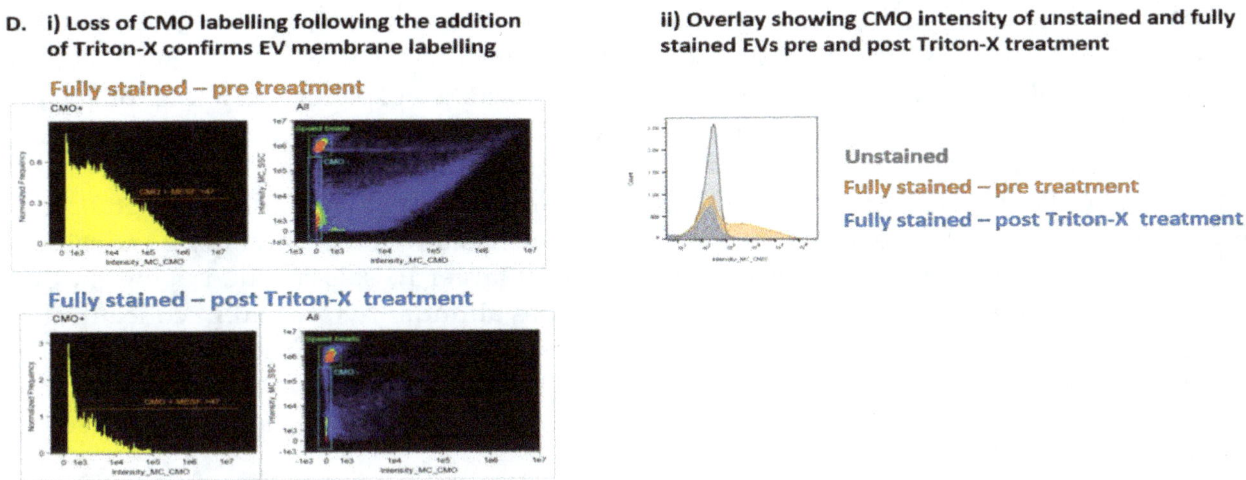

Figure 2. Cell Mask Orange can label EVs to enable standardisation across experiments. (**A**) Cell Mask Orange (CMO) labelling was optimised by titration using parent cells. Serial dilutions from 1 in 1000–100,000 were used to label parent UW228-2 cells and analysed by flow cytometry. A clear relationship between CMO concentration and fluorescence intensity is shown. (**B**) ISX: Objects/mL reduce with increased sample dilution (i) whilst CMO intensity remains constant (ii). Sample dilution was used to validate EV staining protocol. Using a CMO dilution of 1 in 2000 (2.5 µg/mL), the Imagestream acquired fewer CMO positive objects per ml with increasing dilution of CMO labelled EVs (i), whilst the CMO fluorescence intensity in channel 03 was maintained (ii). (**C**) Quantibrite PE Beads were run at the time of each experiment and used to assign molecules of equivalent soluble fluorochrome (MESF) values to the CMO intensity of labelled EVs. (Representative example from a single experiment shown) Quantibrite PE beads were acquired at the time of each experiment to assign MESF value. Quantibrite PE beads were separated on a bivariate plot of intensity against side scatter (i). The fluorescence intensity of each bead set was gated on the histogram (ii) and assigned a MESF value according to the number of PE molecules per bead as provided by the manufacturer (iii). Log MFI against Log MESF provided a standard curve for each acquisition (iv). Regression analysis was used to extrapolate the equivalent MESF value for CMO intensity for unstained and labelled samples (v) which enabled lower threshold for CMO intensity to be set to distinguish between unstained and CMO labelled EVs. Representative experiment shown. MESF 50 was subsequently used to set the lower CMO+ threshold to standardise across experiments. (**D**) CMO labelling was diminished following the addition of detergent. CMO positive (CMO+) EVs could be distinguished from unstained EVs by ISX (i). Treatment of the same sample with Triton-X 100 diminished the CMO labelling (ii). Exported .fcs files from the same acquisition were overlaid using FlowJo 10.6 (Ashland). The fully stained sample (orange) shows increased fluorescence intensity in channel 3. The same sample post Triton–X treatment (blue) showed a reduction in fluorescence intensity to a similar level of the unstained sample (grey).

2.2. Large EVs Size Distribution and Quantity was Assessed Using Tunable Resistance Pulse Sensing (TRPS)

At the time of each preparation, an aliquot of freshly isolated EVs was assessed using the qNano GOLD particle counter to ascertain EV size distribution (diameter) in terms of percentage population (%) and concentration (particles/mL) prior to labelling for ISX analysis. Using a series of 3 overlapping Nanopores (Figure 3A(i)), each with an optimal size range which spans a total size of approximately 275 nm to 5.7 µm, we assessed the size range of particles in the EV preparations. Representative profiles of diameter against percentage population are shown for a single sample using NP600, NP1000, and NP2000 Nanopores (Figure 3A(ii–iv)). By selecting overlapping Nanopore sizes, the same calibrator beads (1000 nm) could be used and therefore it was possible to overlay the resulting profiles to visualise the total large EV population within each sample (Figure 3A(v)). As expected, the EVs present in each biological replicate were heterogeneous in size. However, the size range of EVs across biological replicates was consistent (250 nm to 6 µm); with some variation in the median diameter per sample. Size is presented with a bin of 100 (nm) and in each case the most prevalent large EVs were detected using the NP600. The median size across the replicates using the NP600 for comparison were 250–350 nm 5.15×10^7; 450–550 nm 1.7×10^7, and 250–350 nm 2.8×10^8. The starting volume of EV-containing media used per replicate was 100 mL. The resulting pellet was re-suspended in a total of 700 μL PBS (prior to labelling). As an example, applying this dilution factor (142.86) to the first replicate equates to 7.4×10^9 EVs specifically of the 250–350 nm size range that were released by 7.9×10^6 UW228-2 cells (NB cell count at harvest) in 24 h into 100 mL serum-free media. It should be noted that due to the number of washes and centrifugation steps during the isolation, this can only be considered as an illustration of the quantity of EVs produced. The total larger EVs as counted using the NP2000 (935 nm – 5.7 μm) were less common but are nevertheless abundantly present in concentrations of 2.3×10^8, 6.9×10^8 and 2.1×10^9 in the 100 mL harvested media from this particular cell line. The total particle count for each Nanopore size, across 3 biological replicates is shown in Table 1. It should be noted that the counts shown are not cumulative as they represent the same sample analysed across 3 Nanopore sizes.

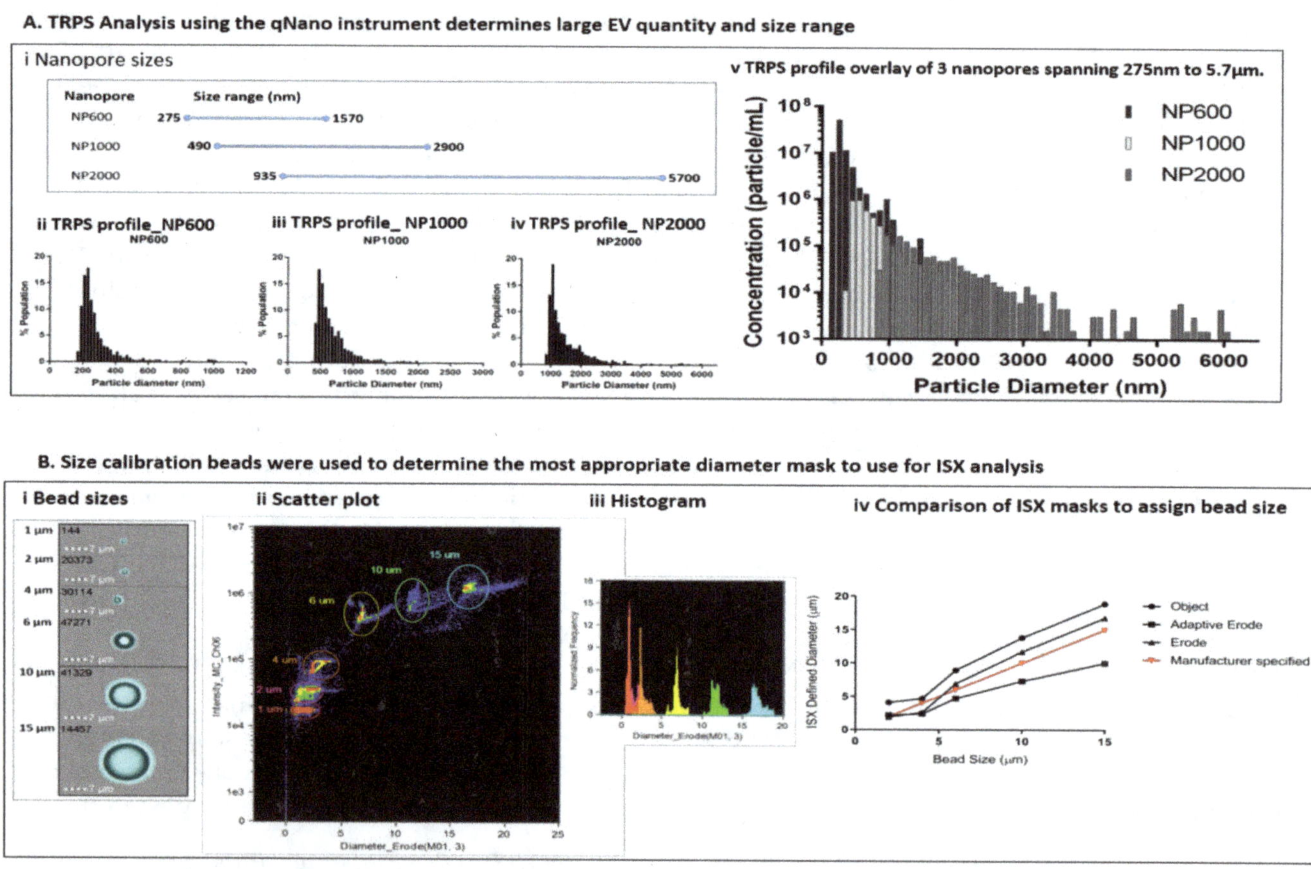

Figure 3. EV size and quantity was assessed by Tunable Resistance Pulse Sensing (TRPS) using the qNANO instrument (iZON Science) and compared to diameter masks in the ISX IDEAS software. (**A**) Three Nanopores of over-lapping size ranges were used to determine the particle diameter and concentration in each EV preparation (i). Size profiles were established for each individual Nanopore (ii NP600; iii NP1000; iv NP2000). The overlaid profiles of diameter by concentration provided by the 3 Nanopores show a size range of 250 nm and 6μm against a common calibrator bead of 1000 nm (representative sample shown) (v). (**B**) Calibration beads of known size were used to determine the most accurate diameter mask. The masks available within the ISX IDEAS software for diameter analysis were compared. The 3 which assigned the most accurate sizes according to manufacturer specified beads sizes were the Object, Erode and Adaptive Erode masks. These were compared across bead size by applying the mask (ii) visualising a scatter plot (iii) and histogram for each mask type. For the size range within our EV preparations, we found the Erode mask to be the most accurate (iv).

Table 1. Total particle count using different Nanopore sizes (qNANO) in biological replicates.

-	NP600	NP1000	NP2000
1	8.4×10^8	3.8×10^6	1.6×10^6
2	4.9×10^7	1.4×10^7	4.8×10^6
3	5.0×10^8	4.4×10^7	1.5×10^7

Size calibration beads were used to determine the most appropriate diameter mask to use for ISX analysis of EV diameter. The bead sizes provided by the manufacturer were 1, 2, 4, 6, 10, and 15 μm. We used the feature tool in IDEAS to apply a range of different diameter masks to the bright field image of the beads (Figure 3B(i)). We selected bright field for this using non-labelled beads because the Quantibrite beads showed over exaggeration of EV size, possibly due to saturation and flare of fluorescence. The beads were visualised on bivariate plots of diameter versus side scatter (channel 06). Density plots (Figure 3B(ii)) allowed the bead populations to be gated individually and subsequent histograms (Figure 3B(iii)) to be viewed. We compared the diameters for each bead population assigned using the Object, Adaptive Erode and Erode Masks against the manufacturer specified size (Figure 3B(iv)) and found the Erode mask on the bright field image, using 3 pixel reduction to be the most comparable.

2.3. *Fluorescence Membrane Labelling Can Be Used to Distinguish EVs from Background and Speed Beads*

We devised an analysis template within IDEAS which comprised a hierarchical gating strategy aimed at characterising heterogeneous large EVs (Figure 4A). All buffers were filtered using a 0.2 μm filter and samples were acquired for 5 min to avoid recording different amounts of background per acquisition. Speed beads were included. The speed beads and CMO^- events were defined using density plots of channel 03 (CMO) versus channel 06 (side scatter) intensity (Gate i). Running acquisition buffer only (left side) and unstained EVs (centre) showed the instrument detected a high level of background with low to medium side scatter. Labelling with the cell membrane dye (CMO) therefore helped to distinguish between the low CMO intensity/low side scatter EVs and background detected in both the acquisition buffer and unstained sample (right side). Applying a gate to capture low Raw Max Pixel events (channel 06—SSC: Gate ii) eliminated those events with saturated side scatter, including the speed beads. Outliers in the side scatter versus CMO intensity plots were individually inspected and found to be dual events comprising a speed bead (CMO^-, high SSC) and an EV (CMO+, low SSC) which occupied the flow chamber at the same time. This resulted in aberrant events with both high CMO intensity and high SSC and was therefore excluded from further analysis.

Events classed as CMO+ at this point were included in the initial CMO+ gate (Gate iii). Within the same acquisition, we ran the Quantibrite PE beads which enabled a minimum threshold for CMO+ events to be calculated and converted to MESF units, as described. The lower MESF cut off for this experiment was defined at >50 and a further gate on the CMO intensity histogram (Gate iv) captured all CMO+ events with an MESF >50 for downstream analysis. The proportion of CMO+ events as a percentage of all events acquired over the 5 min period varied across 10 replicates (mean 18.73% ± 1.2 SEM) CMO+ MESF > 50; Figure 4B). A final gate was included to eliminate outliers which were either high CMO intensity but appeared on the bright field image to be membrane fragments (Gate v) or were not assigned a diameter due to lack of bright field image focus. This stringent gating strategy was designed to eliminate any events which could not be further analysed for EV marker expression or diameter.

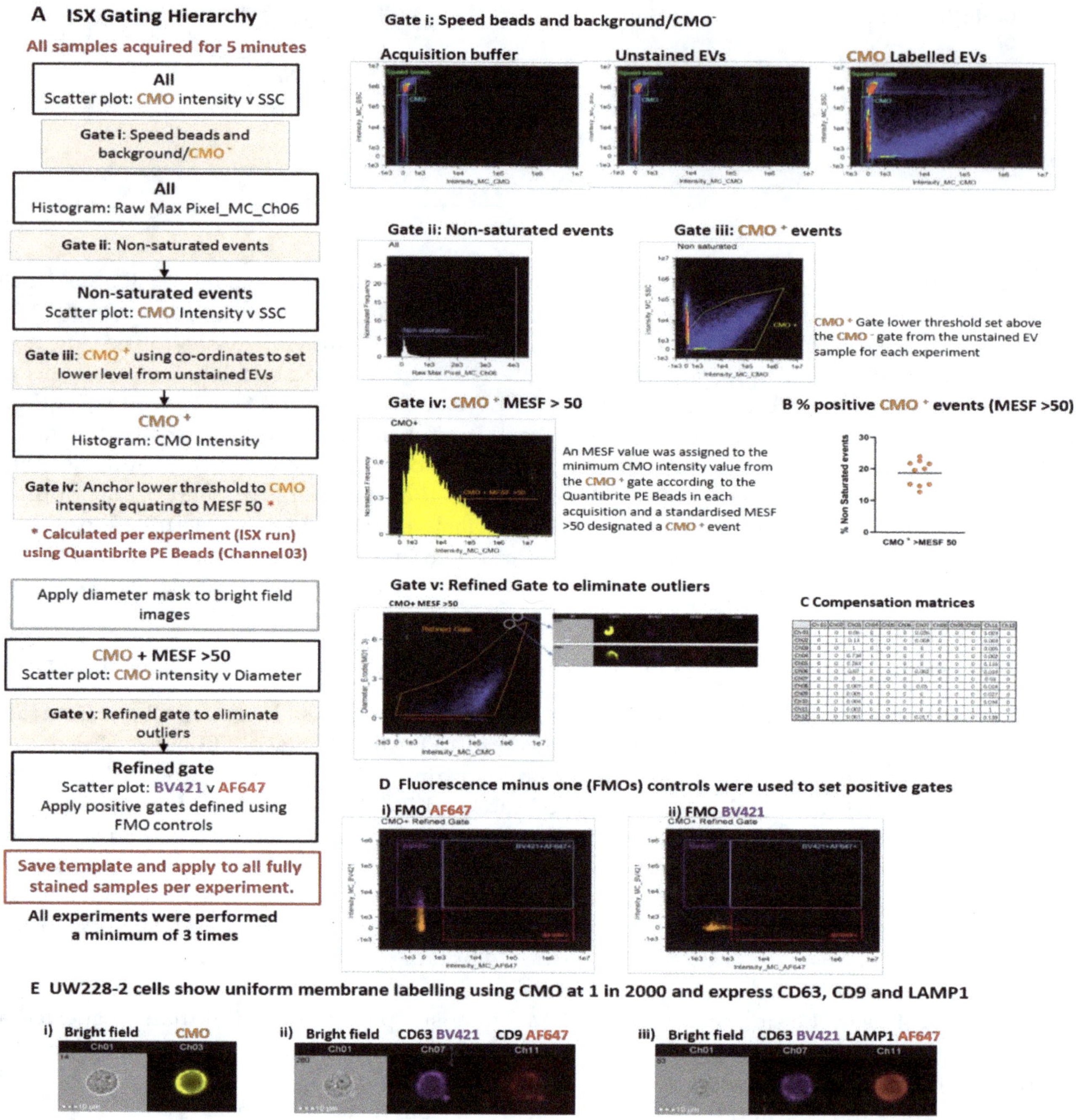

Figure 4. Optimised acquisition by Imaging flow cytometry and florescence membrane labelling can distinguish large extracellular vesicles from background and enable multiplex labelling. (**A**) Hierarchical gating can confidently refine the EV population for characterisation. Samples were acquired for 5 min using the ISX INSPIRE software and all events visualised using bivariate plots for fluorescence intensity in channel 03 (CMO) and channel 06 (side scatter). Initial gates for speed beads and CMO$^-$ events were set using the Acquisition buffer only and unstained EV samples included in each run (Gate i). These gates were applied to the labelled samples. Saturated events were excluded from the analysis by gating on the histogram plot: Raw Max Pixel for Channel 06 (Gate ii) thus removing the speed beads and very high side scatter events. An initial CMO + gate was placed (Gate iii). A lower fluorescence threshold of MESF > 50 in channel 3 (CMO) was set in each experiment using the regression analysis of Quantibrite PE Beads as described. The CMO + gate was anchored using co-ordinates for the equivalent fluorescence intensity and a further CMO + MESF > 50 gate applied to the histogram (Gate iv). Finally, a refined gate was used to eliminate outliers (Gate v). The analysis template was set for each experiment

and applied to all samples. (**B**) Percentage positive CMO+ events (MESF > 50). The percentage of CMO+ events (MESF > 50) is represented by an orange dot for 10 replicate experiments. Black bar represents mean 18.73% (± 1.2 SEM). (**C**) A compensation matrix was applied to all samples. The compensation wizard was used to create a matrix which was applied to all samples. (**D**) Single, double, and triple labelling protocols were used for compensation and accurate gating. Dual labelled EVs provided Fluorescence minus one (FMO) controls for (i) AF647; EVs labelled with CMO and CD63 BV421) or (ii) BV421; (EVs labelled with CMO and LAMP1 AF647) and used to set positive gates. (**E**) UW228-2 cells were used to optimise multiplex labelling. Cell mask orange (CMO) provides a general membrane label (i) and could be used to co-label with EV markers CD63 BV421 and CD9 AF647 (ii) or CD63 BV421 and LAMP1 AF647 (iii). Representative gallery images from IDEAS software are shown.

We tested a number of approaches to apply compensation to our data in consultation with the manufacturer's specialists. Initially, we attempted to construct a matrix using single stained EVs; however, the single pixel fluorescence was insufficient for the in-built compensation Wizard to assign a matrix. Next, we tried single stained parent cells, but the compensation matrix formulated by the wizard resulted in over-compensation and negative fluorescence intensity events. Over-compensation was thought to be due to the imbalance between the strong fluorescence signal resulting from cell mask orange, a membrane marker which indiscriminately labels lipids, and relatively weak fluorescence signal from target specific antibodies. These experiments were originally performed using a FITC conjugated CD63 antibody, however we found the level of adjustment required between channel 02 (FITC) and channel 03 (CMO) contributed to the negative populations. We re-optimised using the same CD63 antibody clone (HSC6) conjugated to BV421 (channel 07) which is spectrally more distinct from CMO. Finally, we found that using commercially available compensation beads labelled with our antibodies and acquired in the channels used for our study, in conjunction with assisted manual adjustment, provided the most reproducible compensation matrix.

We combined CMO with BV421 and AF647 and Fluorescence Minus One (FMO) controls for each fluorophore were used to set positive gates (Figure 4D). The parent UW228-2 cells were screened for expression of the EV markers chosen for this study: CD63, CD9, and LAMP1 as defined in the latest MISEV guidelines [1]. Multiplex labelling was performed using the following combinations: CMO with CD63 BV421 and CD9 AF647; or CMO with CD63 BV421 and LAMP1 AF647.

2.4. *Multiplex Labelling Reveals Heterogeneity in EV Marker Expression*

The proportion of CMO+ only events varied across the 10 replicates (Figure 5A). The mean percentage positive CMO+ (MESF > 50) events (63.3% ± 4.6 SEM) which did not demonstrate expression of any EV markers included in this study were designated CMO only. Considering the single EV markers, the proportion of EVs positive for each EV marker varied CD63 19.0%, (± 2.2 SEM), CD9 mean 10.3% (± 0.5 SEM); LAMP1 47.6% (± 3.8 SEM). These data shown LAMP1 to be the most prevalent EV marker screened in this study.

A Within the CMO⁺ (MESF >50) gate: The percentage positive CMO only events which did not express the EV markers screened in this study varied across replicates. The proportion of CMO⁺ events expressing a single EV marker also varied across replicate experiments with LAMP1 being the most prevalent.

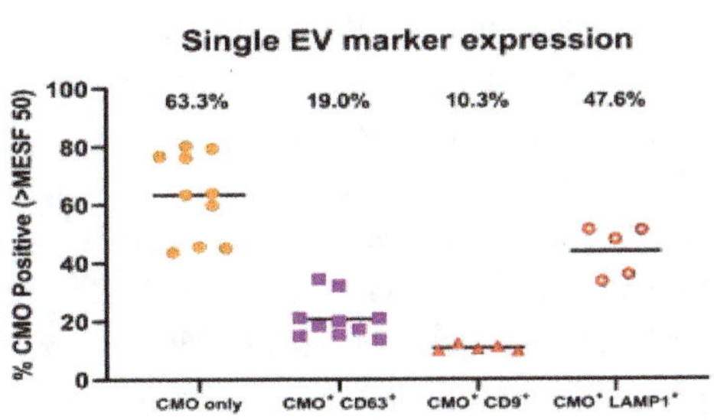

Figure 5. *Cont.*

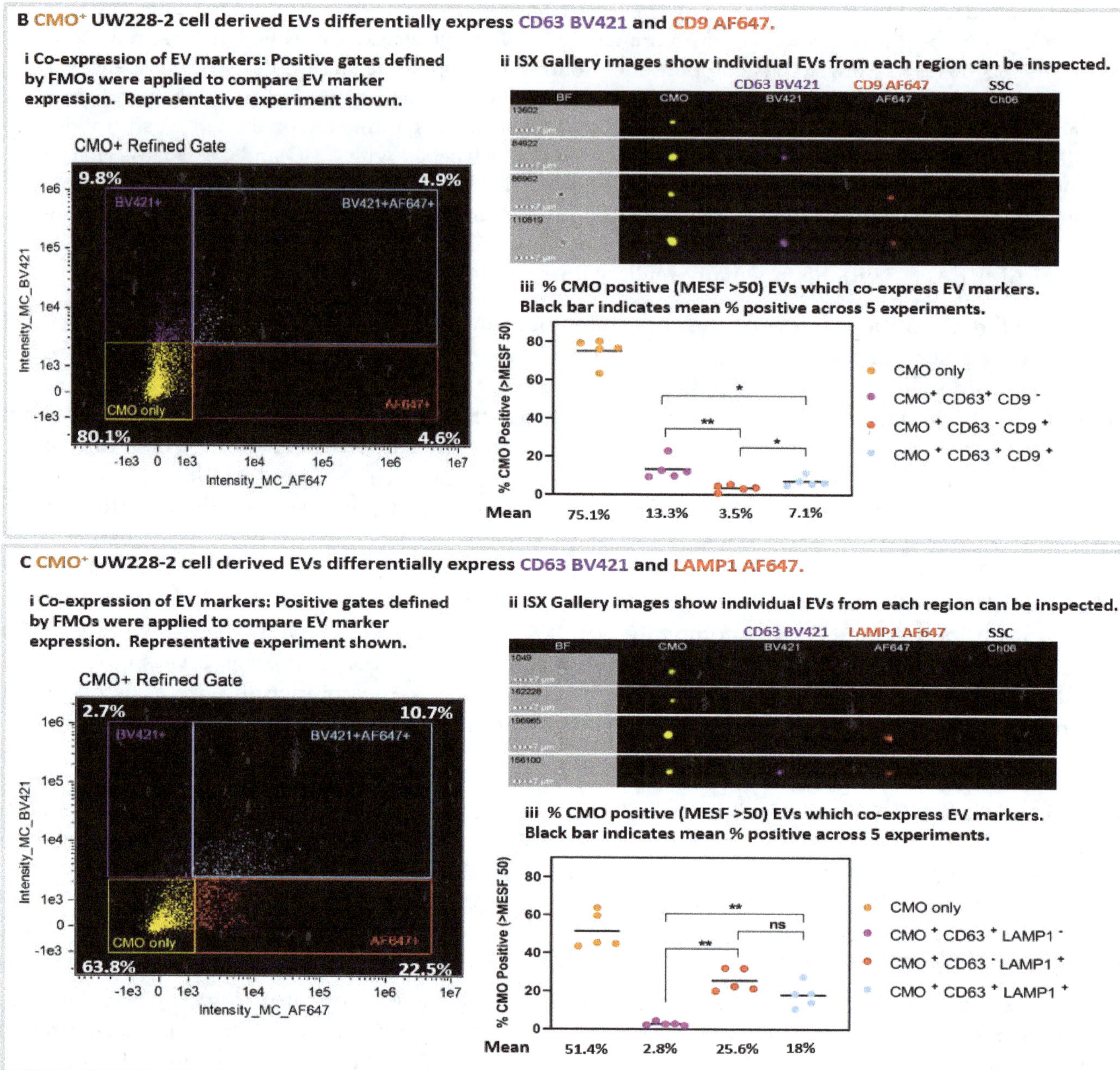

Figure 5. Isolated large EVs can be triple-labelled and individual events can be scrutinised post-acquisition. (**A**) The proportion of CMO+ events expressing a single EV marker varied across replicates. The percentage positive events within the CMO+ MESF > 50 gate which expressed either no EV marker (CMO only), CD63, CD9, or LAMP1 varied across replicates. Black bar represents the mean across 10 (CMO or CD63) or 5 (CD9 or LAMP1) replicates. CMO only 63.3% (± 4.6 SEM), CD63 only 19% (± 2.2 SEM), CD9 (± 0.5 SEM) only 10.3% and LAMP1 only 47.6% (± 3.8 SEM). (**B**) UW228-2 cell derived EVs differentially express EV markers CD63 and CD9, or both. EVs were labelled with CMO, CD63 BV421 and CD9 AF647. CMO $^{+}$ events (MESF > 50) were analysed for co-expression with CD63, CD9 or both. Quadrant gating of bivariate plots for fluorescence intensity in channel 07 (BV421) and channel 11 (AF647) demonstrated single labelled (CMO^{+} only), dual and triple labelled EV populations (i). Percentage positive in each quadrant is shown. Representative experiment. Gallery images display examples of individual events from each of the 4 quadrants (ii). Bright Field (BF) and side scatter (SSC) channels are shown alongside CMO, CD63 BV421 and CD9 AF647. Of the CMO+ MESF >50: 75.1% (± 3.0 SEM) were CMO $^{+}$ only; 13.3% (± 2.4 SEM) expressed CD63 BV421 and 3.5% (± 0.8 SEM)

expressed CD9 AF647. 7.1% (± 1.1 SEM) of the CMO^+ events expressed both CD63 and CD9. **(C)** UW228-2 cell derived EVs differentially express EV markers CD63 and LAMP1, or both. EVs were labelled with CMO, CD63 BV421 and LAMP1 AF647. CMO^+ events were analysed for co-expression with CD63, LAMP1 or both. Quadrant gating of bivariate plots for fluorescence intensity in channel 07 (BV421) and channel 11 (AF647) demonstrated single labelled (CMO^+ only), dual and triple labelled EV populations (i). Percentage positive in each quadrant is shown. Representative experiment. Gallery images display examples of individual events from each of the 4 quadrants (ii). Bright Field (BF) and side scatter (SSC) channels are shown alongside CMO, CD63 BV421 and LAMP1 AF647. Of the CMO+ MESF >50: 54.1% (± 4.3 SEM) were CMO^+ only; 2.8% (± 0.4 SEM) expressed CD63 BV421 and 25.6% (± 2.7 SEM) expressed LAMP1 AF647 whilst 18.0% (± 2.8 SEM) of the CMO^+ events expressed both CD63 and LAMP1. (* $p = 0.0159$ or 0.0317; ** $p = 0.0079$). Mann Whitney U test.

Differential expression and co-expression of EV markers was examined between five replicate experiments. Representative bivariate plots of AF647 intensity against BV421 intensity (Figure 5B(i) and Figure 5C(i) for CD63/CD9 and CD63/LAMP1, respectively), and representative galleries (Figure 5B(ii) and Figure 5C(ii)) of EVs displaying each labelling combination taken from the quadrant plots are shown. The proportion of EVs which co-express EV markers within each gate for the five replicates is shown in Figure 5B(iii) and Figure 5C(iii). Where the EVs were labelled for CMO, CD63 BV421, and CD9 AF647 (Figure 5B(iii)): 75.1% (± 3.0 SEM; yellow) were CMO only, 13.3% (± 2.4 SEM; purple) were CMO^+ $CD63^+$ $CD9^-$; 3.5% (± 0.8 SEM; red) were CMO^+ $CD63^-$ $CD9^+$ and 7.1% (± 1.1 SEM; blue) were CMO^+ $CD63^+$ $CD9^+$. There were significantly fewer $CD63^-$ $CD9^+$ EVs compared with $CD63^+$ ($p = 0.0079$) only or $CD63^+$ $CD9^+$ EVs ($p = 0.0159$). Similarly, there were fewer EV co-expressing CD63 and CD9 compared with $CD63^+$ alone ($p = 0.0317$). (Mann Whitney U test). Considering the EVs labelled with CMO, CD63 BV421 and LAMP1 AF647 (Figure 5C(iii)). 51.4% (+/− 4.3 SEM; yellow) were CMO+ only, a comparatively low 2.8% (± 0.4 SEM; purple) were CMO^+ $CD63^+$ $LAMP1^-$; whilst 25.6% (± 2.7 SEM; red) were CMO^+ $CD63^-$ $LAMP1^+$ and 18% (± 2.8 SEM; blue) co-expressed both CD63 and LAMP1. There were significantly more EVs which expressed LAMP1 compared with CD63 alone ($p = 0.0079$) and more expressing both CD63 and LAMP1 compared with CD63 alone ($p = 0.0079$).

2.5. ISX Allow Correlative Analyses of Diameter with EV Marker Expression for Phenotyping

We applied the Diameter Mask; Erode (M01, 3) in IDEAS software to all fully labelled replicates using the Batch analysis function and compared the range of diameters present within the CMO+ MESF> 50 gate. Figure 6A shows single event diameters within the 10 replicates. This median diameter (indicated with the black line) varied significantly across the replicates (922–1129 nm: Kruskal-Wallis; $p > 0.0001$). However, these data represent EVs isolated as 3 biological replicates and when each replicate is analysed using ANOVA there was no significant difference between the median diameters of each EV preparation. There was a clear correlation between EV diameter and CMO intensity as shown in Figure 6B. Regression analysis was performed on the individual exported feature values from each replicate (2 representative plots are shown from different biological replicates) ($p < 0.0001$).

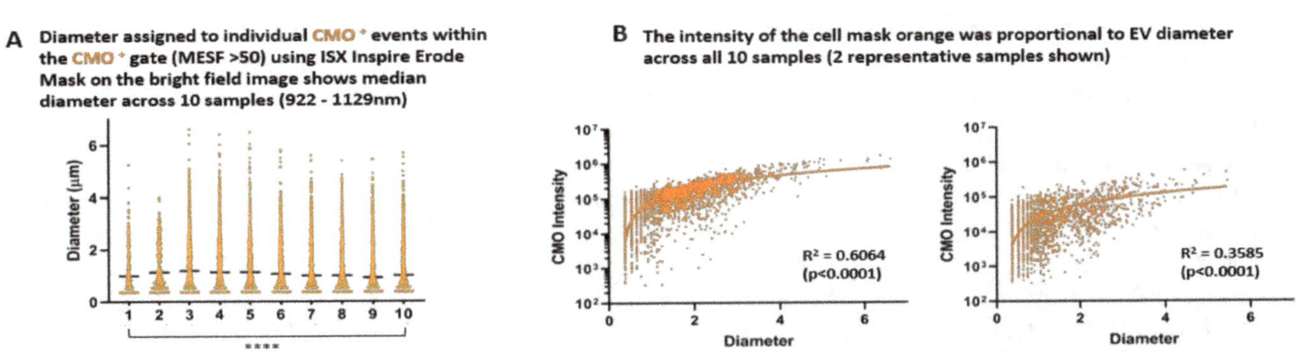

Figure 6. *Cont.*

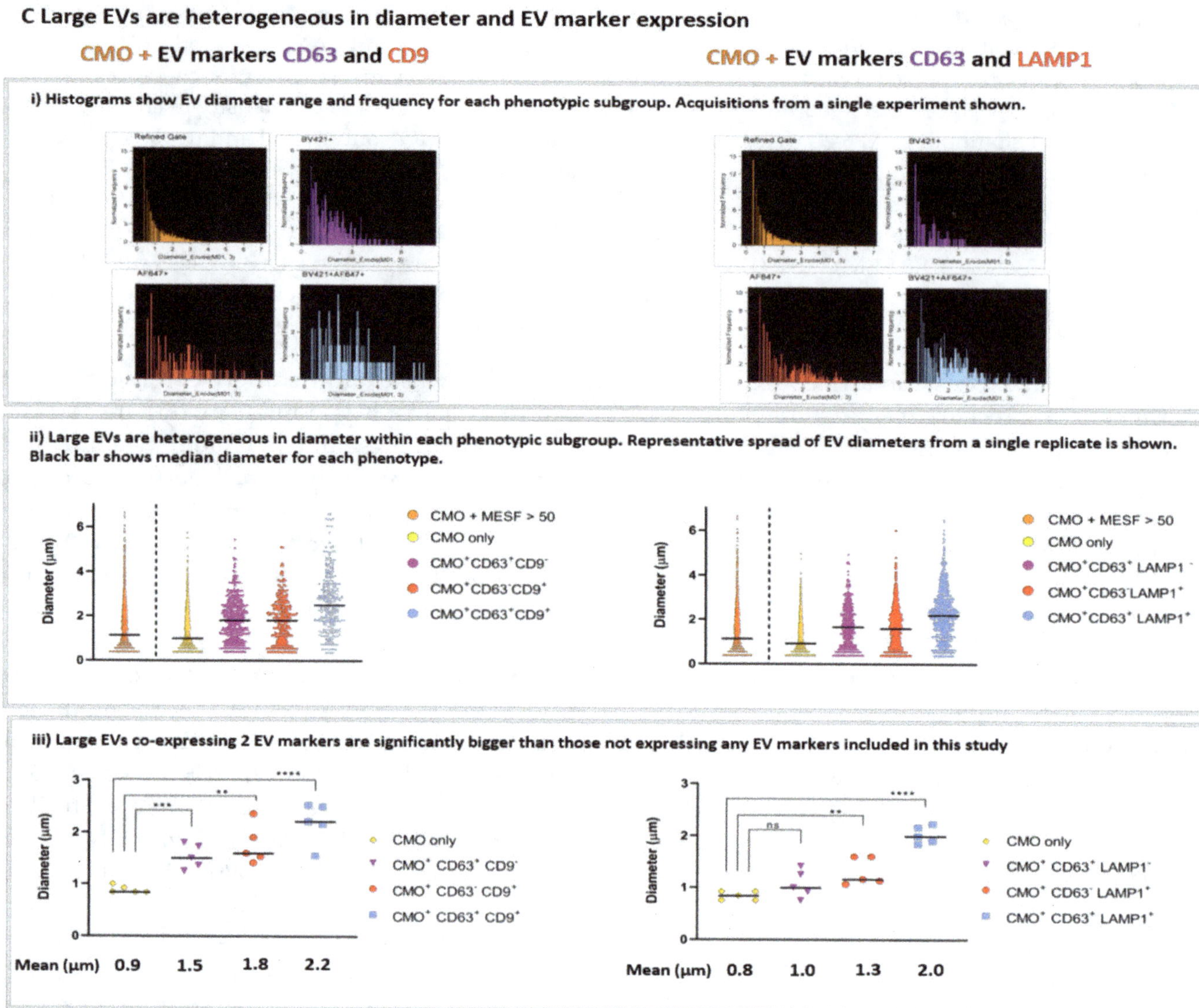

Figure 6. The ISX can accurately assign diameter to large EVs and facilitate individual event analyses to phenotype heterogeneous EV populations. (**A**) The median diameter of CMO $^+$ EVs from UW228-2 cells was comparable across 10 samples. The erode mask was applied to the bright field images of those EVs which were included in the CMO$^+$ >MESF 50 gate. The diameter of individual events within this gate was exported from IDEAS into PRISM for analysis. This figure shows a median diameter (represented by the black bar) of 922–1129 nm across the 10 samples and visualises the broad size range within each sample. (**** $p > 0.0001$ Kruskal-Wallis test). (**B**) The fluorescence intensity of the CMO was proportional to EV diameter. Bivariate plots showing CMO intensity against diameter consistently demonstrates a proportional relationship between EV size and intensity of the membrane label across samples run in triplicate experiments. 2 representative samples shown. (Regression analysis $p < 0.0001$). (**C**) Comparison of EV phenotypes across triplicate experiments. Individual histograms displaying EV diameter for each phenotype demonstrated differential size ranges according to EV marker(s) expression (i a and i b). In each case, those labelled with CMO only and not expressing either CD63, CD9 or LAMP1 have a smaller median diameter compared with EVs expressing CD63 and/or CD9 or LAMP1 (ii a and ii b). In both cases, those EVs co-expressing both CD63 with CD9 (iii a) or CD63 with LAMP1 (iii b) are significantly larger than CMO only. (** $p < 0.005$, *** $p < 0.0004$, **** $p < 0.0001$).

We found that large EVs are heterogeneous in both diameter and EV marker expression. Histogram plots of individual events exported from the quadrant positive gates show the different diameter range and frequency within each phenotypic subgroup (Figure 6C(i)). Representative acquisitions from a single replicate are shown. ISX analysis facilitates the comparison of individual EV diameters from within each phenotype (Figure 6C(ii)). A representative plot for all CMO $^{+}$ (MESF > 50) in a single replicate of each phenotype is shown and the black bars represent the median diameter for each subgroup. Within this typical replicate, the median diameter and range for each subgroup was as follows: All CMO >MESF 50 1.1 μm (362 nm–6.2 μm), CMO only 1.0 μm (362 nm–5.4 μm), CMO+ CD63^{+} CD9^{-} 1.8 μm (362 nm–5.1 μm), CMO+ CD63^{-} CD9^{+} 1.8 μm (362 nm–4.7 μm), and CMO+ CD63^{+} CD9^{+} 2.5 μm (362 nm–6.2 μm). When considering the second phenotype: All CMO >MESF 50 1.4 μm (362 nm–6.2 μm), CMO only 1.1 μm (362 nm–4.6 μm), CMO+ CD63^{+} CD9^{-} 1.7 μm (362 nm–4.5 μm), CMO+ CD63^{-} CD9^{+} 1.6 μm (362 nm–5.6 μm), and CMO+ CD63^{+} CD9^{+} 2.2 μm (362 nm–6.1 μm). In each case the median diameter increases in size with accumulating EV marker expression.

When considering all 5 replicates; we compared the mean diameter for each subgroup (Figure 6C(iii)). For the first group (left panel): CMO only 0.89 μm (±0.03 SEM), CMO+ CD63^{+} CD9^{-} 1.53 μm (± 0.1 SEM), CMO+ CD63^{-} CD9^{+} 1.76 μm (±0.17 SEM) and CMO+ CD63^{+} CD9^{+} 2.19 μm (±0.18 SEM). Those EVs expressing either CD63 ($p < 0.0004$) or CD9 ($p < 0.0010$) were significantly larger than EV which did not (CMO only). For the second subgroup (right panel) the mean diameters were as follows: CMO only 0.84 μm (±0.04 SEM), CMO+ CD63^{+} CD9^{-} 1.06 μm (±0.12 SEM), CMO+ CD63^{-} CD9^{+} 1.31 μm (±0.12 SEM), and CMO+ CD63^{+} CD9^{+} 2.02 μm (±0.07 SEM). In this group, those EVs expressing CD63 only were not significantly larger than those which did not. However, EVs expressing LAMP1 were larger ($p < 0.005$). In both cases, those CMO $^{+}$ EVs co-expressing 2 EVs markers: either CD63 and CD9, or CD63 and LAMP1 (blue) were significantly larger than those which do not carry the markers screened in this study. (**** $p < 0.0001$. Un-paired t-test).

3. Materials and Methods

3.1. Cell Line

UW228-2 cells were kindly provided by DTW Jones (DKFZ, Heidelberg, Germany). Cells were grown in DMEM with L-glutamine (Lonza, Manchester, UK; cat: R8758) supplemented with 10% FCS (SIGMA, Gillingham, UK; cat: F9665) in Corning 225 cm^2 Angled Neck Cell Culture Flask with Vent Cap (Fisher Scientific, Leicestershire, UK; cat: 431082) at 37 °C with 5% CO_2 in normoxia. Cell lines were passaged with 1 × Trypsin-EDTA (Lonza, UK; cat: T3924). All cell lines tested negative for Mycoplasma and all were authenticated in-house (CRUK-Manchester Institute) by examination of a total of 21 loci across the genome using the Powerplex 21 System (Promega, Southampton, UK).

3.2. Cell Culture

For each experiment, cells were seeded at 2.5 × 10^6 cells in 50 mL DMEM 10% FBS (complete media) per 225 cm^3 tissue culture flask and allowed to adhere overnight. On day 2 media was switched to 50 mL serum-free DMEM for 24 h prior to EV isolation and cell preparation. Conditioned media (containing EVs) was removed, and the cells washed x2 with PBS before trypsinisation using 1 x Trypsin-EDTA (Lonza, Manchester, UK; cat: T3924). Cell counts and viability were checked at the time of EV harvest using the trypan blue exclusion assay (0.4% Trypan blue solution; SIGMA; Gillingham, UK; cat T8154).

3.3. Vesicle Isolation

Large EVs were harvested using a standard operating procedure (SOP) as previously reported (Figure 1F) [15]. For each experiment, EVs were isolated from the serum-free, conditioned media from 2 × 225 cm^2 flasks (pooled; total 100 mL). Cell culture supernatant (conditioned media: CM) was centrifuged in 2 tubes to remove cells (300× g 5 min, ×2) and filtered using a double layered 5 μm pore nylon Sieve (Fisher Scientific, Leicestershire, UK; cat 12994257). The supernatant was collected and centrifuged at 2000× g for 30 min and prepared for ISX analysis. Centrifugation steps were performed using an Eppendorf 5702 bench top centrifuge with an A-4-38 rotor. All EV preparations were performed on the day of analysis and not stored. Experimental procedures for EV isolation have been submitted to the EV-TRACK database (EV TRACK ID: EV190013) [22].

3.4. Chemicals and Reagents Including Antibodies

3.4.1. Apoptosis Assay

UW228-2 cells were seeded at 1×10^{5}/well into 6 well plates (Corning; Fisher Scientific, Leicestershire, UK; cat CL S3516) and incubated at 37 °C overnight in DMEM containing 10% FBS. Triplicate wells were cultured in complete media or switched to serum-free DMEM for 24 h before screening with Annexin V APC/PI using the Apoptosis detection kit (Biolegend UK, London, UK; cat 640932) according to the manufacturer's instructions. 30,000 events were acquired using the LSR II flow cytometer with lasers for APC (640 nm laser, emission captured at 660 nm) and PI (488 nm laser with emission captured at 575 nm). Positive controls were generated to inform accurate gating: cells were treated with 200 µM (UW228-2) Etoposide (SIGMA; Gillingham, UK; cat E1383) for 24 h (Apoptotic cells: Annexin V), or heated at 56 °C for 10 min prior to labelling (Dead cells: PI).

3.4.2. Immunofluorescence Microscopy

Cells and EVs were immobilised onto CellCarrier 96 well plates (Perkin Elmer; Bucks, UK; Cat 6005550), fixed with 3.7% paraformaldehyde, permeabilised with 0.2% Triton X in PBS and probed using anti-human CD63 antibody (Clone: H5C6, Biolegend UK, London, UK; Cat: 353005) directly conjugated to FITC and counterstained for polymerised actin using 0.2 × Alexa Fluor 555 Phalloidin (Fisher Scientific, Leicestershire, UK; cat A34055) and 300 nM DAPI (Biolegend UK, London, UK, 422801). Images were captured using the Perkin Elmer Operetta system (Perkin Elmer; Bucks, UK) at ×40 magnification.

3.4.3. Transmission Electron Microscopy (TEM)

EVs were immobilised onto ACLAR (poly-chloro-tri-fluoro-ethylene (PCTFE) film) coated with Corning CellTak (Fisher Scientific, Leicestershire, UK; cat 354240) and fixed with glutaraldehyde in sodium cacodylate buffer (pH 7.2) followed by post-fix staining with osmium tetroxide and uranyl acetate (supplied in-house by the FBMH Core Facility). Preparations were dehydrated and embedded in resin to allow serial 60–200 µm sections to be taken. Images were captured using a Biotwin Philips TECNAI G2 transmission electron microscope.

3.4.4. Tunable Resistance Pulse Sensing (TRPS)

Size and quantity were determined by Tunable Resistance Pulse Sensing (TRPS) using the qNano GOLD instrument (iZON Science, Christchurch, New Zealand) as per manufacturer's instructions. The principles are discussed elsewhere [23]. We analysed an aliquot of isolated EVs alongside downstream analyses using overlapping sizes of Nanopores (NP600, NP1000, and NP2000) to provide a full picture of EV size distribution and quantity.

3.4.5. Cell Mask Orange Labelling

100 mL (2 × 50 mL) EV containing media was used to harvest large EVs for each experiment as described and the 2000× *g* pellets re-suspended in either (1) 2 mL serum-free DMEM or (2) 2 mL Cell Mask Orange (CMO: 2.5 µg/mL in serum-free DMEM) (Fisher Scientific, Leicestershire, UK; cat C10045) and both were incubated at 37 °C for 10 min.

3.5. Antibody Labelling

See Table 2 for antibody details and manufacturer information. Antibody titrations were performed for each antibody using parent cells and EVs. In all cases, the maximum recommended volume (5 µL) provided the greatest fluorescence signal from the EVs. Antibody only controls (no EVs) were included in the ISX analysis and showed no fluorescence events above the unstained gate in each case. Both CMO labelled and unlabelled EVs preparations were washed by addition of 5 mL 1% BSA/PBS and centrifuged at 2000× g for 30 min. The pellets were re-suspended in 700 μL 0.2 μm filtered 1% BSA/PBS and split into 7 × 100 μL aliquots. For each labelling combination, 5 μL directly conjugated primary antibodies were added simultaneously as follows: Non-CMO labelled EVs were used for unstained, single labelled CD63 BV421, CD9 AF647, or LAMP1 AF647 and CMO FMO controls (×2) (BV421 + AF647: both antibodies were screened). The final aliquot was used to establish EV concentration and diameter range using TRPS analysis on the qNANO. The CMO labelled EV pellet was re-suspended in 700 μL 0.2 μm filtered 1% BSA/PBS and split into 7 × 100 μL aliquots. CMO labelled EVs were used as single stained (CMO+ only), AF647 FMO (CMO + BV421), BV421 FMO (CMO + AF647), and multiplexed CMO + CD63 BV421 + either CD9 AF647 or LAMP1 AF647. Antibodies were incubated for 1 h on ice in the dark. EVs were washed by addition of 500 μL 0.2 μm filtered 1% BSA/PBS and centrifuged at 2000× g for 30 min. Resulting pellets were re-suspended in 75 μL 0.2 μm filtered 1% BSA/PBS for ISX analysis. Fully labelled EV preparations were treated post-acquisition with 0.1% Triton-X 100 and acquired again to demonstrate loss of fluorescence due to antigen degradation.

Table 2. Chemicals and Reagents Including Antibodies.

<table>
<tr><th colspan="2">Reagent</th><th colspan="2">Manufacturer</th><th colspan="2">Catalogue No.</th></tr>
<tr><td colspan="2">Cell mask orange: plasma membrane marker</td><td colspan="2">Thermo Fisher Scientific</td><td colspan="2">C10045</td></tr>
<tr><td colspan="2">Annexin V APC/PI: Apoptosis assay</td><td colspan="2">Biolegend</td><td colspan="2">640932</td></tr>
<tr><td colspan="2">Alexa Fluor 555 phalloidin: Polymerised actin cytoskeleton</td><td colspan="2">Thermo Fisher Scientific</td><td colspan="2">A34055</td></tr>
<tr><td colspan="2">Speed beads: Imagestream flow calibration</td><td colspan="2">Amnis</td><td colspan="2">400041</td></tr>
<tr><td colspan="2">Quantibrite PE Beads</td><td colspan="2">Beckton Dickinson UK</td><td colspan="2">340495</td></tr>
<tr><th>Antibody</th><th>Clone</th><th>Fluorophore</th><th>Isotype</th><th>Manufacturer</th><th>Catalogue No.</th></tr>
<tr><td>Anti-human CD 63</td><td>H5C6</td><td>BV421</td><td>Mouse IgG1, k</td><td>Biolegend</td><td>353030</td></tr>
<tr><td>Anti-human CD9</td><td>MEM-61</td><td>AF647</td><td>Mouse IgG1</td><td>Fisher Scientific</td><td>15317424</td></tr>
<tr><td>Anti-human LAMP 1</td><td>H4A3</td><td>AF647</td><td>Mouse IgG1, k</td><td>Biolegend</td><td>328611</td></tr>
</table>

3.6. *Imagestream Acquisition*

Sheath buffer (PBS without calcium and magnesium: SIGMA, Gillingham, UK; Cat D5652) was filtered using 0.2 μm bottle top filters (SIGMA, Gillingham, UK, Nalgene: FIL8184) to minimise background signal. Internal instrument calibrations were performed before every run according to manufacturer's guidelines using the ASSIST Calibrations to include: camera synchronisation, spatial offsets, dark current, bright field crosstalk coefficient, core stage position, horizontal laser, side scatter, and a retro illumination scheme to maximise the amount of light incident. This was followed by a series of internal operations designed to measure performance including excitation laser power, bright field uniformity, and focus. Specific laser powers used for this study are detailed in Table 3.

Table 3. Settings used for Imagestream.

Laser	Channel/Filter	Power mW	Parameter
405	Ch07/435–80 nm	120	BV421
561	Ch03/577–35 nm	200	CMO
785	Ch06/762–35 nm	70	SSC
642	Ch11/702–85	150	AF647

Speed beads with an exaggerated irregular surface were incorporated into every analysis for internal calibration. A dedicated laser was assigned to assess side scatter (CH 06; SSC: 785 nm laser). For each experiment, a separate readout was obtained from 0.2 μm filtered 1% BSA/PBS acquisition buffer alone. All events were acquired for 5 min and visualised using bivariate plots of side scatter against fluorescence intensity.

For cells, 10,000–30,000 total events were acquired. EVs were acquired for 5 min at ×60 magnification using lasers as described (Table 3) (Image stream, Amnis, Seattle, WA, USA). The ×60 objective provides a Numerical Aperture of 0.9 enabling resolution of 0.3 μm^2/pixel [19].

4. Compensation

Antibody labelled compensation beads (anti-mouse compensation beads: BD Biosciences, San Jose, CA, USA; cat 552843) were used to acquire single colour controls within the channels used for this study. The final compensation matrices were constructed by the wizard (INSPIRE) with manual adjustment in consultation with the manufacturer's specialist adviser and applied to the .rif files of all controls, dual and triple labelled EVs. Data were analysed using the IDEAS software (v. 6.2, Amnis, Seattle, WA, USA). The compensation matrices and analysis template were applied using the batch processing tool to all .rif files to produce .daf files for each sample. FCS files were exported and uploaded onto the Flow Repository according to the requirements.

5. Mask Selection for Assessment of EV Diameter

We investigated which of the diameter masks available within the IDEAS software would be most accurate for EVs. We used non-fluorescence size calibration beads (Fisher Scientific—UK Ltd., Loughborough, UK; cat F13838) to validate the masks. The beads were acquired using the same laser powers and settings as the EV preparations and analysis templates were constructed to identify which mask fitted most closely to the bead diameter according to the manufacturer's instructions. We found that applying the diameter mask Erode (03; indicating 3 pixel erosion) to the bright field channel most closely assigned the correct diameter. This mask formed part of an analysis template which was applied to all samples using the batch analysis tool.

6. Molecules of Equivalent Soluble Fluorochrome (MESF) Calculation

To enable comparisons between experiments, Molecules of Equivalent Soluble Fluorochrome (MESF) values were calculated as previously described [19,24]. Quantibrite PE beads (BD Biosciences, San Jose, CA, USA; Cat: 340495. Lot: 90926) were the closest available calibration beads for the fluorescent channel used to detect Cell Mask Orange labelled EVs. A fresh aliquot of lyophilised Quantibrite beads was reconstituted for each run, and 5000 events were acquired using the identical laser settings for each fluorophore as described. The SSC laser (channel 06) was adjusted to ensure the beads could be visualised on the bivariate plots and therefore each bead could be gated as a separate population and the median fluorescence intensity recorded. The CMO+ events were then analysed for expression of the EV markers included in this study: CD63, CD9 and LAMP1.

7. Discussion

Our principal aim was to develop a standardised method for the isolation and characterisation of individual large EVs, which could be further developed for phenotyping large EVs from clinical samples. The value of the Imagestream to the field of EV characterisation has been explored elsewhere [19,24,25]. However, reports focussing on the large EV populations, frequently discarded during small EV isolation protocols, are rarely present in the current literature.

By using in vitro cultures, we were able to use the ideal conditions to generate EVs and optimise experiments. Specifically, by culturing in serum-free media, we eliminated contamination from bovine EVs present in FBS [18] and subsequent false positive fluorescence signals from serum lipoproteins [20]. Nevertheless, harvesting large EVs from any source presents challenges as cells or cell debris, including intracellular vesicles released due to parent cell membrane rupture from early centrifugation steps, can contaminate the subsequent EV pellets. For the EVs to be truly extracellular prior to isolation, the outer membrane of accompanying cells must not be ruptured by mechanical or chemical means during initial harvest. The centrifugation speeds we have used here are low compared to some commonly reported EV isolation protocols [26] to specifically preserve large EV membrane integrity. Electron microscopy remains the sole technique that can examine individual EVs and EV preparations for sheared cell fragments, but it is neither quantitative nor high throughput. In addition, whilst TEM is widely used for EV investigation, it cannot perhaps distinguish between bone fide EVs and other particles. In this case, cryoEM or CLEM (correlative light and electron microscopy) which facilitates overlaid fluorescence or gold labelled antibody binding and EM would prove more informative. It is necessary to use a combination of techniques to explore the quantity, quality, and biology of EVs. All techniques, many originally designed for analysing cells, have technical challenges when applied to considerably smaller entities. For flow cytometry, background scatter events due to particles in the sheath fluid are an anticipated phenomenon which is rarely reported. In the work we report here, a high level of background appeared within the same gate as unstained EVs and persisted despite 0.2 μm filtration of sheath fluid. Our protocol is therefore reliant on strong and uniform, membrane-bound fluorescence labelling in order to assign an initial gate that separates potential EVs from speed beads or background scatter events. However, we found that the lower fluorescence intensity threshold to define CMO positive events was not clearly distinct from the instrument background. This was likely due to a combination of EV size and relative fluorescence intensity. As recommended elsewhere [19] we used commercially available fluorescent beads of known intensity (Quantibrite) to provide a means to assign standardised units (MESF) and therefore a mathematical cut-off for our CMO+ gate. We acknowledge that calibration beads differ in the physical properties of EVs, in particular the refractive index, which can affect the intensity of scattered light, a particularly important consideration for the study of smaller EVs. PE was the closest available fluorophore to CMO and used as a standard for channel 03 on the ISX. CMO is a membrane label incorporated into the plasma membrane, and emits a greater fluorescence compared with a target-specific, conjugated antibody. This hampered the use of FITC alongside CMO as the spectrally close fluorophores led to overcompensation between channels 02 (FITC) and channel 03 (CMO). BV421 was a successful alternative but the difficulties encountered raised concerns about trying to further multiplex with additional fluorophores using this platform.

The challenges for choosing the correct technique(s) to analyse EVs have been well described elsewhere [27]. For detailed advice on considerations for EV separation or enrichment methodologies and recommended steps for EV characterisation, we recommend referring to the MISEV guidelines; a position paper prepared by the international EV community to support EV research [1]. A major challenge now is to adapt the protocol we describe for the analysis of large EVs from biological fluids. Clinical samples are more complex: EVs from a single cell type as described here offer the ideal model for characterisation, however clinical samples contain heterogeneous EVs from numerous cells [10]. EV isolation has been reported from peripheral blood with some correlations to clinical outcome in other cancers [28,29]. We have previously identified leukaemic EVs in patients' bone marrow plasma. This was possible using a marker which identified the cell of origin (B cell marker CD19) and understanding that the bone marrow of a leukemia patient is primarily composed of malignant B cells [9]. However, surface markers for Medulloblastoma cells are less well known, rather molecular signatures define subgroups in this cancer type [30]. Investigating surface marker expression on medulloblastoma cell-derived EVs would help to develop a more rapid screening tools. The value of a liquid biopsy to diagnose brain tumours in the clinic is clear and some suitable candidates have been identified in Glioblastoma [31,32]. A next step could be to investigate EVs released by primary medulloblastoma cells although the requirements for optimal culture conditions to generate sufficient material will add additional complexity. In the case of medulloblastoma, ideally, EVs circulating in either the cerebrospinal fluid or peripheral blood for comparison to matched primary tumour-derived EVs would be of considerable interest.

Our SOP is likely to exclude most small EVs and exosomes, expected to be present in the supernatant discarded at the final step (2000× g). Experiments to isolate these for comparison are on-going. One classification we examined in this study was the distinction from apoptotic bodies. Demonstration of intact EV membranes, a lack of fragmented nuclei staining (DAPI) and evidence that EVs were derived from viable cells supported our assertion that the EVs analysed were not apoptotic bodies or cell debris [33]. It might be possible to further distinguish these populations using molecular profiling. Other studies suggest that these EV subgroups display distinct RNA profiles [34] an approach which is reliant on pure populations and therefore robust and meticulous isolation protocols.

Our data show that large EVs are ubiquitous and whilst absolute quantification is not yet within reach, we demonstrated similar size profiles using two independent techniques: TRPS and ISX. We have previously identified EVs of up to 6 μM using immunofluorescence, ISX and TEM [15]. However, validating the quantity and size range has only been possible using the qNANO instrument. The qNANO employs TRPS technology to quantify EV count in a given sample and assign a size relative to a calibration bead of known diameter. It is currently the only instrument which can provide this information across the large EV population which spans 250 nm up to 6 μm (from our cells). Other platforms are restricted to small EVs (<1 μM) due to the measurements being reliant on Brownian movement (e.g., Nanosight and Zetaview). We found the most prevalent EV populations to be around 250–450 nm; however, we consistently detect EVs with a much larger diameter range in every cell line we have screened to date. We remain cautious not to define these as oncosomes; as although derived from cancer cells, we have not yet demonstrated their oncogenic potential [4].

Whilst fluorescence intensity alone cannot be used to quantify protein expression levels due to low level antigen expression on EVs, we did observe patterns of differential expression. In EV literature, 3 principle markers are used to define EVs: CD63, CD9, and CD81. CD63, however, has been identified as a pan-EV marker, present in all defined EV subgroups to date [6] and therefore CD63 was our preferred initial marker. However, large EVs are as yet poorly characterised and we found that CD63 was not the most abundant EV marker in our study. CD63 is a tetraspanin which has been used as a selection tool for immuno-capture experiments and also for tracking EV release [35,36]. Based on our observations, if a full EV repertoire was of interest, then a cocktail of multiple markers should be considered as screening with CD63 alone will likely fail to capture a significant proportion of EVs. We found LAMP1, previously identified on exosomes and EVs from a variety our cell types and biological fluids [37], was significantly more prevalent.

We also show here that the majority of large EVs did not express any of the three markers screened. We did observe a significant increase in the median diameter of individual EVs which expressed 1 or more markers compared to none (CMO only), in each of the experiments performed. Further, across all replicates, those EVs which co-expressed 2 markers (CD63 + CD9, or CD63 + LAMP1) were significantly larger than those without. Others have suggested that larger EVs are likely to accommodate a greater number of tumour-derived molecules than exosomes [13] and data presented here would support that hypothesis. Further investigations to define a broader panel of large EV markers followed by more comprehensive techniques, such as proteomic profiling, would help to fully phenotype the large EVs population. From a clinical perspective, it is likely that large EVs will be a rich source of biomarkers of benefit to the study of human disease. Standardised protocols and instruments capable of measuring multiple markers are key to moving the field forward and expanding the interest from exosomes only.

The research we report here demonstrates that high resolution, high throughput imaging flow cytometry is an exceptional tool offering the unique ability to quantify and analyse individual events within heterogeneous EV populations. We set out to develop an isolation protocol consisting of minimal manipulation and processing which may abrogate, mask, or indeed elicit changes in EV structure or biology, which could impact on any functional read outs in downstream experiments. Indeed, fully understanding the biological consequence(s) of EV release or uptake by recipient cells is an essential part of the field and will advance with new technologies and innovation.

Author Contributions: Conceptualization, S.M.J. and A.B.; Data curation, S.M.J., A.B., C.S. and A.M.; Formal analysis, S.M.J., A.B. and M.G.M.; Funding acquisition, S.M.J. and M.G.M.; Investigation, S.M.J. and C.S.; Methodology, C.S. and A.M.; Resources, A.B. and A.M.; Validation, S.M.J.; Writing—original draft, S.M.J. Writing—review & editing, A.B. and M.G.M. All authors have read and agreed to the published version of the manuscript.

References

1. Thery, C.; Witwer, K.W.; Aikawa, E.; Alcaraz, M.J.; Anderson, J.D.; Andriantsitohaina, R.; Antoniou, A.; Arab, T.; Archer, F.; Atkin-Smith, G.K.; et al. Minimal information for studies of extracellular vesicles 2018 (MISEV2018): A position statement of the International Society for Extracellular Vesicles and update of the MISEV2014 guidelines. *J. Extracell. Vesicles* **2018**, *7*, 1535750. [CrossRef]
2. Cocucci, E.; Meldolesi, J. Ectosomes and exosomes: Shedding the confusion between extracellular vesicles. *Trends Cell Biol.* **2015**, *25*, 364–372. [CrossRef] [PubMed]
3. Van Niel, G.; Porto-Carreiro, I.; Simoes, S.; Raposo, G. Exosomes: A common pathway for a specialized function. *J. Biochem.* **2006**, *140*, 13–21. [CrossRef] [PubMed]
4. Meehan, B.; Rak, J.; Di Vizio, D. Oncosomes—Large and small: What are they, where they came from? *J. Extracell. Vesicles* **2016**, *5*, 33109. [CrossRef] [PubMed]
5. Yanez-Mo, M.; Siljander, P.; Andreu, Z.; Zavec, A.B.; Borràs, F.E.; Buzas, E.I.; Buzas, K.; Casal, E.; Cappello, F.; Carvalho, J.; et al. Biological properties of extracellular vesicles and their physiological functions. *J. Extracell. Vesicles* **2015**, *4*, 27066. [CrossRef]
6. Kowal, J.; Arras, G.; Colombo, M.; Jouve, M.; Morath, J.P.; Primdal-Bengtson, B.; Dingli, F.; Loew, D.; Tkach, M.; Théry, C. Proteomic comparison defines novel markers to characterize heterogeneous populations of extracellular vesicle subtypes. *Proc. Natl. Acad. Sci. USA* **2016**, *113*, E968–E977. [CrossRef]
7. Simpson, R.J.; Mathivanan, S. Extracellular Microvesicles: The Need for Internationally Recognised Nomenclature and Stringent Purification Criteria. *J. Proteomics Bioinform.* **2012**, *5*. [CrossRef]
8. Tkach, M.; Thery, C. Communication by Extracellular Vesicles: Where We are and Where We Need to Go. *Cell* **2016**, *164*, 1226–1232. [CrossRef]
9. Johnson, S.M.; Dempsey, C.; Chadwick, A.; Harrison, S.; Liu, J.; Di, Y.; McGinn, O.J.; Fiorillo, M.; Sotgia, F.; Lisanti, M.P.; et al. Metabolic reprogramming of bone marrow stromal cells by leukemic extracellular vesicles in acute lymphoblastic leukemia. *Blood* **2016**, *128*, 453–456. [CrossRef]
10. Gyorgy, B.; Szabó, T.G.; Pásztói, M.; Pál, Z.; Misják, P.; Aradi, B.; László, V.; Pállinger, É; Pap, E.; Kittel, Á.; et al. Membrane vesicles, current state-of-the-art: Emerging role of extracellular vesicles. *Cell. Mol. Life Sci.* **2011**, *68*, 2667–2688. [CrossRef]
11. Colombo, M.; Raposo, G.; Thery, C. Biogenesis, secretion, and intercellular interactions of exosomes and other extracellular vesicles. *Annu. Rev. Cell Dev. Biol.* **2014**, *30*, 255–289. [CrossRef] [PubMed]
12. Xu, R.; Rai, A.; Chen, M.; Suwakulsiri, W.; Greening, D.W.; Simpson, R.J. Extracellular vesicles in cancer—Implications for future improvements in cancer care. *Nat. Rev. Clin. Oncol.* **2018**, *15*, 617–638. [CrossRef] [PubMed]
13. Minciacchi, V.R.; Freeman, M.R.; di Vizio, D. Extracellular vesicles in cancer: Exosomes, microvesicles and the emerging role of large oncosomes. *Semin. Cell Dev. Biol.* **2015**, *40*, 41–51. [CrossRef] [PubMed]
14. Morello, M.; Minciacchi, V.R.; De Candia, P.; Yang, J.; Posadas, E.; Kim, H.; Griffiths, D.; Bhowmick, N.; Chung, L.W.K.; Gandellini, P.; et al. Large oncosomes mediate intercellular transfer of functional microRNA. *Cell Cycle* **2013**, *12*, 3526–3536. [CrossRef] [PubMed]
15. Johnson, S.M.; Dempsey, C.; Parker, C.; Mironov, A.; Bradley, H.; Saha, V. Acute lymphoblastic leukaemia cells produce large extracellular vesicles containing organelles and an active cytoskeleton. *J. Extracell. Vesicles* **2017**, *6*, 1294339. [CrossRef]
16. Vagner, T.; Spinelli, C.; Minciacchi, V.R.; Balaj, L.; Zandian, M.; Conley, A.; Zijlstra, A.; Freeman, M.R.; Demichelis, F.; De, S.; et al. Large extracellular vesicles carry most of the tumour DNA circulating in prostate cancer patient plasma. *J. Extracell. Vesicles* **2018**, *7*, 1505403. [CrossRef]
17. Pezzicoli, G.; Tucci, M.; Lovero, D.; Silvestris, F.; Porta, C.; Mannavola, F. Large Extracellular Vesicles—A New Frontier of Liquid Biopsy in Oncology. *Int. J. Mol. Sci.* **2020**, *21*, 6543. [CrossRef]
18. Lehrich, B.M.; Liang, Y.; Khosravi, P.; Federoff, H.J.; Fiandaca, M.S. Fetal Bovine Serum-Derived Extracellular Vesicles Persist within Vesicle-Depleted Culture Media. *Int. J. Mol. Sci.* **2018**, *19*, 3538. [CrossRef]

19. Lannigan, J.; Erdbruegger, U. Imaging flow cytometry for the characterization of extracellular vesicles. *Methods* **2017**, *112*, 55–67. [CrossRef]
20. Simonsen, J.B. Pitfalls associated with lipophilic fluorophore staining of extracellular vesicles for uptake studies. *J. Extracell. Vesicles* **2019**, *8*, 1582237. [CrossRef]
21. Osteikoetxea, X.; Benke, M.; Rodriguez, M.; Pálóczi, K.; Sódar, B.W.; Szvicsek, Z.; Szabó-Taylor, K.; Vukman, K.V.; Kittel, Á.; Wiener, Z.; et al. Detection and proteomic characterization of extracellular vesicles in human pancreatic juice. *Biochem. Biophys. Res. Commun.* **2018**, *499*, 37–43. [CrossRef] [PubMed]
22. Consortium, E.-T.; Van Deun, J.; Mestdagh, P.; Agostinis, P.; Akay, Ö.; Anand, S.; Anckaert, J.; Martinez, Z.A.; Baetens, T.; Beghein, E.; et al. EV-TRACK: Transparent reporting and centralizing knowledge in extracellular vesicle research. *Nat. Methods* **2017**, *14*, 228–232.
23. Vogel, R.; Coumans, F.A.W.; Maltesen, R.G.; Böing, A.N.; Bonnington, K.E.; Broekman, M.L.; Broom, M.F.; Buzás, E.I.; Christiansen, G.; Hajji, N.; et al. A standardized method to determine the concentration of extracellular vesicles using tunable resistive pulse sensing. *J. Extracell. Vesicles* **2016**, *5*, 31242. [CrossRef] [PubMed]
24. Gorgens, A.; Bremer, M.; Ferrer-Tur, R.; Murke, F.; Tertel, T.; Horn, P.A.; Thalmann, S.; Welsh, J.A.; Probst, C.; Guerin, C.; et al. Optimisation of imaging flow cytometry for the analysis of single extracellular vesicles by using fluorescence-tagged vesicles as biological reference material. *J. Extracell. Vesicles* **2019**, *8*, 1587567. [CrossRef] [PubMed]
25. Headland, S.E.; Jones, H.R.; D'Sa, A.S.V.; Perretti, M.; Norling, L.V. Cutting-edge analysis of extracellular microparticles using Image Stream (X) imaging flow cytometry. *Sci. Rep.* **2014**, *4*, 5237. [CrossRef] [PubMed]
26. Witwer, K.W.; Buzás, E.I.; Bemis, L.T.; Bora, A.; Lässer, C.; Lötvall, J.; Hoen, E.N.N.T.; Piper, M.G.; Sivaraman, S.; Skog, J.; et al. Standardization of sample collection, isolation and analysis methods in extracellular vesicle research. *J. Extracell. Vesicles* **2013**, *2*, 2. [CrossRef] [PubMed]
27. Erdbrugger, U.; Lannigan, J. Analytical challenges of extracellular vesicle detection: A comparison of different techniques. *Cytom. Part A* **2016**, *89*, 123–134. [CrossRef]
28. Menck, K.; Bleckmann, A.; Wachter, A.; Hennies, B.; Ries, L.; Schulz, M.; Balkenhol, M.; Pukrop, T.; Schatlo, B.; Rost, U.; et al. Characterisation of tumour-derived microvesicles in cancer patients' blood and correlation with clinical outcome. *J. Extracell. Vesicles* **2017**, *6*, 1340745. [CrossRef]
29. Taylor, D.D.; Gercel-Taylor, C. MicroRNA signatures of tumor-derived exosomes as diagnostic biomarkers of ovarian cancer. *Gynecol. Oncol.* **2008**, *110*, 13–21. [CrossRef]
30. Juraschka, K.; Taylor, M.D. Medulloblastoma in the age of molecular subgroups: A review. *J. Neurosurg. Pediatr.* **2019**, *24*, 353–363. [CrossRef]
31. Santiago-Dieppa, D.R.; Steinberg, J.; Gonda, D.; Cheung, V.J.; Carter, B.S.; Chen, C.C. Extracellular vesicles as a platform for 'liquid biopsy' in glioblastoma patients. *Expert Rev. Mol. Diagn.* **2014**, *14*, 819–825. [CrossRef] [PubMed]
32. Saenz-Antonanzas, A.; Auzmendi-Iriarte, J.; Carrasco-Garcia, E.; Moreno-Cugnon, L.; Ruiz, I.; Villanua, J.; Egaña, L.; Otaegui, D.; Samprón, N.; Matheu, A. Liquid Biopsy in Glioblastoma: Opportunities, Applications and Challenges. *Cancers (Basel)* **2019**, *11*, 950. [CrossRef] [PubMed]
33. Ihara, T.; Yamamoto, T.; Sugamata, M.; Okumura, H.; Ueno, Y. The process of ultrastructural changes from nuclei to apoptotic body. *Virchows Arch.* **1998**, *433*, 443–447. [CrossRef] [PubMed]
34. Crescitelli, R.; Lässer, C.; Szabó, T.G.; Kittel, A.; Eldh, M.; Dianzani, I.; Buzás, E.I.; Lötvall, J. Distinct RNA profiles in subpopulations of extracellular vesicles: Apoptotic bodies, microvesicles and exosomes. *J. Extracell. Vesicles* **2013**, *2*, 2. [CrossRef]
35. Verweij, F.J.; Bebelman, M.P.; Jimenez, C.R.; Garcia-Vallejo, J.J.; Janssen, H.; Neefjes, J.; Knol, J.C.; Haas, R.D.G.D.; Piersma, S.R.; Baglio, S.R.; et al. Correction: Quantifying exosome secretion from single cells reveals a modulatory role for GPCR signaling. *J. Cell. Biol.* **2018**, *217*, 1157. [CrossRef]
36. Cashikar, A.G.; Hanson, P.I. A cell-based assay for CD63-containing extracellular vesicles. *PLoS ONE* **2019**, *14*, e0220007. [CrossRef]

12

More than Nutrition: Therapeutic Potential of Breast Milk-Derived Exosomes in Cancer

Ki-Uk Kim [1], Wan-Hoon Kim [1], Chi Hwan Jeong [1], Dae Yong Yi [2,3,*] and Hyeyoung Min [1,*]

[1] College of Pharmacy, Chung-Ang University, Seoul 06974, Korea; hakiukah@cau.ac.kr (K.-U.K.); vessel37@cau.ac.kr (W.-H.K.); tolast12@cau.ac.kr (C.H.J.)
[2] Department of Pediatrics, Chung-Ang University College of Medicine, Seoul 06974, Korea
[3] Department of Pediatrics, Chung-Ang University Hospital, Seoul 06973, Korea
* Correspondence: meltemp2@cau.ac.kr (D.Y.Y.); hymin@cau.ac.kr (H.M.);

Abstract: Human breast milk (HBM) is an irreplaceable source of nutrition for early infant growth and development. Breast-fed children are known to have a low prevalence and reduced risk of various diseases, such as necrotizing enterocolitis, gastroenteritis, acute lymphocytic leukemia, and acute myeloid leukemia. In recent years, HBM has been found to contain a microbiome, extracellular vesicles or exosomes, and microRNAs, as well as nutritional components and non-nutritional proteins, including immunoregulatory proteins, hormones, and growth factors. Especially, the milk-derived exosomes exert various physiological and therapeutic function in cell proliferation, inflammation, immunomodulation, and cancer, which are mainly attributed to their cargo molecules such as proteins and microRNAs. The exosomal miRNAs are protected from enzymatic digestion and acidic conditions, and play a critical role in immune regulation and cancer. In addition, the milk-derived exosomes are developed as drug carriers for delivering small molecules and siRNA to tumor sites. In this review, we examined the various components of HBM and their therapeutic potential, in particular of exosomes and microRNAs, towards cancer.

Keywords: human milk; nutrient; microbiota; exosomes; microRNAs; cancer

1. Introduction

Human breast milk (HBM) is an irreplaceable source of nutrition for early infant growth and development. For this reason, the World Health Organization and the American Academy of Pediatrics recommend exclusive breastfeeding for at least 6 months and to continue breastfeeding until the age of 2 years [1–3]. It is widely known that HBM provides advantages for cognition and development in the short and long term [3,4]. Breast-fed children are known to have a low prevalence of necrotizing enterocolitis (NEC), gastroenteritis, otitis media, respiratory diseases, and acute diseases, as well as obesity, inflammatory bowel disease, and diabetes [3,5,6]. In addition, breastfeeding for more than 6 months reduces the risk of acute lymphocytic leukemia and acute myeloid leukemia, and there is evidence of reduced morbidity associated with lymphoma and other tumors [3,5–7]. Furthermore, breastfeeding mothers show various short-term benefits and long-term positive effects on cardiovascular disease, diabetes, and bone density [5]. In particular, breastfeeding is known to lower the risk of breast and ovarian cancer.

Although various multinational companies are conducting research to develop breast milk replacements, it has not yet been possible to completely replace breast milk by any method [8]. This is because HBM contains various ingredients that have not yet been identified, so it is difficult to artificially reproduce them. HBM contains not only nutritional components, such as macronutrients and micronutrients, but also various non-nutritional proteins or cellular components, including immune components, hormones, and growth factors [9–11]. In recent years, studies on HBM components, such as human milk oligosaccharides (HMOs) and fatty acids, have been actively conducted. The HBM components that have not been well appreciated in the past, such as the microbiome, extracellular vesicles (EVs) or exosomes, and microRNAs (miRNAs), are being examined, assisted by the development of various testing techniques [11–13]. Some of these components consistently show no considerable differences between regions and races, such as protein components and with regard to the energy content. However, there are differences according to the diet or weight of the breastfeeding mother in components such as vitamin A, vitamin D, water-soluble vitamins, and the composition of fatty acids, and also in other components according to the underlying condition of the breastfeeding mother [3,11,12]. In addition, the maturation, colonization, and immunity acquisition of immature intestinal mucosa are obtained through breastfeeding, and the incidence of various diseases is lower in breastfed infants than that in formula-fed infants. Considering these aspects, the HBM components, such as the immune components and exosomes, would be helpful in predicting and treating diseases in actual clinical practice.

In this article, we examined the various well-known components of HBM and those that are currently being studied, focusing on exosomes. In addition, we described how these ingredients affect the health of children and breastfeeding mothers and especially their potential use in diseases such as cancer.

2. Human Breast Milk Components

2.1. Nutritional Components (Macronutrients and Micronutrients)

As already known, HBM is mainly composed of macronutrients such as carbohydrates, proteins, and fats, along with 87–88% water [11]. Among them, carbohydrates are the most important component in HBM and play a critical role in infant nutrition. Although lactose is a major carbohydrate constituent of HBM, HMOs, which have recently been attracting attention, are unique components of HBM and are the third largest constituent of HBM. HMOs play a prebiotic role that affect the development of intestinal colonization and gut microbiota immediately after birth, directly affecting immunity, and also play a role in the production of short chain fatty acids [14,15]. The proteins in HBM are composed of various peptides along with a mixture of casein and whey. These proteins play an essential role in growth and development by being involved in the functionalization and organization of cells in the human body [11]. As HBM is known to play an essential role in the development of the early human immune system, whey proteins, such as alpha-lactalbumin, lactoferrin, and secretory IgA, play an important role in the immune system and have antibacterial properties [10,16,17]. Fat accounts for 50% of the total nutrition supplied by HBM, is the second largest macronutrient, and plays an important role in the development of the nervous system [18]. Fat in HBM varies according to the maternal diet and body weight during pregnancy, so there are large regional differences [18].

In addition, HBM contains various vitamins and micronutrients, such as iron, calcium, zinc, and copper. Although most of these micronutrients are sufficiently contained in breast milk, vitamin D or vitamin K may require supplementation if exclusively breastfeeding due to insufficiencies [19,20]. The nutrient content of HBM varies depending on whether it is colostrum or mature milk, and the ratio changes as lactation progresses. In addition, HBM fed to premature infants may have different components from that fed to mature infants [11].

2.2. Non-Nutritional Components and Clinical Applications

In addition to the nutrient components, HBM contains various non-nutritional components, bioactive proteins, and peptides. HBM is known to contain various hormones, such as parathyroid hormone, insulin, and leptin, but their functions are not fully understood [11]. However, the roles of many growth factors are well known. Epidermal growth factors play a crucial role in the maturation and healing of intestinal mucosa, and neuronal growth factors are known to be necessary for the development of the enteric nervous system and immature intestine in newborns [10,11]. Moreover, the insulin-like growth factor superfamily and vascular endothelial growth factor regulate erythropoiesis and angiogenesis, respectively [10,11].

Considering the development of the immune system through breastfeeding, the formation of a functional microbial community through colonization of microorganisms in the intestinal tract, and the low disease incidence in breast-fed infants, various components of HBM have been studied for their applicability in the clinic and infection [21]. Attempts have been made to use lactoferrin to treat colon cancer, advanced stage non-small-cell lung carcinoma, newly diagnosed lung cancer, and breast cancer in connection with regulating cellular growth and differentiation, playing an important role in immune response [22–26]. Another immune component, alpha-lactalbumin, is known as HAMLET (human alpha-lactalbumin made lethal to tumor cells) for its clinical applicability in oncologic diseases. Through the intratumoral administration of HAMLET, damage to the intact brain in human glioblastoma is minimized and tumor cell apoptosis is induced [27]. The effects of alpha-lactalbumin on apoptosis and resolution have also been confirmed in human skin papilloma and bladder cancer [28,29]. The transforming growth factor β (TGF-β) contained in milk acts on immunomodulation and cell proliferation and differentiation, particularly in pediatric Crohn's disease, and is used for enteral nutrition [21].

2.3. Microbiomes and Their Derived Extracellular Vesicles

Until the 20th century, HBM was considered sterile, and bacterial species identified in HBM were believed to be due to contamination or infection. However, it was known to some researchers that HBM also contains commensal bacteria, and rich and diverse communities have been identified by new testing techniques such as Next Generation Sequencing [30–32]. Through these advanced techniques, *Staphylococcus* and *Streptococcus* were found to be the predominant core genera in HBM, despite differences in regions and test techniques [11]. In addition, Togo et al. confirmed the diversity of human milk microbiota through a systematic review and identified commonly found species: *Staphylococcus aureus*, *Staphylococcus epidermidis*, *Streptococcus agalactiae*, *Cutibacterium acnes*, *Enterococcus faecalis*, *Bifidobacterium breve*, *Escherichia coli*, *Streptococcus sanguinis*, *Lactobacillus gasseri*, and *Salmonella enterica* [33].

The origin of the HBM microbiota remains unknown. The possibility of contamination of the skin or oral cavity cannot yet be ruled out. In fact, Ramsay et al. confirmed milk-ejection reflux from the mouth of the infant to mammary ducts during breastfeeding through ultrasound [34]. However, an introduction of microbiota from an endogenous origin, such as glands, rather than contamination is possible, since anaerobic gut-associated microbiota has been identified in breast milk. The maternal gut microbiota may be absorbed through the maternal intestinal epithelium and present in the mammary glands [35,36]. It is thought that these HBM microbes act as prebiotics with HMOs and indirectly affect the infant gut and various extra-intestinal environments [37–40]. Through many studies conducted so far, vertical transmission of HBM microbiota has been shown to play an important role in the initial formation of infant gut microbiota. It is already known that there is a difference between the stool microbiota of the infant and adulthood periods in breast-fed and non-breast-fed individuals, and a role for HBM microbiota in human health is possible, as it is associated with a low incidence of various diseases such as infectious diseases and allergies in breast-fed infants [41,42].

In addition to these microbiomes, HBM was also found to contain EVs of bacterial origin [12]. EVs are nanometer-sized membrane vesicles that have various bioactive functions in intercellular communication. EVs were generally classified into exosomes (endocytic pathway), microvesicles (plasma membrane), and apoptotic bodies (plasma membrane) according to their cargos, biogenesis, and size [43]. Bacterial origin EVs are present in a variety of body fluids such as blood, urine, and stool, but their role in breast milk remains unknown. At the genus level, HBM bacteria and bacterial EVs show a significant difference. It is believed that bacterial origin EVs act on the host receptor to move bioactive molecules to host cells or to move the EVs themselves to host cells to play a role in infant gut colonization and immunity [12].

3. Breast Milk-Derived Exosomes

Exosomes are small membranous extracellular vesicles secreted by most eukaryotic cells into surrounding body fluids, such as blood, saliva, urine, cerebrospinal fluid, lymphatic fluid, and amniotic fluid [44–46]. These vesicles are 40~100 nm in diameter and are involved in cell–cell communication by transporting the bioactive cargo molecules, including DNAs, mRNAs, microRNAs, lipids, and proteins, derived from the cells of origin to the target cells. Exosomes are also present in breast milk, and the diverse health benefits of breastfeeding are attributed to exosomes, as well as to well-known immunoregulatory components such as lactoferrin, lactalbumin, lysozyme, and sIgA [47,48].

Exosomes were first extracted from human colostrum and breast milk and characterized in 2007 [47], and subsequent studies have reported the isolation and characterization of milk exosomes from cows, camels, buffalos, pigs, sheep, and pigeons [49–55]. Milk exosomes are secreted by mammary gland epithelial cells and also released from milk fat globules during lactation [56]. Higher concentrations of exosomes were detected in early milk collected at day 3–8 postpartum than in mature milk collected at 2 months, whereas no significant differences were found in the expression of marker proteins, such as the tetraspanins CD63, CD81, and the scavenger receptor CD36, between early and mature milk [57].

3.1. The Role of Milk Exosomes in Cell Proliferation and Inflammation

Studies have reported that milk exosomes function in regulating cell proliferation and inflammation. Hock et al. showed that rat milk-derived exosomes promote the viability and proliferation of intestinal epithelial cells and intestinal stem cell activity [58], supporting the previous observations that HBM reduces the incidence of NEC in infants [59–62]. Consistently, Martin et al. have also reported that human milk-derived exosomes protect intestinal epithelial cells from cell death upon exposure to oxidative stress [63]. Furthermore, a neonatal mouse intestinal organoid model and experimental NEC in C57BL/6 mouse pups supported the protective activities of HBM exosomes by decreasing inflammation and intestinal damage [64]. A reduction in intestinal damage has also been observed by exosomes derived from pasteurized breast milk and raw breast milk, indicating that milk exosomes are resistant to the pasteurization process. In addition, an increasing number of studies have reported the protective effects of milk exosomes in various in vitro and in vivo NEC models, confirming the critical role of exosomes in the prevention of NEC development in premature infants [65,66].

The anti-inflammatory effects of bovine milk-derived exosomes have also been shown in IL-1Ra(-/-) and collagen-induced arthritis mouse models [67]. When administered orally by oral gavage or drinking water, milk exosomes delay the onset of arthritis and reduce the swelling of ankle joints, cartilage depletion, and bone marrow inflammation. In addition, the circulating levels of MCP-1, IL-6, and anti-collagen IgG_{2a} declined, accompanied by a decrease in mRNA expression of T-bet (Th1) and ROR-γT (Th17) in the primary splenocytes. These results suggest the therapeutic potential of milk exosomes in the treatment of autoimmune and inflammatory diseases.

3.2. Immunomodulatory Function of Milk Exosomes

Breast milk exosomes also show immunomodulatory effects. The Gabrielsson group demonstrated that HBM-derived exosomes inhibit CD3-induced production of IL-2, IFN-γ, and TNF-α in peripheral blood mononuclear cells (PBMCs), while increasing the number of $FoxP3^+$ $CD4^+$ $CD25^+$ regulatory T cells in PBMCs [47]. They also reported that MUC-1 expressed on HBM-derived exosomes binds to DC-SIGN on monocyte-derived dendritic cells (MDDCs) and blocks HIV-1 infection and viral transfer from MDDCs to $CD4^+$ T cells [68]. In addition, human milk exosomes possess different phenotypes, depending upon maternal sensitization and lifestyle, and differentially influence allergy development in children [57]. For example, exosomes derived from mothers with an anthroposophic lifestyle are associated with a lower prevalence of allergic sensitization than those from non-anthroposophic mothers. Cow milk-derived exosomes upregulate CD69 expression in NK cells and IFN-γ production in NK cells and $\gamma\delta$ T cells in the presence of IL-2 and IL-12 [69]. Furthermore, bovine milk exosomes induce proliferation of macrophages and RAW264.7 cells and protect macrophages against cisplatin-induced cell death [70]. These data suggest that milk exosomes potentially influence the immune system.

A proteomic analysis has been performed to compare the exosomal protein contents between bovine colostrum and mature milk (mid-lactation period) [71]. The results revealed that exosomes from colostrum are highly enriched in proteins implicated in the innate immune response, inflammatory response, acute-phase response, platelet activation, cell growth, and complement activation, whereas proteins implicated in transport and apoptosis are enriched in exosomes from mature milk [71]. A further proteomic analysis of bovine milk exosomes also reported that exosomal proteins map to the Kyoto Encyclopedia of Genes and Genomes (KEGG) pathways with immunological functions such as Fc gamma receptor-mediated phagocytosis, antigen processing and presentation, T cell receptor signaling, B cell receptor signaling, and NK cell-mediated cytotoxicity [72]. Proteome studies supported the immunoregulatory function of milk exosomes and suggested the importance of colostrum in immune defense and immune system development during the early period after birth.

3.3. Milk Exosomes and Cancer

The chemopreventive effects of breast milk have been well studied in childhood leukemia and lymphoma [73–77]. Martin et al. performed a systemic review and meta-analysis and showed that ever having been breast-fed is inversely associated with the incidence of acute lymphoblastic leukemia, Hodgkin's disease, and neuroblastoma in childhood [78]. Similarly, a recent meta-analysis and systemic review reported that 14–19% of all childhood leukemia cases might be prevented by breastfeeding for 6 months or more, whereas breastfeeding for a short duration or non-breastfeeding may be associated with a slightly increased risk of acute childhood leukemia [7,79]. Given the severity and detrimental effects of these illnesses, any factor lowering the risk of childhood cancer would be useful.

Reif et al. reported that HBM-derived exosomes promote the proliferation of normal colon epithelial cells without affecting the growth of colonic cancer cells [80]. Although this study does not show the direct antitumor effects of human milk exosomes, it suggests the beneficial function of human milk exosomes through differential effects on normal cells compared to cancer cells. Bovine milk-derived exosomes have revealed their intrinsic antitumor activity by inhibiting the proliferation of various types of cancers, such as lung, prostate, colon, pancreatic, breast, and ovarian cancers [81]. Treatment of cancer cells with 50 μg/mL exosomal proteins for 72 h reduces cell growth by 8–47% as assessed by MTT assay, suggesting the benefit of exosomes as delivery vehicles for anticancer drugs. In addition, camel milk and its components have also been shown to exert antitumor effects in human hepatoma (HepG2), human breast cancer cells (MCF7), and murine hepatoma (Hepa 1c1c7) [82,83]. Camel milk inhibits cell growth and induces apoptosis by activating caspase-3 and death receptor DR4 and accumulates intracellular reactive oxygen species in HepG2 and MCF7 cells [83]. Badawy et al. demonstrated that camel milk and its exosomes inhibit the proliferation of MCF7 cells, accompanied by a decrease in MCF7 migration, as measured by a wound-healing assay [53]. In tumor-bearing rats, oral or local (in the tumor tissue) administration of camel milk and its exosomes significantly reduces tumor weight and progression by inducing apoptosis and inhibiting oxidative stress, inflammation, angiogenesis, and metastasis. Moreover, the numbers of $CD4^{+}$, $CD8^{+}$, and $NKT1.1^{+}$ T cells increase in the spleen following treatment with camel milk and its exosomes. Although exosomes show better overall antitumor effects, the increase in the numbers of splenic T cells is more potent in milk-treated rats, suggesting that camel milk possesses more immune-stimulating constituents than exosomes.

In contrast, breast milk exosomes may negatively influence some type of cancers. TGF-β isolated from breast milk exosomes promotes the proliferation of breast cancer cells and epithelial–mesenchymal transition (EMT), as demonstrated by changes in the actin cytoskeletal structure and loss of E-cadherin expression [84]. Since epithelial cells acquire migratory and invasive properties during EMT, TGF-β in exosomes may lead to the transformation of normal cells and a change in breast cancer cells into a more aggressive and invasive tumor.

4. Breast Milk-Derived Exosomal MicroRNAs

MicroRNAs are endogenous small non-coding RNA molecules that are 19~24 bp in length [85–87]. miRNAs interact with the 3′ untranslated region (UTR) of target mRNAs and control gene expression post-transcriptionally. Ample studies have reported the function of miRNAs in diverse biological processes, such as proliferation, differentiation, cell cycle, and cell death [88,89], and dysregulated miRNAs are implicated in the development and progression of many human diseases [90,91]. In relation to cancer, miRNAs may function as tumor suppressor genes or oncogenes, and accordingly, aberrantly expressed miRNAs in particular types of cancer have been suggested as useful biomarkers for cancer diagnostics and therapeutic targets [92–95]

HBM is highly enriched in miRNAs, with more than 1400 identified miRNAs [96–100]. Breast milk miRNAs are thought to originate from the mammary epithelium and are present in the cells, skimmed fractions, and lipid fractions of breast milk [98]. More miRNAs are detected in the cell and lipid fractions of breast milk than the skimmed milk fraction, suggesting the importance of cell and lipid fractions in the analysis of breast milk miRNAs [96,100]. Exosomes are mostly present in the skimmed milk fraction, and exosomal miRNAs are often reported within breast milk miRNA if it has not been specifically mentioned that exosomes were isolated from milk before miRNA extraction.

Studies have reported that milk exosomes protect exosomal miRNAs from enzymatic, chemical, or mechanical degradation. When exposed to acidic conditions that mimic gastric and pancreatic digestion, milk exosomes prevent the degradation of vulnerable miRNAs [101]. Commercial bovine milk also protects miRNAs from acidic environments and RNase treatment, safely delivering miRNAs into the digestive tract [102]. Subsequently, milk exosomes are taken up by intestinal epithelial cells by endocytosis and moved into systemic circulation [103,104]. In addition, studies have shown that milk exosomes containing miRNAs are absorbed into macrophages, PBMCs, and kidney cells, and regulate gene expression in the target cells [67,105]. For example, upon incubation of normal intestinal CRL 1831 cells, K562 leukemia cells, and Lim 1215 colon cancer cells with milk exosomes, the expression of miR-148a-3p, the most abundant miRNA in breast milk, is increased accompanied by a decrease in the expression of DNA methyltransferase1 (DNMT1), a target of miR-148a-3p [106]. Although the composition of breast milk and miRNAs vary depending on the maternal health and nutrition status and lactation stage [107,108], miR-148a-3p is highly expressed in breast milk and conserved across mammalian species, and other highly expressed miRNAs, such as miR-320, miR-375, and miR-99, are also frequently found in different species [106].

4.1. *Immune-Regulating miRNAs*

Exosomal miRNAs contribute to the physiological and therapeutic functions of milk exosomes. Several studies have reported that breast milk is rich in immune-regulating miRNAs, such as miR-125b, miR-146b, miR-155, miR-181a, and miR-181b [108–110]. These miRNAs are known to regulate B cell, T cell, and monocyte development, and control the innate immune response and cytokine production (Table 1). Zhou et al. demonstrated that immune-regulating miRNAs are abundant in milk exosomes [99]. Among the 453 pre-miRNAs detected in milk exosomes, 59 pre-miRNAs were immune related, based on annotation in the Pathway Central database. For example, miR-30b-5p promotes cellular invasion by directly targeting GalNAc transferase and immunosuppression by increasing IL-10 [111], whereas miR-182-5p promotes helper T cell-mediated immune response upon induction by IL-2 [112]. Other frequently found immune-related exosomal miRNAs are miR-148a-3p, miR-146a, miR-146b-5p, miR-200a-3p, and miR-29a-3p [99,106,113]. These immune-regulating miRNAs may be transferred into the infant body during breastfeeding and function in the development of the immune system [99]. Future studies extending beyond in silico analysis are needed to elucidate the mechanisms by which individual miRNAs control immune responses in infants.

Table 1. Immune-regulating miRNAs in milk.

miRNA	Immune Function	Reference
miR-17 and miR-92 cluster	B-cell, T-cell, and monocyte development	[108]
miR-29a-3p	Suppression of immune responses to intracellular pathogens by targeting IFN-γ	[99]
miR-30b-5p	Promotion of cellular invasion by directly targeting GalNAc transferase, immunosuppression by increasing IL-10	[99,111]
miR-106	Regulation of IL-10 production	[113]
miR-125b	Negative regulation of TNF-α production, activation, and sensitivity	[108]
miR-146b-5p	Negative regulation of the innate immune response by targeting NF-κB signaling, control of TLR and cytokine signaling	[108,109]
miR-150	Control of B cell differentiation, pre- and pro-B cell formation or function	[109]
miR-155	T- and B-cell maturation, the innate immune response	[108,109]
miR-181a, miR-181b	B-cell differentiation, $CD4^+$ T-cell sensitivity and selection	[108,109,113]
miR-182-5p	Promotion of helper T cell-mediated immune responses upon induction by IL-2	[99,112]
miR-223	Neutrophil proliferation and activation	[108,109]
miR-451	Regulation of Macrophage migration inhibitory factor (MIF)	[113]
let-7i	Toll-like receptor 4 expression in human cholangiocytes	[108]

IFN-γ, interferon- γ; GalNAc, N-acetylgalactosamine; IL, interleukin; NF-κB, nuclear factor kappa-light-chain-enhancer of activated B cells.

4.2. Tumor-Associated miRNAs

In addition to immune-regulating miRNAs, milk exosomes also contain tumor-suppressive or oncogenic miRNAs (oncomiRs) [114]. miR-148a-3p has been shown to function as a tumor suppressor by targeting DNMT1, ERBB3, and ROCK1, which are all involved in the development, proliferation, and metastasis of tumors [106,115–117]. Given that miR-148a-3p is highly expressed in breast milk but less expressed in leukemia, miR-148a-3p in breast milk may be able to protect infants against childhood leukemia [79,118,119].

In contrast, some well-known oncomiRs, such as the miR-21, miR-155, miR-223, and miR-17-92 clusters, have been found in breast milk [120]. Of note, Melnik suggested that miRNA-21-5p, a highly expressed miRNA in human and cow milk, is one of the major environmental factors eliciting melanomagenesis through exosomal transfer of miRNAs [99,121]. miR-21-5p downregulates the expression of tumor suppressor genes, such as PTEN (phosphatase and tensin homolog), Sprouty1 and Sprouty2, and PDCD4 (programmed cell death protein 4), thereby promoting the initiation and progression of malignant melanoma [122–124]. It has been proposed that an exogenous supply of exosomal miR-21 through breastfeeding or milk consumption may enhance oncogenic miR-21 signaling and lead to the transition of benign melanocytes to malignant melanoma [121]. Additionally, miR-155, an oncomiR that enhances STAT3 expression by targeting SOCS1 (suppressor of cytokine signaling 1) and facilitates STAT3-mediated tumorigenesis, is also found in exosomes derived from bovine colostrum, implying the association of milk exosomal miR-155 with tumorigenesis [113,125,126].

Nonetheless, studies that provide a direct evidence of the role of breast milk miRNAs in the prevention or development of cancer are limited. Most studies reporting the function of milk exosomal miRNAs are in silico analyses accompanied by Gene Ontology and KEGG pathway analyses. Further investigations merging in silico analysis and in vitro and/or in vivo experimental validation are required to determine any significant association between breast milk miRNAs and cancer.

5. Breast Milk-Derived Exosomes as Natural Carriers for Drug Delivery

As natural carriers of endogenous biomolecules, exosomes have remarkable advantages over other drug delivery vehicles. Exosomes are biocompatible and low in toxicity and immunogenicity and have long circulating half-lives [44,45]. Furthermore, they can cross many biological barriers such as the blood–brain barrier and cytoplasmic membrane [127]. Accordingly, numerous exosome-based delivery systems have been developed for targeted drug delivery by transporting small molecules, proteins, and small interfering RNA (siRNA) to target tissues. In particular, milk exosomes possess unique benefits compared with those of other origins [128,129]. Milk exosomes can be obtained from bovine milk enabling low-cost and mass production and show cross-species tolerance [128]. Moreover, milk exosomes are stable in acidic environments and can be absorbed from the digestive tract in humans, suggesting the potential use of milk exosomes as oral delivery vehicles of drugs that are conventionally administered intravenously [130]. With additional modification for target-specific drug delivery, milk exosomes have been widely investigated as promising carriers to transport diverse biomolecules and chemotherapeutic agents.

5.1. Delivery of Small Molecules

Milk exosomes have been shown to deliver natural compounds isolated from plant resources, such as celastrol, anthocyanidin, and curcumin [81,131,132]. When loaded onto cow milk-derived exosomes, curcumin with poor oral bioavailability is taken up and shows enhanced antiproliferative effects compared with free curcumin in Caco-2 cells [131]. Exosomal formulation of anthocyanidin (ExoAnthos) also exerts greater antiproliferative effects than that of free Anthos in lung, breast, ovarian, colon, pancreatic, and prostate cancers in vitro, and has higher antitumor efficacy in mice bearing A549 xenografts upon oral administration [81]. In addition to the improvement in oral bioavailability, ExoAnthos is well tolerated with no systemic toxicity, as determined by biochemical and hematological parameters. Similarly, celastrol encapsulated in milk exosomes inhibits the proliferation of the non-small-cell lung carcinoma cell lines A549 and H1299 in vitro, and shows greater antitumor effects in vivo without toxicity than that of free celastrol [132].

Agrawal et al. demonstrated that paclitaxel-loaded milk exosomes (ExoPACs) inhibit the growth of human lung tumor in athymic nude mice bearing A549 xenografts [133]. The ExoPACs are highly stable in simulated gastrointestinal fluids and under cold storage conditions at −80 °C for 1 month, allowing oral delivery of paclitaxel. Furthermore, oral administration of ExoPACs maintains the number and function of immune cells with reduced systemic and tissue toxicity in mice compared with intravenous administration of the drug, indicating the safety and efficacy of drug-loaded milk exosomes. Considering the discomfort and inconvenience of the intravenous route, the use of milk exosomes would be beneficial for the oral delivery of chemotherapeutic agents with low oral bioavailability.

5.2. Delivery of Nucleic Acid

Since the discovery of RNA interference mediated by siRNA, siRNAs have attracted much attention as a new generation of therapeutics based on nucleic acid. Despite the promising therapeutic potential of silencing gene expression in a sequence-specific manner, siRNAs have some intrinsic undruggable properties [134,135]. They are readily degraded by nuclease in serum and tissue and cleared by renal excretion or phagocytes in the reticuloendothelial system (RES). Negatively charged siRNAs are not easily taken up by cells due to electrostatic repulsion, and endosomal escape is required for siRNAs to reach mRNA targets, once internalized. To overcome the obstacles that limit the safe and effective application of siRNAs, the chemical modification of their nucleotide components, including ribose, phosphate, and base, as well as the special siRNA delivery systems, such as polymeric nanoparticles and liposomes, have been extensively investigated [134,135].

In addition to the oral delivery of small molecule drugs, milk exosomes have also been used for the delivery of siRNAs. siRNAs were loaded onto milk exosomes by electroporation and chemical transfection, and their gene silencing activities were tested in multiple cancers in vitro [129]. The siRNA-loaded exosomes are resistant to RNase and taken up by cancer cells accompanied by silencing of the target genes. Exo-siKRAS (siRNA against KRAS-loaded exosomes) suppresses the proliferation of A549 in vitro and exhibits antitumor effects in A549 tumor-bearing mice upon intravenous administration.

Similarly, siRNAs against *bcl-2*, an oncogene that inhibits apoptosis and promotes proliferation, were loaded onto milk exosomes using an ultrasonic approach, and their antitumor activity was tested in vitro and in vivo [136]. The results demonstrated that exosiBcl-2 (siRNA against *bcl-2*-loaded exosomes) inhibits *bcl-2* expression followed by an increase in apoptosis and a decrease in cell growth, migration, and invasion of cancer cells in vitro. An in vivo study also confirmed that intravenous administration of exosiBcl-2 suppresses tumor growth in xenograft nude mice. In view of the advantages of milk exosomes over other siRNA delivery vehicles, future studies are expected to establish optimal methods for siRNA loading, surface functionalization, and administration routes to attain effective antitumor therapies.

5.3. Targeted Dug Delivery

Despite excellent biocompatibility and low toxicity, exosomes from bovine milk may require further modification for targeted drug delivery. Folate receptors are highly expressed on many types of tumor cells, and accordingly, folic acid (FA)-conjugated milk exosomes have been generated for enhanced delivery of siRNAs to tumor sites [129]. FA-functionalized milk exosomes show higher accumulation in tumor tissue than non-FA milk exosomes and significantly inhibit tumor growth in vivo.

CD44, a receptor for hyaluronic acid, is overexpressed in some types of cancer such as pancreatic, lung, ovarian, and breast cancer [137]. For targeted cancer therapy, attempts have been made to generate conjugates of membrane incorporating molecule DSPE-PEG$_{200}$ and Hyaluronan (HA), a CD44-specific ligand, which are then self-assembled into the phospholipid bilayer of milk exosomes [138]. The resulting HA-incorporated milk exosomes (HA-mExo) expose the HA ligand on the surface of the exosomes and are loaded with doxorubicin to obtain HA-mExo-Dox, a doxorubicin-loaded milk exosome with an HA ligand. The results showed that HA-mExo-Dox selectively delivers doxorubicin to CD44-overexpressing cancer cells and exerts enhanced antitumor activity.

Another example of a tumor-targeting ligand is iRGD (CRGDK/RGPD/EC), a 9-amino acid cyclic peptide that binds to $\alpha v\beta 3$ and $\alpha v\beta 5$ integrin on the endothelial cells of tumor vessels to promote transcytosis across tumor vasculature [139]. After translocation, the iRGD ligand is cleaved by an endogenous protease to yield a CRGDK/R peptide that serves as a ligand for the neuropilin-1 receptor to activate the transport of co-administered anticancer drugs into tumors [140]. When iRGD peptides are incorporated into milk exosomes, the oral administration of iRGD exosomes exhibits an increase in penetration and accumulation into tumors compared with intravenously administered control exosomes [130]. In contrast, the accumulation of iRGD exosomes in other organs, such as liver, heart, lung, and spleen, is decreased. Considering the increased tumor accumulation and decreased systemic distribution, iRGD exosomes might be used as promising delivery vehicles for chemotherapeutic agents with a high antitumor efficacy and decreased off-target effects.

In addition, a milk exosome-based pH and light-sensitive drug carrier has been developed for targeting an acidic and hypoxic tumor microenvironment (TME) [141]. Given that most solid tumors exhibit a pH from 6.5 to 7.4, milk exosomes are conjugated to doxorubicin by a pH-sensitive imine bond, which gets cleaved and releases doxorubicin in an acidic TME. Furthermore, photodynamic therapy is introduced by encapsulating photosensitizer-chlorin e6 (Ce6) and the anthracene endoperoxide derivative (EPT1) within milk exosomes. Upon exposure to 808 nm near-infrared light, Ce6 releases plasmonic heat that generates singlet oxygen from EPT1 and corrects hypoxia in TME. This Exo@Dox-EPT1 nanocarrier shows enhanced antitumor effects and biocompatibility with less cardiotoxicity caused by doxorubicin in an in vitro and in vivo mouse model of oral squamous cell carcinoma.

6. Conclusions

With the significant advancements in technology analyzing nucleic acids, proteins, and microbiota, HBM and animal-derived milk are now known to contain distinct bioactive molecules along with renowned nutritional components, and their therapeutic roles are becoming appreciated. In particular, milk-derived exosomes have gathered much attention owing to their intrinsic antitumor activities.

The milk exosomes modulate immune function and suppress the proliferation of various cancer cells in vitro and in vivo, and their chemotherapeutic function is mainly due to the exosomal miRNAs with immunoregulatory and tumor-suppressive activities. The milk exosomes can also serve as oral delivery vehicles for chemotherapeutic agents and siRNAs, and further modification of the exosomes with a tumor-targeting ligand enables the targeted delivery of drugs to the tumor sites.

Given the physiological activity, safety, biocompatibility, and drug delivery potential, the application of milk exosomes in cancer therapeutics are innumerable. Future studies are required to demonstrate how individual bioactive components within exosomes exert biological function, including antitumor activity, and to clarify any possible unwanted effects upon exosome treatment. In addition, it would be important to establish a cost effective and standardized method to isolate, purify, and manipulate exosomes from milk to ensure the quality of the exosomes for clinical and industrial implementation.

Author Contributions: Conceptualization, D.Y.Y. and H.M.; original draft preparation, K.-U.K., W.-H.K., C.H.J., D.Y.Y. and H.M.; editing, K.-U.K., D.Y.Y. and H.M.; funding acquisition, D.Y.Y. and H.M.; supervision, D.Y.Y. and H.M. All authors have read and agreed to the published version of the manuscript.

Abbreviations

HBM	human breast milk
NEC	necrotizing enterocolitis
HMOs	human milk oligosaccharides
EVs	extracellular vesicles
TGF-β	transforming growth factor β
TME	tumor microenvironment

References

1. Section on Breastfeeding. Breastfeeding and the Use of Human Milk. *Pediatrics* **2012**, *129*, e827–e841. [CrossRef]
2. Gartner, L.M.; Morton, J.; Lawrence, R.A.; Naylor, A.J.; O'Hare, D.; Schanler, R.J.; Eidelman, A.I. Breastfeeding and the use of human milk. *Pediatrics* **2005**, *115*, 496–506. [CrossRef]
3. Agostoni, C.; Braegger, C.; Decsi, T.; Kolacek, S.; Koletzko, B.; Michaelsen, K.F.; Mihatsch, W.; Moreno, L.A.; Puntis, J.; Shamir, R.; et al. Breast-feeding: A commentary by the ESPGHAN Committee on Nutrition. *J. Pediatr. Gastroenterol. Nutr.* **2009**, *49*, 112–125. [CrossRef]
4. Bernard, J.Y.; De Agostini, M.; Forhan, A.; Alfaiate, T.; Bonet, M.; Champion, V.; Kaminski, M.; de Lauzon-Guillain, B.; Charles, M.A.; Heude, B.; et al. Breastfeeding duration and cognitive development at 2 and 3 years of age in the EDEN mother-child cohort. *J. Pediatr.* **2013**, *163*, 36–42. [CrossRef] [PubMed]
5. van Rossum, C.T.M.; Buchner, F.L.; Hoekstra, J. *Quantification of Health Effects of Breastfeeding—Review of the Literature and Model Simulation*; RIVM: Bilthoven, The Netherlands, 2006; pp. 19–21.
6. Ip, S.; Chung, M.; Raman, G.; Chew, P.; Magula, N.; DeVine, D.; Trikalinos, T.; Lau, J. Breastfeeding and maternal and infant health outcomes in developed countries. *Evid Rep. Technol. Assess. (Full Rep.)* **2007**, *153*, 1–186.
7. Güngör, D.; Nadaud, P.; Dreibelbis, C.; LaPergola, C.C.; Wong, Y.P.; Terry, N.; Abrams, S.A.; Beker, L.; Jacobovits, T.; Järvinen, K.M.; et al. Infant milk-feeding practices and childhood leukemia: A systematic review. *Am. J. Clin. Nutr.* **2019**, *109*, 757S–771S. [CrossRef] [PubMed]
8. Koletzko, B.; Baker, S.; Cleghorn, G.; Neto, U.F.; Gopalan, S.; Hernell, O.; Hock, Q.S.; Jirapinyo, P.; Lonnerdal, B.; Pencharz, P.; et al. Global standard for the composition of infant formula: Recommendations of an ESPGHAN coordinated international expert group. *J. Pediatr. Gastroenterol. Nutr.* **2005**, *41*, 584–599. [CrossRef] [PubMed]

9. Kim, M.H.; Shim, K.S.; Yi, D.Y.; Lim, I.S.; Chae, S.A.; Yun, S.W.; Lee, N.M.; Kim, S.Y.; Kim, S. Macronutrient Analysis of Human Milk according to Storage and Processing in Korean Mother. *Pediatr. Gastroenterol. Hepatol. Nutr.* **2019**, *22*, 262–269. [CrossRef]
10. Ballard, O.; Morrow, A.L. Human milk composition: Nutrients and bioactive factors. *Pediatr. Clin. N. Am.* **2013**, *60*, 49–74. [CrossRef]
11. Kim, S.Y.; Yi, D.Y. Components of human breast milk: From macronutrient to microbiome and microRNA. *Clin. Exp. Pediatr.* **2020**, *63*, 301–309. [CrossRef]
12. Kim, S.Y.; Yi, D.Y. Analysis of the human breast milk microbiome and bacterial extracellular vesicles in healthy mothers. *Exp. Mol. Med.* **2020**. [CrossRef] [PubMed]
13. Hegar, B.; Wibowo, Y.; Basrowi, R.W.; Ranuh, R.G.; Sudarmo, S.M.; Munasir, Z.; Widodo, A.D.; Kadim, M.; Suryawan, A.; Diana, N.R.; et al. The Role of Two Human Milk Oligosaccharides, 2′-Fucosyllactose and Lacto-N-Neotetraose, in Infant Nutrition. *Pediatr. Gastroenterol. Hepatol. Nutr.* **2019**, *22*, 330–340. [CrossRef] [PubMed]
14. Walker, A.W.; Ince, J.; Duncan, S.H.; Webster, L.M.; Holtrop, G.; Ze, X.; Brown, D.; Stares, M.D.; Scott, P.; Bergerat, A.; et al. Dominant and diet-responsive groups of bacteria within the human colonic microbiota. *ISME J.* **2011**, *5*, 220–230. [CrossRef] [PubMed]
15. Plaza-Díaz, J.; Fontana, L.; Gil, A. Human Milk Oligosaccharides and Immune System Development. *Nutrients* **2018**, *10*, 1038. [CrossRef] [PubMed]
16. Lönnerdal, B.; Woodhouse, L.R.; Glazier, C. Compartmentalization and quantitation of protein in human milk. *J. Nutr.* **1987**, *117*, 1385–1395. [CrossRef]
17. Lönnerdal, B.; Lien, E.L. Nutritional and physiologic significance of alpha-lactalbumin in infants. *Nutr. Rev.* **2003**, *61*, 295–305. [CrossRef]
18. Martin, C.R.; Ling, P.R.; Blackburn, G.L. Review of Infant Feeding: Key Features of Breast Milk and Infant Formula. *Nutrients* **2016**, *8*, 279. [CrossRef]
19. Perrine, C.G.; Sharma, A.J.; Jefferds, M.E.; Serdula, M.K.; Scanlon, K.S. Adherence to vitamin D recommendations among US infants. *Pediatrics* **2010**, *125*, 627–632. [CrossRef]
20. Committee on Fetus and Newborn. Controversies concerning vitamin K and the newborn. American Academy of Pediatrics Committee on Fetus and Newborn. *Pediatrics* **2003**, *112*, 191–192. [CrossRef]
21. Hill, D.R.; Newburg, D.S. Clinical applications of bioactive milk components. *Nutr. Rev.* **2015**, *73*, 463–476. [CrossRef]
22. Parodi, P.W. A role for milk proteins and their peptides in cancer prevention. *Curr. Pharm. Des.* **2007**, *13*, 813–828. [CrossRef] [PubMed]
23. Kozu, T.; Iinuma, G.; Ohashi, Y.; Saito, Y.; Akasu, T.; Saito, D.; Alexander, D.B.; Iigo, M.; Kakizoe, T.; Tsuda, H. Effect of orally administered bovine lactoferrin on the growth of adenomatous colorectal polyps in a randomized, placebo-controlled clinical trial. *Cancer Prev. Res.* **2009**, *2*, 975–983. [CrossRef] [PubMed]
24. Parikh, P.M.; Vaid, A.; Advani, S.H.; Digumarti, R.; Madhavan, J.; Nag, S.; Bapna, A.; Sekhon, J.S.; Patil, S.; Ismail, P.M.; et al. Randomized, double-blind, placebo-controlled phase II study of single-agent oral talactoferrin in patients with locally advanced or metastatic non-small-cell lung cancer that progressed after chemotherapy. *J. Clin. Oncol.* **2011**, *29*, 4129–4136. [CrossRef] [PubMed]
25. Digumarti, R.; Wang, Y.; Raman, G.; Doval, D.C.; Advani, S.H.; Julka, P.K.; Parikh, P.M.; Patil, S.; Nag, S.; Madhavan, J.; et al. A randomized, double-blind, placebo-controlled, phase II study of oral talactoferrin in combination with carboplatin and paclitaxel in previously untreated locally advanced or metastatic non-small cell lung cancer. *J. Thorac. Oncol.* **2011**, *6*, 1098–1103. [CrossRef]
26. Sun, X.; Jiang, R.; Przepiorski, A.; Reddy, S.; Palmano, K.P.; Krissansen, G.W. "Iron-saturated" bovine lactoferrin improves the chemotherapeutic effects of tamoxifen in the treatment of basal-like breast cancer in mice. *BMC Cancer* **2012**, *12*, 591. [CrossRef]
27. Fischer, W.; Gustafsson, L.; Mossberg, A.K.; Gronli, J.; Mork, S.; Bjerkvig, R.; Svanborg, C. Human alpha-lactalbumin made lethal to tumor cells (HAMLET) kills human glioblastoma cells in brain xenografts by an apoptosis-like mechanism and prolongs survival. *Cancer Res.* **2004**, *64*, 2105–2112. [CrossRef] [PubMed]
28. Mossberg, A.K.; Hun Mok, K.; Morozova-Roche, L.A.; Svanborg, C. Structure and function of human α-lactalbumin made lethal to tumor cells (HAMLET)-type complexes. *FEBS J.* **2010**, *277*, 4614–4625. [CrossRef]

29. Gustafsson, L.; Aits, S.; Onnerfjord, P.; Trulsson, M.; Storm, P.; Svanborg, C. Changes in proteasome structure and function caused by HAMLET in tumor cells. *PLoS ONE* **2009**, *4*, e5229. [CrossRef]
30. Heikkilä, M.P.; Saris, P.E. Inhibition of Staphylococcus aureus by the commensal bacteria of human milk. *J. Appl. Microbiol.* **2003**, *95*, 471–478. [CrossRef]
31. Hunt, K.M.; Foster, J.A.; Forney, L.J.; Schütte, U.M.; Beck, D.L.; Abdo, Z.; Fox, L.K.; Williams, J.E.; McGuire, M.K.; McGuire, M.A. Characterization of the diversity and temporal stability of bacterial communities in human milk. *PLoS ONE* **2011**, *6*, e21313. [CrossRef]
32. Fitzstevens, J.L.; Smith, K.C.; Hagadorn, J.I.; Caimano, M.J.; Matson, A.P.; Brownell, E.A. Systematic Review of the Human Milk Microbiota. *Nutr. Clin. Pract.* **2017**, *32*, 354–364. [CrossRef] [PubMed]
33. Togo, A.; Dufour, J.C.; Lagier, J.C.; Dubourg, G.; Raoult, D.; Million, M. Repertoire of human breast and milk microbiota: A systematic review. *Future Microbiol.* **2019**, *14*, 623–641. [CrossRef] [PubMed]
34. Ramsay, D.T.; Kent, J.C.; Owens, R.A.; Hartmann, P.E. Ultrasound imaging of milk ejection in the breast of lactating women. *Pediatrics* **2004**, *113*, 361–367. [CrossRef] [PubMed]
35. Civardi, E.; Garofoli, F.; Tzialla, C.; Paolillo, P.; Bollani, L.; Stronati, M. Microorganisms in human milk: Lights and shadows. *J. Matern. Fetal Neonatal. Med.* **2013**, *26* (Suppl. S2), 30–34. [CrossRef]
36. Rodríguez, J.M. The origin of human milk bacteria: Is there a bacterial entero-mammary pathway during late pregnancy and lactation? *Adv. Nutr.* **2014**, *5*, 779–784. [CrossRef] [PubMed]
37. Gomez-Llorente, C.; Plaza-Diaz, J.; Aguilera, M.; Muñoz-Quezada, S.; Bermudez-Brito, M.; Peso-Echarri, P.; Martinez-Silla, R.; Vasallo-Morillas, M.I.; Campaña-Martin, L.; Vives-Piñera, I.; et al. Three main factors define changes in fecal microbiota associated with feeding modality in infants. *J. Pediatr. Gastroenterol. Nutr.* **2013**, *57*, 461–466. [CrossRef] [PubMed]
38. Martín, V.; Maldonado-Barragán, A.; Moles, L.; Rodriguez-Baños, M.; Campo, R.D.; Fernández, L.; Rodríguez, J.M.; Jiménez, E. Sharing of bacterial strains between breast milk and infant feces. *J. Hum. Lact.* **2012**, *28*, 36–44. [CrossRef]
39. Coppa, G.V.; Bruni, S.; Morelli, L.; Soldi, S.; Gabrielli, O. The first prebiotics in humans: Human milk oligosaccharides. *J. Clin. Gastroenterol.* **2004**, *38*, S80–S83. [CrossRef]
40. Oozeer, R.; van Limpt, K.; Ludwig, T.; Ben Amor, K.; Martin, R.; Wind, R.D.; Boehm, G.; Knol, J. Intestinal microbiology in early life: Specific prebiotics can have similar functionalities as human-milk oligosaccharides. *Am. J. Clin. Nutr.* **2013**, *98*, 561S–571S. [CrossRef]
41. Zimmermann, P.; Curtis, N. Factors Influencing the Intestinal Microbiome During the First Year of Life. *Pediatr. Infect. Dis. J.* **2018**, *37*, e315–e335. [CrossRef]
42. Moossavi, S.; Sepehri, S.; Robertson, B.; Bode, L.; Goruk, S.; Field, C.J.; Lix, L.M.; de Souza, R.J.; Becker, A.B.; Mandhane, P.J.; et al. Composition and Variation of the Human Milk Microbiota Are Influenced by Maternal and Early-Life Factors. *Cell Host Microbe* **2019**, *25*, 324–335. [CrossRef] [PubMed]
43. Yoo, J.Y.; Rho, M.; You, Y.A.; Kwon, E.J.; Kim, M.H.; Kym, S.; Jee, Y.K.; Kim, Y.K.; Kim, Y.J. 16S rRNA gene-based metagenomic analysis reveals differences in bacteria-derived extracellular vesicles in the urine of pregnant and non-pregnant women. *Exp. Mol. Med.* **2016**, *48*, e208. [CrossRef] [PubMed]
44. Rashed, M.H.; Bayraktar, E.; Helal, G.K.; Abd-Ellah, M.F.; Amero, P.; Chavez-Reyes, A.; Rodriguez-Aguayo, C. Exosomes: From Garbage Bins to Promising Therapeutic Targets. *Int. J. Mol. Sci.* **2017**, *18*, 538. [CrossRef] [PubMed]
45. Hessvik, N.P.; Llorente, A. Current knowledge on exosome biogenesis and release. *Cell Mol. Life Sci.* **2018**, *75*, 193–208. [CrossRef]
46. Vlassov, A.V.; Magdaleno, S.; Setterquist, R.; Conrad, R. Exosomes: Current knowledge of their composition, biological functions, and diagnostic and therapeutic potentials. *Biochim. Biophys. Acta* **2012**, *1820*, 940–948. [CrossRef]
47. Admyre, C.; Johansson, S.M.; Qazi, K.R.; Filén, J.J.; Lahesmaa, R.; Norman, M.; Neve, E.P.; Scheynius, A.; Gabrielsson, S. Exosomes with immune modulatory features are present in human breast milk. *J. Immunol.* **2007**, *179*, 1969–1978. [CrossRef]
48. Galley, J.D.; Besner, G.E. The Therapeutic Potential of Breast Milk-Derived Extracellular Vesicles. *Nutrients* **2020**, *12*, 745. [CrossRef]
49. Hata, T.; Murakami, K.; Nakatani, H.; Yamamoto, Y.; Matsuda, T.; Aoki, N. Isolation of bovine milk-derived microvesicles carrying mRNAs and microRNAs. *Biochem. Biophys. Res. Commun.* **2010**, *396*, 528–533. [CrossRef]

50. Gu, Y.; Li, M.; Wang, T.; Liang, Y.; Zhong, Z.; Wang, X.; Zhou, Q.; Chen, L.; Lang, Q.; He, Z.; et al. Lactation-related microRNA expression profiles of porcine breast milk exosomes. *PLoS ONE* **2012**, *7*, e43691. [CrossRef]
51. Chen, Z.; Xie, Y.; Luo, J.; Chen, T.; Xi, Q.; Zhang, Y.; Sun, J. Milk exosome-derived miRNAs from water buffalo are implicated in immune response and metabolism process. *BMC Vet. Res.* **2020**, *16*, 123. [CrossRef]
52. Baddela, V.S.; Nayan, V.; Rani, P.; Onteru, S.K.; Singh, D. Physicochemical Biomolecular Insights into Buffalo Milk-Derived Nanovesicles. *Appl. Biochem. Biotechnol.* **2016**, *178*, 544–557. [CrossRef]
53. Badawy, A.A.; El-Magd, M.A.; AlSadrah, S.A. Therapeutic Effect of Camel Milk and Its Exosomes on MCF7 Cells In Vitro and In Vivo. *Integr. Cancer Ther.* **2018**, *17*, 1235–1246. [CrossRef] [PubMed]
54. Ma, Y.; Feng, S.; Wang, X.; Qazi, I.H.; Long, K.; Luo, Y.; Li, G.; Ning, C.; Wang, Y.; Hu, S.; et al. Exploration of exosomal microRNA expression profiles in pigeon 'Milk' during the lactation period. *BMC Genom.* **2018**, *19*, 828. [CrossRef] [PubMed]
55. Quan, S.; Nan, X.; Wang, K.; Jiang, L.; Yao, J.; Xiong, B. Characterization of Sheep Milk Extracellular Vesicle-miRNA by Sequencing and Comparison with Cow Milk. *Animals* **2020**, *10*, 331. [CrossRef]
56. Gallier, S.; Vocking, K.; Post, J.A.; Van De Heijning, B.; Acton, D.; Van Der Beek, E.M.; Van Baalen, T. A novel infant milk formula concept: Mimicking the human milk fat globule structure. *Colloids Surf. B Biointerfaces* **2015**, *136*, 329–339. [CrossRef] [PubMed]
57. Torregrosa Paredes, P.; Gutzeit, C.; Johansson, S.; Admyre, C.; Stenius, F.; Alm, J.; Scheynius, A.; Gabrielsson, S. Differences in exosome populations in human breast milk in relation to allergic sensitization and lifestyle. *Allergy* **2014**, *69*, 463–471. [CrossRef] [PubMed]
58. Hock, A.; Miyake, H.; Li, B.; Lee, C.; Ermini, L.; Koike, Y.; Chen, Y.; Määttänen, P.; Zani, A.; Pierro, A. Breast milk-derived exosomes promote intestinal epithelial cell growth. *J. Pediatr. Surg.* **2017**, *52*, 755–759. [CrossRef]
59. Maffei, D.; Schanler, R.J. Human milk is the feeding strategy to prevent necrotizing enterocolitis! *Semin. Perinatol.* **2017**, *41*, 36–40. [CrossRef]
60. Cortez, J.; Makker, K.; Kraemer, D.F.; Neu, J.; Sharma, R.; Hudak, M.L. Maternal milk feedings reduce sepsis, necrotizing enterocolitis and improve outcomes of premature infants. *J. Perinatol.* **2018**, *38*, 71–74. [CrossRef]
61. O'Connor, D.L.; Gibbins, S.; Kiss, A.; Bando, N.; Brennan-Donnan, J.; Ng, E.; Campbell, D.M.; Vaz, S.; Fusch, C.; Asztalos, E.; et al. Effect of Supplemental Donor Human Milk Compared with Preterm Formula on Neurodevelopment of Very Low-Birth-Weight Infants at 18 Months: A Randomized Clinical Trial. *JAMA* **2016**, *316*, 1897–1905. [CrossRef]
62. Quigley, M.; Embleton, N.D.; McGuire, W. Formula versus donor breast milk for feeding preterm or low birth weight infants. *Cochrane Database Syst. Rev.* **2018**, *6*, CD002971. [CrossRef] [PubMed]
63. Martin, C.; Patel, M.; Williams, S.; Arora, H.; Sims, B. Human breast milk-derived exosomes attenuate cell death in intestinal epithelial cells. *Innate Immun.* **2018**, *24*, 278–284. [CrossRef] [PubMed]
64. Miyake, H.; Lee, C.; Chusilp, S.; Bhalla, M.; Li, B.; Pitino, M.; Seo, S.; O'Connor, D.L.; Pierro, A. Human breast milk exosomes attenuate intestinal damage. *Pediatr. Surg. Int.* **2020**, *36*, 155–163. [CrossRef] [PubMed]
65. Wang, X.; Yan, X.; Zhang, L.; Cai, J.; Zhou, Y.; Liu, H.; Hu, Y.; Chen, W.; Xu, S.; Liu, P.; et al. Identification and Peptidomic Profiling of Exosomes in Preterm Human Milk: Insights Into Necrotizing Enterocolitis Prevention. *Mol. Nutr. Food Res.* **2019**, *63*, e1801247. [CrossRef] [PubMed]
66. Li, B.; Hock, A.; Wu, R.Y.; Minich, A.; Botts, S.R.; Lee, C.; Antounians, L.; Miyake, H.; Koike, Y.; Chen, Y.; et al. Bovine milk-derived exosomes enhance goblet cell activity and prevent the development of experimental necrotizing enterocolitis. *PLoS ONE* **2019**, *14*, e0211431. [CrossRef] [PubMed]
67. Arntz, O.J.; Pieters, B.C.; Oliveira, M.C.; Broeren, M.G.; Bennink, M.B.; de Vries, M.; van Lent, P.L.; Koenders, M.I.; van den Berg, W.B.; van der Kraan, P.M.; et al. Oral administration of bovine milk derived extracellular vesicles attenuates arthritis in two mouse models. *Mol. Nutr. Food Res.* **2015**, *59*, 1701–1712. [CrossRef]
68. Näslund, T.I.; Paquin-Proulx, D.; Paredes, P.T.; Vallhov, H.; Sandberg, J.K.; Gabrielsson, S. Exosomes from breast milk inhibit HIV-1 infection of dendritic cells and subsequent viral transfer to CD4+ T cells. *Aids* **2014**, *28*, 171–180. [CrossRef]
69. Komine-Aizawa, S.; Ito, S.; Aizawa, S.; Namiki, T.; Hayakawa, S. Cow milk exosomes activate NK cells and $\gamma\delta$T cells in human PBMCs in vitro. *Immunol. Med.* **2020**, 1–10. [CrossRef]

70. Matic, S.; D'Souza, D.H.; Wu, T.; Pangloli, P.; Dia, V.P. Bovine Milk Exosomes Affect Proliferation and Protect Macrophages against Cisplatin-Induced Cytotoxicity. *Immunol. Investig.* **2020**, 1–15. [CrossRef]
71. Samuel, M.; Chisanga, D.; Liem, M.; Keerthikumar, S.; Anand, S.; Ang, C.S.; Adda, C.G.; Versteegen, E.; Jois, M.; Mathivanan, S. Bovine milk-derived exosomes from colostrum are enriched with proteins implicated in immune response and growth. *Sci. Rep.* **2017**, *7*, 5933. [CrossRef]
72. Reinhardt, T.A.; Lippolis, J.D.; Nonnecke, B.J.; Sacco, R.E. Bovine milk exosome proteome. *J. Proteomics* **2012**, *75*, 1486–1492. [CrossRef]
73. Amitay, E.; Keinan-Boker, L. Breastfeeding and childhood leukemia and lymphoma. *Harefuah* **2014**, *153*, 273–279. [PubMed]
74. Amitay, E.L.; Dubnov Raz, G.; Keinan-Boker, L. Breastfeeding, Other Early Life Exposures and Childhood Leukemia and Lymphoma. *Nutr. Cancer* **2016**, *68*, 968–977. [CrossRef]
75. Küçükçongar, A.; Oğuz, A.; Pınarlı, F.G.; Karadeniz, C.; Okur, A.; Kaya, Z.; Çelik, B. Breastfeeding and Childhood Cancer: Is Breastfeeding Preventative to Childhood Cancer? *Pediatr. Hematol. Oncol.* **2015**, *32*, 374–381. [PubMed]
76. Mathur, G.P.; Gupta, N.; Mathur, S.; Gupta, V.; Pradhan, S.; Dwivedi, J.N.; Tripathi, B.N.; Kushwaha, K.P.; Sathy, N.; Modi, U.J. Breastfeeding and childhood cancer. *Indian Pediatr.* **1993**, *30*, 651–657.
77. Shu, X.O.; Clemens, J.; Zheng, W.; Ying, D.M.; Ji, B.T.; Jin, F. Infant breastfeeding and the risk of childhood lymphoma and leukaemia. *Int. J. Epidemiol.* **1995**, *24*, 27–32. [CrossRef] [PubMed]
78. Martin, R.M.; Gunnell, D.; Owen, C.G.; Smith, G.D. Breast-feeding and childhood cancer: A systematic review with metaanalysis. *Int. J. Cancer* **2005**, *117*, 1020–1031. [CrossRef]
79. Amitay, E.L.; Keinan-Boker, L. Breastfeeding and Childhood Leukemia Incidence: A Meta-analysis and Systematic Review. *JAMA Pediatr.* **2015**, *169*, e151025. [CrossRef] [PubMed]
80. Reif, S.; Elbaum Shiff, Y.; Golan-Gerstl, R. Milk-derived exosomes (MDEs) have a different biological effect on normal fetal colon epithelial cells compared to colon tumor cells in a miRNA-dependent manner. *J. Transl. Med.* **2019**, *17*, 325. [CrossRef]
81. Munagala, R.; Aqil, F.; Jeyabalan, J.; Agrawal, A.K.; Mudd, A.M.; Kyakulaga, A.H.; Singh, I.P.; Vadhanam, M.V.; Gupta, R.C. Exosomal formulation of anthocyanidins against multiple cancer types. *Cancer Lett.* **2017**, *393*, 94–102. [CrossRef]
82. Korashy, H.M.; El Gendy, M.A.; Alhaider, A.A.; El-Kadi, A.O. Camel milk modulates the expression of aryl hydrocarbon receptor-regulated genes, Cyp1a1, Nqo1, and Gsta1, in murine hepatoma Hepa 1c1c7 cells. *J. Biomed. Biotechnol.* **2012**, *2012*, 782642. [CrossRef] [PubMed]
83. Korashy, H.M.; Maayah, Z.H.; Abd-Allah, A.R.; El-Kadi, A.O.; Alhaider, A.A. Camel milk triggers apoptotic signaling pathways in human hepatoma HepG2 and breast cancer MCF7 cell lines through transcriptional mechanism. *J. Biomed. Biotechnol.* **2012**, *2012*, 593195. [CrossRef] [PubMed]
84. Qin, W.; Tsukasaki, Y.; Dasgupta, S.; Mukhopadhyay, N.; Ikebe, M.; Sauter, E.R. Exosomes in Human Breast Milk Promote EMT. *Clin. Cancer Res.* **2016**, *22*, 4517–4524. [CrossRef]
85. Bartel, D.P.; Chen, C.Z. Micromanagers of gene expression: The potentially widespread influence of metazoan microRNAs. *Nat. Rev. Genet.* **2004**, *5*, 396–400. [CrossRef] [PubMed]
86. Bartel, D.P. MicroRNAs: Genomics, biogenesis, mechanism, and function. *Cell* **2004**, *116*, 281–297. [CrossRef]
87. Friedman, R.C.; Farh, K.K.; Burge, C.B.; Bartel, D.P. Most mammalian mRNAs are conserved targets of microRNAs. *Genome Res.* **2009**, *19*, 92–105. [CrossRef]
88. Ambros, V. MicroRNA pathways in flies and worms: Growth, death, fat, stress, and timing. *Cell* **2003**, *113*, 673–676. [CrossRef]
89. Carrington, J.C.; Ambros, V. Role of microRNAs in plant and animal development. *Science* **2003**, *301*, 336–338. [CrossRef]
90. Chang, T.C.; Mendell, J.T. microRNAs in vertebrate physiology and human disease. *Annu. Rev. Genom. Hum. Genet.* **2007**, *8*, 215–239. [CrossRef]
91. Bushati, N.; Cohen, S.M. microRNA functions. *Annu. Rev. Cell Dev. Biol.* **2007**, *23*, 175–205. [CrossRef]
92. Mollaei, H.; Safaralizadeh, R.; Rostami, Z. MicroRNA replacement therapy in cancer. *J. Cell Physiol.* **2019**, *234*, 12369–12384. [CrossRef] [PubMed]
93. Rupaimoole, R.; Slack, F.J. MicroRNA therapeutics: Towards a new era for the management of cancer and other diseases. *Nat. Rev. Drug Discov.* **2017**, *16*, 203–222. [CrossRef] [PubMed]

94. Garzon, R.; Calin, G.A.; Croce, C.M. MicroRNAs in Cancer. *Annu. Rev. Med.* **2009**, *60*, 167–179. [CrossRef] [PubMed]
95. Ling, H.; Fabbri, M.; Calin, G.A. MicroRNAs and other non-coding RNAs as targets for anticancer drug development. *Nat. Rev. Drug Discov.* **2013**, *12*, 847–865. [CrossRef] [PubMed]
96. Alsaweed, M.; Hepworth, A.R.; Lefèvre, C.; Hartmann, P.E.; Geddes, D.T.; Hassiotou, F. Human Milk MicroRNA and Total RNA Differ Depending on Milk Fractionation. *J. Cell Biochem.* **2015**, *116*, 2397–2407. [CrossRef]
97. Alsaweed, M.; Lai, C.T.; Hartmann, P.E.; Geddes, D.T.; Kakulas, F. Human milk miRNAs primarily originate from the mammary gland resulting in unique miRNA profiles of fractionated milk. *Sci. Rep.* **2016**, *6*, 20680. [CrossRef]
98. Alsaweed, M.; Hartmann, P.E.; Geddes, D.T.; Kakulas, F. MicroRNAs in Breastmilk and the Lactating Breast: Potential Immunoprotectors and Developmental Regulators for the Infant and the Mother. *Int. J. Environ. Res. Public Health* **2015**, *12*, 13981–14020. [CrossRef]
99. Zhou, Q.; Li, M.; Wang, X.; Li, Q.; Wang, T.; Zhu, Q.; Zhou, X.; Gao, X.; Li, X. Immune-related microRNAs are abundant in breast milk exosomes. *Int. J. Biol. Sci.* **2012**, *8*, 118–123. [CrossRef]
100. Munch, E.M.; Harris, R.A.; Mohammad, M.; Benham, A.L.; Pejerrey, S.M.; Showalter, L.; Hu, M.; Shope, C.D.; Maningat, P.D.; Gunaratne, P.H.; et al. Transcriptome profiling of microRNA by Next-Gen deep sequencing reveals known and novel miRNA species in the lipid fraction of human breast milk. *PLoS ONE* **2013**, *8*, e50564. [CrossRef]
101. Lönnerdal, B. Human Milk MicroRNAs/Exosomes: Composition and Biological Effects. *Nestle Nutr. Inst. Workshop Ser.* **2019**, *90*, 83–92. [CrossRef]
102. Izumi, H.; Kosaka, N.; Shimizu, T.; Sekine, K.; Ochiya, T.; Takase, M. Bovine milk contains microRNA and messenger RNA that are stable under degradative conditions. *J. Dairy Sci.* **2012**, *95*, 4831–4841. [CrossRef] [PubMed]
103. Liao, Y.; Du, X.; Li, J.; Lönnerdal, B. Human milk exosomes and their microRNAs survive digestion in vitro and are taken up by human intestinal cells. *Mol. Nutr. Food Res.* **2017**, *61*, 1700082. [CrossRef] [PubMed]
104. Kahn, S.; Liao, Y.; Du, X.; Xu, W.; Li, J.; Lönnerdal, B. Exosomal MicroRNAs in Milk from Mothers Delivering Preterm Infants Survive in Vitro Digestion and Are Taken Up by Human Intestinal Cells. *Mol. Nutr. Food Res.* **2018**, *62*, e1701050. [CrossRef] [PubMed]
105. Zempleni, J.; Aguilar-Lozano, A.; Sadri, M.; Sukreet, S.; Manca, S.; Wu, D.; Zhou, F.; Mutai, E. Biological Activities of Extracellular Vesicles and Their Cargos from Bovine and Human Milk in Humans and Implications for Infants. *J. Nutr.* **2017**, *147*, 3–10. [CrossRef]
106. Golan-Gerstl, R.; Elbaum Shiff, Y.; Moshayoff, V.; Schecter, D.; Leshkowitz, D.; Reif, S. Characterization and biological function of milk-derived miRNAs. *Mol. Nutr. Food Res.* **2017**, *61*, 1700009. [CrossRef]
107. Nojiri, K.; Kobayashi, S.; Higurashi, S.; Takahashi, T.; Tsujimori, Y.; Ueno, H.M.; Watanabe-Matsuhashi, S.; Toba, Y.; Yamamura, J.; Nakano, T.; et al. Maternal Health and Nutrition Status, Human Milk Composition, and Growth and Development of Infants and Children: A Prospective Japanese Human Milk Study Protocol. *Int. J. Environ. Res. Public Health* **2020**, *17*, 1869. [CrossRef]
108. Kosaka, N.; Izumi, H.; Sekine, K.; Ochiya, T. microRNA as a new immune-regulatory agent in breast milk. *Silence* **2010**, *1*, 7. [CrossRef]
109. Na, R.S.; E, G.X.; Sun, W.; Sun, X.W.; Qiu, X.Y.; Chen, L.P.; Huang, Y.F. Expressional analysis of immune-related miRNAs in breast milk. *Genet. Mol. Res.* **2015**, *14*, 11371–11376. [CrossRef]
110. Chen, X.; Gao, C.; Li, H.; Huang, L.; Sun, Q.; Dong, Y.; Tian, C.; Gao, S.; Dong, H.; Guan, D.; et al. Identification and characterization of microRNAs in raw milk during different periods of lactation, commercial fluid, and powdered milk products. *Cell Res.* **2010**, *20*, 1128–1137. [CrossRef]
111. Gaziel-Sovran, A.; Segura, M.F.; Di Micco, R.; Collins, M.K.; Hanniford, D.; Vega-Saenz de Miera, E.; Rakus, J.F.; Dankert, J.F.; Shang, S.; Kerbel, R.S.; et al. miR-30b/30d regulation of GalNAc transferases enhances invasion and immunosuppression during metastasis. *Cancer Cell* **2011**, *20*, 104–118. [CrossRef]
112. Stittrich, A.B.; Haftmann, C.; Sgouroudis, E.; Kühl, A.A.; Hegazy, A.N.; Panse, I.; Riedel, R.; Flossdorf, M.; Dong, J.; Fuhrmann, F.; et al. The microRNA miR-182 is induced by IL-2 and promotes clonal expansion of activated helper T lymphocytes. *Nat. Immunol.* **2010**, *11*, 1057–1062. [CrossRef] [PubMed]
113. Sun, Q.; Chen, X.; Yu, J.; Zen, K.; Zhang, C.Y.; Li, L. Immune modulatory function of abundant immune-related microRNAs in microvesicles from bovine colostrum. *Protein Cell* **2013**, *4*, 197–210. [CrossRef]

114. Melnik, B.C.; Schmitz, G. Exosomes of pasteurized milk: Potential pathogens of Western diseases. *J. Transl. Med.* **2019**, *17*, 3. [CrossRef] [PubMed]
115. Long, X.R.; He, Y.; Huang, C.; Li, J. MicroRNA-148a is silenced by hypermethylation and interacts with DNA methyltransferase 1 in hepatocellular carcinogenesis. *Int. J. Oncol.* **2014**, *44*, 1915–1922. [CrossRef] [PubMed]
116. Li, J.; Song, Y.; Wang, Y.; Luo, J.; Yu, W. MicroRNA-148a suppresses epithelial-to-mesenchymal transition by targeting ROCK1 in non-small cell lung cancer cells. *Mol. Cell Biochem.* **2013**, *380*, 277–282. [CrossRef]
117. Yu, J.; Li, Q.; Xu, Q.; Liu, L.; Jiang, B. MiR-148a inhibits angiogenesis by targeting ERBB3. *J. Biomed. Res.* **2011**, *25*, 170–177. [CrossRef]
118. Zhang, H.; Luo, X.Q.; Zhang, P.; Huang, L.B.; Zheng, Y.S.; Wu, J.; Zhou, H.; Qu, L.H.; Xu, L.; Chen, Y.Q. MicroRNA patterns associated with clinical prognostic parameters and CNS relapse prediction in pediatric acute leukemia. *PLoS ONE* **2009**, *4*, e7826. [CrossRef]
119. de Oliveira, J.C.; Brassesco, M.S.; Scrideli, C.A.; Tone, L.G.; Narendran, A. MicroRNA expression and activity in pediatric acute lymphoblastic leukemia (ALL). *Pediatr. Blood Cancer* **2012**, *59*, 599–604. [CrossRef]
120. Tomé-Carneiro, J.; Fernández-Alonso, N.; Tomás-Zapico, C.; Visioli, F.; Iglesias-Gutierrez, E.; Dávalos, A. Breast milk microRNAs harsh journey towards potential effects in infant development and maturation. Lipid encapsulation can help. *Pharmacol. Res.* **2018**, *132*, 21–32. [CrossRef]
121. Melnik, B.C. MiR-21: An environmental driver of malignant melanoma? *J. Transl. Med.* **2015**, *13*, 202. [CrossRef]
122. Meng, F.; Henson, R.; Wehbe-Janek, H.; Ghoshal, K.; Jacob, S.T.; Patel, T. MicroRNA-21 regulates expression of the PTEN tumor suppressor gene in human hepatocellular cancer. *Gastroenterology* **2007**, *133*, 647–658. [CrossRef] [PubMed]
123. Sayed, D.; Rane, S.; Lypowy, J.; He, M.; Chen, I.Y.; Vashistha, H.; Yan, L.; Malhotra, A.; Vatner, D.; Abdellatif, M. MicroRNA-21 targets Sprouty2 and promotes cellular outgrowths. *Mol. Biol. Cell* **2008**, *19*, 3272–3282. [CrossRef] [PubMed]
124. Asangani, I.A.; Rasheed, S.A.; Nikolova, D.A.; Leupold, J.H.; Colburn, N.H.; Post, S.; Allgayer, H. MicroRNA-21 (miR-21) post-transcriptionally downregulates tumor suppressor Pdcd4 and stimulates invasion, intravasation and metastasis in colorectal cancer. *Oncogene* **2008**, *27*, 2128–2136. [CrossRef] [PubMed]
125. Cao, Q.; Li, Y.Y.; He, W.F.; Zhang, Z.Z.; Zhou, Q.; Liu, X.; Shen, Y.; Huang, T.T. Interplay between microRNAs and the STAT3 signaling pathway in human cancers. *Physiol. Genom.* **2013**, *45*, 1206–1214. [CrossRef]
126. Huang, C.; Li, H.; Wu, W.; Jiang, T.; Qiu, Z. Regulation of miR-155 affects pancreatic cancer cell invasiveness and migration by modulating the STAT3 signaling pathway through SOCS1. *Oncol. Rep.* **2013**, *30*, 1223–1230. [CrossRef]
127. Ha, D.; Yang, N.; Nadithe, V. Exosomes as therapeutic drug carriers and delivery vehicles across biological membranes: Current perspectives and future challenges. *Acta Pharm. Sin. B* **2016**, *6*, 287–296. [CrossRef]
128. Munagala, R.; Aqil, F.; Jeyabalan, J.; Gupta, R.C. Bovine milk-derived exosomes for drug delivery. *Cancer Lett.* **2016**, *371*, 48–61. [CrossRef]
129. Aqil, F.; Munagala, R.; Jeyabalan, J.; Agrawal, A.K.; Kyakulaga, A.H.; Wilcher, S.A.; Gupta, R.C. Milk exosomes—Natural nanoparticles for siRNA delivery. *Cancer Lett.* **2019**, *449*, 186–195. [CrossRef]
130. Betker, J.L.; Angle, B.M.; Graner, M.W.; Anchordoquy, T.J. The Potential of Exosomes from Cow Milk for Oral Delivery. *J. Pharm. Sci.* **2019**, *108*, 1496–1505. [CrossRef]
131. Carobolante, G.; Mantaj, J.; Ferrari, E.; Vllasaliu, D. Cow Milk and Intestinal Epithelial Cell-derived Extracellular Vesicles as Systems for Enhancing Oral Drug Delivery. *Pharmaceutics* **2020**, *12*, 226. [CrossRef]
132. Aqil, F.; Kausar, H.; Agrawal, A.K.; Jeyabalan, J.; Kyakulaga, A.H.; Munagala, R.; Gupta, R. Exosomal formulation enhances therapeutic response of celastrol against lung cancer. *Exp. Mol. Pathol.* **2016**, *101*, 12–21. [CrossRef] [PubMed]
133. Agrawal, A.K.; Aqil, F.; Jeyabalan, J.; Spencer, W.A.; Beck, J.; Gachuki, B.W.; Alhakeem, S.S.; Oben, K.; Munagala, R.; Bondada, S.; et al. Milk-derived exosomes for oral delivery of paclitaxel. *Nanomedicine* **2017**, *13*, 1627–1636. [CrossRef] [PubMed]
134. Dong, Y.; Siegwart, D.J.; Anderson, D.G. Strategies, design, and chemistry in siRNA delivery systems. *Adv. Drug Deliv. Rev.* **2019**, *144*, 133–147. [CrossRef] [PubMed]
135. Das, M.; Musetti, S.; Huang, L. RNA Interference-Based Cancer Drugs: The Roadblocks, and the "Delivery" of the Promise. *Nucleic Acid Ther.* **2019**, *29*, 61–66. [CrossRef] [PubMed]

136. Tao, H.; Xu, H.; Zuo, L.; Li, C.; Qiao, G.; Guo, M.; Zheng, L.; Leitgeb, M.; Lin, X. Exosomes-coated bcl-2 siRNA inhibits the growth of digestive system tumors both in vitro and in vivo. *Int. J. Biol. Macromol.* **2020**, *161*, 470–480. [CrossRef]
137. Naor, D.; Nedvetzki, S.; Golan, I.; Melnik, L.; Faitelson, Y. CD44 in cancer. *Crit. Rev. Clin. Lab. Sci.* **2002**, *39*, 527–579. [CrossRef]
138. Li, D.; Yao, S.; Zhou, Z.; Shi, J.; Huang, Z.; Wu, Z. Hyaluronan decoration of milk exosomes directs tumor-specific delivery of doxorubicin. *Carbohydr. Res.* **2020**, *493*, 108032. [CrossRef]
139. Sugahara, K.N.; Teesalu, T.; Karmali, P.P.; Kotamraju, V.R.; Agemy, L.; Greenwald, D.R.; Ruoslahti, E. Coadministration of a tumor-penetrating peptide enhances the efficacy of cancer drugs. *Science* **2010**, *328*, 1031–1035. [CrossRef]
140. Teesalu, T.; Sugahara, K.N.; Kotamraju, V.R.; Ruoslahti, E. C-end rule peptides mediate neuropilin-1-dependent cell, vascular, and tissue penetration. *Proc. Natl. Acad. Sci. USA* **2009**, *106*, 16157–16162. [CrossRef]
141. Zhang, Q.; Xiao, Q.; Yin, H.; Xia, C.; Pu, Y.; He, Z.; Hu, Q.; Wang, J.; Wang, Y. Milk-exosome based pH/light sensitive drug system to enhance anticancer activity against oral squamous cell carcinoma. *RSC Adv.* **2020**, *10*, 28314–28323. [CrossRef]

Permissions

All chapters in this book were first published by MDPI; hereby published with permission under the Creative Commons Attribution License or equivalent. Every chapter published in this book has been scrutinized by our experts. Their significance has been extensively debated. The topics covered herein carry significant findings which will fuel the growth of the discipline. They may even be implemented as practical applications or may be referred to as a beginning point for another development.

The contributors of this book come from diverse backgrounds, making this book a truly international effort. This book will bring forth new frontiers with its revolutionizing research information and detailed analysis of the nascent developments around the world.

We would like to thank all the contributing authors for lending their expertise to make the book truly unique. They have played a crucial role in the development of this book. Without their invaluable contributions this book wouldn't have been possible. They have made vital efforts to compile up to date information on the varied aspects of this subject to make this book a valuable addition to the collection of many professionals and students.

This book was conceptualized with the vision of imparting up-to-date information and advanced data in this field. To ensure the same, a matchless editorial board was set up. Every individual on the board went through rigorous rounds of assessment to prove their worth. After which they invested a large part of their time researching and compiling the most relevant data for our readers.

The editorial board has been involved in producing this book since its inception. They have spent rigorous hours researching and exploring the diverse topics which have resulted in the successful publishing of this book. They have passed on their knowledge of decades through this book. To expedite this challenging task, the publisher supported the team at every step. A small team of assistant editors was also appointed to further simplify the editing procedure and attain best results for the readers.

Apart from the editorial board, the designing team has also invested a significant amount of their time in understanding the subject and creating the most relevant covers. They scrutinized every image to scout for the most suitable representation of the subject and create an appropriate cover for the book.

The publishing team has been an ardent support to the editorial, designing and production team. Their endless efforts to recruit the best for this project, has resulted in the accomplishment of this book. They are a veteran in the field of academics and their pool of knowledge is as vast as their experience in printing. Their expertise and guidance has proved useful at every step. Their uncompromising quality standards have made this book an exceptional effort. Their encouragement from time to time has been an inspiration for everyone.

The publisher and the editorial board hope that this book will prove to be a valuable piece of knowledge for researchers, students, practitioners and scholars across the globe.

List of Contributors

Giulia Chiabotto, Chiara Gai and Giovanni Camussi
Department of Medical Sciences, University of Torino, Torino 10126, Italy

Maria Chiara Deregibus
2i3T Business Incubator and Technology Transfer, University of Torino, Torino 10126, Italy

Hans A. Schlößer, Zhefang Wang, Jie Qin, Jiahui Li, Felix Popp, Marie Christine Popp, Hakan Alakus, Seung-Hun Chon and Christiane J. Bruns
Department of General, Visceral und Tumor Surgery, University Hospital Cologne, Kerpener Straße 62, 50937 Cologne, Germany

Karl-Walter Jauch
Department of General, Visceral und Vascular Surgery, Ludwig-Maximilian-University (LMU), 81377 Munich, Germany

Jiangang Zhao
Department of General, Visceral und Tumor Surgery, University Hospital Cologne, Kerpener Straße 62, 50937 Cologne, Germany
Department of General, Visceral und Vascular Surgery, Ludwig-Maximilian-University (LMU), 81377 Munich, Germany

Hinrich P. Hansen
Department I of Internal Medicine, University Hospital of Cologne, Center for Integrated Oncology Cologne-Bonn, CECAD Center of Excellence on "Cellular Stress Responses in Aging-Associated Diseases", Center for Molecular Medicine Cologne, University of Cologne, 50931 Cologne, Germany

Wolfram F. Neiss
Department of Anatomy I, Medical Faculty, University of Cologne, 50937 Cologne, Germany

Yue Zhao
Department of General, Visceral und Tumor Surgery, University Hospital Cologne, Kerpener Straße 62, 50937 Cologne, Germany
Department of General, Visceral und Vascular Surgery, Otto von Guericke University, 39120 Magdeburg, Germany

Clodagh P. O'Neill, Katie E. Gilligan and Róisín M. Dwyer
Discipline of Surgery, Lambe Institute for Translational Research, National University of Ireland Galway (NUIG), Galway H91 YR71, Ireland

Sibylle Ohl and Martina Seiffert
Division of Molecular Genetics, German Cancer Research Center (DKFZ), 69120 Heidelberg, Germany

Marie Bordas
Division of Molecular Genetics, German Cancer Research Center (DKFZ), 69120 Heidelberg, Germany
Faculty of Biosciences, University of Heidelberg, 69120 Heidelberg, Germany

Géraldine Genard
Division of Biomedical Physics in Radiation Oncology, German Cancer Research Center (DKFZ), 69120 Heidelberg, Germany

Michelle Nessling and Karsten Richter
Central Unit Electron Microscopy, DKFZ, 69120 Heidelberg, Germany

Tobias Roider and Sascha Dietrich
Department of Medicine V, Hematology, Oncology and Rheumatology, University of Heidelberg, 69120 Heidelberg, Germany

Kendra K. Maaß
Hopp-Children's Cancer Center Heidelberg (KiTZ), 69120 Heidelberg, Germany
Division of Pediatric Neurooncology, German Cancer Research Center (DKFZ), 69120 Heidelberg, Germany

Maria Luisa Fiani, Valeria Barreca, Massimo Sargiacomo, Flavia Ferrantelli, Francesco Manfredi and Maurizio Federico
National Center for Global Health, Istituto Superiore di Sanità, 00161 Rome, Italy

Banuja Balachandran and Chrit T. Moonen
Imaging Division, University Medical Center Utrecht, 3584 CX Utrecht, The Netherlands

Yuana Yuana
Imaging Division, University Medical Center Utrecht, 3584 CX Utrecht, The Netherlands
Department of Biomedical Engineering, TU Eindhoven, 5600 MB Eindhoven, The Netherlands

Kim M. G. van der Wurff-Jacobs
Department of Pharmaceutical Sciences, University of Utrecht, 3584 CG Utrecht, The Netherlands

Raymond M. Schiffelers
Laboratory of Clinical Chemistry and Haematology, University Medical Center Utrecht, 3584 CX Utrecht, The Netherlands

Ancuta Jurj and Cecilia Pop-Bica
Research Center for Functional Genomics, Biomedicine and Translational Medicine, "Iuliu Hat,ieganu" University of Medicine and Pharmacy, 400337 Cluj-Napoca, Romania

Ondrej Slaby
Central European Institute of Technology, Masaryk University, 625 00 Brno, Czech Republic
Department of Pathology, Faculty Hospital Brno and Faculty of Medicine, Masaryk University, 625 00 Brno, Czech Republic

Cristina D. Ştefan
SingHealth Duke-NUS Global Health Institute, Singapore 169857, Singapore

William C. Cho
Department of Clinical Oncology, Queen Elizabeth Hospital, Hong Kong, China

Schuyler S. Korban
Department of Natural Resources and Environmental Sciences, University of Illinois at Urbana-Champaign, Urbana, IL 61801, USA

Ioana Berindan-Neagoe
Research Center for Functional Genomics, Biomedicine and Translational Medicine, "Iuliu Hat,ieganu" University of Medicine and Pharmacy, 400337 Cluj-Napoca, Romania
Department of Functional Genomics and Experimental Pathology, "Prof. Dr. Ion Chiricuta" Oncology cInstitute, 400015 Cluj-Napoca, Romania

Kerstin Menck and Suganja Sivaloganathan
Department of Medicine A, Hematology, Oncology, and Pneumology, University Hospital Münster, 48149 Münster, Germany

Claudia Binder
Department of Hematology/Medical Oncology, University Medical Center Göttingen, 37075 Göttingen, Germany

Annalen Bleckmann
Department of Medicine A, Hematology, Oncology, and Pneumology, University Hospital Münster, 48149 Münster, Germany
Department of Hematology/Medical Oncology, University Medical Center Göttingen, 37075 Göttingen, Germany

Agnieszka Szyposzynska, Aleksandra Bielawska-Pohl, Agnieszka Krawczenko, Olga Doszyn, Maria Paprocka and Aleksandra Klimczak
Laboratory of Biology of Stem and Neoplastic Cells, Hirszfeld Institute of Immunology and Experimental Therapy, Polish Academy of Sciences, R.Weigla 12, 53-114Wroclaw, Poland

Eliane Ebnoether and Laurent Muller
Department of Biomedicine, University of Basel, 4031 Basel, Switzerland
Department of Otorhinolaryngology, Head and Neck Surgery, University Hospital of Basel, 4051 Basel, Switzerland

SuzanneM Johnson and Christopher Smith
Children's Cancer Group, Division of Cancer Sciences, School of Medical Sciences, Faculty of Biology Medicine and Health, University of Manchester, Oglesby Cancer Research Building, Manchester Academic Health Science Centre, Manchester Cancer Research Centre, Manchester M20 4GJ, UK

Antonia Banyard
Flow Cytometry Core Facility, Cancer Research UK Manchester Institute, University of Manchester, Alderley Park, Macclesfield, UK

Aleksandr Mironov
Electron Microscopy Core Facility, School of Biological Sciences, Faculty of Biology Medicine and Health, University of Manchester, Manchester M13 9PT, UK

Martin G. McCabe
Children's Cancer Group, Division of Cancer Sciences, School of Medical Sciences, Faculty of Biology Medicine and Health, University of Manchester, Oglesby Cancer Research Building, Manchester Academic Health Science Centre, Manchester Cancer Research Centre, Manchester M20 4GJ, UK
Manchester Children's Brain Tumour Research Network Royal Manchester Children's Hospital, Manchester M13 9WL, UK

Ki-Uk Kim, Wan-Hoon Kim, Chi Hwan Jeong and Hyeyoung Min
College of Pharmacy, Chung-Ang University, Seoul 06974, Korea

Dae Yong Yi
Department of Pediatrics, Chung-Ang University College of Medicine, Seoul 06974, Korea
Department of Pediatrics, Chung-Ang University Hospital, Seoul 06973, Korea

Index

M

N

O

P

R

S

T

U

V

www.ingramcontent.com/pod-product-compliance
Lightning Source LLC
LaVergne TN
LVHW081019210926
848485LV00004B/900